# EARTH'S NATURAL RESOURCES

# JOHN V. WALTHER

Southern Methodist University

## JONES & BARTLETT
LEARNING

*World Headquarters*
Jones & Bartlett Learning
5 Wall Street
Burlington, MA 01803
978-443-5000
info@jblearning.com
www.jblearning.com

Jones & Bartlett Learning books and products are available through most bookstores and online booksellers. To contact Jones & Bartlett Learning directly, call 800-832-0034, fax 978-443-8000, or visit our website, www.jblearning.com.

**Production Credits**
Executive Publisher: Kevin Sullivan
Senior Acquisitions Editor: Erin O'Connor
Editorial Assistant: Michelle Bradbury
Production Editor: Keith Henry
Senior Marketing Manager: Andrea DeFronzo
V.P., Manufacturing and Inventory Control: Therese Connell
Composition: diacriTech
Cover and Title Page Design: Kristin E. Parker
Rights and Photo Research Associate: Lauren Miller
Illustrations: Electronic Publishing Services Inc.
Cover Image: (mine image) © iStockphoto/Thinkstock, (top background texture) © Mironov56/ShutterStock, Inc.
Printing and Binding: Courier Companies
Cover Printing: Courier Companies

*About the cover*: A bucket wheel excavator operating in a strip mine near Lausitz, Germany, removing overburden layers of soil to expose lignite coal below. Bucket wheel excavators have a rotating wheel with a series of buckets attached. The upswing movement of the wheel causes the buckets to scoop soil or lignite, carrying it to the back of the wheel where it falls on a conveyor belt for removal. These 13,000 metric ton vehicles move on caterpillar tracks, cost more than U.S. $100 million each, can remove more than 75,000 meters$^3$ of material a day, and require four to five operators.

**Library of Congress Cataloging-in-Publication Data**
Walther, John.
  Earth's natural resources / John Walther. — 1st ed.
      p. cm.
  Includes index.
  ISBN 978-1-4496-3234-2 (alk. paper)
  1. Natural resources.  I. Title.
  HC85.W35 2013
  333.7—dc23
                                            2012028195

6048
Printed in the United States of America
17 16 15 14 13    9 8 7 6 5 4 3 2 1

# Brief Contents

# Contents

## Chapter 11    Chemicals from Evaporation of Water and Gaseous Elements from Air . . . . . . . . . . . . . . . . 263

# Preface

"Man shapes himself through decisions that shape his environment."

*– René Dubos*

Because of the intellect we possess, our evolutionary rise is different from other species on Earth. As civilization advances, humankind can decide about how to use the earth's natural resources it requires. Rather than taking the evolutionary path of easiest maximum exploitation, humankind can make choices. Potentially, humans can exploit natural resources so they can supply needs far into the future. To do this, they must understand resource availability and the extent of demand.

Exploitation of natural resources creates employment and wealth. Today, worldwide production of natural resources typically is undertaken with a capitalistic system of supply and demand. This works reasonably well when resources are abundant and widely distributed, but as the 1973–1974 Arab oil embargo demonstrated, political concerns can be problematic when resources are limited or confined to certain countries. As demand increases and the supply of a key natural resource becomes restricted, increased political disruption can occur. Knowledge of resource location and availability becomes more important to anticipate and address these concerns.

The rise in resource prices indicates a limited supply. Whether this is predicted to be temporary or permanent depends on an understanding of future supply sources. This knowledge is important for the informed citizen. This understanding cannot be divorced from an understanding of future population growth and, therefore, demand as well as the environmental impact of resource extraction and usage. I hope that *Earth's Natural Resources* will impart the needed information by considering where natural resources occur, how they are concentrated and extracted, and the extent of their supply and usage.

The book has a U.S. bias in that many of the examples are derived from the United States, but the development should be of interest to a wider audience. The book is appropriate for a student with a scientific background equivalent to a strong U.S. high school education and some lower division college courses in physical geology. To assist those with less experience, Chapter 1 provides a short review of some terms. Important terms, when first used, are *italicized* and provided in a glossary at the end of the book. In many U.S. curriculums, the book fits into a lower division college major's course in Earth or Environmental Science or as a background resource course in Civil Engineering. Each chapter provides problems to help the reader develop a deeper understanding of the material covered. Answers to these problems are available from the publisher.

## *Instructor's Resources*

A downloadable **PowerPoint® Image Bank** is available to instructors. This resource provides the book's illustrations, photographs, and tables (to which Jones & Bartlett holds the copyright or has permission to reproduce digitally) inserted into PowerPoint slides. These images and tables can be easily copied into existing lecture slides.

Also available for download is the **PowerPoint® Lecture Outline** presentation package. This provides lecture notes and images for each chapter of *Earth's Natural Resources*. Instructors with Microsoft PowerPoint software can customize the outlines, art, and order of each presentation.

# Acknowledgments

I would like to thank my colleagues at Southern Methodist University, Kurt Ferguson, Ian Richards, Maria Richards, and Neil Tabor for their helpful chapter reviews. I would also like to thank those who reviewed chapters throughout the editorial process:

- Diane M. Burns, Eastern Illinois University
- Jim Constantopoulos, Eastern New Mexico University
- Winton Cornell, University of Tulsa
- Rónadh Cox, Williams College
- Udo Fehn, University of Rochester
- Emilio Mutis-Duplat, The University of Texas of the Permian Basin
- Kent Ratajeski, University of Kentucky
- Martin Schoonen, Stony Brook University
- Brandon E. Schwab, Humboldt State University
- Stephen Van Horn, Muskingum University

John V. Walther
Southern Methodist University

# Table of Elements

Listed for the given element are the chapter where it is discussed, its atomic number, atomic weight, common valence states, and average abundance in the continental crust and seawater.w

| ELEMENT | CHAPTER | SYMBOL | ATOMIC NUMBER | ATOMIC WEIGHT | COMMON VALENCE STATES | AVERAGE CONTINENTAL CRUST | SEAWATER |
|---|---|---|---|---|---|---|---|
| Actinium | - | Ac | 89 | 227.03 | +3 | - | - |
| Aluminum | 6 | Al | 13 | 26.98 | +3 | 8.2% | $8 \times 10^{-4}$ ppm |
| Americium | - | Am | 95 | (243) | +3,+4,+5,+6 | - | - |
| Antimony | 9 | Sb | 51 | 121.75 | +3,+5,−3 | 0.2 ppm | 0.15 ppb |
| Argon | 11 | Ar | 18 | 39.95 | 0 | - | - |
| Arsenic | 9 | As | 33 | 74.92 | +3,+5,−3 | 2.1 ppm | 1.7 ppb |
| Astatine | - | At | 85 | (210) | −1 | - | - |
| Barium | 10 | Ba | 56 | 137.34 | +2 | 340 ppm | 0.014 ppm |
| Berkelium | - | Bk | 97 | (247) | +3,+4 | - | - |
| Beryllium | 9 | Be | 4 | 9.01 | +2 | 1.9 ppm | $2 \times 10^{-7}$ ppm |
| Bismuth | 9 | Bi | 83 | 208.98 | +3,+5 | 25 ppb | 0.02 ppb |
| Boron | 11 | B | 5 | 10.81 | +3 | 8.7 ppm | 4.5 ppm |
| Bromine | 11 | Br | 35 | 79.91 | +1,+5,−1 | 3 ppm | 67 ppm |
| Cadmium | 8 | Cd | 48 | 112.30 | +2 | 0.15 ppm | 0.08 ppb |
| Calcium | 6 | Ca | 20 | 40.08 | +2 | 5.0% | 413 ppm |
| Californium | - | Cf | 98 | (251) | +3 | - | - |
| Carbon | 2–3, 10 | C | 6 | 12.01 | +2,+4,−4 | 0.18% | 28 ppm |
| Cerium | 9 | Ce | 58 | 140.12 | +3,+4 | 60 ppm | 0.0035 ppb |
| Cesium | 9 | Cs | 55 | 132.90 | +1 | 1.9 ppm | 0.29 ppb |

| ELEMENT | CHAPTER | SYMBOL | ATOMIC NUMBER | ATOMIC WEIGHT | COMMON VALENCE STATES | AVERAGE CONTINENTAL CRUST | SEAWATER |
|---|---|---|---|---|---|---|---|
| Chlorine | 11 | Cl | 17 | 35.45 | +1,+5,+7,−1 | 170 ppm | 1.95% |
| Chromium | 7 | Cr | 24 | 52.00 | +2,+3,+6 | 140 ppm | 0.2 ppb |
| Cobalt | 7 | Co | 27 | 58.93 | +2,+3 | 30 ppm | 0.002 ppb |
| Copper | 8 | Cu | 29 | 63.54 | +1,+2 | 68 ppm | 0.3 ppb |
| Curium | - | Cm | 96 | (247) | +3 | - | - |
| Dysprosium | 9 | Dy | 66 | 162.50 | +3 | 6.2 ppm | 0.0011 ppb |
| Einsteinium | - | Es | 99 | (254) | +3 | - | - |
| Erbium | 9 | Er | 68 | 167.26 | +3 | 3.0 ppm | $9.2 \times 10^{-4}$ ppb |
| Europium | 9 | Eu | 63 | 151.96 | +2,+3 | 1.8 ppm | $1.5 \times 10^{-4}$ ppb |
| Fermium | - | Fm | 100 | (257) | +3 | | - |
| Fluorine | 8 | F | 9 | 19.00 | −1 | 540 ppm | 1.3 ppm |
| Francium | - | Fr | 87 | (223) | +1 | - | - |
| Gadolinium | 9 | Gd | 64 | 157.25 | +3 | 5.2 ppm | 0.001 ppb |
| Gallium | 9 | Ga | 31 | 69.72 | +3 | 19 ppm | 0.02 ppb |
| Germanium | 9 | Ge | 32 | 72.59 | +2,+4 | 1.4 ppm | 0.005 ppb |
| Gold | 9 | Au | 79 | 196.97 | +1,+3 | 3.1 ppb | 0.0049 ppb |
| Hafnium | 5 | Hf | 72 | 178.49 | +4 | 3.3 ppm | < 0.007 ppb |
| Helium | 11 | He | 2 | 4.00 | 0 | - | - |
| Holmium | 9 | Ho | 67 | 164.93 | +3 | 1.2 ppm | $2.8 \times 10^{-4}$ ppb |
| Hydrogen | 3 | H | 1 | 1.01 | +1,−1 | 0.15% | 10.82% |
| Indium | 9 | In | 49 | 114.82 | +3 | 0.16 ppm | $1 \times 10^{-4}$ ppb |
| Iodine | 11 | I | 53 | 126.90 | +1,+5,+7,−1 | 0.49 ppm | 0.056 ppm |
| Iridium | 9 | Ir | 77 | 192.20 | +3,+4 | 0.4 ppb | $1 \times 10^{-5}$ ppb |
| Iron | 6 | Fe | 26 | 55.85 | +2,+3 | 6.3% | 0.06 ppb |
| Krypton | 11 | Kr | 36 | 83.80 | 0 | - | - |
| Lanthanum | 9 | La | 71 | 174.97 | +3 | 34 ppm | 0.0045 ppb |
| Lead | 8 | Pb | 82 | 207.19 | +2,+4 | 10 ppm | 0.002 ppb |
| Lithium | 11 | Li | 3 | 6.94 | +1 | 17 ppm | 0.17 ppm |
| Lutetium | 9 | Lu | 71 | 174.97 | +3 | 350 ppb | $1.4 \times 10^{-4}$ ppb |
| Magnesium | 6 | Mg | 12 | 24.31 | +2 | 2.90% | 0.129% |
| Manganese | 6 | Mn | 25 | 54.94 | +2,+3,+4,+7 | 0.11% | 0.3 ppb |
| Mendelevium | - | Md | 101 | (258) | +2,+3 | - | - |

| ELEMENT | CHAPTER | SYMBOL | ATOMIC NUMBER | ATOMIC WEIGHT | COMMON VALENCE STATES | AVERAGE CONTINENTAL CRUST | SEAWATER |
|---|---|---|---|---|---|---|---|
| Mercury | 8 | Hg | 80 | 200.59 | +1,+2 | 67 ppb | 0.001 ppb |
| Molybdenum | 7 | Mo | 42 | 95.95 | +4,+6 | 1.1 ppm | 0.011 ppm |
| Neodymium | 9 | Nd | 60 | 144.24 | +3 | 33 ppm | 0.0042 ppb |
| Neon | 11 | Ne | 10 | 20.18 | 0 | - | - |
| Neptunium | - | Np | 93 | 237.05 | +3,+4,+5,+6 | - | - |
| Nickel | 7 | Ni | 28 | 58.71 | +2,+3 | 980 ppm | 0.5 ppb |
| Niobium | 9 | Nb | 41 | 92.91 | +3,+5 | 17 ppm | < 0.005 ppb |
| Nitrogen | 11 | N | 7 | 14.01 | +1,+2,+3,+4, | 20 ppm | 15.5 ppm |
| Nobelium | - | No | 102 | (259) | +2,+3 | - | - |
| Osmium | 9 | Os | 76 | 190.20 | +3,+4 | 1.8 ppb | $1 \times 10^{-5}$ ppb |
| Oxygen | 11 | O | 8 | 16.00 | -2 | 46% | 85.84% |
| Palladium | 9 | Pd | 46 | 106.40 | +2,+4 | 6.3 ppb | $2 \times 10^{-4}$ ppb |
| Phosphorus | 11 | P | 15 | 30.97 | +3,+5,-3 | 0.10% | 0.071 ppm |
| Platinum | 9 | Pt | 78 | 195.09 | +2,+4 | 3.7 ppb | 0.05 ppb |
| Plutonium | 5 | Pu | 94 | (244) | +3,+4,+5,+6 | - | - |
| Polonium | - | Po | 84 | (209) | +2,+4 | - | - |
| Potassium | 11 | K | 19 | 39.10 | +1 | 1.5% | 399 ppm |
| Praseodymium | 9 | Pr | 59 | 140.91 | +3 | 8.7 ppm | 0.001 ppb |
| Promethium | 9 | Pm | 61 | (145) | +3 | - | - |
| Protactinium | - | Pa | 91 | 231.04 | +4,+5 | - | - |
| Radium | 5 | Ra | 88 | 226.03 | +2 | - | - |
| Radon | 5 | Rn | 86 | (222) | 0 | - | - |
| Rhenium | 9 | Re | 75 | 186.20 | +4,+6,+7 | 2.6 ppb | 0.004 ppb |
| Rhodium | 9 | Rh | 45 | 102.90 | +3 | 0.7 ppb | 7 ppb |
| Rubidium | - | Rb | 37 | 85.47 | +1 | 60 ppm | 0.12 ppm |
| Ruthenium | 9 | Ru | 44 | 101.07 | +3 | 1 ppb | $7 \times 10^{-4}$ ppb |
| Samarium | 9 | Sm | 62 | 150.35 | +2,+3 | 6 ppm | $8.4 \times 10^{-4}$ ppb |
| Scandium | 9 | Sc | 21 | 44.96 | +3 | 26 ppm | $6.7 \times 10^{-4}$ ppb |
| Selenium | 10 | Se | 34 | 78.96 | +4,+6,-2 | 50 ppb | 0.13 ppb |
| Silicon | 6 | Si | 14 | 28.09 | +2,+4,-4 | 27% | 2.8 ppm |
| Silver | 9 | Ag | 47 | 107.87 | +1 | 80 ppb | 0.0027 ppb |
| Sodium | 11 | Na | 11 | 22.99 | +1 | 2.30% | 1.08% |

| ELEMENT | CHAPTER | SYMBOL | ATOMIC NUMBER | ATOMIC WEIGHT | COMMON VALENCE STATES | AVERAGE CONTINENTAL CRUST | SEAWATER |
|---------|---------|--------|---------------|---------------|------------------------|---------------------------|----------|
| Strontium | 11 | Sr | 38 | 87.62 | +2 | 360 ppm | 7.6 ppm |
| Sulfur | 11 | S | 16 | 32.06 | +4,+6,−2,−1,0 | 420 ppm | 900 ppm |
| Tantalum | 9 | Ta | 73 | 180.95 | +5 | 1.7 ppm | < 0.0025 ppb |
| Technetium | - | Tc | 43 | 98.91 | +4,+6,+7 | - | - |
| Tellurium | 9 | Te | 52 | 127.60 | +4,+6,−2 | 1 ppb | $8 \times 10^{-4}$ ppb |
| Terbium | 9 | Tb | 65 | 158.92 | +3 | 0.95 ppm | $1.7 \times 10^{-4}$ ppb |
| Thallium | 9 | Tl | 81 | 204.37 | +1,+3 | 0.53 ppm | 0.01 ppb |
| Thorium | 5 | Th | 90 | 232.04 | +4 | 6 ppm | $6 \times 10^{-5}$ ppb |
| Thulium | 9 | Tm | 69 | 168.93 | +3 | 0.45 ppm | $1.3 \times 10^{-4}$ ppb |
| Tin | 8 | Sn | 50 | 118.69 | +2,+4 | 2.2 ppm | $5 \times 10^{-4}$ ppb |
| Titanium | 6 | Ti | 22 | 47.90 | +2,+3,+4 | 0.66% | < 0.96 ppb |
| Tungsten | 7 | W | 74 | 183.85 | +6 | 190 ppm | 0.1 ppb |
| Uranium | 5 | U | 92 | 238.03 | +3,+4,+5,+6 | 1.8 ppm | 3.1 ppb |
| Vanadium | 7 | V | 23 | 50.94 | +2,+3,+4,+5 | 190 ppm | 1.2 ppb |
| Xenon | 11 | Xe | 54 | 131.30 | 0 | - | - |
| Ytterbium | 9 | Yb | 70 | 173.04 | +2,+3 | 2.8 ppm | $9.0 \times 10^{-4}$ ppb |
| Yttrium | 9 | Y | 39 | 88.90 | +3 | 29 ppm | 0.007 ppb |
| Zinc | 8 | Zn | 30 | 65.37 | +2 | 79 ppm | 0.4 ppb |
| Zirconium | 5 | Zr | 40 | 91.22 | +4 | 130 ppm | 0.03 ppb |

Values in parentheses are the atomic weight of the longest-lived isotope. Crustal abundance from: www.webelements.com/periodicity/abundance_crust/ and seawater composition from www.seafriends.org.nz/oceano/seawater.htm.

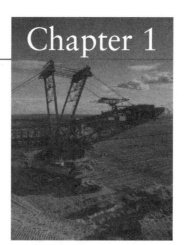

## Chapter 1

# Understanding the Earth's Natural Resources: An Introduction

*"The conservation of our natural resources and their proper use constitute the fundamental problem which underlies almost every other problem of our national life"*

*— Theodore Roosevelt*

*Resource* issues are central to the important challenges facing the world today. They are woven into society at every level as materialistic lifestyles compete with each other and with subsistence living for limited resources. A resource is something that can be used, an asset. On earth these can be divided into living and nonliving resources. This text considers the nonliving ones as given in the grey boxes in **Figure 1.1**. The arrows indicate that energy must be added to rocks, the atmosphere, and water to produce the natural resources shown. Humankind adds most of this energy before a resource is useful but, as shown by the open arrows, some resource production is dominated by the natural input of energy.

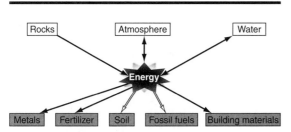

*Figure 1.1* Resource relationships on the earth.

Energy resources are anything used by society as a source for the ability to do work and include coal, crude oil, and natural gas as well as wind and the flow of water, among others. The first three energy resources are termed *fossil fuels* because they have been formed from the organic remains of prehistoric plants and animals.

Two ways to compare resources are shown in **Figure 1.2**. On the top in (a) are the top 10 mineral resources by quantity taken from the earth and on the bottom in (b) are resources ordered by their values. As might be expected, sand plus gravel and aggregate dominate the volume produced, but crude oil is by far the natural resource of greatest value.

Resources can be renewable or nonrenewable. *Renewable resources* are sources of energy or other natural material that are replenished shortly after being used. Renewable resources can depend on the rate of consumption. For instance, the amount of fish consumed on the earth is close to its maximum sustainable yield. If the consumption exceeds this value fish become a nonrenewable resource. Renewable resources include the following:

> solar energy,
> organic matter and its derivatives (food),
> water,
> wind,
> forests, and
> fish.

Nonrenewable resources are natural resources that cannot be remade, regrown, or regenerated

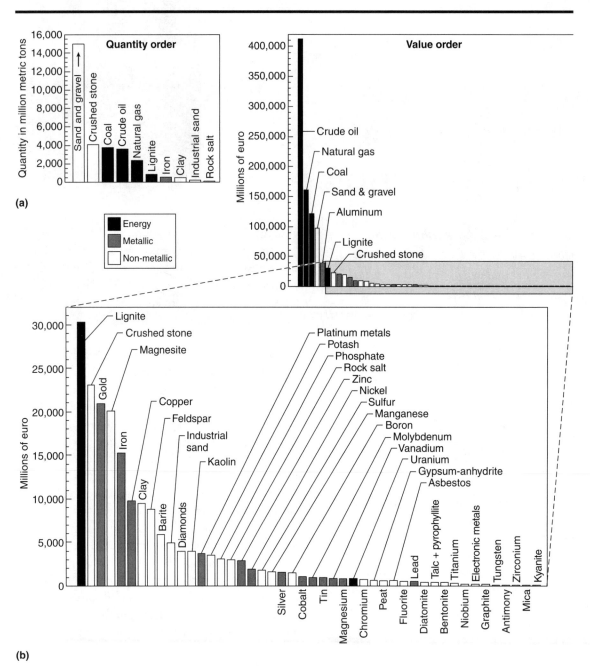

*Figure* 1.2 Resource production in 1998 (a) by quantity (iron given in metal equivalent and natural gas in 1,000 million m⁻³) and (b) by value. Inset in (b) is an expansion of the low value resources. (Data from: Wellmer, F. W. and Becker-Platen, J. D., 2002, Sustainable development and the exploitation of mineral and energy resources: a review, *Inter. Jour. Earth Sci.,* v. 91, pp. 723–745.)

on a time scale comparative to its consumption. These resources are consumed faster by humankind than they are produced by nature and therefore the amounts decrease with time. Nonrenewable resources include fossil fuels and metals extracted from the earth as shown for copper in **Figure** 1.3. Geothermal energy, which is heat extracted from the earth, is also nonrenewable as the extraction

*Figure 1.3* Bingham Canyon copper mine in Utah is the largest excavation (Over 4 km wide) and the deepest open pit mine (Over 1.2 km deep) in the world.

cools the earth locally over human time scales. However, if the total heat output from the earth is considered then the potential to develop geothermal energy is almost limitless.

# Energy and Resources

The earth is in a dynamic state powered by *energy*. This energy comes from a *flux* of sunlight through the earth's atmosphere and a heat flux through rocks from the earth's hot interior. By flux what is meant is the flow of mass or energy through a unit surface area per unit time. The energy flux average over a year at the top of the atmosphere from the sun is 1,360 joules per meter squared per second ($J\ m^{-2}\ s^{-1}$). Because 1 watt = 1 joule per second, the average flux of energy from the sun is 1,360 $W\ m^{-2}$. This is much larger than the average heat flux from the interior of the earth through the top of the crust, which typically varies from 25 to 150 $mW\ m^{-2}$ ($mW$ = milliwatt = $10^{-3}$ watt) and averages 75 $mW$. Therefore, the interior heat flux is too small to affect the temperature or the earth's weather.

The average amount of energy from the interior of the earth that fluxes through an area of 36.5 m × 36.5 m is on average only about

$$36.5\ m \times 36.5\ m \times 0.075\ watts\ m^{-2} = 100\ watts. \quad [1.1]$$

If all this heat was captured and completely converted to electricity it could power a 100-watt light bulb. Not a lot of energy. However, the energy from this internal heat flux is not evenly distributed across the earth. As a result there are tectonically active areas where the heat flux is much higher. Heat in these high-energy areas, as well as energy from the sun, cause reactions that can concentrate minerals of interest and other natural resources. This text explains how.

# Mineral Resources

Modern industrial societies are dependent on energy, water, and mineral resources to produce the goods and services needed. Informed citizens understand their dependence on energy and water as they are used directly and fluctuations in their price are felt immediately. Mineral resources on the other hand are incorporated into finished goods and the connections are not as obvious but the dependence is just as great. Every American born in 2008 is estimated to use the amount of nonfuel mineral resources given in **Table 1.1** in their lifetime. Again, these are generally not used directly but appear in finished products, some of which are outlined in the table.

The estimated average amount of energy by source used every year by an American is outlined in **Table 1.2**. **Figure 1.4** shows what the energy source was used for. The petroleum obtained from crude oil is consumed dominantly in passenger transportation. An average U.S. passenger car or light truck is driven about 12,000 miles (~19,300 km) a year and averages about 20 mpg (~32 km per gallon) so the vehicle consumes 600 gallons of petroleum per year.

The rest of the petroleum consumed in the U.S. is used as jet fuel, to produce heating oil, to make plastics, and as the asphalt base for roads. The coal and uranium are consumed dominantly to make electricity. However, some coal is used to produce heat for industrial applications. Much of the electricity is used by industries that make finished products. About 1/3 of natural gas goes into the production of electricity, 1/3 for heating

*Table 1.1* ESTIMATED NONENERGY MINERAL RESOURCES USED BY AN AMERICAN OVER A LIFETIME.

| MINERAL COMMODITY | AMOUNT REQUIRED OVER A LIFETIME | USES |
|---|---|---|
| Aluminum (bauxite) | 5,677 pounds | Building supports, beverage containers, autos, airplanes |
| Cement | 65,480 pounds | Roads, sidewalks, buildings |
| Clays | 19,245 pounds | Floor and wall tile, bricks and cement, paper, dinnerware |
| Copper | 1,309 pounds | Plumbing, electrical wire |
| Gold | 1,576 ounces | Jewelry, electronic products |
| Iron ore | 29,608 pounds | Mainly steel |
| Lead | 928 pounds | Batteries, TV screens |
| Phosphate rock | 19,815 pounds | Fertilizer, animal feed supplements |
| Stone, sand, and gravel | 1.61 million pounds | Roads, concrete, asphalt, building blocks |
| Zinc | 671 pounds | Metal rust inhibiter, paint, skin creams |

*Data from:* U.S. Geological Survey and U.S. Energy Information Administration; statistical analysis from the National Mining Association.

*Table 1.2* ESTIMATED AVERAGE AMOUNT OF ENERGY USED BY EACH AMERICAN BY SOURCE FOR 2010.

| SOURCE | PETROLEUM | COAL | NATURAL GAS | URANIUM (0.72% = U$^{235}$) |
|---|---|---|---|---|
| Volume or weight | 1,055 gallons | 7,540 lbs | 72,980 cu. ft | 1/3 lb |
| Energy content | $1.2 \times 10^{11}$ J | $7.0 \times 10^{10}$ J | $8.3 \times 10^{10}$ J | $3.0 \times 10^{10}$ J |

J = joules.

*Data from:* Mineral Information Institute, Golden, Colorado.

buildings, and 1/3 for industrial uses. Natural gas vehicle fuel at present accounts for only 0.15% of the natural gas used.

# Determination of Resource Prices

In general, a resource sold as liquid or gas is measured by a standard volume. In the case of liquids this is typically a barrel = 159 liters = 42 U.S. gallons. For gases such as natural gas a volume given by cubic feet or meters at a standard pressure and temperature is used. Solids, like minerals, use a standard weight. This varies from carats for gemstones to *metric tons* for industrial minerals. A metric ton is 1,000 kg while a carat is 0.0002 kg.

For a given mineral commodity prices occur at various stages of production. Consider bauxite, the rock material from which aluminum metal is obtained. Bauxite contains the minerals gibbsite, Al(OH)$_3$, boehmite, AlO(OH), and diaspore,

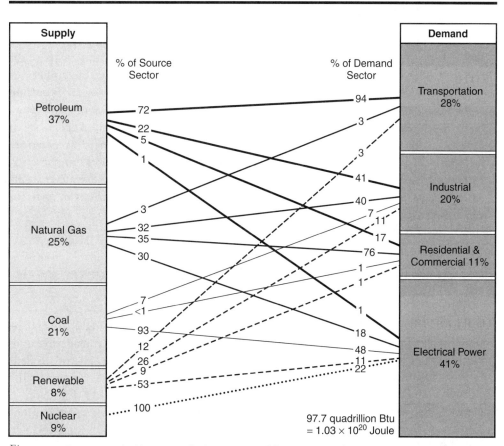

*Figure 1.4* Energy supplied by type and what it was used for in the U.S. for 2010. Note renewable includes hydroelectric power. (Modified from USA Energy Information Administration, Department of Energy.)

AlO(OH). The price of bauxite with a content of about 50 wt% alumina, $Al_2O_3$, was about U.S. $65 per metric ton in 2012. Bauxite is processed and converted to nearly pure alumina that costs about U.S. $365 per metric ton. To produce 1 metric ton of alumina requires about 2 metric tons of 50 wt% alumina bauxite and about $1.3 \times 10^{10}$ J of energy. The alumina is then processed to make Al metal. A ton of Al metal requires about 2.5 tons of alumina and 16 thousand kWh (kilowatt hours) of electricity or $5.8 \times 10^{10}$ joules of energy. A metric ton of Al metal costs about U.S. $2,200. Prices can change for any of the three Al commodities depending on demand and the cost of energy.

Prices are determined by different techniques. Each is an effort to stabilize prices with changes

in demand. The producer can announce the price as well as the terms and conditions for meeting a demand. These are producer prices. Resource prices are often set by a trade association or periodical that determines prices by recent transactions that have taken place. Markets exist for metals such as the London Metal Exchange (LME), and the United States Commodities Exchange (COMEX) for metal's futures and option trading. Alternatively, the buyer and seller can negotiate a price directly. These are then contract prices that typically extend over a number of years with provisions for price adjustments depending on the changes in costs of energy needed, etc. Prices can also be established by auctions open to anyone on future markets such as the Shanghai Futures Exchange (SHFE).

## Resource Classification

There are many ways to classify resources. In this text a classification of resources based on their desired properties is used. Therefore, broad divisions of energy, metals, building and industrial materials, water, and soil are considered and then these are subdivided to produce the following topics:

**Energy Resources**
> Petroleum *(Chapter 2)*
> Natural Gas, Coal, and Related Resources *(Chapter 3)*
> Alternative Energy Resources *(Chapter 4)*
> Nuclear Power *(Chapter 5)*

**Metal Resources**
> Abundant metals *(Chapter 6)*
> Scarce Metals
>> Ferroalloy Metals *(Chapter 7)*
>> Base Metals *(Chapter 8)*
>> Precious and Specialty Metals *(Chapter 9)*

**Life Supporting Resources**
> Building and Industrial Minerals *(Chapter 10)*
> Chemicals from Evaporation of Water and Gaseous Elements from Air *(Chapter 11)*

**Water and Soil Resources**
> The Distribution and Movement of Water *(Chapter 12)*
> Water Quality, Usage, and Law *(Chapter 13)*
> Soil as a Resource *(Chapter 14)*

Given in the Table of Elements, following the Preface, are the elements considered in this text and the chapter where they are discussed. This can be found in the beginning pages of the book. Appendix A gives metric multipliers used for the resource units and Appendix B outlines some common ore minerals. Energy and power unit conversions are tabulated in Appendix C. A glossary of the terms introduced as given in italics when first used can be found in the back of the text and a geological time scale that gives the names of various times periods in earth history is presented on the inside back cover.

## Mineral Resources and Reserves

The importance of energy and mineral resources to humankind is clear by considering a historical perspective of civilizations. Historians define civilizations based on their use of energy and mineral resources as given in **Table 1.3**. Note that we are presently making the transformation to the nuclear plus renewable energy age and petroleum will no longer define our existence. How will we make this transition? This text will consider our use of petroleum, nuclear, and renewable energy going forward as well as the other resources modern society depends on.

*Table 1.3* HISTORICAL AGE AND ITS APPROXIMATE TIME PERIOD.

| AGE | APPROXIMATE PERIOD |
|---|---|
| Paleolithic (Old Stone) | 500,000 to 9,500 BC |
| Neolithic (New Stone) | 9,500 to 5,000 BC |
| Bronze | 5,000 to 700 BC |
| Iron | 700 BC to 200 AD |
| Coal | 200 to 1,850 AD |
| Petroleum | 1,850 AD to present |
| Nuclear/Renewable Energy | Future |

## Resource Evaluation

Understanding the extent of fossil fuel and mineral resources and reserves in a particular property is the basis of determining its value. This includes production cost estimates as well as the varieties of fossil fuel and mineral products contained in the property. Fossil fuel and mineral reserves are then determined by a combination of the economics of extraction and processing operations, and specifics of the market for the fossil fuel/mineral products. Extractions of fossil fuel/mineral reserves are limited by either their physical exhaustion or loss of economic viability.

For a resource to become a *reserve* the location, concentration, quality, and quantity of the resource must be known or estimated using geological insight. It must also be extractable economically under current market conditions. To reflect varying degrees of geological certainty, resources can be subdivided into measured, indicated, inferred, and undiscovered categories as given in **Figure 1.5**.

**Measured reserves**: The size of measured reserves is estimated from examination of outcrops, trenches, road cuts, and/or drill holes. The amount present is determined by physical and chemical analysis of samples. The sampling and geological observations are so closely done that the size, shape, depth, and changes in concentration of the resource are well established. Geophysical methods such as seismic sections and magnetic, electrical, and gravity surveys can be used to confirm the extent of the reserve.

**Indicated reserves**: The size and concentration of indicated reserves are estimated from information similar to that used for measured resources, but the sampling and observations are less frequently spaced. The degree of assurance, although lower than that for measured resources, is high enough to reasonably assume geological continuity between sampling and observations.

**Inferred resources**: For inferred resources estimates are based on geological evidence and assumed continuity in the geological processes operating in the area but there is less confidence than for measured or indicated reserves. Because of the uncertainty these are considered a resource rather than dependable reserves. Inferred resources need not be based on sampling or other measurements. However, the inference needs to be supported by a geological understanding of the resource formation process and particulars of the area considered.

**Possible resources**: Estimates of possible resources are based on broad geological knowledge and an economic model. There is less confidence than for inferred resources. The time lines for possible production are much longer so economic changes over time become more important.

Often the term *reserve base* is used in considering resource availability. In resource analysis this is typically the sum of measured reserves + indicated reserves + marginally economic reserves + a portion of subeconomic reserves (see Figure 1.5).

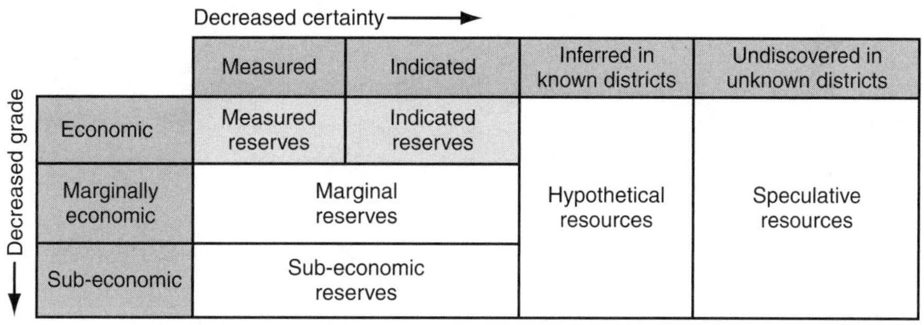

| | Decreased certainty → | | | |
|---|---|---|---|---|
| | Measured | Indicated | Inferred in known districts | Undiscovered in unknown districts |
| Economic | Measured reserves | Indicated reserves | Hypothetical resources | Speculative resources |
| Marginally economic | Marginal reserves | | | |
| Sub-economic | Sub-economic reserves | | | |

*Decreased grade ↓*

*Figure 1.5* Relation of reserves to resources for materials found in the earth.

# Geochemical Cycles

An important broad way to view resources is in the context of *geochemical cycles*. A geochemical cycle indicates chemical changes in terms of fluxes between reservoirs on a particular time scale and generally considers the whole earth. These reservoirs can include the solid earth, ocean, and atmosphere.

## Water Cycle

As an example of a geochemical cycle, consider the present day *water cycle*. This outlines the changes in the amount of $H_2O$ in reservoirs on the present earth, as given in **Figure** 1.6. Note that a reservoir is in *steady state* and, therefore, does not change its size with time when the flux of $H_2O$, as given by the arrows, into a reservoir is equal to the flux out. Therefore, as shown in the figure the amount of water on the continents is in steady state.

The amount of $H_2O$ in the atmosphere is also in steady state as the $H_2O$ that precipitates as rain and snow on the continents, $110 \times 10^{15}$ kg per year, together with that which precipitates on the ocean, $380 \times 10^{15}$ kg per year, is equal to the sum of $H_2O$ that transpires from plants and evaporates from wet surfaces on the continents, $70 \times 10^{15}$ kg per year, together with that which evaporates from the ocean. Whether a reservoir is in steady state or not depends on the time frame considered. Clearly the size of the continental $H_2O$ reservoir has changed over a 100,000-year time frame as the amount of ice in ice sheets has expanded and contracted during the earth's most recent ice ages. Also, if global warming occurs the amount of water in the atmosphere will increase if the average relative humidity in the atmosphere stays constant.

## Carbon Cycle

A particularly important geochemical cycle for resource considerations is that for carbon. This is because humankind—through fossil fuel and biomass burning—is interfering with the natural balances of carbon in the cycle as shown by the long dashed arrows in **Figure** 1.7. The cycle is more complex than the hydrological cycle because carbon can exist in a variety of compounds, both organic and inorganic.

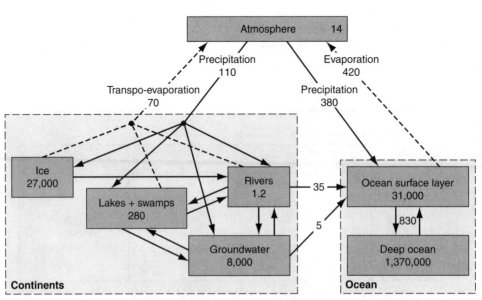

*Figure* 1.6 Present day water cycle with boxes denoting reservoirs of the indicated size in Eg (Eg = exagram = $10^{18}$ g) and arrows showing the fluxes in Eg $yr^{-1}$ between the reservoirs.

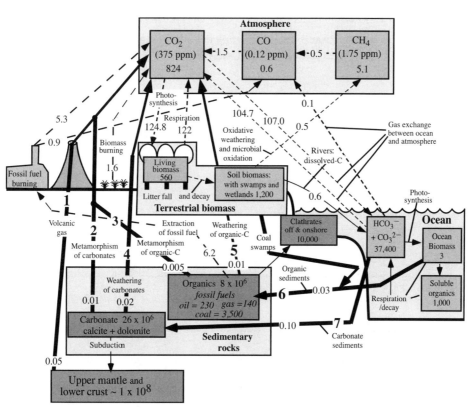

*Figure* 1.7 Carbon cycle in the year 2005 for the whole earth showing the reservoirs of carbon (units of Pg) as boxes and major fluxes (units of Pg yr$^{-1}$) between reservoirs given as arrows (Pg = 10$^{15}$ g).

**Table** 1.4 tabulates the flux of grams of carbon into and out of the atmospheric $CO_2$ reservoir as given in Figure 1.7. Note that this reservoir is not in steady state as 3.4 Pg (Pg = petagram = 10$^{15}$ g) per year more carbon fluxes into the atmosphere than fluxes out. Judging from the anthropogenic flux this increase is due to humankind's input of carbon into the atmosphere.

### $CO_2$ in the Atmosphere

Shown in **Figure 1.8** are measurements of the $CO_2$ concentration of air as a function of time at the Mauna Loa Observatory in Hawaii, the "Keeling curve." The curve is named for the late Charles Keeling of the Scripps Institution of Oceanography, who had been undertaking the measurements until his death in 2005. Note the yearly cycle of $CO_2$ that declines in the spring when photosynthesis,

and therefore the growth of plants, is at its maximum in the Northern Hemisphere. In the autumn $CO_2$ increases as photosynthesis decreases and dead vegetation starts to decay, releasing $CO_2$ and $CH_4$ to the atmosphere. Written in terms of equations, these observations are as follows:

Spring: $H_2O + CO_2 + sunlight \rightarrow CH_2O_{(organic\ matter)} + O_2$ [1.2]

Fall: $CH_2O_{(organic\ matter)} + O_2 \rightarrow H_2O + CO_2.$ [1.3]

Burning of carbon containing fossil fuels in the autumn and winter to generate heat also contribute significantly to the yearly cycling of $CO_2$. This can also be represented by reaction [1.3].

Figure 1.8 shows an increase in $CO_2$ on a year-to-year basis. Anthropogenic contributions to this increased concentration of $CO_2$ are due to the clearing of forests, which decreases the photosynthetic $CO_2$ sequestering effect, labeled

*Table 1.4* PRESENT-DAY CARBON FLUXES INTO AND OUT OF THE ATMOSPHERIC $CO_2$ RESERVOIR*.

| ATMOSPHERIC $CO_2$ RESERVOIR (UNITS OF Pg CARBON $YR^{-1}$) | |
|---|---|
| **FLUX IN** | **FLUX OUT** |
| 122.0 = Respiration | 124.8 = Land photosynthesis |
| 104.7 = Ocean degassing | 107.0 = Ocean absorption |
| 5.3 = Fossil fuel burning | **231.8 = Total** |
| 1.5 = Oxidation of CO | |
| 1.6 = Biomass burning | |
| 0.1 = Long-term fluxes from mantle + crust (shown with thick lines) | In – Out = 235.2 – 231.8 = 3.4 Pg/yr |
| **235.2 = Total** | |

*As given in Figure 1.7 in Pg (= $10^{15}$ g).

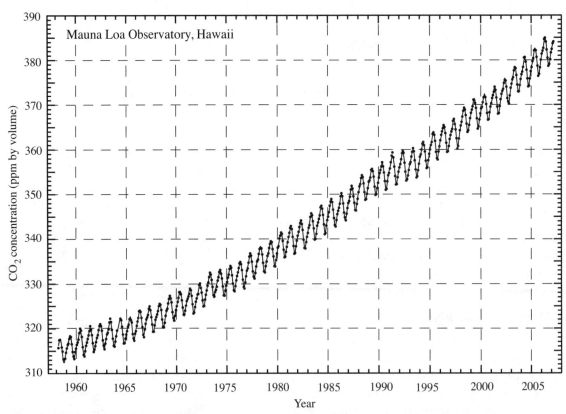

*Figure 1.8* Monthly average $CO_2$ concentration in dry air measured in Hawaii for the indicated year. (Data from: Keeling, C. D. and Whorf, T. P., 2003, Atmospheric $CO_2$ records from sites in the SIO air sampling network. In Trends: A Compendium of Data on Global Change. Carbon Dioxide Information Analysis Center, Oak Ridge National Laboratory, U.S. Department of Energy, Oak Ridge, TN.)

"photosynthesis" in Figure 1.7 as well as the burning of fossil fuels. The $CO_2$ increase to the atmosphere is mitigated to some extent by an increased absorption by the ocean. A "biologic pump" in the ocean helps in this removal of $CO_2$ from the atmosphere. $CO_2$ is taken up by phytoplankton at the ocean's surface because of reaction [1.2]. When the phytoplankton die, they sink to the deep ocean where they are increasingly unstable and decay by reaction [1.3] and thus transfer $CO_2$ from shallow to deep water. This then promotes a greater flux of atmospheric $CO_2$ to the shallow ocean.

To determine the $CO_2$ concentration in the atmosphere before direct measurements were made the concentration of $CO_2$ in air trapped in ice can be measured. Those formed from atmospheric precipitation of $H_2O$ in annual layers in the Arctic and Antarctica give some records greater than 10,000 years. Determinations from Siple Station in West Antarctica along with the Keeling curve are given in **Figure 1.9**.

The increases of concentration of $CO_2$ with time in the atmosphere outlined in Figures 1.7 and 1.9 can be compared with the increases given in Figure 1.8. From Figure 1.8 the average increase of $CO_2$ in the atmosphere with time is about 1.55 ppm by volume per year for the past few years. This is determined by measuring the slope of the line connecting points at the same time of year. With the atmosphere modeled as an ideal gas, this ppm by volume increase is equal to its mole fraction increase per year of $1.55 \times 10^{-6} \ yr^{-1}$. The increase in carbon is then given by this mole fraction times the molecular weight of carbon, $12.01 \ g \ mol^{-1}$, times the mass of the atmosphere in moles. This molar mass of carbon equals the mass of the atmosphere in grams, $5.3 \times 10^{21} \ g$ (Campbell, 1977), divided by the grams of carbon in a mole of air, $28.97 \ g \ mol^{-1}$ or

$$\frac{1.55 \times 10^{-6} \ yr^{-1} \times 12.01 \ g \ mol^{-1} \times 5.3 \times 10^{21} \ g}{28.97 \ g \ mol^{-1}}$$

$$= 3.4 \ Pg \ yr^{-1}. \qquad [1.4]$$

This is consistent with the value outlined in Table 1-4. Measurements in ice cores from ice layers older than the year 1750 give an atmospheric $CO_2$ concentration of $280 \pm 3$ ppm for hundreds of years. Therefore, the increase in atmospheric $CO_2$ started with the advent of the Industrial Revolution.

### Methane Clathrate Stability

An examination of Figure 1.7 indicates the large size of the *methane clathrate* reservoir of carbon containing methane gas, $CH_4$. Methane in clathrates occurs with bacterial decomposition of organic matter in a low oxygen environment, similar to the formation of "swamp gas." Under low enough temperatures and high enough pressures a solid methane clathrate phase forms in the sediments. Methane clathrates are cage-like structures of cubic ice with a methane gas atom within the cage. In the ocean, methane clathrate is not stable in sediments until the water above reaches a thickness of 1 km. At these depths sediments are cool enough and under high enough pressure.

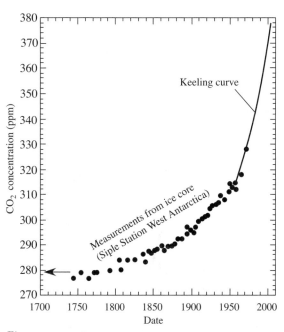

*Figure 1.9* Yearly average $CO_2$ concentration in the atmosphere for the given date from the Keeling curve and ice core determinations. (Data from: Friedli, H., H. Lötscher, H. Oeschger, U. Siegenthaler, and B. Stauffer, 1986, Ice core record of 13C/12C ratio of atmospheric $CO_2$ in the past two centuries. *Nature* v. 324, pp. 237–238.)

With increasing depth in the earth, the increasing temperature along the geothermal gradient makes clathrate unstable. As shown in **Figure 1.10**, this occurs about 1/2 km below the ocean floor or at a depth 3/4 km below the land surface.

Because the size of the methane clathrate reservoir is greater than all the carbon residing in other fossil fuel deposits, researchers have investigated the possibility of obtaining methane from clathrates as a potential fuel source. However, the technical problems of large-scale development have been intractable to date. Some investigators have suggested natural large-scale release of methane from clathrates limits the extent of ice ages. Sea level is lowered in an ice age as more water is put on the continents as ice. These investigators argue the lower sea level lowers the pressure put on the methane clathrates at the bottom of the ocean and they become unstable. The clathrates release $CH_4$ to the atmosphere that reacts with oxygen producing $CO_2$. As greenhouse gasses, $CH_4$ and $CO_2$ cause global warming, which ends the ice age.

# Common Rocks

In order to understand the formation of mineral resources it is helpful to have some background with terms used to describe rocks and minerals. Rocks can be classified as igneous, sedimentary, or metamorphic.

*Igneous rocks* are formed by the crystallization of molten *magma*. A *sedimentary rock* is produced from solid grains that have weathered from material at the earth's surface, then settled and accumulated or they have formed from solid grains that have precipitated directly out of water. *Metamorphic rocks* are previously formed igneous or sedimentary rocks that have changed their appearance by undergoing significant changes in mineralogy, structure, and/or chemistry in response to changes in temperature and/or pressure as they are buried in the earth.

Rocks are made up of minerals, glass, and organic material. There are many minerals of importance. Given in **Table 1.5** are some common rock-forming minerals divided into those that are produced in igneous, sedimentary, and metamorphic processes.

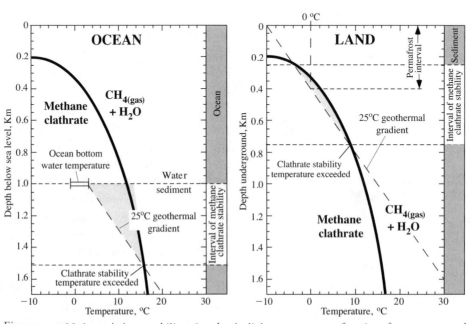

*Figure 1.10* Methane-clathrate stability, given by the light grey areas, as a function of temperature and depth both in the ocean and on land.

*Table 1.5* COMMON ROCK-FORMING MINERALS.

| MINERALS FORMED IN IGNEOUS ROCKS | | |
|---|---|---|
| MINERAL | COMPOSITION | COLOR |
| Quartz | $SiO_2$ | Translucent, white, grey |
| Feldspar: Plagioclase | $NaAlSi_3O_8$ to $CaAl_2Si_2O_8$ | Light to dark in color |
| K-feldspar | $KAlSi_3O_8$ | Typically pink |
| Hornblende | $Ca_2(Mg,Fe)_4Al(AlSi_7O_{22})(OH)_2$ | Black |
| Pyroxene | $Ca(Mg,Fe)Si_2O_6$ or $(Mg,Fe)_2Si_2O_6$ | Green to black |
| Mica: Muscovite | $KAl_2(AlSi_3)O_{10}(OH)_2$ | Translucent |
| Biotite | $K(Mg,Fe)_3AlSi_3O_{10}(OH)_2$ | Black |
| Olivine | $(Mg,Fe)_2SiO_4$ | Olive green to black |
| Magnetite | $Fe_3O_4$ | Black, shiny |
| **MINERALS FORMED IN SEDIMENTARY ROCKS** | | |
| MINERAL | COMPOSITION | COLOR |
| Calcite | $CaCO_3$ | Translucent, white |
| Dolomite | $CaMg(CO_3)_2$ | White, grey, pink |
| Kaolinite | $Al_2Si_2O_5(OH)_4$ | White |
| Halite | $NaCl$ | Translucent |
| Gypsum | $CaSO_4 \cdot 2H_2O$ | Translucent, white |
| Hematite | $Fe_2O_3$ | Red, shiny grey |
| Limonite | $FeOOH \cdot nH_2O$ | Yellowish red |
| **MINERALS FORMED IN METAMORPHIC ROCKS** | | |
| MINERAL | COMPOSITION | COLOR |
| Talc | $Mg_3Si_4O_{10}(OH)_2$ | White |
| Chlorite | $Mg_5Al_2Si_3O_{10}(OH)_8$ | Green |
| Garnet | $(Fe,Mg,Ca)_3Al_2Si_3O_{12}$ | Pink, red, green, black |
| Alumino-silicates: Andalusite | $Al_2SiO_5$ | Often pink |
| Kyanite | $Al_2SiO_5$ | Blue to white, grey, green, black |
| Sillimanite | $Al_2SiO_5$ | Transparent to white |

# Igneous Rocks

Igneous rock that has formed within the earth is termed plutonic and that crystallized from lavas as well as eruptive gas and airborne magma at the earth's surface is called volcanic. **Figure** 1.11 gives the names and mineral content of common volcanic and plutonic igneous rocks. Note that these

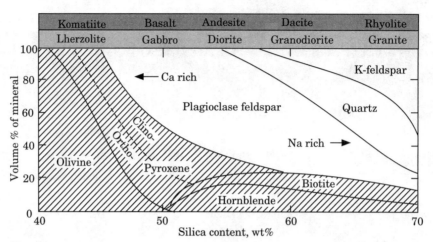

*Figure 1.11* Volume percent of minerals in plutonic igneous rocks as a function of their silica content. The lined region of the diagram indicates the mafic (dark-colored) minerals, whereas the unlined area gives the felsic (light-colored) minerals in the rock. The light grey bar on the top of the diagram outlines the name of the intrusive plutonic igneous rock for the indicated range of $SiO_2$ content. The darker grey bar specifies the equivalent volcanic rocks of similar chemical composition.

common igneous rocks are defined based on their $SiO_2$ content. Rocks, magma, or minerals that are rich in magnesium and iron are termed *mafic*. They tend to be dark in color. Mafic rocks and magmas have 45 wt% < $SiO_2$ < 52 wt%. This includes both basalt and gabbro as shown in Figure 1.11. *Ultramafic* is a term used for rocks and magma with less than 45 wt% $SiO_2$. They occur much less frequently than mafic rocks in the crust. The upper mantle of the earth is, however, composed of the ultramafic rock peridotite.

## Sedimentary Rocks

Sedimentary rocks cover the continental crust to an extensive depth in some locations. However, on a global scale they are a thin veneer over the metamorphic and igneous rocks that make up 95% of the earth's crust. Sedimentary rocks are formed from deposited sediments in layers termed strata that produce beds of rock. A bed is the smallest unit in sedimentary rocks, ranging in thickness from a centimeter to several meters that is distinguishable from beds above and below it. Given in

Table 1.6 are some sedimentary rock names and their characteristics.

## Metamorphic Rocks

When rocks, formed at the earth's surface, are buried in the earth they are subjected to higher temperatures and pressures. This is because heat is escaping from the earth. This heat was produced by the conversion of gravitational energy to heat energy when the earth was formed. Added to this is heat produced by radioactive decay of some elements such as radioactive potassium and uranium. As a result, near the earth's surface the increase in temperature with depth, the *geothermal gradient*, is generally between 15° and 40°C per kilometer.

With burial at temperatures of about 150°C and above, sediments and volcanic material produced at the earth's surface undergo notable transformations leading to their recrystallization that are termed metamorphic. Given in **Table 1.7** are names for some metamorphic rocks and their characteristics.

*Table 1.6* COMMON SEDIMENTARY ROCKS.

| TYPE | PARTICLE SIZE | DESCRIPTION | ROCK NAME |
|------|---------------|-------------|-----------|
| Clastic (Fragments of preexisting rocks) | Coarse grain | Round clasts | Conglomerate |
| | | Angular clasts | Breccia |
| | Fine grain (visible to naked eye) | Predominately quartz and/or feldspar | Sandstone |
| | | Type of sandstone of quartz with >25% K-feldspar | Arkose |
| | | Predominately rock fragments, mica & clay | Graywacke |
| | Very fine grain (invisible to naked eye) | Some grains can be seen with hand lens | Siltstone |
| | | Grains can't be seen with hand lens, non-laminated | Mudstone |
| | | Grains can't be seen with hand lens, laminated | Shale |
| Organic | Varies | Calcite with or without fossils | Limestone |
| | | Soft, porous carbonaceous plant material | Peat |
| | | Blocky, black carbonaceous plant material | Lignite/Coal |
| Chemical | Generally, fine grain | Composed of dolomite ($CaMg(CO_3)_2$) | Dolostone |
| | | Composed of chalcedony ($SiO_2$) | Chert |
| | | Composed of halite (NaCl) or gypsum ($CaSO_4 \bullet 2H_2O$) | Evaporite |

*Table 1.7* COMMON METAMORPHIC ROCKS.

| TEXTURE | COMPOSITION | DESCRIPTION | ROCK NAME |
|---------|-------------|-------------|-----------|
| Foliated or banded | Pelite (originally clay-rich) | Fine grained with dense, thin pieces | Slate |
| | | Fine grained with satiny luster | Phyllite |
| | | Medium grained with plainer aligned mica | Schist |
| | | Medium-coarse grained, alternating light and dark bands | Gneiss |
| | | Fine grained, dense and dark | Hornfels |
| Non-foliated | Basalt | Medium-coarse grained, black with prismatic amphibole | Amphibolite |
| | Carbonate | Medium-coarse grained calcite or dolomite | Marble |
| | Quartz | Medium-coarse grained quartz | Quartzite |
| | Organic carbon | Black, shiny, conchoidal fractures | Anthracite |

# Population Growth

There are two major drivers of increased resource use, population growth, and increased *per capita* use. Estimated past world population as a function of time from 0 AD is given in **Figure 1.12** along with the dates of some historic events. It is estimated that over 55 million people lived in the Eastern and Western Roman Empire at the time the city of Constantinople was built. The *Black Death plague* between 1347 and 1351 likely reduced the population of the world from 450 to 350 million. It is probable that the world population reached 1 billion in about 1810, stood at 2.5 billion in 1950, and increased to over 3 billion by 1960. In 1999 the world population was estimated to be 6 billion, a doubling of the population in less than 40 years.

There were 7 billion people on the earth at the end of 2011. World births have leveled off at about 135 million per year. Deaths are now about 60 million per year resulting in 75 million people added per year. The growth rate of the population is then equal to 75 million people per year/7.0 billion people = 1.07% per year. This 2011 rate of population growth is less than half of its peak of 2.2% per year, which was reached in 1963 (see **Figure 1.13**). A dip in the growth rate in 1959–1960 occurred because of both natural disasters and decreased agricultural output in China due to a massive social reorganization termed the "Great Leap Forward." China's death rate rose sharply and its fertility rate fell by almost half.

**Table 1.8** gives the populations and annual growth rates of the world's 20 largest countries. Most of the people in the world live in only a few countries. For instance, in 2009, 37% of the world's population lived in China and India while 22% of the population lived in the next

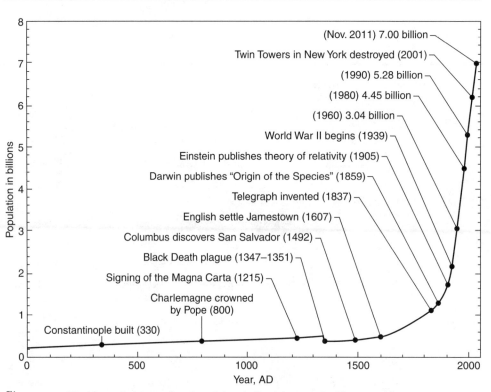

*Figure 1.12* World population as a function of time indicating some significant world events.

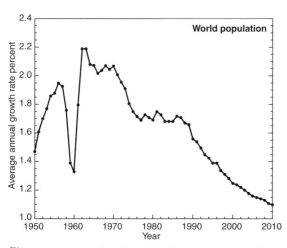

*Figure 1.13* Annual mid-year world population growth rate from the U.S. Census Bureau International Data Base (IDS). (Data from: U. S. Census Bureau International Data Base.)

eight largest countries, which in order of decreasing population are the United States, Indonesia, Brazil, Pakistan, Bangladesh, Nigeria, Russian Federation, and Japan. The 230 countries that have populations less than 20 million people contain only 11% of the world's population.

Note in Table 1.8 that Russia and Japan have a negative growth rate so they are decreasing populations as a function of time. The growth rate of India is greater than that of the People's Republic of China so that if rates stayed the same India will become the most populous country in the world. To determine when this will happen consider the mathematics of growth.

## Mathematics of Growth

If the annual compounded growth rate stabilizes at 1.1% what is the consequence for world population growth into the future? The equation for compounded (exponential) growth of variable $x$ is

$$\frac{dx}{dt} = kx \qquad [1.5]$$

where k is a constant, called the growth rate, giving the rate of increase of $x$ with time, $t$. Consider starting with a population of 6.83 billion in 2010 (year = 0) with a rate of growth of

k = 1.1% per year. Rearranging and writing the integral of equation [1.5] for $x$ = *population* gives

$$\int_{6.83\ \text{billion}}^{population\ at\ t} \frac{d(population)}{population} = \int_{year = 0}^{year = t} k\ dt. \qquad [1.6]$$

With k = 0.011, performing the integration of both sides of equation [1.6] and evaluating the limits results in

$$\ln\,(population\ \text{at}\ t) - \ln\,(6.83\ \text{billion}) = 0.011t - 0. \quad [1.7]$$

Taking the exponential of both sides, gives

$$(population\ \text{at}\ t) = 6.83\ \text{billion} \times e^{0.011t}. \qquad [1.8]$$

The evaluation of equation [1.8] as a function of time to 2110 is shown in **Figure 1.14**. Note that in 100 years of growth the population on the earth is calculated to be 20.5 billion people or a tripling of the present population.

To determine when the countries of the People's Republic of China, with a growth rate of 0.60%, and India, with a growth rate of 1.44%, have the same population, equation [1.8] can be written for each country replacing 6.83 billion with their current populations and equating the populations at a time, $t$, in the future to give

$$\underset{\text{India}}{1,181,263,000 \times (e^{0.0144t})} = \underset{\text{Peoples Republic of China}}{1,337,700,000 \times (e^{0.0060t}).} \quad [1.9]$$

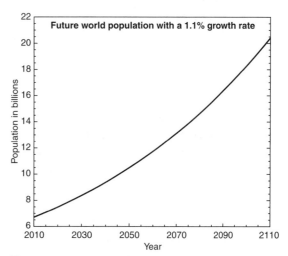

*Figure 1.14* Predicted future world population with a 1.1% yearly growth rate as a function of time for the indicated year.

*Table 1.8* POPULATION AND GROWTH RATE RANGE OF THE 20 LARGEST COUNTRIES IN 2010.

| COUNTRY | POPULATION | ANNUAL GROWTH RATE |
|---|---|---|
| People's Rep. of China | 1,337,700,000 | 0.58–0.66 |
| India | 1,181,263,000 | 1.41–1.46 |
| United States | 309,345,000 | 0.97–0.98 |
| Indonesia | 231,370,000 | 1.14–1.16 |
| Brazil | 192,976,000 | 1.20–1.26 |
| Pakistan | 169,580,000 | 1.56–1.84 |
| Bangladesh | 162,221,000 | 1.29–1.67 |
| Nigeria | 154,729,000 | 2.00–2.27 |
| Russia | 141,927,000 | −0.47–0.51 |
| Japan | 127,390,000 | −0.02–0.19 |
| Mexico | 107,551,000 | 1.12–1.13 |
| Philippines | 92,227,000 | 1.72–1.96 |
| Vietnam | 85,790,000 | 1.14–1.32 |
| Germany | 81,758,000 | −0.05–0.07 |
| Ethiopia | 79,221,000 | 2.51–3.21 |
| Egypt | 78,308,000 | 1.76–2.03 |
| Iran | 74,196,000 | 0.88–1.35 |
| Turkey | 72,561,000 | 1.26–1.31 |
| Dem Rep. of Congo | 66,020,000 | 2.11–3.22 |
| France | 65,447,000 | 0.49–0.55 |

Taking the natural logarithm of both sides of equation [1.9] and solving for $t$ gives $t = 15$ years. Therefore, in 2025 India will have a greater population than China if the present growth rates stay constant.

## Fertility

Humankind presently has an average *fertility rate* of 2.56. The fertility rate is the average number of children a woman will bear in her lifetime. The replacement rate in a population where there is zero population growth (ZPG) is somewhat greater than 2.0 because of a significant infant mortality rate before women are of childbearing age. For the world as a whole it takes a fertility rate of 2.1 to 2.2 to have ZPG. In countries with low life expectancies, the fertility rate for zero population growth is even higher, 2.2 to 3.0. However, even a current fertility rate as low as 2.1 may not ensure zero population growth over time.

If during a particular period of time in the past a population has an unusually large number of children exceeding its ZPG, the added population will pass through their childbearing years increasing the population even if their fertility is 2.1 or less. Most childbearing is done by women between the ages of 15 and 49. If a population has a large number of young people just entering their reproductive years, the rate of growth of that population is sure to rise.

# Distribution of People on the Earth as a Function of Time

By 2050, Bangladesh, Ethiopia, and the Democratic Republic of the Congo will be among the 10 most populous countries in the world according to the United Nations Population Division. **Table 1.9** gives estimates of the past world population

*Table 1.9* DISTRIBUTION OF THE WORLD'S POPULATION (%) IN MAJOR AREAS ACCORDING TO DIFFERENT MODELS TO 2050.

| MAJOR AREA | 1950 | 1975 | 2009 | 2050 LOW | 2050 MEDIUM | 2050 HIGH | 2050 CONSTANT |
|---|---|---|---|---|---|---|---|
| More-developed regions | 32.1 | 25.8 | 18.1 | 14.2 | 13.9 | 13.8 | 11.4 |
| Less-developed regions | 67.9 | 74.2 | 81.9 | 85.8 | 86.1 | 86.2 | 88.6 |
| Least-developed countries | 7.9 | 8.8 | 12.2 | 18.4 | 18.3 | 18.1 | 22.4 |
| Other less-developed countries | 60.0 | 65.4 | 69.7 | 67.5 | 67.8 | 68.1 | 66.2 |
| Africa | 9.0 | 10.3 | 14.8 | 22.0 | 21.8 | 21.7 | 27.2 |
| Asia | 55.5 | 58.6 | 60.3 | 57.0 | 57.2 | 57.4 | 54.5 |
| Europe | 21.6 | 16.6 | 10.7 | 7.6 | 7.6 | 7.5 | 6.0 |
| Latin America and Caribbean | 6.6 | 8.0 | 8.5 | 7.9 | 8.0 | 8.1 | 7.6 |
| Northern America | 6.8 | 6.0 | 5.1 | 5.0 | 4.9 | 4.8 | 4.2 |
| Oceania | 0.5 | 0.5 | 0.5 | 0.6 | 0.6 | 0.6 | 0.5 |

*Data from*: Population Division of the Department of Economic and Social Affairs of the United Nations Secretariat (2009). *World Population Prospects: The 2008 Revision.*

by region and a prediction for the year 2050. Note that 32% of the population lived in more-developed countries in 1950 but this percentage is decreasing every year and by 2050 is predicted to be somewhere between 14.2% and 13.8%. This is because the population of Africa with its less-developed regions is predicted to grow from 9% of the world population in 1950 to about 22% in 2050.

Figure 1.15 shows the predicted changes in world population by region as a function of time plotted on a logarithmic scale. Note that European population is expected to decrease into the future due to a low birth rate. As given in Table 1.9 the African population is expected to increase rapidly as will the Latin American population. The Latin American population will increase past that in Europe in about the year 2030.

## The Demographic Transition

In general, countries become more developed with time. In less-developed countries this development causes a shift from high birth and death rates to low birth and death rates. This is called the *demographic transition*. Slowly declining birth rates following an earlier sharp decline in death rates are today characteristic of most of the less-developed regions of the world. This is due to many factors including the following:

better nutrition,
greater access to medical care,
improved sanitation, and
more widespread immunization.

This decline in death rates with nearly as high a birthrate has caused significant increase in developing nations' populations.

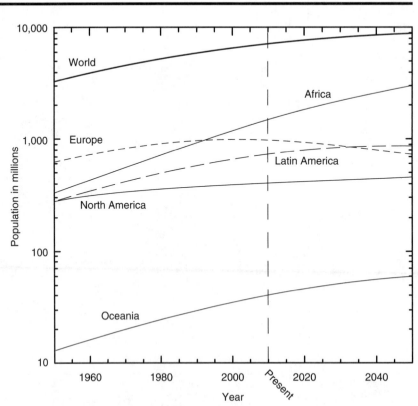

*Figure 1.15* Past and estimated future population of the indicated region plotted on a log scale. (Data from: United Nations, Department of Economics and Social Affairs)

## Age Distribution of the Human Population

**Figure** 1.16 gives *population pyramids* for Uganda and Sweden. These indicate the percentage of population in 5-year increments of age for males and females in the country. Uganda with its broad base that rapidly narrows as the population's age increases is characteristic of a population with a high birthrate and high mortality rate in all age groups to age 50 where the relative mortality rate is lower. On the other hand, the barrel-shaped population pyramid for Sweden is characteristic of a low birthrate and low mortality rate population. Variations to the population pyramids point to significant population events. Large permutations in immigration or past high birthrates can lengthen a bar relative to its neighbors. Relative shortening of bars can be the result of war or epidemics that are prevalent in a particular age range of the population.

**Figure** 1.17 gives population pyramids for China and the United States in terms of total population. China had a classic high birthrate and high mortality rate population distribution like Uganda until 1979 when it instituted a one-child-per-couple policy. Since then the percentage of children has decreased. Note that if the birthrate stabilizes at its current values and the morality rate decreases, a barrel-shaped population distribution given by the 0–14 year groups will develop.

In China population spikes occur in the 40–44 and 20–24 age groups while in the United States the spikes occur at ages of 45–49 and 20–24. The older age spike in population is the *baby boomers*, the demographic boom in births after World War II. This includes those born between 9 months after the war (1946) and 19 years later (1965). This would make them 64 to 45 years old in 2010. Note that the boom in births in China was delayed by about five years from that in the U.S. because of the slower change to a post-war industrial society in China. Baby Boomers are entering their senior years. Many are now retiring and leaving the labor force. With a shrinking working-age population, who will take care of the country's retirees? This is a question that needs to be answered not only in the U.S., but also in the world at large.

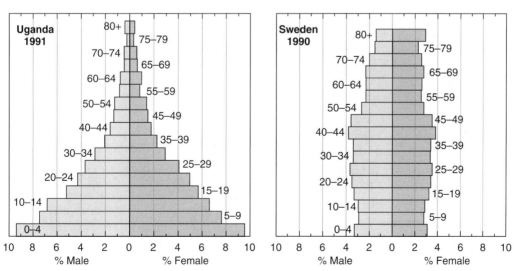

*Figure* 1.16 Age distribution and male/female ratio of people in Uganda (1991) and Sweden (1990) from U.S. Census Bureau International Data Base. (Data from: U. S. Census Bureau International Data Base.)

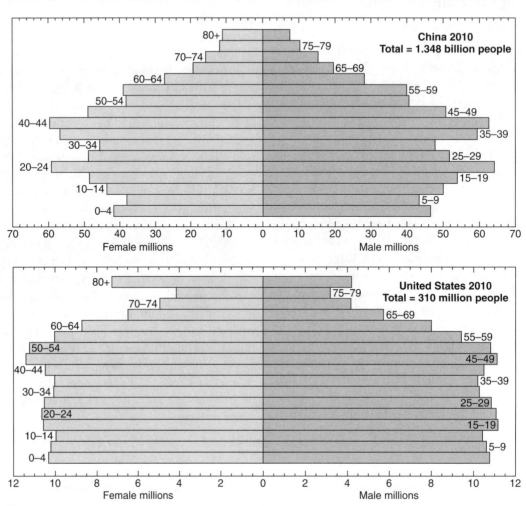

*Figure 1.17* Age distribution and male/female ratio in China and the United States in 2010 from data in the U.S. Census Bureau International Data Base. (Data from: U.S. Census Bureau International Data Base.)

The 20–24 age population spike, born between 1990 and 1995 is part of an "Echo Boomer" generation as they are offspring of the Baby Boomers. In the U.S. and Europe they are also referred to as late *Millennials* (or Generation Y, born ~1980–1997) because they were brought up using digital technology and exposed to mass media. Millennials are new voters of which many political parties wish to embrace. In the United States they are the first generation to grow up in a society with both desegregation and sexual equality by law.

# The Earth's Human Carrying Capacity

Can the world sustain increased population? A concept developed from population dynamics is the *carrying capacity*. The carrying capacity within a given habitat gives the maximum sustainable abundance of a species in the habitat. When a species population is at its carrying capacity the birth and death rates are equal, and the size of the population does not change with time. Populations that overshoot the carrying capacity are not sustainable, and the environment will adjust to bring the population back to its carrying capacity. In nature, populations vary with time for reasons that may be complex and difficult to understand. The notion of a carrying capacity is useful as it highlights that for all species, including humans, there are habitat limitations to the sizes of populations that can be sustained.

**Figure 1.18** shows how a typical species reaches its carrying capacity. When a species is introduced into a habitat its population grows exponentially with time. At some point negative feedbacks such as limits on food or increases in the number of predators slows the increase until the population reaches a steady state, the carrying capacity. The equation for development of a steady-state population is given by

$$\frac{d(population)}{dt} = k \times population \times \left( \frac{K - population}{K} \right) \quad [1.10]$$

where K is the carrying capacity and k is the growth rate as discussed above. Note the similarity of equation [1.10] to equation [1.5] but the right-hand side is multiplied by an added term in parenthesis involving K. Because the population is always less than K, this term is less than 1 and acts as a negative feedback against exponential growth. As the population approaches K the term in parenthesis becomes zero and the steady state of the carrying capacity is reached as the change in population with time is zero. In Figure 1.18 equation [1.10] is plotted as a dashed line with k = 1.1% per time unit (years) and K = 9 billion people.

Given the likely increase in the world population one can ask what the carrying capacity of humans on the earth is? That is, what is the number of people the earth can sustain and what negative feedbacks could limit the population? It has been argued that the earth may be able to support 40 to 50 billion people ~ 3 doublings. However, of the 7 billion people alive today, 0.5 to 1.1 billion are presently undernourished.

Does carrying capacity increase with time because humans have the ability to alter their environment and can make rational choices? It can be argued that humankind will apply advances in

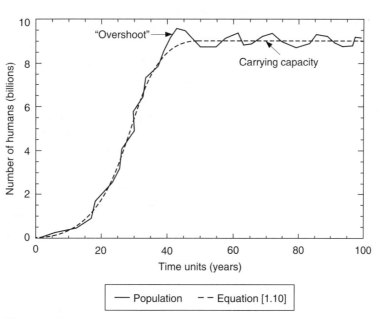

*Figure 1.18* A possible model of population growth with time for a system that reaches its carrying capacity plotted as a dashed line as given by equation [1.10] with k = 1.1% and K = 9 billion.

technology and agricultural production techniques to increase the food supply. Also the carrying capacity could increase from improvements in public health and development of vaccines for diseases that decreases the negative feedback in equation [1.10]. Is it possible, however, that humankind will be limited by the resources needed to make these advancements? In the case of humans what could K, the carrying capacity, be controlled by: limited resources of energy, minerals, or fresh water? This text helps one understand the natural resource supplies available to humankind.

In regard to resources, a worldwide *economic stagnation* could occur if we run out of a resource because it is nonrenewable and the world can't satisfy increased demand by substitution or recycling. It could also occur if a resource is renewable but only at a slower rate than required, leading to decreased supply relative to demand. Finally, economic stagnation could occur if problems of pollution and resource degradation occur because industry is free of government intervention (*laissez-faire policy*) leading to world health problems or the inability to obtain available resources, economically. Knowledge of resource availability is required to anticipate these problems.

## SUMMARY

Energy is considered in fluxes, an amount per unit area per time. The average flux of energy from the sun at the top of the atmosphere is 1,360 W m$^{-2}$ and from the interior of the earth only 75 $m$W m$^{-2}$. Both these fluxes are important for resource formation. Resources can be renewable or nonrenewable and their use increases both with population growth and increased per capita use.

Resources such as mineral resources can be classified based on how well their size and concentration are known. Resources are generally considered reserves if they are of high enough concentration to be extracted at current prices and known with reasonable certainty to exist in the earth.

Geochemical cycles are used to characterize the flux of resources through the various reservoirs on the earth. The carbon cycle is not in steady state and more carbon is going into the atmosphere than leaving. It can be shown this increase started in the middle of the eighteenth century and is continuing to the present. Large amounts of methane clathrates exist on the ocean floor and under permafrost. This resource does not appear to be exploitable for its methane gas but if released would lead to extensive global warming.

Rocks are made up of minerals, glass, and organic material. Common rocks contain silicate or carbonate minerals. They are classified as igneous, sedimentary, or metamorphic. Igneous rocks are the most common rock type found in the earth's crust with sedimentary rock occurring on the top of the crust.

World population is growing at about 1.1% per year and with exponential growth will triple the population in 100 years. This is consistent with the world fertility rate, which currently is 2.56. It is not clear what the carrying capacity of the earth with regard to humankind is but the exponential increase is not likely to be sustained much longer.

At their present rates of growth India will surpass China as the world's most populous country in about 2025. A demographic transition occurs as countries develop from high birth and death rate to low birth and death rate. The age distribution in population pyramids of less-developed countries tends to have a broad base and decreases in number significantly with increasing ages. This is contrasted with developed countries where the number of people in a particular age bracket is similar to the others.

Civilizations have been defined by the resources they use. It appears at present that humankind is making the transformation from the petroleum age to the nuclear plus renewable energy age.

## KEY TERMS

| | |
|---|---|
| baby boomers | metamorphic rock |
| Black Death plague | methane clathrate |
| carrying capacity | metric ton |
| demographic transition | millennials |
| economic stagnation | per capita |
| energy | population pyramid |
| fertility rate | renewable resource |
| flux | reserve |
| fossil fuels | reserve base |
| geochemical cycle | resource |
| geothermal gradient | sedimentary rock |
| igneous rock | steady state |
| laissez-faire policy | ultramafic |
| mafic | water cycle |
| magma | |

## PROBLEMS

1. *Working with units of measure*

   The most widely used systems of measurements are S.I. (Systéme International d'Unités) and CGS (Centimetre–Gram–Second) units. Given below in **Table 1.10** and **Table 1.11** are conversion factors between the two systems and some other commonly used units.

   CGS
   dyne = force to accelerate a mass of 1 g by 1 cm s$^{-2}$

   S.I. (MKS)
   newton = force to accelerate a mass of 1 kg by 1 m s$^{-2}$

   a. Give a possible unit of both *energy* and *pressure*.
   b. What unit would you have to multiply your pressure unit by to get your energy unit?
   c. If pressure changes on a volume of 10 cm$^3$ from 10 to 100 bars, how many thermo calories of energy have been added to the volume? How many joules?
   d. How many joules in a barrel of oil?

*Table 1.10* PRESSURE UNIT CONVERSION FACTORS.

| PRESSURE UNIT | BAR | ATMOSPHERE | psi | mm Hg | PASCAL |
|---|---|---|---|---|---|
| CGS: 1 μbar = dyne cm$^{-2}$ = | $10^{-6}$ | $0.98692 \times 10^{-6}$ | $14.504 \times 10^{-6}$ | $7.50 \times 10^{-4}$ | 0.1 |
| S.I.: 1 pascal = newton m$^{-2}$ = | $10^{-5}$ | $0.98692 \times 10^{-5}$ | $14.504 \times 10^{-5}$ | $7.50 \times 10^{-3}$ | 1 |
| 1 bar = | 1 | 0.98692 | 14.504 | 750.1 | $10^5$ |
| 1 atmosphere = | 1.01325 | 1 | 14.696 | 760 | $1.01325 \times 10^5$ |

*Table 1.11* ENERGY UNIT CONVERSION FACTORS.

| ENERGY UNIT | ERG | JOULE | THERMO CALORIE | cm$^3$ × BAR | BTU |
|---|---|---|---|---|---|
| CGS: 1 erg = dyne cm = | 1 | $10^{-7}$ | $2.3890 \times 10^{-8}$ | $10^{-6}$ | |
| S.I.: 1 joule = newton meter = | $10^7$ | 1 | 0.23901 | 10.00 | |
| 1 thermo calorie = | $4.184 \times 10^7$ | 4.1840 | 1 | 41.84 | $3.968 \times 10^{-3}$ |
| 1 Btu (British thermal unit) = | | 1055 | | | 1 |
| 1 Bbl (barrel of oil) = | | | | | $5.8 \times 10^6$ |
| 1 kWh (kilowatt hour) = | | $3.6 \times 10^6$ | $8.601 \times 10^5$ | | 3412 |

2. If the growth rate of a population, *P*, is 2% per year how long does it take for the population to double? Time to increase by a factor of 10?

## REFERENCES

Archer, A. A., Lutting, G. W., and I. I. Snezhko, eds. 1987. *Man's dependence on the earth: The role of the geosciences in the environment*. Paris: United Nations Educational, Scientific and Cultural Organization (UNESCO).

Walther, J. V. 2009, *Essentials of Geochemistry*, 2nd ed. Sudbury, MA: Jones and Bartlett Learning.

Wellmer, F. W. and Becker-Platen, J. D. 2002. Sustainable development and the exploitation of mineral and energy resources: A review, *Inter Jour Earth Sci*, 91: 723–745.

# Part One

# Energy Resources

*The energy available to humankind limits <u>what can</u> be done, and influences <u>what will</u> be done.*

Energy can be defined as the amount of work that can be performed by a force. Energy resources are then resources that can be used to do useful work. The types of energy are defined by the force: gravitational, heat, electrical, light, etc. Any form of energy can be transformed to another, but during the process some energy is converted to heat. Heat energy has a decreased ability to do useful work so greater work efficiency occurs when it is minimized during an energy conversion. We quantify this more work-depleted energy state as a higher entropy state.

Starting with nuclear and gravitational energy **Figure PT1.1** outlines the types of energy humankind uses to produce the work, heat and *electricity* needed in modern society. What is electricity? The typical answer is electrons flowing in a wire that produces an electric current. In this context electricity is a set of charges and not energy. In the context of energy usage it is best to think of electricity as the movement of an *electromagnetic force field* (electric force + magnetic force field) produced by the motion of charged electrons. A force field is a region of space where a force operates. Moving a force (e.g., Newton) through a distance (meter) in a field (electromagnetic) is then energy (e.g., Joule).

The properties of an electromagnetic field are used to make electrical energy with an electric generator. In an electric generator, a voltage is generated between the ends of an electrical wire when a magnetic field is fluxed across the wire. In this way mechanical energy needed for the movement of the magnetic field in the generator is transformed to electrical energy (See Figure 4.3 in the chapter on alternative energy).

As outlined in Figure PT1.1 most of the sources of energy humankind uses are derived ultimately from nuclear reactions. In the sun this is the fusion of hydrogen to create helium and on earth it is humankind's fission of uranium and plutonium in nuclear

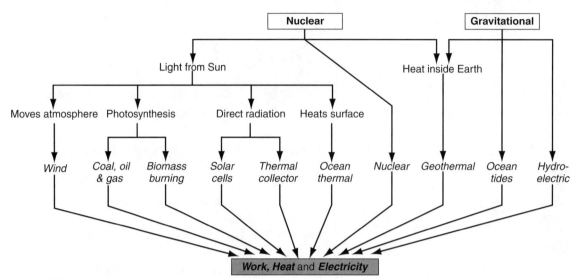

*Figure PT1.1* Types of energy and their flux that humankind captures to perform work as well as produce heat and electricity.

reactors. Also the natural nuclear reaction of radioactive decay in rocks produces heat in the earth that humankind taps as geothermal energy. The other ultimate source of energy is gravitational. Mankind utilizes gravitational energy by harnessing tides and constructing dams to produce hydroelectric power.

When turning on a switch to allow the transport of electricity to illuminate a light bulb, the energy source that produced the electricity is not one's first concern. The important thing is that the light goes on when the switch is activated. The debate about which sources of energy to use is the great "Energy Crisis" or "Energy Debate." Each energy source has its own set of pluses and minuses. As humankind's use of energy touches all aspects of society, understanding the choices in this debate are important. There are positive and negative consequences with the different sources of energy that humankind uses. As stated in the first law of thermodynamics, energy can't be created or destroyed, only changed from one state to another. From a scientific point of view the energy crisis is really a crisis of increased entropy production as energy is always conserved.

## World Energy Consumption

Global energy consumption has grown by nearly 2.2% per year over the past few years, about twice the rate of population growth. In 2012 it

stands at a bit over 12 billion metric *tons of oil equivalent* (toe) per year (2012 BP Statistical Review of World Energy). A ton oil equivalent, or "toe," is the amount of heat released by burning 1,000 kilograms = 1 metric ton of crude oil, approximately 42 GJ (= gigajoules = trillion joules). As outlined in **Figure PT1.2**, in the next 25 years the world's consumption of energy is predicted to grow by 1/3, less than the current 2.2% rate that would increase consumption by 70%.

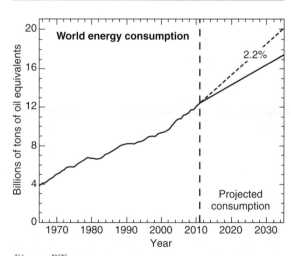

*Figure PT1.2* Total world energy consumption by year in oil equivalents. (Adapted from Year of Energy 2009, Sigma Xi, The Scientific Research Society.)

**Figure PT1.3** outlines the amount of purchased energy in the world by fuel type. It is clear that modern humankind exists on energy obtained from petroleum, coal, and natural gas, the fossil fuels. Petroleum is the energy that is most widely used at present. It is also predicted to be the most used in the foreseeable future. Note, however, that renewable energy (e.g., wind, solar, hydroelectric) will likely make up an increasing percentage of the world's energy use portfolio in the future.

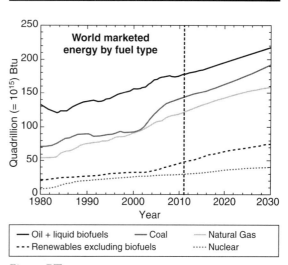

*Figure PT1.3* World energy use by fuel type in British thermal units (Btu) from the U.S. Energy Information Administration. (Modified from International Energy Outlook 2011.)

# Petroleum

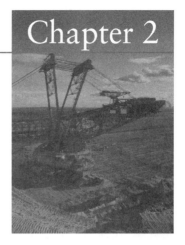

**Chapter 2**

The most important role of petroleum in human society is to supply a highly concentrated, easily distributed fuel that releases its energy by simple, clean combustion for transportation of people and goods. This makes possible a mobile, highly productive society where goods need to be transported and people need to interact.

## Organic Matter and Fossil Fuels

Fossil fuels like petroleum are hydrocarbon (hydrogen + carbon) deposits produced from organic matter (OM) in once living organisms that were deposited in rocks at an earlier geological time. The main constituent of OM is organic carbon and the second is hydrogen. Carbon can be inorganic as in a $CO_3^{2-}$ group in a carbonate mineral such as calcite, $CaCO_3$. Organic carbon has four bonding electrons that can form four single bonds as with hydrogen in methane, $CH_4$, or a smaller number of multiple electron bonds with other atoms. Hydrocarbon OM is divided into two different groups: OM containing *benzene-like rings* of six carbon atoms and OM consisting of long chains and branches of carbon atoms without rings of carbon as shown for two simple organic compounds in **Figure 2.1**.

Cellulose $(C_6H_{10}O_5)_n$, where n is a large whole number, is a molecule made up of multiple linked glucose $(C_6H_{10}O_6)$ six-member benzene-like rings as shown in **Figure 2.2**. It is the most abundant

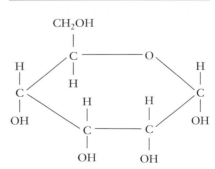

*Figure* 2.1 **(a)** Trichlorobenzene, an organic compound based on a closed ring of six carbons with three carbon double-bonds, which in this case has three attached chlorine atoms. **(b)** Octane, an organic compound made up of a chain of eight bonded carbon atoms with attached hydrogen atoms.

*Figure* 2.2 Structural formula of a glucose molecule.

31

compound in plants. Note the approximate ratio of carbon to hydrogen to oxygen of $CH_2O$, that is 1:2:1. The total mass of organic matter on the earth is dominated by land plants and their buried remains with much less of the earth's organic matter in the ocean or oceanic sediments. Often $CH_2O$ is used as shorthand for specifying OM on the earth. In contrast, petroleum, which includes compounds like octane (see Figure 2.4b), contains little or no oxygen and more hydrogen per carbon. In general, with their extensive amount of oxygen, cellulose and therefore land plants do not break down and produce OM from which petroleum is produced. Land plants produce coal instead (see *Chapter 3* on natural gas and coal).

## Petroleum Formation

Petroleum is typically produced by the breakdown of algae and *zooplankton* for which up to 40% of their body mass is fatty acid. These are small drifting organisms that grow in lake and ocean surface environments where light can penetrate. Fatty acid has the general formula of $C_nH_{2n+1}COOH$ where *n* is a large whole number. Fatty acids, therefore, have more H and less oxygen ($O_2$) relative to carbon than in $CH_2O$ found in vascular land plants. Therefore, it is algae and zooplankton rather than vascular plants that contain a significant amount of fatty acids and other low oxygen lipid-like carbon chain compounds that can be transformed to petroleum in the subsurface.

The quantity of OM in a sediment or rock is typically expressed as *total organic carbon* (TOC) in wt%. Most sandstones have low TOC because they are rapidly deposited relative to any OM that is fluxing downward in the water and any of the limited OM present is destroyed in oxygenated waters that flow through the deposited sands. Most OM that forms petroleum accumulates in shales as the time frame of clay deposition is much longer than sands and, therefore, shale deposits have more time to capture the organic matter fluxed from the surface layer of the ocean or lake. However, some limestones that accumulate slowly can also contain significant quantities of OM that form petroleum.

When algae and zooplankton in the top of the water column die the OM fluxes to the sediments below. There is typically a correlation between grain size and TOC. Finer grain rocks have higher TOC because the mineral particle accumulation rate is less allowing relatively more OM to be deposited for the same OM flux from surface water. The darker the color of the shale or limestone generally the greater is its TOC. TOC in typical sandstone is 0.03 wt%. An average shale has about 0.4 wt% TOC. A reasonable petroleum producing shale, termed a "source rock," has a minimum of ~0.6 wt% TOC. Therefore, somewhat less than half but still a large amount of shale in the subsurface can be source rock. Black shales are those that have about 15 wt% or more TOC. The U.S. Green River shale of Wyoming is black shale with 18 wt% TOC.

The OM in source rocks is composed of *bitumen* and *kerogen*. The distinction is based on its solubility in normal organic solvents such as acetone, $C_3H_6O$, and ethanol, $C_2H_6O$. The soluble portion of OM is termed bitumen and the insoluble portion, kerogen. The bitumen is made up of tar-like organic molecules containing large quantities of highly condensed benzene-like carbon rings. These are bonded together as shown in **Figure 2.3**. There is also a small quantity of short-chained carbon molecules in the bitumen fraction.

### Kerogen

Kerogen is composed of very long carbon chains with attached functional groups. Because of their length kerogen molecules have very high

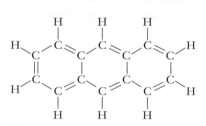

*Figure 2.3* Structural formula of anthracene, a polycyclic aromatic hydrocarbon made up of three bonded benzene rings, often found in bitumen.

molecular weight, making them insoluble in low molecular weight organic solvents. Kerogens can contain compounds like palmitic acid (**Figure 2.4**). Palmitic acid is a fatty acid that is found in both animal and plant material (palm oil). With increased heating on burial these long chains break down to produce a shorter chain liquid hydrocarbon like octane, a constituent of gasoline.

Depending on the atomic hydrogen-to-carbon ratio, H/C, and atomic oxygen to carbon ratio, O/C, kerogen is classified as

Type I (very petroleum prone),
Type II (petroleum prone),
Type III (natural gas prone), and
Type IV (inert).

Higher hydrogen and lower oxygen contents in kerogen correspond to greater petroleum-generating potential. Type I kerogen has an atomic H/C ratio of at least 1.35 (high) and O/C ratio of less than 0.15 (low). Type II kerogen has a H/C between 1.0 and 1.35 and O/C ratios generally between 0.03 and 0.18. Some investigators put the boundary between Type I and II at H/C = 1.5. Type III kerogen has a H/C ratio

between 1.0 and 0.5 and high O/C (up to ~0.3). In Type IV kerogen the H/C ratio is less than 0.5 with relatively high O/C of 0.2 to 0.3. It has no potential to form petroleum. The low hydrogen containing organic matter is due to the presence of carbon rings and is derived mostly from higher plants. Type II kerogens account for most petroleum source rocks. If it has high sulfur content it is often referred to as Type II sulfur (Peters and Moldowan, 1993).

While most coal is *humic coal* that has passed through a peat stage and does not produce kerogens on burial, there is an uncommon type of coal that does. *Sapropelic coal*, also called boghead or cannel coal, is coal formed from algal material in oxygen-depleted ponds, lakes, and lagoons. On burial sapropelic coal, which contains waxes and fats, with its higher hydrogen content can produce kerogen.

As the kerogen is buried in the subsurface it heats up. At about 2 km depth (~65°C) the long kerogen chains start to breakdown. The liquid produced is referred to as *crude oil*. The composition of crude oil is given in **Box 2.1**. As seen from the percentage range of the various components given in parentheses, crude oil compositions vary widely.

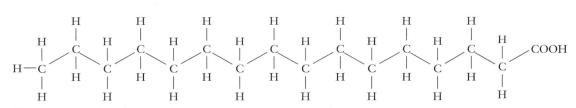

*Figure 2.4* Structural formula of hexadecanoic acid [$CH_3(CH_2)_{14}COOH$], which is also referred to as palmitic acid.

*Box 2.1* Major components of crude oil

1) **Alkanes** also called paraffins (15% to 30%): Chains of single-bonded carbon atoms with attached hydrogens ($C_nH_{2n+2}$) from $CH_4$ (methane) to $C_{70}H_{142}$. "Sweet oil" has many short-chained alkanes.
2) **Cycloalkanes** also called naphthenes (30% to 60%): Alkanes in a circular structure, primarily cyclopentane and cyclohexane both of which are rather unreactive and used for lubricating oils.
3) **Aromatic hydrocarbons** (3% to 30%): Compounds with six-member carbon rings that are separated by boiling the crude oil.
4) Original **bitumens** also called asphaltics (0% to 40%): If concentrated enough the petroleum is not liquid but rather is tar-like.

### Crude Oil and Natural Gas Windows

Because the kerogen breaks down with increasing depth in the earth's crust there is a depth region where crude oil is produced referred to as the *oil window*. The oil window at a particular place on the earth depends on both the temperature gradient and the rate the organic-rich sediments were buried and, therefore, time. Given in **Figure** 2.5 is a typical oil window, existing from about 2 to 5 km below the surface. Above 2 km depth crude oil is not released from the kerogen structure and below 5 km depth (~135°C) any crude oil produced is not stable and is broken down to methane gas instead. At 3 km methane gas, $CH_4$, as well as crude oil is released from the kerogen. The natural gas window extends from about 3 to 6.5 km in depth. In most locations on the earth by a depth of 10 km only a carbon residue remains of any original organic matter.

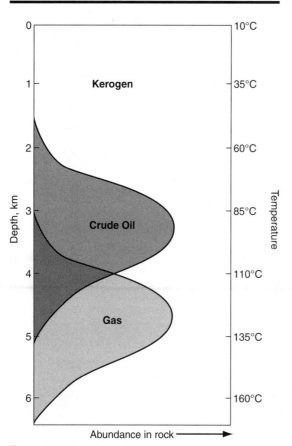

*Figure 2.5* Typical crude oil and natural gas "windows" showing where they occur as a function of depth in the subsurface.

### Fingerprinting Oil

Fingerprinting oil is a technique where crude oil is identified by the uniqueness of its composition of different organic molecules. This helps oil companies identify the source reservoirs from which oil is taken. When a new well intersects an oil producing region it can be determined whether a new reservoir has been located or whether the well intersected an extension of a previously known reservoir. When a well intersects multiple producing horizons the fingerprinting of oil can determine the amount of production from each horizon.

Oil fingerprinting provides forensic evidence in oil spill cases to prove or disprove a particular tanker or oil platform as the source of a spill. In using fingerprinting to identify the source of oil in the environment one of the problems is that the oil degrades with time and the identification become more difficult.

# Methods of Petroleum Source Rock Evaluation

## Vitrinite Reflectance

Vitrinite reflectance (VR) is the most commonly used organic maturation indicator used in the petroleum industry to quantify the state of development of petroleum source rocks. It is accurate, quick, nondestructive, and inexpensive. *Vitrinite* forms in organic matter during burial of rocks by the thermal alteration of the six-member carbon rings found in coals and to a smaller extent in most kerogens. Vitrinite macerals (compounds) are not strongly prone to crude oil and gas formation, and become increasingly reflective with time and temperature. The reflectance, $\%R_o$, is the percentage of light reflected from the sample surface, calibrated against a mirror (100% reflectance). Typically the onset of crude oil generation (~60°C) corresponds with a vitrinite reflectance of 0.5 to 0.6 $\%R_o$ while at a reflectance of 0.85 to 1.1 $\%R_o$ (~120°C) the strata will no longer produce crude oil.

## Van Krevelen Diagram

A *Van Krevelen* (atomic H/C vs. O/C) *diagram* as given in **Figure** 2.6 is probably the best way to evaluate the quality and maturation state of kerogen as well as coal in the subsurface and relate it

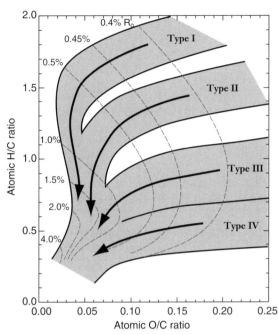

*Figure* 2.6 Van Krevelen diagram showing the pathways of maturation for types I to IV kerogens with burial. The dashed lines give percentage of vitrinite reflectance, $R_o$. (Adapted from Hunt, J. M., 1996, *Petroleum Geochemistry and Geology*, Freeman and Co., New York, 743 p.)

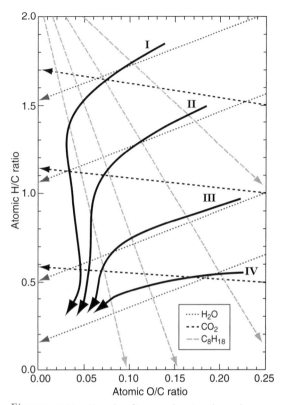

*Figure* 2.7 Van Krevelen diagram showing the pathways of maturation for types I to IV kerogens with burial. The effects of loss of $H_2O$, $CO_2$ and $C_8H_{18}$ from kerogen during maturation are shown by the dotted and dashed lines.

to vitrinite reflectance. General maturation paths for the four different types of kerogen in shale outlined above are shown with arrows in the figure. The dashed lines give percent of vitrinite reflectance, which increases with increased maturation of the oil-forming layer.

What causes the changes in H/C and O/C ratios as a function of increasing maturation? Bacteria present in the source rock transform hydrogen and oxygen in the kerogen to $H_2O$. Being insoluble in oily kerogen the generated $H_2O$ is released, causing the H/C and O/C ratio of the remaining kerogen to decrease towards the left, parallel to the dotted grey lines given in **Figure** 2.7. $CO_2$ is also produced by bacterial activity. Loss of $CO_2$ causes a decrease in O/C with a slight increase in H/C parallel to the black dashed lines. If crude oil is represented as octane, $C_8H_{18}$, the production of crude oil from the kerogen rapidly decreases the H/C ratio while increasing the O/C ratio as given by the grey dashed lines in the figure. The maturation process can

be considered an initial loss of $H_2O$ and $CO_2$ before crude oil is produced. Clearly, Type I and II kerogen have the high H content necessary to produce significant crude oil as their H/C ratio decreases on their maturation path.

## Rock-Eval Pyrolysis

Often *Rock-Eval pyrolysis* (**Box** 2.2) is used to identify the type and extent of maturity of the OM in a rock by programmed heating of the rock in a pyrolysis (heating) oven. Helium gas is passed through pulverized rock at 300°C for 3 minutes. Free hydrocarbons are volatilized during this process and measured as the $S_1$ peak as shown in **Figure** 2.8. Temperature is then increased 25°C per minute to 550°C. Very heavy hydrocarbons (> C40) are volatilized during this heating and some nonvolatile organic matter is volatilized by *cracking* and produces the $S_2$ peak.

---

*Box 2.2* ROCK-EVAL PYROLYSIS

$S_1$ peak is the amount of free hydrocarbons (gas and oil) in the sample. If $S_1 > 1$ mg per g of rock, it generally indicates significant hydrocarbons are present. $S_1$ normally increases with depth as OM is naturally cracked with greater time and temperature.

$S_2$ peak is considered to give the amount of hydrocarbons that could be generated through thermal cracking of nonvolatile OM. That is, $S_2$ indicates the quantity of hydrocarbons a rock has the potential of producing should burial and maturation continue. Below 1 km $S_2$ normally decreases with burial depth.

$S_3$ peak is the amount of $CO_2$ produced during pyrolysis of kerogen. $S_3$ is an indication of the amount of oxygen in the kerogen.

$T_{max}$ is the temperature of maximum hydrocarbon release rate from cracking of the kerogen (top of $S_2$ peak). $T_{max}$ indicates the stage of full maturation of the OM.

**TOC** the total organic carbon in the rock is determined by oxidizing (at 600°C) the organic matter remaining in the sample after pyrolysis and adding this to the amount released by pyrolysis.

---

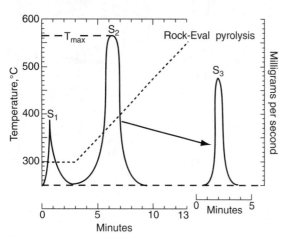

*Figure 2.8* Typical Rock-Eval pyrolysis plot showing loss of hydrocarbons as a function of timed heating.

Cracking is the breakdown of large hydrocarbon molecules into smaller molecules of lower molecular weight. Cracking occurs naturally as more heat is put into the hydrocarbon as temperature increases. The temperature of the maximum rate of milligrams of hydrocarbon volatilized is recorded as $T_{max}$. The amount of $CO_2$ produced from kerogen cracking in the 300° to 390°C range is trapped and determined on cooling and is given as the $S_3$ peak.

As shown in **Figure 2.9** the free hydrocarbons, given by the Rock-Eval $S_1$ peak, increase with depth as the kerogen is naturally cracked. The hydrocarbons available for cracking, given by the $S_2$ peak, therefore decrease. Note the production index, given by $S_1/(S_1 + S_2)$, and the $S_2$ peak temperature ($T_{max}$) increase with maturation.

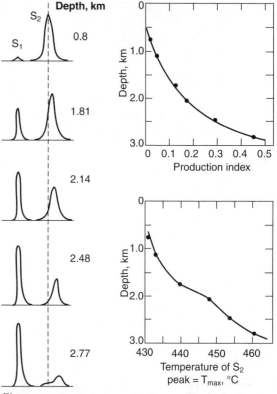

*Figure 2.9* Pyrolysis determinations as a function of depth on tertiary-aged core samples from a drill hole in West Africa. (Modified from Espitalié, J. J., La Porte, L., Madec, M., Marquis, F., Le Plat, P., Paulet, J., and Boutefeu, A., 1977, Méthode rapid de caractérisation des roches méres de leur potentiel pétrolier et de leur degré d'évolution. *Rev. l'Inst. Français pétrole*, v. 32(1), pp. 23–42.)

# Age of Source Rocks

Source rocks for petroleum are not evenly distributed through the sedimentary rock column, that is, as a function of the strata laid down or their age. Six time intervals account for 91.5% of recoverable petroleum (**Figure 2.10**). A cyclic geologic process is not responsible for the distribution but rather a variety of factors working together control it. Some of the most important are the paleolatitude of the depositional areas, the time of evolution of biota on the earth, burial mechanisms of the source rocks, and presence or absence of a variety of traps for any crude oil produced.

# Crude Oil Deposits

With time the crude oil formed by thermal and bacterial processes in the subsurface is released from the shale and carbonate layers where it is formed. Because the crude oil is less dense than any water in the pores of the surrounding rocks it migrates upward through the water in permeable strata. When a nonpermeable fault or sealing impermeable layer, called a cap, is reached it collects in the porosity of the rock beneath the cap, displacing the typically salty water present initially in the pores. This concentration of hydrocarbons is considered a *reservoir*.

Given in **Figure 2.11** are various kinds of reservoir traps labeled by their environment. These include an anticlinal structure where a highly porous layer like sandstone is folded convex up with the oldest beds at the interior. Another kind of structural petroleum trap is a fault. Grains along the fault can be ground to a powder making it a seal for the movement of petroleum. Stratigraphic traps of petroleum are also formed when impervious beds above seal a reservoir bed or when permeability changes along a bed. Porous coral reef limestone can make excellent reservoirs if impermeable strata cap them. Often reservoirs exist at angular unconformities where the upper layer is impermeable shale. A reservoir type, which is

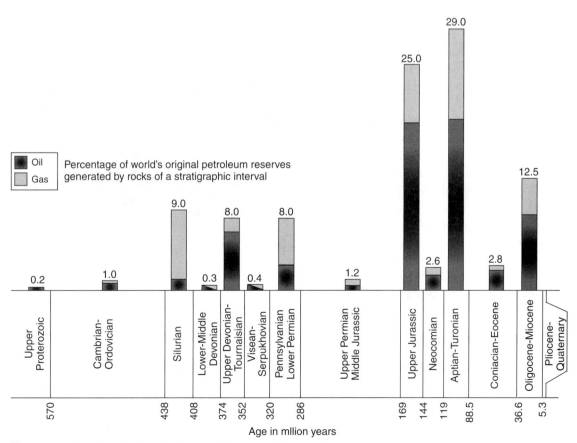

*Figure 2.10* Stratigraphic (age) distribution of source rocks for petroleum.

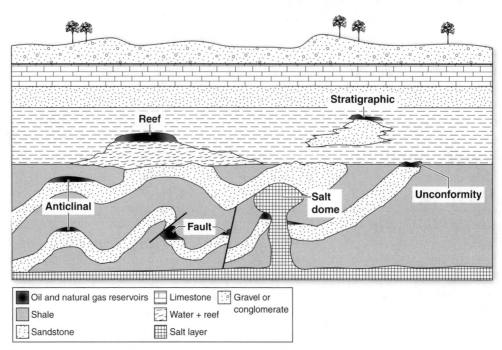

*Figure* 2.11 Types of petroleum reservoirs in the subsurface labeled by their mechanism of trapping their oil and natural gas.

common in the U.S. Gulf Coast, is produced by a salt dome. Less dense salt layers in the subsurface can become unstable bending layers upward as they rise as diapirs. Because salt accumulations are impermeable to petroleum a trap is formed.

In most profitable reservoirs 40% or more of the pore fluids must be hydrocarbons. If this reservoir is a porous, permeable sandstone or limestone a significant petroleum and/or natural gas reservoir can be present as shown in **Figure 2.12**. Generally a porosity of 10% or greater is needed in sandstones, but can be less in limestone if it is highly fractured. Of limestone, the skeletal remains preserved in ancient coral reefs often have high porosity and permeability and make excellent reservoirs. If these reservoirs can be found much of the petroleum and/or natural gas they contain can be recovered. Problems can occur with recovery if the crude oil is highly viscous or the reservoir has low permeability.

Crude oil is usually characterized by its density as given by the *API* (American Petroleum Institute) *gravity* number expressed in degrees (°). This is a measure of how heavy or light the oil is

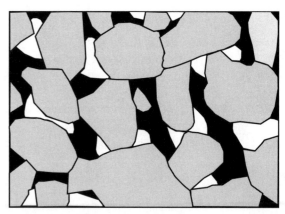

*Figure* 2.12 Microscopic section of an oil deposit showing original sand grains in grey, silica cement in white, and black crude oil.

relative to water. With API gravity greater than 10°, the oil floats on water while less than 10° it sinks. The lighter an oil the higher is its API gravity and the more of the short-chained carbon molecules that can be made into gasoline it contains. The API number is determined from

$$API = \frac{141.5}{\rho} - 131.5 \qquad [2.1]$$

where API stands for the degrees of API gravity and ρ gives the crude oil density in g/cm³ at 60°F (= 15.6°C). Light crude oil is defined as having an API > 31.1°, medium crude oil has an API between 22.3° and 31.1°, and heavy crude oil has an API < 22.3°. The benchmark Brent crude oil from the North Sea has an API = 38.3° and West Texas Intermediate's API is 39.6°. Sweet crude oil has a sulfur < 0.5 wt% and a higher sulfur content is considered sour crude oil.

**Figure 2.13** is a cross-section through the earth showing the subsurface reservoir of petroleum and natural gas at Prudhoe Bay, Alaska. This occurs in the Sadlerochit Group, a large deep-sea fan-delta system that contains sands and gravels. The high porosity of these rocks makes them a good reservoir to hold petroleum and natural gas. The rocks above the Sadlerochit Group are shales with low permeability that seal the reservoir to escaping crude oil and natural gas. In the petroleum industry these shales are termed *caprocks* as they cap the reservoir.

# Crude Oil Recovery

The earliest oil wells were drilled by driving rotating pieces of pipe into the ground. This drilling technique was replaced by cable-tool drilling where a heavy bit is attached to a long cable and rotated. Presently rotary drills are used where the bit is attached to sections of connected hollow rigid metal pipe to produce what is called the drill stem. The drill stem is rotated by a turntable attached near the stem's top as shown in **Figure 2.14**.

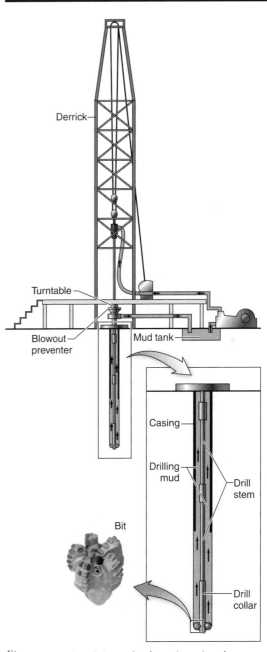

*Figure 2.14* An oil rig used to bore through rocks to produce a well from which crude oil and/or natural gas at depth can be extracted.

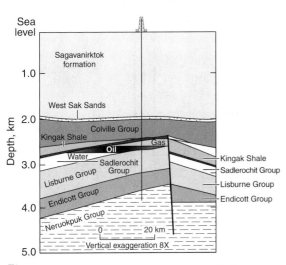

*Figure 2.13* North-south cross-section through the Prudhoe Bay petroleum and gas field, Alaska. (Adapted from Jamison, H. C., L. D. Brockett, and R. A. McIntosh, 1980, Prudhoe Bay—a 10-year perspective, in M. T. Halbouty, ed., Giant oil and gas fields of the decade 1968–1978: *AAPG Memoir* 30, pp. 289–314.)

The hollow drill stem allows water, infused with the clay, *bentonite*, and barite, $BaSO_4$, or cesium formate, $CsCHO_2$, termed *drilling mud*, to be sent down the hole to the drill bit. The bentonite increases fluid viscosity so it is not lost to permeable layers in the subsurface. The barite with a density of ~ 4.5 g cm$^{-3}$ and the cesium formate with a density of ~ 2.4 g cm$^{-3}$ are used to control fluid density. It is important to control the mud density and therefore bottom hole mud pressure. If formation fluids are at a higher pressure than mud pressure they can enter the drill hole and lift the mud rapidly leading to a well blow as described below. The drilling mud travels back to the surface around the outside of the drill pipe in what is called the *annulus*. In doing so the drilling mud stabilizes the wall of the borehole and removes drill cuttings. The drilling mud also lubricates, cools by carrying away heat, and cleans the drilling bit.

## Operation of a Drill Rig

Examining rock chips extracted from drilling mud returning to the surface is known as *mud logging* and the downhole record produced is known as a mud log. An examination of these rock chips reveals the type of rock and formations being drilled and estimates the oil and gas content of the section of rock through which the drill bit is cutting. Mud loggers also connect various sensors to the drill stem to characterize rock formation properties as a function of depth. For instance, downhole changes in gamma ray production can identify shale layers, changes in voltage with depth can locate permeable formations, and their boundaries and electrical resistance measurements can distinguish water from hydrocarbons in porous formations.

As the hole is drilled, sections of steel tubing or casing, slightly smaller in diameter than the borehole, are set in the hole and these are cemented to the sides of the hole. This casing provides structural integrity and isolates potentially dangerous high-pressure fluid zones from each other and from the surface. A large valve at the wellhead, a blowout preventer, is used to seal off the well if overpressured formation fluids enter the well below.

After drilling and casing the well, it must be *completed*. Completion is the process where the well is enabled to produce petroleum and/or natural gas. Small perforations are made in a portion of the casing that passes through the production zones where the hydrocarbons reside. In open-hole completion to a reservoir section, often "sand screens" or a "gravel pack" is installed in the uncased bottom section. Finally, the area above the reservoir section of the well is sealed off inside the casing, and connected to the surface via a smaller diameter pipe called tubing as shown in **Figure 2.15**.

### Directional Drilling

Despite more widespread use of directional drilling, most oil reservoirs are still tapped directly from above. Horizontal boreholes cost twice as much to drill as vertical ones thus only 5% to 8% of all U.S. land wells have been drilled at an angle. However, directional drilling is presently becoming more common as many horizontal wells can be drilled from a single vertical drill site. They can expose a lot more formation to the borehole (**Figure 2.16**). This advantage makes horizontal

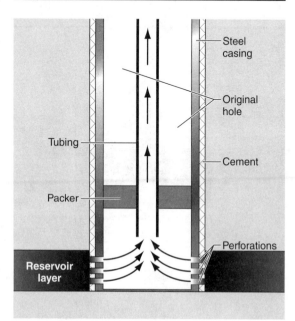

*Figure 2.15* A completed well showing the flow of crude oil or gas from the reservoir layer to the production tubing in the bottom of the well.

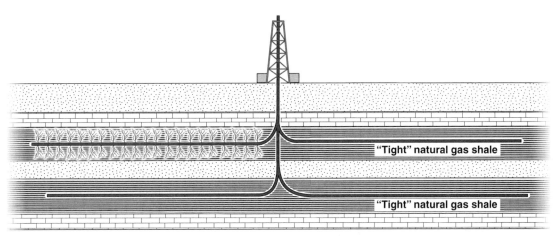

"Tight" natural gas shale

"Tight" natural gas shale

*Figure 2.16* Single drilling platform with multiple directionally drilled production wells. In the top well on the left the shale has been "fracked" (see Chapter 3).

drilling ideal for reservoirs that are shallow, spread out, fractured, and/or in sensitive environments, such as the Arctic tundra or in populated areas. Multiple directional wells are drilled from many offshore drilling platforms, particularly as the cost of the platforms increase with increasing water depth. Extensive horizontal drilling is being used to extract tightly held natural gas from the Barnett shale in Texas, Marcellus shale in New York and Pennsylvania, and the Utica shale in New York and the province of Quebec, Canada (See *Chapter 3* on natural gas and coal production).

## Deepwater Production

Because oil is formed dominantly from organisms that lived in the ocean and are buried in sedimentary rock it makes sense that potential oil reservoirs exist in ocean sediments on the continental shelves. As a function of time offshore oil and gas drilling has been done at deeper and deeper water depths in the ocean as the drilling and production technology has improved and the price of and demand for oil increases. At present deepwater drilling and production is generally considered to be in water depths of 300 m or greater and ultra-deepwater at depths of 1.5 km or more. Drilling and production platforms built on the ocean floor are constructed with steel or concrete legs to depths near 520 m. This depth is extended

to 910 m by employing compliant flexible steel towers. At greater depths floating platforms are exploited. These are semisubmersible structures resting on buoyant columns and pontoons.

Tension-leg floating platforms, which are attached to the ocean floor with long nearly vertical cables grouped to produce "tension legs" at each of the structure's corners, are used between depths of 300 m and 1.5 km. Spar floating platforms which are similar to tension-leg platforms but have a single large 90% submerged buoyant cylinder supporting the platform's decks are also used (**Figure 2.17**). The cables that attach it to the sea floor are placed at some distance from the platform rather than directly under it. As of 2012 the deepest Spar platform is in 2,438 m of water in the Gulf of Mexico.

The Deepwater Horizon was the name of an ultra-deepwater Spar-type drilling rig platform. In April 2010 while drilling a well 66 km off the coast of Louisiana in 1.5 km of water it experienced a blowout, oil gusher, and fire. This destroyed the rig and killed 11 workers on the platform deck. It is estimated that 4.9 million barrels of oil was released into the Gulf of Mexico from the well before it was capped 86 days later. This is almost 20 times greater than the Exxon Valdez oil tanker spill in Alaska in 1989 and makes it the largest accidental marine oil spill in history. Because British Petroleum was the operator and principle developer of the well the event is often referred to as the BP oil spill.

## Primary Production

In primary production natural pressure on the crude oil and natural gas brings it to the surface. Typically this recovers about 5% to 15% of the hydrocarbon in the subsurface. A gusher occurs when crude oil is sent out of the top of the drill rig typically from expanding $CO_2$ or natural gas at high pressure in the reservoir section. The crude oil can shoot 60 m or more above the earth's surface as shown in the classic picture from January 10, 1901 of the Lucas Gusher

at Spindletop, Texas (**Figure 2.18**). Prior to the development of blowout preventers in the 1920s, gushers were an unavoidable consequence of drilling into high-pressure petroleum reservoirs. With the current use of blowout preventers gushers rarely happen.

## Secondary Recovery

As a well continues to release crude oil and natural gas the pressure on the petroleum falls. At some point there will be insufficient underground pressure to force the crude oil to the surface. To recover crude oil after natural pressure subsides, secondary recovery methods are used. Fluids are introduced at an injection well nearby to move the crude oil in the reservoir to the production well by increasing the reservoir fluid pressure. This fluid can be either water or natural gas. In the case of water because of its greater density it floods the lower portions of the reservoir and drives the crude oil upward. Natural gas reacts with the oil, lowering its viscosity and allowing the crude oil to flow more freely. These recovery techniques generally

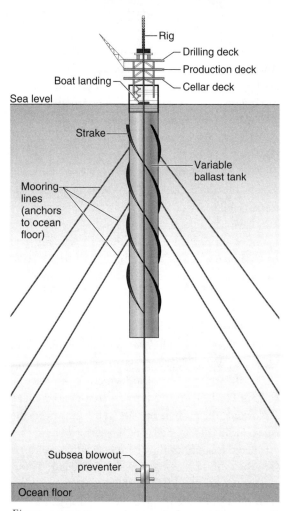

*Figure 2.17* A Spar production platform used to extract oil and natural gas from rocks below deep water.

*Figure 2.18* Lucas Gusher at Spindletop, Texas (Jan. 10, 1901). At a depth of 347 m a high-pressure petroleum reservoir was encountered which released crude oil 46 m in the air at a rate of 100,000 barrels per day.

allow another 30% of the crude oil to be extracted from the reservoir. Ease of recovery depends on the extent the pores that contain the oil-water mixture have surfaces that are water-wet as opposed to oil-wet. If water-wet the surfaces are coated with water and the oil remains as droplets in the water. If oil-wet then the opposite is true. The difference occurs because mineral surfaces carry a charge that can attract either polar water molecules or polar asphaltene organic molecules in crude oil depending on local conditions. If asphaltenes coat the pore surfaces so oil congregates there, it is more difficult to extract it from the reservoir. If water molecules are attached to the pore surfaces then the oil is more easily transported out of the rock. Most reservoirs are intermediate in character in that some pores are water-wet while others are oil-wet.

## Tertiary Recovery

As hydrocarbons are extracted the remaining hydrocarbons become more and more viscous and reservoir pressure continues to fall. Tertiary crude oil recovery techniques can be used to lower the remaining oil's viscosity and therefore increase this oil's ability to flow to the production well. These techniques include steam flooding (heat lowers viscosity), chemical flooding (uses light hydrocarbon, detergents, or $CO_2$ that react with the crude oil, breaking long hydrocarbon chains apart and lowering its viscosity), and injection of microbes + nutrients (microbes grow at the interface between the crude oil and the mineral grains, which helps release the crude oil from the surfaces of grains in the reservoir rocks). Tertiary recovery can extract an additional 5% to 15% of the reservoir's petroleum. Tertiary methods are used as long as petroleum can still be extracted profitably. Previously unprofitable wells can be brought back into production when prices are high enough and shut down when prices are too low.

In low producing wells a *pumpjack* can be installed as shown in **Figure** 2.19. These reciprocating piston pumps mechanically lift 5 to 40 liters of an emulsion of crude oil and water out of the well with each stroke by closing a

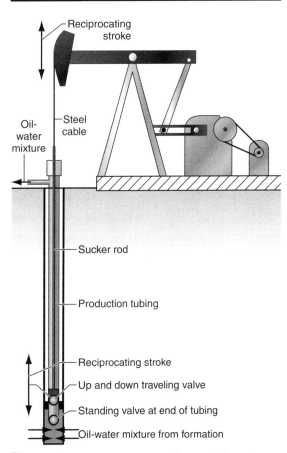

*Figure 2.19* Reciprocating stroke of a pumpjack lifting petroleum plus water to the surface from underground.

traveling valve and opening a standing valve on the upstroke. Fluid above the closed valve is lifted. On the downstroke the traveling valve is open and the standing valve is closed filling the tubing with more fluid. The crude oil is then separated from the water in the lifted fluid.

## Abandonment

If a well produces so poorly that it is unprofitable, it is abandoned. Often 50% of the crude oil is left in the ground because it is too difficult/expensive to obtain. Tubing is removed from the well and sections of well bore are filled with cement to isolate the flow path between gas and water zones so hydrocarbons can't contaminant aquifers. Wellhead and casing are cut off at ground level and a cap is welded in place.

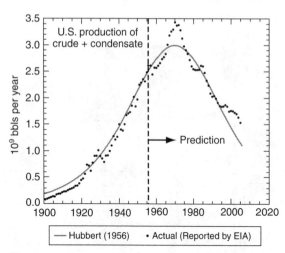

*Figure 2.20* Hubbert's 1956 analysis and prediction given by the solid line and actual production for the year as reported by the U.S. Energy Information Administration.

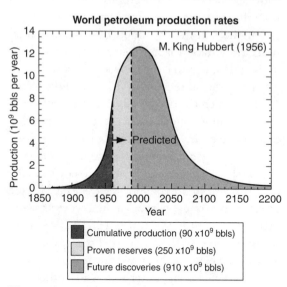

*Figure 2.21* Hubbert's 1956 prediction of world production of petroleum based on known reserves and likely future discoveries. (Adapted from Nuclear Energy and the Fossil Fuels, M. K. Hubbert, Fig. 20 of publication No. 95 Shell Development Company, June 1956.)

## Peak Oil

In 1956, M. King Hubbert, a geologist with Shell Oil, based on the information to date correctly predicted a peak in U.S. petroleum production, or *peak oil*, would occur between 1965 and 1970 as shown by the solid line in **Figure 2.20**. He also predicted world "peak oil" would occur in about half a century (~2006) based on his curve shown in **Figure 2.21**. While the increase in world production has slowed we have yet to reach peak oil. The importance of peak oil is that it implies that demand will not stimulate any additional supply and production enters a terminal decline.

Most petroleum industry experts believe that the modern industrial world is heavily dependent on the availability of relatively low-cost petroleum and the post-peak production decline will severely increase prices and have very large negative implications for the global economy. There is a great debate about when world peak petroleum production will occur but nearly every expert believes it will occur. Many recent estimations of peak production forecast the global decline will begin by 2020 or later. If so, major investments

in alternative energy sources can still occur and major rapid changes in the lifestyle of heavily petroleum-consuming nations possibly need not occur. It is difficult to predict peak world petroleum because there is no consensus on how to assess reserves.

Oilfield reserves are confidential in every country except in the U.K., Norway, and those available to the U.S. government. Reserve and production "estimates" given by countries in the Organization of Petroleum Exporting Countries (OPEC) have political overtones and are known to be unreliable as quotas rule OPEC production. World's proven conventional petroleum reserves were estimated to be 1.317 trillion barrels as of January 1, 2007.

Some estimates of the timing of peak oil are given in **Figure 2.22**, which appear to be increasing into the future the more recent the forecast is made. Most experts agree that "easy" crude oil will soon peak. More sophisticated techniques can be used to extract the immobile crude oil in a reservoir, up from the current average of 35%

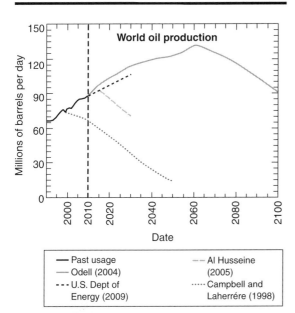

*Figure* 2.22 Past usage and prediction of petroleum production into the future from a number of sources. (Adapted from Maugeri, Leonardo. 2009. Squeezing more oil out of the ground. *Scientific American*, April 1.)

of the crude oil present, although this adds to the expense. Drilling techniques such as using deep offshore platforms and horizontal drilling have also recently opened up reservoirs of petroleum and natural gas that previously were impossible or uneconomic to drill. Exploration tools are becoming more sophisticated. 3-D and 4-D (3-D plus time) imaging techniques using seismic waves have been developed to locate many smaller petroleum reservoirs. Additionally, a large resource base in oil sands and oil shales is starting to be brought into production. Unfortunately the environmental cost of extracting crude oil from oil sands and oil shales is high.

# Oil Shale and Tar Sand

## *Oil Shale*

Oil shales are petroleum source rocks that have not released all their hydrocarbons. They are fine-grained and contain significant concentrations of kerogen. Kerogen requires more processing to produce the short-chained hydrocarbons desired than does crude oil. These rocks are present in sedimentary basins across the globe including major deposits in the United States. Oil shale is starting to be mined as the price of conventional crude oil has risen. Brazil, China, Estonia, Germany, Israel, and Russia currently utilize oil shale as an energy source.

The global amount of crude oil equivalent estimated to be recoverable from oil shale ranges from 2.8 to 3.3 trillion barrels. The Green River Formation, covering parts of Colorado, Utah, and Wyoming, is the largest known deposit in the world. Its estimated reserves are thought by some to total 1.5 trillion barrels of oil. This is more than five times the stated reserve of Saudi Arabia. However, it is likely that only 0.8 trillion barrels is economically recoverable. Most of the oil shale reserves are on U.S. government-owned land and they will control any leases to recover it.

Oil shale is mined with surface mining techniques, either in an open pit or strip mine (**Figure 2.23**). In either type of mine overburden must be removed. An open pit mine is typically used where the overburden is greater. *Open pit mines* are conical chasms with a set of winding levels called benches that lead from the earth's surface down to the pit floor at the bottom (see Figure 1.3 in the introductory chapter). The benches serve both as working terraces that are drilled, blasted, and the rock loaded in haulage trucks, and as haulage roads. A *strip mine* is employed where the oil shale is flat-lying and not far below the surface. The overburden is first removed by large earthmoving equipment in long strips. This can be a dragline excavator, which is a large bucket suspended from a boom, or a bucket wheel excavator (See **Figure 2.24** and the book's cover), which consists of a large wheel with a continuous set of buckets. After displacement of the overburden, the oil shale is removed and loaded into haulage trucks or open-topped gondola train cars. After the oil shale is excavated, piles of overburden waste rock, the spoilage mounds, are resurfaced with soil and revegetated to produce reclaimed land.

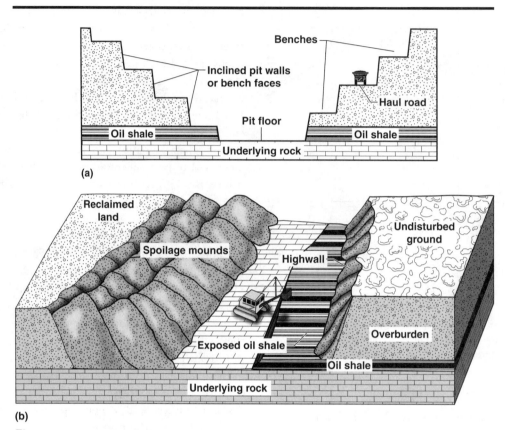

Figure 2.23 (a) Cross-section through an open pit mine for recovery of oil shale. (b) Overhead view of a strip mine used to recover oil shale.

Figure 2.24 Bucket-wheel excavator used for removing large quantities of rock in flat lying terrain. They can stand taller than the Statue of Liberty on its base (93 meters).

After mining the oil shale can be burned to produce energy or processed to remove the hydrocarbons. Hydrocarbon liquids and gasses are removed by heating the kerogen containing shale to a temperature greater than 300°C in the absence of $O_2$ in a process called pyrolysis or retorting. The kerogen decomposes into hydrocarbon gas, condensable petroleum liquids, and a solid residuum. The products can be upgraded by hydrogenation (see below). Compared to conventional fossil fuels, oil shale requires larger investments of energy relative to energy it generates. Much of this is used during retorting.

Oil shale mining and its processing involve a number of environmental concerns. These include problems with disposal of processed shale, acidic groundwater production, erosion, sulfur gas production during retorting of sulfur containing shale, and particulate air pollution from processing and oil shale transport. The production of petroleum from oil shale also uses significant amounts of water. This is a particular concern in arid regions where the local water supply is limited.

As the price of petroleum increases the interest in oil shale development also increases. In the United States environmentally sensitive mining and retorting methods are required. While aboveground retorting is most often used to extract the useful hydrocarbons, in situ methods are being developed where oil shale is heated underground. While in situ methods are more energy intensive they are also more environmentally friendly.

Shell Oil is developing an in situ retorting method involving drilling holes up to 600 meters deep into the oil shale, inserting electrical resistance heaters, and heating the shale to 340° to 370°C over a period of three to four months. The heated kerogen releases liquid and gaseous hydrocarbon, which can be extracted by a conventional type oil well as it collects in a reservoir. It is argued the process results in the production of about 65% to 70% of the original carbon in the subsurface. The in situ processing does pose a risk for groundwater contamination. Shell's technique to minimize groundwater contamination involves freezing the perimeter of the extraction zone through massive belowground refrigeration systems in what is called freeze-wall technology. Its ability to protect against leakage once the site is abandoned is not well established.

Adding up the capital investment to install the retorting mechanisms, cost of heat and refrigeration energy required, costs of refining, as well as the cost of needed water, transport, and environmental compliance, these types of production only make sense with the elevated price of petroleum. For each unit of energy used in production about 3.5 units of energy are produced for the consumer market.

Raytheon Company has developed a technology for in situ retorting using microwaves to heat the underground kerogen to release its liquid and gas components. By tuning the microwave frequencies to the kerogen vibrational modes, most of the energy can be directed to the kerogen rather than the surrounding silicate minerals. This is similar to microwave ovens that heat water by tuning the frequency to water vibrational modes that are not present in the container in which the water resides. Because of the more rapid heating the retorting can be accomplished in a shorter amount of time than resistance heating.

## Tar Sand

Tar sands are concentrations of very dense and viscous bitumen in sand. Most appear to be formed in normal petroleum reservoirs by biodegradation, reaction with oxidizing formation waters, and/or loss of their volatile fraction through time. The tar has a 6° to 12° API. They are found in significant amounts in many countries but Canada and Venezuela have particularly large deposits: the Athabasca tar sand in Alberta, Canada and the Orinoco River Basin tar sand in Venezuela. The deposits in Alberta, Canada have an API gravity of around 8°. It is estimated that within the Athabasca and Orinoco tar sands are 3.6 trillion barrels of crude oil while 1.75 trillion barrels of conventional oil reserves are thought to exist worldwide making oil sands the major resource base for petroleum. Higher petroleum prices and new production technology enable hydrocarbons from tar sands to be profitably extracted and upgraded to usable products.

Open pits are used to mine tar sands. The tar sand is crushed to reduce its size. Hot water between 50° and 80°C is added to the tar sand to release the bitumen in a slurry. The bitumen is recovered by flotation from the top of the slurry. Depending on the bitumen content of the tar sand 90% to 100% of the bitumen can be recovered. Molecules in the bitumen are broken down by hydrocracking to produce shorter chained molecules and then hydrogenated to produce gasoline and kerosene. Because of the needed hydrogen, a large amount of natural gas $CH_4$ is required in the processing of tar sands to petroleum. After bitumen extraction, the spent sand is returned to the open pits with the land eventually being reclaimed.

Concerns have been raised both with oil shale and sand production because the volume of the spent shale and sand with its high porosity is greater than the original volume excavated. This makes reclaiming the mined area difficult. Also the high porosity allows rainwater to penetrate deeply into the waste material and to react with it, potentially contaminating groundwater. The high porosity also decreases surface runoff and stream formation in the area.

## Strategic Petroleum Reserve

The United States Strategic Petroleum Reserve (SPR) is the world's largest supply of emergency crude oil. Petroleum is stored in huge underground *salt domes* along the U.S. coastline of the Gulf of Mexico. The U.S. SPR current capacity is 727 million barrels and is generally kept at this level ready to use. The second largest emergency supply of petroleum is Japan's SPR with a reported capacity of 579 million barrels.

Salt (NaCl) originally formed bedded deposits by continuous evaporation of seawater in the ancestral Gulf of Mexico during the Jurassic period, 160 to 170 million years ago. After many layers of rock were deposited on top of the salt layer, the salt became unstable. Because the density of salt ($\sim 2.16$ g cm$^{-3}$) is less than the density of rock ($\sim 2.7$ g cm$^{-3}$) the salt rose as vertical elongate cylinders into the rocks above forming a dome-like diapiric structure of salt a few kilometers across and about 5 km deep with a limestone plus anhydrite cap as shown in **Figure 2.25**. The limestone (CaCO$_3$) plus anhydrite (CaSO$_4$) cap is the result of impurities of carbonate and gypsum in the original ocean evaporate deposit that remain after dissolution of salt due to interaction of formation waters in the top of the salt diapir.

To produce an SPR reservoir, caverns are carved out of underground salt domes by solution mining. A well is drilled into the salt dome and fresh water is injected dissolving the salt. By controlling the freshwater injection process, salt caverns with precisely controlled dimensions are produced. Each SPR cavern is about 600 meters deep and holds about 10 million barrels of petroleum.

## Crude Oil Refining

Crude oil is processed to produce fractions dictated by the market for different oils. This market can change in terms of the amount of a particular fraction desired. Typically more of the gasoline fraction, of carbon chain length from 4 to 10 (C$_4$–C$_{10}$), is wanted than is present in the crude oil.

The crude oil is first separated into fractions in a distillation tower as given in **Figure 2.26**. The crude oil is heated to 400°C. At this temperature all the hydrocarbons of chain length 20 or lower are vaporized to a gas. Hydrocarbons of chain lengths greater than 20 remain liquid and pass downward. If this heavy oil can't be sold directly it undergoes more refining (see below).

Starting at 400°C the distillation tower is kept at lower and lower temperatures with increasing height. The gases of chain lengths C-20 or lower rise in the tower. The gas mixture encounters a barrier trap through which there are only openings into a liquid hydrocarbon of C-20 at its condensation temperature. The gas mixture is then forced to go through the liquid at this temperature. The C-20 fraction in the gas will condense into liquid while the lighter fractions stay gaseous and are transported upward to the next barrier trap and the process is continued.

The heavy fraction of chain length greater than 20 typically undergoes vacuum distillation. The presence of low pressures causes the boiling temperature of these heavy fractions to be lowered by around 150°C allowing their separation in a distillation tower.

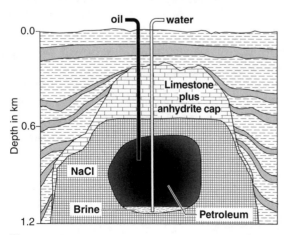

*Figure 2.25* Salt dome in the earth containing a cavity used to store petroleum.

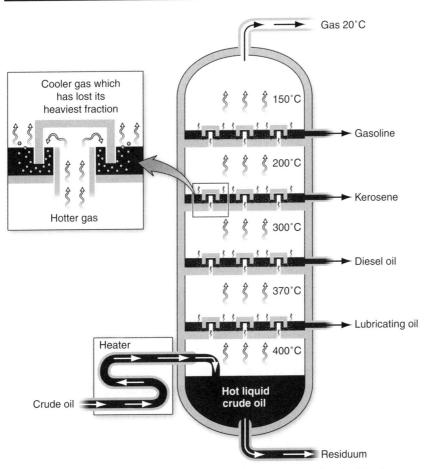

*Figure* 2.26 Distillation tower used for separating the constituents of crude oil by heating to the boiling point of each component.

Shorter-length hydrocarbons are in great demand for their use in vehicles. The heavy fraction is therefore typically cracked and hydrogenated. Thermal cracking applies heat and pressure to heavy oil to break it apart at temperatures of 455° to 540°C and pressures between 700 to 6,900 kPa in the presence of a catalyst. Originally this was alumina but now *zeolites* are used (See Figure 10.24 in the chapter on industrial minerals). Hydrocracking is a catalytic cracking process with an elevated partial pressure of hydrogen gas. This saturates the broken bonds between carbon atoms with hydrogen. This process can also be used by itself to increase the state of hydrogen saturation of hydrocarbon by adding more hydrogen to the carbon where it

is referred to as hydrogenation. In hydrogenation sulfur can be removed and hydrogen added by a reaction like

$$C_2H_5SH + H_2 \rightarrow C_2H_6 + H_2S \qquad [2.2]$$

where ethane and hydrogen sulfide are produced.

The residual heavy oil from a distillation tower can be sold to a producer of plastics or coked. In coking, temperatures above 480°C are maintained until the residual oil cracks into heavy oil, gasoline, and some lighter fractions. Most, however, becomes an almost pure carbon residue, petroleum coke. This coke is sold mainly for use as a fuel for steel and other processing mills but is also used for the manufacture of dry cell batteries and carbon electrodes.

The liquid petroleum products are stored onsite at the production plant in large tanks until they are distributed to customers such as gas distribution centers, airports, and chemical processing plants. Most of this distribution is through pipelines.

## Consumption and Production of Petroleum

Of the approximately 84 million barrels of petroleum processed per day worldwide in 2009, **Table 2.1** gives the yearly per capita consumption of countries with the largest consumption of petroleum. Note that the United States consumes 22% of the world's petroleum and China 10%. There is a large per capita consumption rate in North America whereas in India it is presently about 42 U.S. gallons per capita per year.

The yearly consumption of the four largest users of petroleum as a function of time is plotted in **Figure 2.27**. Note that China has surpassed Japan to become the world's second largest user of petroleum after the U.S. Together with India,

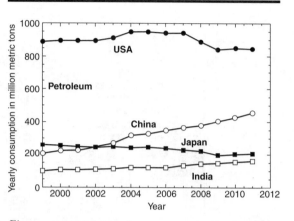

*Figure* 2.27 Yearly petroleum consumption as a function of time for the U.S., China, Japan, and India. (Data from: *BP Statistical Review of World Energy*, June 2012.)

China is industrializing and increasing petroleum consumption as a function of time. The U.S. and Japan, on the other hand, have apparently stabilized consumption to some degree. Their decreased consumption for 2008 to 2010 likely reflects in part the effect of a global recession.

*Table 2.1* PETROLEUM CONSUMPTION FOR THE 10 LARGEST CONSUMING COUNTRIES IN 2009.*

| RANK | NATION | MILLION BBL/DAY | POPULATION IN MILLIONS | BBL/YEAR PER CAPITA |
|------|--------|-----------------|------------------------|---------------------|
| 1 | United States | 18.7 | 308 | 22.2 |
| 2 | China | 8.6 | 1,333 | 2.4 |
| 3 | Japan | 4.4 | 128 | 12.5 |
| 4 | India | 3.2 | 1,170 | 1.0 |
| 5 | Russia | 2.7 | 142 | 6.9 |
| 6 | Saudi Arabia | 2.6 | 26 | 32.3 |
| 7 | Germany | 2.4 | 82 | 11.6 |
| 8 | Brazil | 2.4 | 192 | 4.8 |
| 9 | South Korea | 2.3 | 48 | 16.7 |
| 10 | Canada | 2.2 | 34 | 24.7 |
| World totals | | 84.1 | 6,800 | 4.5 |

* Numbers are in millions. bbl = barrel.

*Data from: BP Statistical Review of World Energy*, June 2010.

**Table** 2.2 provides a list of the top petroleum producing countries, their production, and stated proven reserves. Russia produces 12.0%, Saudi Arabia 11.6%, and the U.S. 10.8%. Given their current rate of production and stated reserves in Table 2.2, Saudi Arabia has a 73-year supply of conventional oil, Russia a 16-year supply, and the U.S. a 6-year supply. Clearly unconventional petroleum supplies and their higher costs are going to be a larger and larger part of the petroleum supply picture in the future. One can assume that mankind will partition off of expensive unconventional oil and into other energy sources as these become more competitive and unconventional oil supplies

*Table* 2.2 TOP WORLD PETROLEUM-PRODUCING COUNTRIES AND THEIR STATED PROVEN RESERVES IN 2009.

| RANK | COUNTRY | PRODUCTION $10^6$ BBL/DAY | PROVEN RESERVES $10^9$ BBL |
|---|---|---|---|
| 1 | Russia | 10.1 | 60.0 |
| 2 | Saudi Arabia | 9.8 | 262.3 |
| 3 | United States | 7.2 | 21.0 |
| 4 | Iran | 4.2 | 136.3 |
| 5 | China | 4.0 | 16.0 |
| 6 | Canada | 3.3 | 179.2 |
| 7 | Mexico | 3.0 | 12.4 |
| 8 | United Arab Emirates | 2.8 | 97.8 |
| 9 | Brazil | 2.6 | 11.8 |
| 10 | Kuwait | 2.5 | 101.5 |
| 11 | Venezuela | 2.5 | 80.0 |
| 12 | Iraq | 2.4 | 115.0 |
| 13 | European Union | 2.4 | <10 |
| 14 | Norway | 2.4 | <10 |
| 15 | Nigeria | 1.8 | 36.2 |
| 16 | Algeria | 2.2 | 12.3 |
| 17 | Angola | 1.8 | <10 |
| 18 | Libya | 1.6 | 41.5 |
| 19 | Kazakhstan | 1.4 | <10 |
| 20 | United Kingdom | 1.6 | <10 |
| 21 | Qatar | 1.2 | 15.2 |
| 22 | Indonesia | 1.1 | <10 |
| Totals | | 84.1 | ~1,340 |

bbl = barrels of oil.

*Data from:* U.S. Central Intelligence Agency (CIA) and Energy Information Administration (EIA).

are used up. How and when this will be done is still a matter of debate, but it is clear that it will be underway by the middle of the twenty-first century. Will mankind ever use up all the crude oil on the earth? No, but only because it will become very expensive to obtain the last remaining supply relative to lower-cost energy alternatives.

**Figure 2.28** gives the production of petroleum as a function of time for Saudi Arabia, the Russian Federation, and the United States. Note that Russian Federation petroleum production surpassed U.S. production in 2002 and Saudi Arabia in 2009. Saudi Arabia petroleum production is adjusted depending on market conditions. The economic slowdown after the 9/11/2001 World Trade Center bombing and 2008 to 2010 global recession with its decreasing demand caused Saudi Arabia to decrease production to keep prices higher.

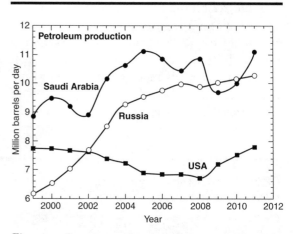

*Figure 2.28* Yearly petroleum production for Saudi Arabia, the Russian Federation, and the U.S. as a function of time. (Data from: BP *Statistical Review of World Energy*, June 2012.)

## SUMMARY

Types of energy are characterized by the force that performs the work. These can be renewable or nonrenewable. Global energy consumption is growing at 2.2% per year, twice the rate of population growth. In 2010 it stood at 12 billion metric tons of crude oil equivalent (toe). Most of this energy now and in the near future will be the fossil fuel petroleum.

Fossil fuels are formed from organic matter (OM) that is made up primarily of carbon with hydrogen the second most abundant element. Land plants are dominantly glucose ($C_6H_{10}O_5$). Because land plants dominate organic matter production on the earth, the formula $CH_2O$ is a reasonable approximation for OM. Petroleum is a hydrocarbon (H + C) and contains little or no oxygen. Because of their oxygen content land plants do not breakdown to produce petroleum, but rather produce coal. Petroleum is produced by the breakdown of algae and zooplankton generally produced in the surface layer of the ocean and some lakes.

Organic matter in petroleum source rocks is either bitumen or kerogen. Kerogen is made up of long chains of carbon atoms with a higher H/C ratio than bitumen. The higher the H/C ratio the more oil prone is the kerogen. The kerogen breaks down and produces petroleum on burial starting at about 2 km depth. The exact depth range over which petroleum is produced is called the oil window.

Vitrinite reflectance, Van Krevelen (Atomic H/C vs. O/C) diagrams, and Rock-Eval pyrolysis are three ways to characterize the maturation state of kerogen in petroleum source rocks. These source rocks occur in distinct time intervals in the geological past. Reservoirs of petroleum occur in rocks of high porosity trapped below caprocks.

Crude oil is recovered by drilling holes into petroleum reservoir rocks with a bit attached to a rotating hollow pipe filled with drilling mud. Natural pressure in the reservoir brings crude oil to the surface in what is called primary production. Fluids are then injected to produce oil in secondary recovery and finally steam and chemicals injected to help move the crude oil out of the reservoir.

There is only a limited amount of petroleum in the earth. Considering production trends in the U.S., the year of peak production of petroleum, or "peak oil," was correctly predicted to be 1970. Based on a worldwide analysis peak oil for the entire world has been predicted. It varies greatly depending on the analysis because reserves are not a matter of public record. The U.S. has a strategic petroleum reserve in cavities produced in salt domes along the Gulf Coast.

Oil shale is petroleum source rock that has not completely released its hydrocarbons. Tar sand is a reservoir rock that has lost its volatile fraction through time. The hydrocarbon can be extracted from these rocks as an energy source.

Crude oil is processed to produce desired hydrocarbons. Much of this is the gasoline fraction of crude oil of carbon chain lengths from 4 to 10 carbon atoms. A distillation tower is used to separate fractions. Longer-length carbon chains can be "cracked," a process for shortening chain lengths. The broken bonds are then hydrogenated.

The U.S. produces 11% of the world's petroleum and consumes 25%. Different countries have different reserves of oil to produce and these are not well known particularly when the unconventional oil in oil shales and sands are considered. It is reasonable to assume that new supplies of petroleum will start to become scarce by the middle of the twenty-first century.

## KEY TERMS

| | |
|---|---|
| annulus | oil window |
| API gravity | open pit mine |
| bentonite | peak oil |
| benzene-like rings | pumpjack |
| bitumen | reservoir |
| caprock | Rock-Eval pyrolysis |
| completed (well) | salt dome |
| cracking | sapropelic coal |
| crude oil | strip mine |
| drilling mud | tons of oil equivalent (toe) |
| electricity | total organic carbon (TOC) |
| electromagnetic force field | Van Krevelen diagram |
| humic coal | vitrinite |
| kerogen | zeolite |
| mud logging | zooplankton |

## PROBLEMS

1.  a.  Is gasoline or electricity a more costly energy source for transportation? Compare first in dollars per joule. Assume the cost of gasoline is US$2.80 per gallon and the cost of electricity is US$0.10 per kWh (kilowatt hour). Gasoline weighs 5.50 lb per gallon and releases 19,000 Btu (British thermal units) per pound (1 joule = $2.78 \times 10^{-7}$ kWh = $9.50 \times 10^{-4}$ Btu).

    b.  If gas engines are 25% efficient in energy use and electric engines are 60% efficient which is more economical for fuel cost?

2.  If a well is drilled 4 km to an oil producing formation what would be the fluid pressure at the bottom of the well if it was: (a) Filled with water (density = 1.0 g cm$^{-3}$)? (b) At the pressure of the overlying column of rocks which have a density of 2.7 g cm$^{-3}$? Note that pressure at depth in the earth equals the density of the material above times the acceleration of gravity times the distance to the surface. (Keep track of your units.)

3.  If the well in problem 2 was a hole 20 cm in diameter how many kg of barite would need to be added to the water in the drilling mud to bring the fluid pressure at the bottom of the hole to the rock pressure? Assume no volume change of barite and water on mixing.

## REFERENCES

Al Husseini, S. 2005. *2015 peak oil forecast*. Interview with the Association for the Study of Peak Oil and Gas, USA (ASPO-USA).

Espitalie´, J. J., La Porte, L., Madec, M., Marquis, F., Le Plat, P., Paulet, J., and Boutefeu, A. 1977. Méthode rapid de caractérisation des roches mères de leur potentiel pétrolier et de leur degré d'évolution. *Rev. l'Inst. Français Pétrole* 32(1):23–42.

Hunt, J. M. 1996. *Petroleum geochemistry and geology*. New York: W. H. Freeman and Company.

U.S. Energy Information Administration. *International energy outlook 2009* (Report #DOE/EIA-0484) Washington, DC: U.S. Department of Energy. Available online at: www.eia.doe.gov/oiaf/ieo/index.html.

Jamison, H. C., Brockett, L. D., and McIntosh, R. A. 1980. Prudhoe Bay—A 10-year perspective. In M. T. Halbouty (ed.) Giant oil and gas fields of the decade 1968–1978. *AAPG Memoir* 30: 289–314.

Klemme, H. D. and Ulmishek, G. F. 1991. Effective petroleum source rocks of the world: Stratigrasphic distribution and controlling depositional factors. *AAPG Bulletin* 75: 1809–1851.

Maugeri L., 2009, Squeezing more oil from the ground, Scientific American, 301(4): 56–63.

North, F. K. 1985. *Petroleum geology*. Boston: Unwin Hyman.

Peters, K. E. and Moldowan, J. M. 1993. *The biomarker guide: Interpreting molecular fossils in petroleum and ancient sediments*. Upper Saddle River, NJ: Prentice Hall, Inc.

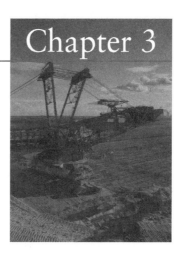

# Chapter 3

# Natural Gas, Coal, and Related Resources

## Natural Gas

The third largest component of the world's supply of energy after crude oil and coal is natural gas (see Figure PT1.3 in the Part 1 introduction). Natural gas is most often found as a product of the generation of petroleum in the earth, referred to as conventional natural gas, but is also associated with some coal (*coal bed methane*) and is formed from bacterial modification of near surface organic matter (*marsh gas*).

### Conventional Natural Gas

The composition of natural gas can vary widely but consists primarily of a mixture of hydrocarbons of short length that remain as gases at ~20°C. Methane ($CH_4$) is the primarily component but ethane ($C_2H_6$), propane ($C_3H_8$), and butane ($C_4H_{10}$) can make up 20% of the natural gas as given in **Table 3.1**. Natural gas made up almost entirely of methane is termed "dry gas."

Natural gas concentrations are typically reported in *cubic feet (cf)* at standard temperature and pressure (for the U.S. and OPEC this is 60°F and 1 atmosphere = 14.73 psi). Natural gas is also reported in energy equivalents, as generally it is the gas's energy production on burning that is of prime importance. These can be *British thermal units (Btu)* where 1 *cf* = 1,027 Btu or metric tons of crude oil equivalent (*toe*), which is ~40,000 *cf* of natural gas. When natural gas is delivered to a residence it is often given in therms. One therm is 100,000 Btu or about 97 *cf* of natural gas depending on its exact composition.

## Unconventional Natural Gas

Besides conventional reservoirs, unconventional reservoirs also exist, which contain natural gas that is difficult and therefore more expensive to extract. Unconventional natural gas resources have been known for quite some time but have been bypassed in favor of more easily obtainable conventional supplies. With higher energy costs and improved technology to exploit these resources unconventional gas deposits are starting to be tapped. These resources appear to be quite extensive and occur in a large number of countries including the U.S., Australia, and China. Given in

*Table 3.1* RANGE OF CONCENTRATIONS OF NATURAL GAS BEFORE REFINING

| GAS | RANGE (%) |
| --- | --- |
| Methane | 70–95 |
| Ethane + propane + butane | 0–20 |
| Carbon dioxide | 0–8 |
| Nitrogen | 0–5 |
| Hydrogen sulfide | 0–5 |
| Oxygen | 0–0.2 |
| Rare gases | Trace |

**Figure 3.1** are the fields in Europe. Until recently only the U.S. exploited its unconventional natural gas resources due to the costs and sophisticated technology required to do so effectively, but this is changing. There are five main categories of unconventional natural gas: deep gas, tight gas, coal bed methane, overpressurized gas, and methane clathrates.

## Deep Gas

Deep natural gas occurs at depths greater than 6 km below the surface. These are deeper than most gas reservoirs, which are generally found at depths of 2 to 6 km. This makes drilling more expensive than for conventional supplies. In the United States deep natural gas is found in the Louisiana-Texas Gulf Basin, Permian Basin in west Texas and southeastern New Mexico,

Anadarko Basin in west Oklahoma and the Texas Panhandle, the Rocky Mountain Basins from Montana to New Mexico, and the Louisiana-Mississippi-Alabama Salt Basin. With the increasing price paid for energy resources these fields are starting to be exploited.

## Tight Gas

Tight gas is present in unusually impermeable rock. This can be an impermeable and low-porosity sandstone, limestone, or shale although most is present in shale. Tight gas containing shales are source beds for petroleum that have retained some of their hydrocarbons as natural gas. The gas has not escaped due to the shales' low permeability. Given in **Figure** 3.2 are the tight gas fields in North America. The Barnett Shale in Texas, Marcellus Shale in New York and Pennsylvania,

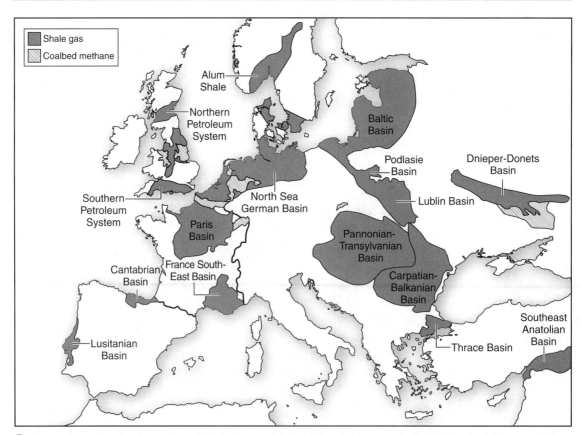

*Figure* 3.1 Unconventional natural gas fields in Europe. (Modified from *World Energy Outlook 2012 - Special Report - Golden Rules for a Golden Age of Gas* © OECD/IEA 2012, figure 3.7, page 121.)

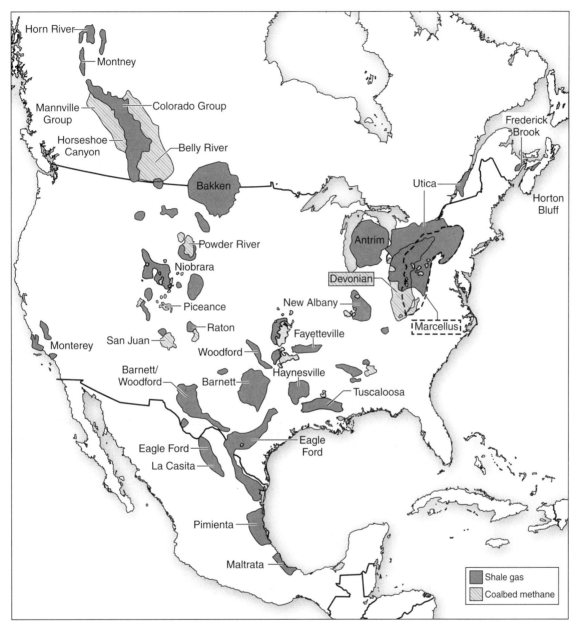

*Figure 3.2* Unconventional natural gas fields in North America. (Modified from *World Energy Outlook 2012 - Special Report - Golden Rules for a Golden Age of Gas* © OECD/IEA 2012, figure 3.1, page 103.)

and the Utica Shale in New York and the province of Quebec, Canada are important tight gas shale locations that have started to be utilized. The Barnett Shale is widely estimated to contain as much as 30 *Tcf* (trillion standard cubic feet) of natural gas. Estimates vary greatly, but the Marcellus Shale is considered by some to contain over 360 *Tcf* of recoverable natural gas. The Utica Shale total recoverable gas resources estimates are from 5 to 25 *Tcf*. To obtain the natural gas from these formations requires a greater density of well sites; about one per square kilometer rather than the one per ~10 square kilometers of conventional reservoirs. In populated areas this means drilling and production activities are considerably more invasive.

**Fracking.**   About 90% of the natural gas wells drilled in the United States are hydrofractured to increase gas flows, termed *fracking* for short. The technique has been used since the late 1940s. Fracking involves injecting a fluid mixture whose composition is about 99.4% water and 0.5% sand, as well as $N_2$, $CO_2$, air, and/or proprietary chemicals down the drill hole. This is done at a rate that increases fluid pressure in the well to a value in excess of the stresses needed to open microfractures in the rocks in producing layers. These microfractures are kept from closing by the injected sand particles allowing the natural gas to flow into the extraction well. In the case of tight shale, the gas is extracted by drilling horizontally along the shale bed and then fracking along the entire horizontal length (see **Figure 2.16** in the chapter on petroleum).

The amount of water used in fracking these horizontally drilled wells is often as much as 7.5 to 15 million liters per well containing up to 200,000 liters of proprietary chemicals. About 25% of the fluid is lost into the shale and about 75% flows back up to the top of the well. Often the backflowing fluid picks up salts from the shale. The used fracking fluid is stored on site for later reuse or is transported to a treatment plant to remove the chemicals in the water.

There is some concern about the proprietary chemicals present in fracking fluids used to help release the gas molecules from the shale mineral surfaces. These fracking fluids can contain chemicals that are toxic to wildlife and humans such as benzene, ethylbenzene, and toluene. Because near-surface aquifer layers are often transected to reach the tight gas reservoirs below, concern about potential drinking water aquifer contamination have been raised by the Environmental Protection Agency. These problems are most likely to occur from poor downhole grouting and misaligned well casings in the aquifer section. Another concern

is in handling the used fracking fluid that is typically stored onsite in pits dug in the ground. Even when lined these storage pits can leak the fluid into surface aquifers if they overflow during a large rainstorm. The environmental impacts of tight gas shale development are clearly a concern but likely manageable with sufficient oversight.

## Coal Bed Methane

Coal bed methane is natural gas produced during the maturation of coal. While much can escape to the atmosphere some methane molecules adsorb to coal particle surfaces. Because of the large internal surface area of coal, large volumes of methane gas can be present. In some cases six to seven times as much methane is present as a similarly sized conventional natural gas reservoir. It can be released by fracturing of the coal when mined. It is a safety concern because at high concentration it can be explosive. A coal bed methane gas explosion occurred in 2010 in the Peak River Mine, New Zealand's largest coal mine, killing 29 people.

Coal bed methane gas is currently captured in wells drilled into fractured coal by first lowering the fluid pressure with removal of formation water. The lower pressure causes methane bubbles to be released from coal surfaces and they make their way to the well following the extracted water. The gas is collected at the top of the well where it is compressed and stored in tanks. Recoverable coal bed methane of 163 *Tcf* is thought to exist in the U.S. according to gas experts. This makes up about 2.3% of the U.S. natural gas resource base. There is some concern because a large amount of formation water must be disposed of. In most cases, the formation water is reinjected back into the subsurface.

## Overpressured Gas

Overpressurized or geopressured natural gas occurs in underground formations where fluid pressures have started to approach the higher pressures experienced by rocks rather than the lower fluid pressure of a standing column of fluid. Overpressurized reservoirs tend to be deep and the high fluid pressures make them difficult and therefore less economic to extract. They are considered a significant resource of natural gas particularly when the price of natural gas increases. In the Gulf of Mexico basin large areas of over-pressured brines saturated with natural gas are present. Many investigators argue these brines are presently economically extractable if the heat from the extracted brines is also used as an energy source.

### Methane Clathrates

Methane clathrates are a lattice of frozen water around a molecule of methane in a cage-like structure. The structure is stable at low temperature under significant pressure as discussed in the introductory chapter. Methane clathrates are found in permafrost below the earth's surface and in cold deep-sea sediments on the continental slopes. Estimates of their abundance worldwide range anywhere from 7,000 *Tcf* to over 73,000 *Tcf*, an immense resource. Technological problems to date, due to their distribution and location, have not allowed methane clathrates to be economically recovered. Whether methane clathrates can be economically recovered in the future is still an open question.

## Production and Reserves

Natural gas is first processed to remove liquid water and any liquid hydrocarbon condensates that form on extraction. The liquid hydrocarbons are sent to an oil refinery for processing. The remaining gas is directed to a gas processing plant where acid gases such as $H_2S$ and $CO_2$ are removed, then any remaining water vapor is removed followed by the removal of Hg and $N_2$. Natural gas liquids (NGL) of ethane, propane, and butane are separated and processed. The remaining nearly pure methane, $CH_4$, gas is then distributed through pipelines to residents, commercial establishments, and industrial users.

Given in **Figure 3.3** is an estimate of conventional natural gas reserves by region of the world. Note the large natural gas reserves in both

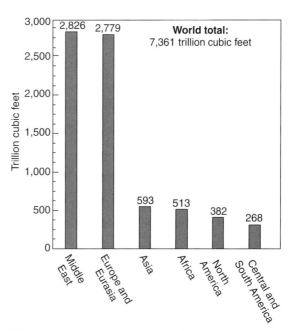

*Figure 3.3* World reserves of conventional natural gas in trillions ($10^{12}$) of standard cubic feet. (Data from: *BP Statistical Review of World Energy,* June 2012.)

the Middle East and Eurasia. It has been argued that with the exploitation of nonconventional reserves *peak natural gas* will not occur until 2090 (**Figure 3.4**).

Consumption of natural gas in the United States in 2011 reached 24.4 trillion standard cubic feet (*Tcf*), a record high due to an increase in heating degree days. In 2011, the U.S. imported 1.8 *Tcf* of natural gas, 8% of the total consumption, mainly from Canada, Mexico, and liquefied natural gas from the Middle East. This was a record low since 1992. Unconventional supplies are becoming a bigger part of the market going forward as shown in **Figure 3.5**. Note that by 2035 shale gas is projected to contribute nearly half of the U.S. natural gas supply. World natural gas consumption is expected to increase by more than 50% from 2010 through 2030. Most experts believe the world's number one natural gas consuming region in 2030, taking over that spot from North America, will be Asia as China's economy is expected to grow at 6.4% annually in the future.

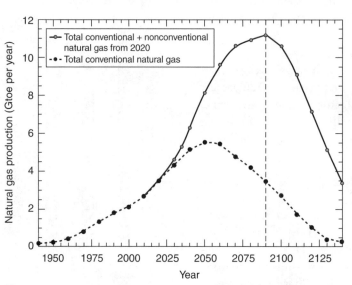

*Figure 3.4* World gas production to 2010 and estimated future world production from conventional and nonconventional sources to 2140. The straight dashed line gives peak natural gas in 2090 of over 11 *Gtoe* ($10^9$ metric tons of oil equivalent) per year. (Adapted from Odell, P. R., 2004, Why carbon fuels will dominate the 21st centuries global energy economy, *Multi-science publishing*, 192 p.)

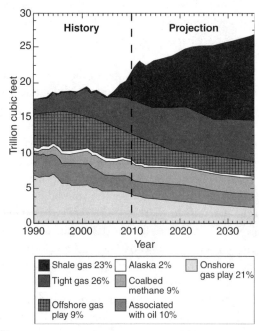

*Figure 3.5* U.S. natural gas production by source from 1990 and predicted to 2035. Legend gives the percentage of supply from each source at the end of 2010. (Modified from the U.S. Energy Information Administration, *AEO2012*, Jan. 23, 2012)

In the United States natural gas for electric power made up 31% of delivered volumes. Residential consumption was 21%. Natural gas for vehicle fuel has increased significantly in the last few years but in 2011 was still less than 0.15% of the total consumed.

## Hydrogen Gas Production

Fossil fuel, in particular natural gas, currently is the main source for hydrogen gas production. At 700° to 1100°C in the presence of a nickel catalyst, steam reacts with methane to yield carbon monoxide and hydrogen gas:

$$CH_4 + H_2O \rightarrow CO + 3H_2. \qquad [3.1]$$

Note that some hydrogen gas is also produced as a byproduct of chorine gas production from NaCl by electrolysis (see chapter on evaporites).

Hydrogen gas is used as a fuel source, for the creation of ammonia ($NH_3$) for fertilizer production (see chapter on evaporites), and to *hydrogenate* broken carbon chains during cracking to produce lighter carbon fractions from heavy petroleum sources (see chapter on petroleum). If the world

*Figure* 3.6 Small seams of black lignite coal exposed by blasting. Note vertical grooves which mark the holes used to set the charges.

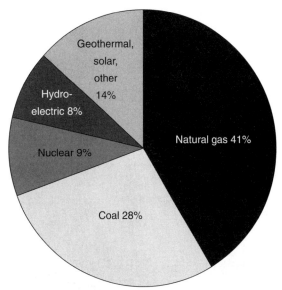

*Figure* 3.7 Percent of electrical power production in the U.S. by type for 2011. (Data from: U.S. EIA.)

embraces fuel-cell powered vehicles using hydrogen gas as a fuel source (see chapter on alternative energy resources) the hydrogen gas supply is likely to have a methane gas source as produced by equation [3.1].

# Coal

Coal is a black to brownish-black colored combustible sedimentary rock made of decomposed organic material of leaves, stems, seeds, and spores, termed humus. The humic material is derived from trees and other land plants, forming tabular bodies of large aerial extent. These occur in layers, beds, veins, or coal seams on vertical cuts through the earth's crust. Shown in **Figure 3.6** are exposures of black beds of coal that extend for kilometers in this sedimentary rock outcrop.

## *Coal Use*

Coal is the largest source of energy for the production of electricity worldwide. Presently 41% of worldwide electrical generation is from coal-powered plants and this is predicted to rise to 44% by 2030. More than 90% of coal production in the U.S. is used to make electricity. Coal-powered plants account for a little over a quarter of the electricity made in the U.S. (see **Figure 3.7**) In China it is about 80%. Coal is also burned to manufacture steel and processed to make liquid

fuel. It is predicted that the liquid fuel segment of production will increase in the future as the price of petroleum increases.

## *Coal Formation*

When plants die, their humic material typically reacts with $O_2$ from the atmosphere and decays. If humus is written in the simple form $CH_2O$, the reaction is

$$CH_2O + O_2 \rightarrow CO_2 + H_2O. \qquad [3.2]$$

$CH_2O$ is an approximation for the complicated composition of organic matter in plants, which gives a good approximation of its relative amount of C, H, and O. However, it needs to be remembered that besides these major constituents a multitude of other elements are required by life and are present in organic matter.

To form coal, humic matter must be buried without reacting with $O_2$ from the atmosphere. This requires a special environment, generally one in water such as a wetland. With water present $O_2$ from air can be excluded from contact with the organic matter. The small amount of $O_2$ that can be dissolved in water will be stripped out by equation [3.2] producing an *anaerobic* environment.

Anaerobic environments, those without air, form in wetlands such as swamps, bogs, and fens (**Figure 3.8**). If the organic matter flux deposited is large compared to other sediment input into the water *peat* is produced. Peat is a volume of partially decomposed plant matter. With time a thick layer of peat can develop in the water-soaked environment.

*Figure 3.8* Conifer marsh in foreground where accumulating organic matter is producing a thick peat-rich layer of sediment.

## Coal Rank

The peat produces humic coal on burial by the deposition of sediments on top of it. Subjected to higher pressures and temperatures it loses much of its water producing a more carbon-rich material (**Figure 3.9**). A soft brown to blackish colored low-rank coal forms termed *lignite* as outlined in **Figure 3.10**. Moisture content is between 40 and 70 wt% in lignite. With greater burial and time first a *subbituminous*, then a *bituminous*, and finally the highest rank coal, *anthracite*, with a moisture content of 2 wt% or less, is created. Note as outlined in Figure 3.10 that the heat content per kilogram increases as the $H_2O$ content decreases making the higher ranked coals more valuable.

Coal is destroyed at great depths in the earth as anthracite is transformed into massive graphite during what is termed granulite facies metamorphism. Granulite facies metamorphism occurs when temperatures increase to greater than about 700°C. This typically occurs at depths exceeding 20 km in the earth's crust but can be much shallower if a magma body is present nearby.

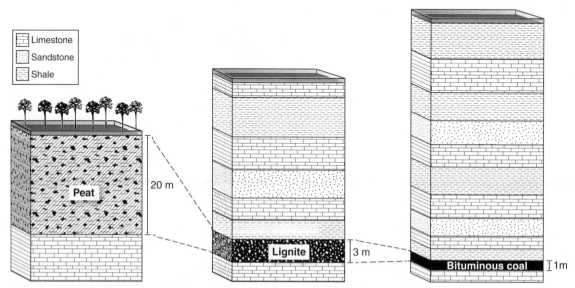

*Figure 3.9* Burial of plant matter in a swamp producing peat. With the added mass of sediments deposited on top and increased temperature with depth in the earth the peat is transformed with the loss of water to lignite and then bituminous coal.

| Stages of coal rank | | Characteristics | H₂O ~wt % | C ~wt % | Heat content |
|---|---|---|---|---|---|
| Peat | | Large pores and free cellulose with details of the original plant material easily recognizable | | | 3,000 kcal/kg |
| | | | 70 | 20 | |
| Soft brown coal | Lignite | No free cellulose | | | 3,500 kcal/kg |
| | | | 40 | 40 | 4,000 kcal/kg |
| Dull brown coal | Hard brown coal "subbituminous" | Plant structures partly recognizable but with obvious compaction | | | 5,500 kcal/kg |
| | | | 30 | | |
| Bright brown coal | | | | | |
| | | | 20 | 55 | 7,000 kcal/kg |
| Bituminous coal | Hard black coal | Plant structures no longer recognizable | | | |
| | | | 2 | 86 | |
| Anthracite | | High percentage of fixed carbon that does not disintegrate when rubbed | | | ~8,650 kcal/kg |
| | | | | 98 | |

*Figure 3.10* Characterization of layers of plant organic matter with the transformations that occur with increasing depth in the earth as it is transformed to coal.

Graphite deposits are mined for a variety of uses including making graphite crucibles, for the production of carbon steel, and use as a lubricant.

## Macerals in Coal

When coal is examined under an optical microscope dehydrated plant fragments are observed. These fragments are called *macerals*. Different macerals have different shapes, morphology, reflectance, and fluorescence. Broadly, the term macerals is equivalent to minerals in rocks. Most coal, when cut thin enough, shows predominantly red colors that are the remains of woody tissues of plants, (e.g., stems and roots) called vitrinite. Coalified spores appear as small, short, yellow blobs termed liptinite. Coalified plant cuticles show up as long orange squiggly lines. Cuticles are protective waxy epidermal plant leaf cells. They prevent moisture loss and resist damage by microorganisms. The black material observed in thin sections can be either a mineral fragment or coalified charcoal termed inertinite (**Figure 3.11**). Because charcoal is burnt wood, the presence of charcoals in coal has been used to say something about the frequency of fires in the fossil record (Scott, 2000).

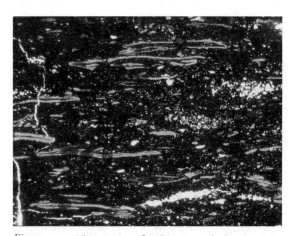

*Figure 3.11* Thin section of coal in greyscale showing macerals. Light grey macerals are coalified spores, medium grey macerals are remains of woody tissues, black material is either mineral fragments or coalified charcoal, and white areas are missing parts of the thin section.

Thin sections are generally not used to examine macerals in coal. Coals are so soft that it is very difficult to make a good thin section. Instead, polished sections of coal are examined. The surface of coal can be polished to a very high degree producing a shiny surface. A *reflectance microscope* is used to examine the surface. The coal shows up as shades of grey in reflected light (**Figure 3.12**). The reflectivity of the maceral's surface is often measured.

As coal rank and the burial of oil producing strata increase, the vitrinite macerals become increasingly reflective. By studying vitrinite reflectance, the temperature history of source beds for petroleum and coal is determined. Given in **Figure 3.13** is a diagram indicating the common maturation parameters used when considering the development of coal and hydrocarbons. Note that a bituminous coal has undergone a state of maturation equal to that of crude oil developed in the oil window.

## Resource Location

Shown in **Figure 3.14** is the distribution of coal reserves as a function of geological time period. Note the absence of coal before the Carboniferous period, 345 million years ago, due to a lack of the evolution of land plants. The extensive shallow inland seas present on the continents during the late Carboniferous period were good environments to accumulate and preserve organic matter as shown by the large coal reserves dating from this geological period.

Given in **Figure 3.15** are the coal regions in the United States. In the lightest grey are lignite deposits of the Gulf of Mexico region and the Northern Great Plains Province in North and South Dakota and Montana. In the West, as shown by medium grey colors, are subbituminous deposits, and in the Midwest and western Pennsylvania are the

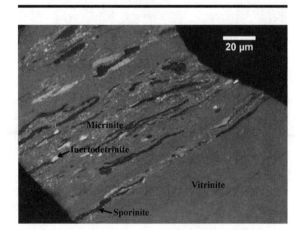

*Figure 3.12* Photomicrograph of a polished section of coal under reflected light showing dominantly compacted macerals of vitrinite, but also micrinite, inertodetrinite, and sporinite.

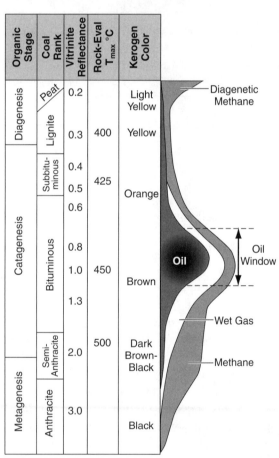

*Figure 3.13* Coal rank and organic maturation parameters with depth related to the amount of hydrocarbon generation. (Adapted from Senftle, J.T., and C.R. Landis, 1991, Vitrinite reflectance as a tool to assess thermal maturity, in R.K. Merrill, ed., Source and migration processes and evaluation techniques: AAPG Treatise of Petroleum Geology, *Handbook of Petroleum Geology*, p. 119-125.)

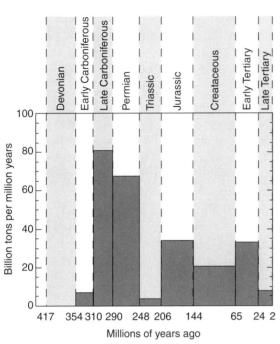

*Figure 3.14* Coal reserves as a function of geological time period.

darkest grey of medium and high-grade bituminous deposits. Small anthracite deposits are found in eastern Pennsylvania and a number of places in the Appalachian Mountains. Most coal in the United States is currently mined in the West particularly in Wyoming's Powder River Basin. This is followed by coal in Pennsylvania and the Midwest.

## Production and Reserves

Around 6.1 billion metric tons of hard coal (anthracite + bituminous) was used worldwide in 2011 and 1 billion metric tons of *brown coal*. Coal resources are more often reported in Btu. This is because the different grades of coal release different amounts of energy and it is the energy that is of most interest when considering coal.

According to the U.S. Energy Information Administration (EIA) international energy outlook, world coal consumption is predicted to increase by ~50% from 127.5 quadrillion ($10^{15}$) Btu in 2006 to 190 quadrillion Btu in 2030 (**Figure 3.16**). The growth rate for coal

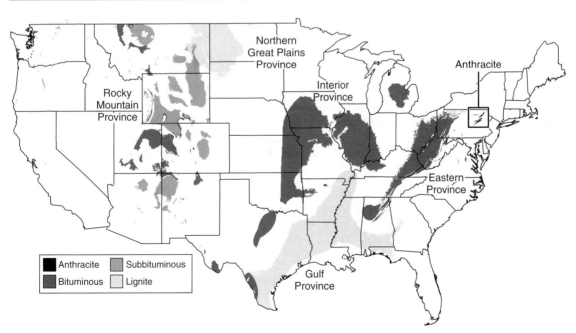

*Figure 3.15* Coal provinces in the continental USA. (Adapted from USGS Open-File Report OF 96-92)

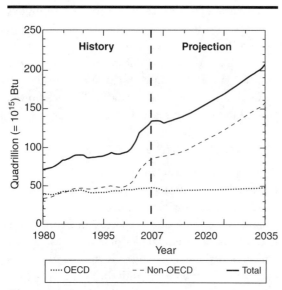

*Figure 3.16* Yearly world coal consumption as a function of time. OECD are those countries in the Organization for Economic Cooperation and Development, a set of 30 high-income countries from Europe and North America that also includes Japan, Australia, New Zealand and Republic of Korea. (Data from: Energy Information Administration (EIA), *International Energy Outlook 2010*.)

consumption will likely increase 1.9% per year from 2012 to 2015 and 1.6% per year from 2015 to 2030 generally reflecting the growth trends for both world gross domestic product and world energy consumption. Gross domestic product (GDP) is the value of all final goods and services produced in a year. In the U.S., however, coal use started to decrease in 2010 as electrical power plants switched to cheaper and cleaner natural gas from unconventional sources for their needed fuel.

The EIA also predicts coal's share of world energy consumption will increase slightly from 27% in 2006 to 28% in 2030. Developing countries will likely account for about 97% of this increase with China and India together accounting for ~85% of the increased demand. This increased demand for coal will be mainly for increased production of electricity.

## Resource Predictions

In many places in this text, such as in the last two paragraphs, numerical predictions are made about future amounts of usage of a resource. It is important to remember that these are predictions subject to error, generally without a certainty associated with them. The size of the prediction depends on the assumptions that were made to do the projections. If you constrain yourself to one source the predictions are typically internally consistent. If you use multiple sources inconsistencies often arise. Even with the same source published at different times new numbers and assumptions modify the prediction. Natural resource use depends on the global economic climate and recessions are notoriously hard to predict, except perhaps in hindsight. A student of resource usage needs to keep these things in mind when considering future resource utilization.

As of the end of 2006 world coal reserves were estimated to be between 800 to 900 gigatons. At the current usage rate this would last 132 years according to the U.S. Energy Information Administration. However, because the growth rate of coal usage is increasing at 1.6% to 1.9% per year this time decreases to about 60 years. Some have argued that the reserves of coal are overstated as calculated world reserves to production from 2000 to 2006 have dropped by a third. The data quality on which coal reserves are determined is poor. *Energy Watch Group* argues coal reserves are overstated and has forecast that *peak coal* will occur as early as 2025 and then coal production will fall into a terminal decline. Given in **Figure 3.17** is their prediction based on a country-by-country analysis. The lines marked "IEA 2006 reference prediction" and "Climate policy prediction" are 2006 International Energy Agency predictions of coal usage unconstrained and constrained by climate policy measures for comparison.

Given in **Table 3.2** are reported coal reserves and production by country. Note that China is the largest producer of coal and the United States has the largest reserves. Most coal is used in the countries where it is mined as transportation costs account for a significant share of the total price of delivered coal. Australia is the world's largest coal exporter accounting for over half of total world exports.

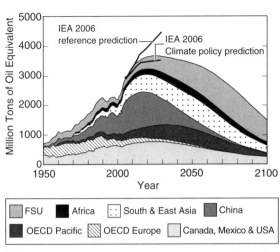

*Figure 3.17* World coal production prediction predicated on extent of reserves by country. (IEA = International Energy Agency; FSU = Former Soviet Union; OECD = Organization for Economic Cooperation and Development, a set of 34 industrialized countries.) (Modified from the Energy Watch Group, 2007, *Coal: Resources and future production*, EWG-Series No 1/2007, Ludwig-Bolkow-Foundation, Ottobrunn. Available at http://go.nature.com/jngfsa.)

## U.S. Coal Production and Consumption

As outlined in **Figure 3.18** little anthracite has been mined in the United States in the last 20 years. Large amounts of subbituminous coal are available in the western U.S. Despite its lower grade, it has a desired low-sulfur content and is used extensively. Given in Figure 3.18a is yearly U.S. coal production as a function of time by coal type. Note that by 1999 subbituminous coal overtook bituminous coal as the largest component of production. As shown in Figure 3.18b U.S. coal production has been increasing except for the economic downturn starting in 2008. Note also that the greater use of subbituminous coal as a function of time correlates with more production from west of the Mississippi River and greater production from surface as opposed to underground mines (Figure 3.18c).

*Table 3.2* COAL RESERVES AND PRODUCTION SHARE FOR COUNTRIES WITH THE GREATEST RESERVES END OF YEAR 2011

| COUNTRY | BITUMINOUS AND ANTHRACITE* | SUBBITU-MINOUS AND LIGNITE* | TOTAL* | WORLD RESERVES | WORLD PRODUC-TION | RESERVE LIFE (YEARS)† |
|---|---|---|---|---|---|---|
| United States | 108,501 | 128,794 | 237,295 | 27.6% | 14.8% | 241 |
| Russian Federation | 49,088 | 107,922 | 157,010 | 18.2% | 4.0% | 495 |
| China | 62,200 | 52,300 | 114,500 | 13.3% | 48.3% | 35 |
| Australia | 37,100 | 39,300 | 76,400 | 8.9% | 6.3% | 180 |
| India | 56,100 | 4,500 | 60,600 | 7.0% | 5.8% | 106 |
| Germany | 99 | 40,600 | 40,699 | 4.7% | 1.2% | 34 |
| Ukraine | 15,351 | 18,522 | 33,873 | 3.9% | 1.0 | 462 |
| Kazakhstan | 21,500 | 12,100 | 33,600 | 3.9% | 1.5 | 303 |
| South Africa | 30,156 | 0 | 30,156 | 3.5% | 3.8% | 119 |
| World Totals | 404,762 | 456,176 | 860,938 | 100% | 100% | 118 |

*In million metric tonnes.

†Reserve life is at present day rates of production.

*Data from: BP Statistical Review of World Energy*, June 2012.

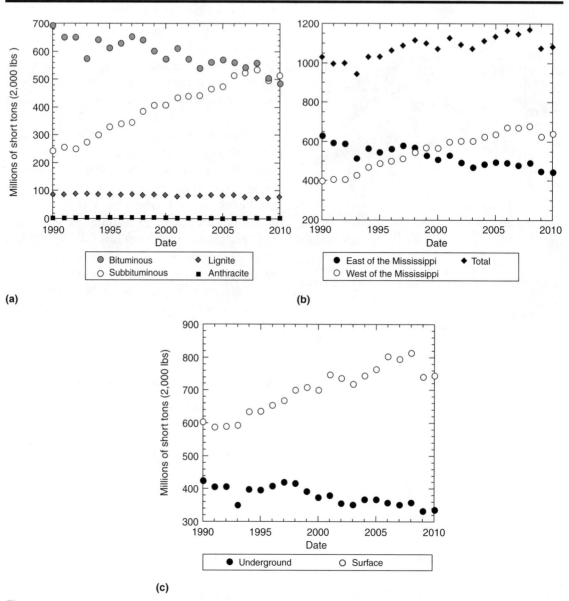

*Figure 3.18* Yearly changes in U.S. coal production by (a) type, (b) location, and (c) method from 1990 to 2010. (Data from: U.S. Energy Information Administration, *Annual Energy Review 2010*.)

**Figure 3.19** shows the flow of coal in the U.S. from production to consumption as determined in 2010. There is a slight increase of exports to import so the U.S. is a net exporter of coal. The figure also indicates the production of coal is used in large part to produce electricity with large sub-equal mining of bituminous and subbituminous coal and smaller productions of lignite and anthracite in 2010 as stated above.

## Environmental Effects

Coal is the dirtiest of all fossil fuels. Environmental problems are associated with coal mining, cleaning,

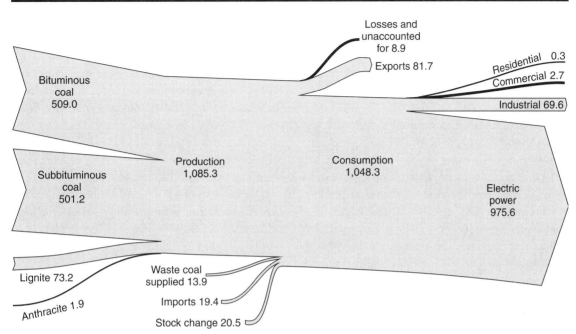

*Figure 3.19* Coal flow from production to consumption in millions of short tons (2,000 pounds) for 2010. Waste coal includes fine coal, coal obtained from a refuse bank or slurry dam, anthracite culm, bituminous gob, and lignite waste that are consumed by the electric power and industrial sectors. (Modified from: U.S. Energy Information Administration, *Annual Energy Review 2010*, p. 211.)

combustion, and disposal of the burnt coal residual. Coal mining is a hazardous occupation. In the United States alone more than 100,000 coal miners have died in coal mining accidents in the last 120 years. Continuous improvements are being made in mining equipment, hazardous gas monitoring, and ventilation. These have reduced rock falls, cave-ins, explosions, and the presence of unhealthy air. As a result fatalities have decreased by two-thirds in the last 20 years. Accidents, however, still happen. In 2010, at the Upper Big Branch coal mine in West Virginia an underground methane gas explosion killed 29 miners.

When strip mines are used to obtain coal it typically degrades a property's commercial and scenic value. While reclaiming and restoring the original landscape after mining, when undertaken, helps minimize these problems, the re-landscaped ecosystems can have unstable soils that easily erode. A variant of strip mining prevalent in the hilly terrain of West Virginia and other areas of the Appalachian Mountains is mountaintop removal (MTR) coal mining. The summit of a mountain is removed to permit easier access to the coal seams below. The removed material is often disposed of in neighboring valleys. This results in radically changed stream drainage patterns in the mined area.

A particular difficulty occurs in the rock-waste dump piles produced from the removed overburden during coal mining. These have very high permeability to water and typically expose pyrite, $FeS_2$, present in the sediments. Oxygenated rainwater can percolate through the waste rock and react with the pyrite:

$$FeS_2 + H_2O + 2O_2 \rightarrow H_2SO_4 + FeO. \qquad [3.3]$$

The sulfuric acid, $H_2SO_4$, produced leads to acid mine drainage. This acid produces toxic water as heavy metals such as Zn, Mn, and Ni are leached from the rock dissolving into the acidified water. When this water enters the local water supply the acid can overwhelm any natural pH buffering reactions in lakes and streams. The pH of the lakes and streams can decrease to such an extent that fish and other aquatic organisms can no longer survive. Mines dug far enough into the earth can cause these effects in underground aquifers. In these mines, water interacts with the coal layers themselves leaching out contaminants such as arsenic, cadmium, and chromium to the aquifer.

Contamination also occurs with the burning of coal. Coal is a chemically complex fuel with significant amounts of uranium, thorium, and toxic metals such as antimony, cadmium, and mercury. The chimney exhaust gas as well as the solid waste residual after combustion can have high concentrations of these metals. Although coal-fired power plants produce a little over one quarter of U.S. electricity they are responsible for 97% of hazardous fine particle soot and sulfur dioxide emissions that produce acid rain, 92% of smog-forming nitrogen oxide emissions, 86% of global warming $CO_2$ emissions, and nearly 100% of toxic mercury emissions.

Fly and bottom ash is produced during the burning of coal. These are fine particles of unburnt $SiO_2$-rich material present in the coal that rise in the exhaust gases given off during combustion or sink to the bottom of the furnace as waste, respectively. Both types of coal ash are typically collected for disposal. Often they have high concentrations of toxic metals such as arsenic, chromium, and cadmium. The coal ash is generally stored at the coal power plants where it is produced or placed in nearby landfills. Some is recycled for use as an additive to Portland cement (See the chapter on building materials).

Often the fine-grained fly ash is washed from the exhaust chimneys of power plants and pumped into a retaining pond where the fly ash settles out. It can then be dredged and put in storage piles. In December 2008, a coal fly ash retaining pond breach occurred in Tennessee at the Kingston Fossil coal-fired power plant. It buried 1.2 km$^2$ of land with up to 1.8 m of fly ash sludge with ash entering tributaries of the Tennessee River killing large numbers of fish. It was the largest fly ash release that has occurred in the United States.

Coal is the greatest contributor of $CO_2$ to the atmosphere of all fossil fuels burned.

Environmentalists argue it is critical to reduce the world's usage in order to slow global warming. To generate 1 kw/hr of electricity requires 154 g of average coal (78% carbon content) and puts 120 g of carbon in the atmosphere, mainly as $CO_2$. Using natural gas to generate 1 kw/hr of electricity requires 0.094 m$^3$ at standard conditions that emits 46 g of carbon into the air. Therefore, the production of electricity by coal rather than natural gas puts 120/46 = 2.6 times as much global warming $CO_2$ into the atmosphere.

The large deposits of bituminous coal in the U.S. Midwest and Pennsylvania tend to have high sulfur content mainly as pyrite ($FeS_2$). The problem with sulfur in coal is that when it is burned the following reaction occurs:

$$FeS_2 + 2.5O_2 \rightarrow 2SO_2 + FeO. \qquad [3.4]$$

The sulfur dioxide gas ($SO_2$), produced reacts with water in the atmosphere to produce sulfuric acid ($H_2SO_4$). This problem makes coal-fired power plants the largest contributor to acid rain. Anthracite deposits with little sulfur occur in the Appalachians of West Virginia and Pennsylvania. While these deposits are attractive, the hilly terrain makes mining more expensive than the strip mining used in the West and Midwest.

Another concern of coal use is coal seam fires. Thousands are burning underground at the present time. They are a particular problem in the People's Republic of China. Forest fires, lightning strikes on outcropping coal seams, and careless human activity in coal mines start most of them. They can smolder underground for centuries fed by oxygen in pores and fractures formed as the coal burns or from oxygen seeping into coal from ventilated mine shafts. They are almost impossible to extinguish. China's coal seam fires are estimated to consume between 20 and 200 million metric tons of coal each year. This is 2% to 20% of their annual production. In Centralia, Pennsylvania, the location of one of hundreds of coal seam fires in the U.S., the fire has been burning since 1962. In 1962 Centralia had a population of over 1,000; now it is a near ghost town as toxic fumes have been seeping from the fire to the surface.

### Clean Coal Technologies

Due to the environmental impact of burning coal a number of technological methods are being developed to reduce emissions of ash, sulfur, heavy metals, and $CO_2$ from coal combustion. These include coal washing, combusting the coal in nearly pure oxygen, and gasification of the coal.

While most coal flue gas exhaust is run through electrostatic precipitators that capture charged ash particles on plates, washing the coal can eliminate these contaminants before the coal is burned. Crushed coal is mixed in a liquid to produce a slurry. Often, but not always, the liquid is water. The mineral impurities, which include flakes of clay, pyrite, and other noncoal material, are allowed to separate either by settling, making a froth to float them to the surface, or using centripetal rotation. If sulfur content is still a problem flue gases can be treated with ground calcite mixed with water to remove the $SO_2$ by the reaction

$$2CaCO_3 + 2SO_2 + 4H_2O + O_2 \rightarrow 2CaSO_4 \bullet 2H_2O + 2CO_2$$
Calcite                              gypsum                 [3.5]

in a process called wet scrubbing.

Although more expensive because of the pure oxygen required, burning of pulverized coal in pure oxygen produces approximately 75% less flue gas emissions than air-fueled combustion. Also, the flue gas produced consists primarily of $CO_2$ suitable for sequestration, that is, permanent storage. Current methods of sequestration of $CO_2$ being considered include pumping it into deep formations underground, dissolving it in deep ocean water masses, or reacting it to form carbonate minerals such as calcite, $CaCO_3$.

Gasification avoids the problems of burning coal altogether. With an integrated gasification combined cycle (IGCC) system, steam together with hot pressurized air or oxygen oxidizes the carbon in the coal to form a CO plus $H_2$ gas, called syngas. CO in the syngas can be separated and reacted with oxygen to produce $CO_2$ for sequestration prior to burning. The syngas or $H_2$ gas is burned to produce an expanding gas that turns a turbine to make electricity. The heat energy from the hot gas is then used to boil water to make steam to power a second turbine. Since IGCC power plants use both gas expansion and heat production, they are very fuel efficient. Alternatively, the hydrogen gas can be used in a fuel cell to generated electricity (see the chapter on alternative energy) instead of being burned to turn a turbine to produce electricity.

To date no commercial-scale coal plants have been designed to capture carbon dioxide for sequestration. Without a price on the greenhouse gas, there has been no economic incentive to do so. However, the separated $CO_2$ from the IGCC power plant can be cooled to −28°C and liquefied. The liquid $CO_2$ is then ready to be injected into an underground depository for permanent disposal. A demonstration project, the FutureGen 2.0 power plant in Morgan County, Illinois, is being constructed to test the oxygen combustion of coal and $CO_2$ sequestration concept.

# Peat Resources

Peat, partially decayed, somewhat compacted vegetation, forms in bogs, moors, and peat swamp forests as outlined above. It can be burned and is, therefore, an energy source with 4 trillion m³ of peat estimated to be available in the world (see **Table 3.3**). The most extensive use of peat is, however, for horticultural purposes. Slightly decomposed moss-rich peats, termed peat moss, is used to improve soil structure and increase acidity.

In Ireland and Finland peat is harvested on a large scale for fuel. In rural areas of Ireland and Scotland bricks of peat are stacked, dried, and used for cooking and domestic heating. They have only about 25% lower heat value than lignite. Research is underway to convert peat into methane gas by bacterial digestion and by thermal decomposition.

The amount of peat in Finland alone is estimated to have twice the energy of the North Sea oil reserves. In Europe, peat fuel is produced in Finland, Ireland, Sweden, Latvia, Estonia, and Lithuania with average annual production exceeding 3,200 ktoe (thousand metric tons of oil equivalent). In 2008, 16% of Ireland's electrical generation

*Table 3.3* PEATLANDS OF THE WORLD

| REGION | ESTIMATED PEATLANDS* | LOCATION OF MAJOR PEAT DEPOSITS |
|---|---|---|
| Former U.S.S.R. | 371 | Russia (Mainly Siberia) |
| Canada | 279 | All provinces have large deposits |
| United States (incl. Alaska) | 124 | Florida, Michigan, Minnesota Wisconsin |
| Africa | 84 | Kenya, Uganda, Burundi, South Africa |
| Europe | 65 | Finland, Sweden |
| South America | 15–25 | Venezuela, Guyana, Brazil, Argentina, Chile |
| Central America and West Indies | 7 | Cuba, Jamaica, Panama |
| Australia & Oceania | < 2.5 | New Zealand, New South Wales, Queensland, Eastern Highlands |

*In millions of acres = 4,047 m².

capacity was by peat of 420 MW in five "peat" generation stations. Peat provides ~6.2% of Finland's energy production where it is used for heating and producing electricity. Peat has low sulfur content, minimal mercury content, low ash content, and is a source of local employment.

## SUMMARY

Natural gas is primarily methane gas ($CH_4$). It is obtained from unconventional as well as conventional petroleum resources. These unconventional sources include deep, tight, or overpressured gas, coal (coal bed methane), and methane clathrates.

Conventional natural gas reserves are present mainly in the Middle East and Eurasia. With the exploitation of nonconventional reserves peak natural gas production will probably not occur until 2090. About 1/3 of natural gas consumption in the U.S. is for electric power production while 1/4 is for residential heating and cooking. Natural gas is the main source to produce hydrogen gas used to hydrogenate carbon chains to make gasoline as well as the primary hydrogen feedstock for ammonia ($NH_3$) production for fertilizers.

Coal is formed from the breakdown of plant material in wetlands such as swamps and bogs. The partially decomposed plant matter is termed peat. With burial, as the peat is subjected to higher temperatures and pressures, the peat loses water and a soft brown to blackish lignite is formed. With additional burial first subbituminous, then bituminous coal and finally anthracite is produced.

When coal or many petroleum precursors are examined under a microscope dehydrated plant fragments termed macerals are observed. The reflectivity of macerals on polished sections of coal or petroleum precursors is used to determine its degree of maturation.

Coal is the largest source of energy for the production of electricity worldwide and used to produce about one quarter of U.S. electricity. More than 90% of coal production in the U.S. is used to make electricity. The absence of any coal before the Carboniferous period occurs because their precursor, land plants, had not yet evolved. Extensive shallow inland seas during the late Carboniferous period preserved organic matter that was transformed on burial to large coal reserves. China is the largest producer of coal and the U.S. has the largest reserves. Presently, the largest coal mining region in the U.S. is Wyoming's Powder River Basin.

Environmental problems are associated with coal mining, cleaning, combustion, and disposal of the burnt coal residual. Particular problems occur with high sulfur coal that produces acid rain on burning and acidic ground water. Clean coal technologies are used to mitigate the harmful effects of burning coal. These include coal washing, combusting the coal in nearly pure oxygen, and gasification of the coal.

Peat is organic matter that has yet to be turned into low-grade coal. It has low sulfur content, minimal mercury content, low ash content, and is a source of local employment. Burning it is used as a source of heat and electricity production in a number of countries.

## KEY TERMS

| | |
|---|---|
| anaerobic | lignite |
| anthracite | maceral |
| bituminous | marsh gas |
| British thermal units (*Btu*) | peak coal |
| brown coal | peak natural gas |
| coal bed methane | peat |
| cubic feet (*cf*) | reflectance microscope |
| Energy Watch Group | subbituminous |
| fracking | |

## PROBLEMS

1. Quantities of natural gas are measured in normal cubic meters at 0°C (273.15 K) and 101.325 kPa or in standard cubic feet at 60°F (288.75 K) and 14.73 psi (= pounds per square inch). If natural gas behaves as an ideal gas (PV = $n$RT) where $n$ is the number of moles, what is the conversion factor between the two units? Show your work (1 psi = 6,894 Pa, 1 foot = 0.3048 meters, and R = 8.31424 J mol$^{-1}$ K$^{-1}$).

2. a. Calculate the heat released in *Btu* from 1 metric ton of lignite and 1 metric ton of anthracite as given in Figure 3.10 (1 *Btu* = 252 calories).

   b. If 127.5 quadrillion ($10^{15}$) *Btu* of coal energy was used in 2008 and on average this was brown coal. How many billion metric tons of coal was used?

3. a. From Figure 3.16 determine the amount of coal predicted to be used in 2035. Convert this number in quadrillion (= $10^{15}$) *Btu* or also called quads to billion of metric tons of oil equivalent (*Btoe*). Use the conversion value of 1 *toe* = 39.68 *MBtu*.

   b. From Figure PT1.2 in the introduction to Part 1 on Energy Resources determine world predicted energy consumption for 2035 in *Btoe* and determine the percentage of 2035 energy consumption that will be from coal using the number derived in a. Given the EIA determined coal percent of total energy given in the text for 2006 and prediction for 2030 does the calculated number for 2035 seem consistent?

## REFERENCES

Doman, L.E. and many others. *International Energy Outlook 2010*. Washington, DC: U.S. Department of Energy, Energy Information Administration. Pub# DOE/EIA-0484(2010). Available online at www.eia.gov.

Levine, J. R. 1987. *Influence of coal composition on the generation and retention of coalbed natural gas: Tuscaloosa, Alabama*. Proceedings of the 1987 Coalbed Methane Symposium, Paper 8711, pp. 15–18.

Odell, P. R. 2004. *Why carbon fuels will dominate the 21st century's global energy economy*. Essex, UK: Multi-Science Publishing Co.

Peters, K. E. and Moldowan, J. M. 1993. *The biomarker guide: Interpreting molecular fossils in petroleum and ancient sediments*. Upper Saddle River, NJ: Prentice Hall, Inc.

Scott, A. C. 2000. The pre-quaternary history of fire. *Palaeogeogr Palaeoclimatol Palaeoecol* 164(1–4):281–329.

Senftle, J. T. and Landis, C. R. 1991. Vitrinite reflectance as a tool to assess thermal maturity. In R. K. Merrill (ed.) Treatise of petroleum geology: Handbook of petroleum geology, source and migration processes and evaluation techniques (pp. 119–125). Tulsa, OK: American Association of Petroleum Geologists (AAPG).

Stach, E., Mackowsky, M.-Th, Teichmüller, M., Taylor, G. H., Chandra, D., and Teichmüller, R. 1982. *Stach's Textbook of Coal Petrology*, 3rd ed. (Plate 1A, p. a4). Berlin: Gebrüder Borntraeger.

Zittel, W. and Schindler, J. of the Energy Watch Group. 2007. *Coal: Resources and future production*, EWG-Series No 1/2007. Ottobrunn, Germany: Ludwig-Bolkow-Foundation. Available at go.nature.com/jngfsa.

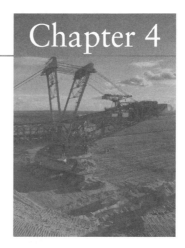

## Chapter 4

# Alternative Energy Resources

Useable energy from any source other than fossil fuels and nuclear energy is usually designated as alternative energy. This energy is typically from a renewable source and includes solar, wind, water, geothermal, and biological sources as given in **Figure** 4.1. Interestingly, the most populous country, China, is presently the number one investor in alternative energy. While solar, wind, tidal, and geothermal energy resources are virtually unlimited, each type has locations where its rate of production is highest and therefore makes the most sense to employ. The largest components of alternative energy resources used, biomass and hydroelectric energy that depend on farmable land area, and the extent of river flow, respectively, are starting to reach their production limits in most areas.

## Solar Energy

Solar energy is energy from a whole spectrum of different wavelengths from short wavelength *ultraviolet radiation* to long wavelength *infrared radiation* from the sun. This is shown in **Figure** 4.2 as light energy per unit area per unit wavelength interval. Note that the species, $O_3$, $H_2O$, and $CO_2$ in the atmosphere absorb parts of the spectrum from the sun transforming it into longer wavelength infrared (heat) energy before it reaches the surface. This infrared energy plus the infrared energy radiated back to space from the earth's surface causes the warming of the earth's atmosphere in what is termed the *greenhouse effect*. Like a greenhouse the

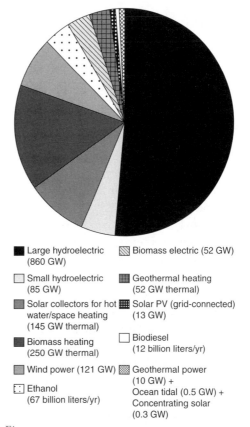

- ■ Large hydroelectric (860 GW)
- □ Small hydroelectric (85 GW)
- ▨ Solar collectors for hot water/space heating (145 GW thermal)
- ■ Biomass heating (250 GW thermal)
- ▨ Wind power (121 GW)
- ⸬ Ethanol (67 billion liters/yr)
- ▨ Biomass electric (52 GW)
- ▦ Geothermal heating (52 GW thermal)
- ▦ Solar PV (grid-connected) (13 GW)
- □ Biodiesel (12 billion liters/yr)
- ▨ Geothermal power (10 GW) + Ocean tidal (0.5 GW) + Concentrating solar (0.3 GW)

*Figure 4.1* Global renewable energy capacity existing at the end of 2011. GW = gigawatts = $10^9$ watts and "thermal" indicates heat energy not used to produce electricity. (Data from: *Renewables Global Status Report*, 2012.)

shorter wavelengths of visible light from the sun pass through the transparent atmosphere. They are

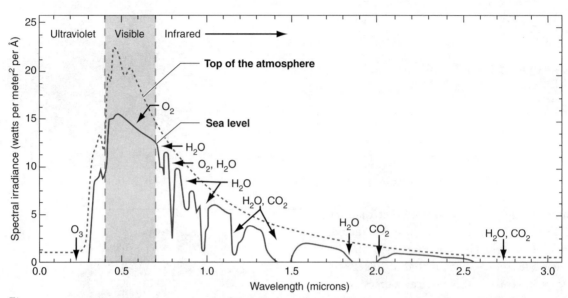

*Figure* 4.2 Spectrum of solar energy at the top of the earth's atmosphere (dashed line) and at sea level (solid line) indicating the dominance of visible light wavelengths. The arrows indicate absorption in the atmosphere of particular wavelengths by the indicated species. The grey area gives the part of the solar spectrum that can be seen with the human eye.

absorbed by the earth's surface producing longer wavelength infrared energy. This infrared energy warms the atmosphere.

Solar energy is the earth's most available energy source and has been used since prehistorical times. In terms of direct usage it can be divided into low-quality solar energy, which uses sunlight to change temperature, and solar power, which uses a sunlight concentrator to produce electricity or to activate a chemical process. The difference between solar energy and *solar power* is that energy is an amount given in an energy unit like joules and power is the rate that the amount of energy is produced or used given in a unit like watts. Typical units of measure for energy and power are given in **Table 4.1**. If considering energy usage over time often the units kilowatt-hr (kWh) or kilowatt-yr (kWyr) are used.

## Low-Quality Solar Energy

The low-quality solar energy of direct sunlight is used primarily for space and water heating, clothes and food drying, cooking, and natural illumination in buildings. For instance, visible light passes through window glass and is converted into infrared energy when it encounters an object in the room. How much the room heats up and stays warm at night or on a cloudy day depends on the heat capacity of the objects in the room and the amount of heat that is transported out of the room. Energy can be transferred by three different mechanisms.

*Radiation* is the way heat or light is transferred through space as waves from a heated or lit object. It is how we obtain light energy from the sun. Standing in front of a fire the infrared radiation can be felt and the visible light radiation that reaches your eyes is observed. An object radiates a spectrum of wavelengths of energy with the wavelengths of the radiation shortening with increased temperature. Because the sun is hotter than a fire it produces shorter wavelength ultraviolet radiation as well as visible light. A human body, which is much cooler than a fire, radiates only longer wavelength infrared energy.

Moving a heated mass of material from one place to another also transfers energy. This is termed *energy convection*. Humankind uses the convection of hot water through pipes to carry heat from a furnace to

*Table 4.1* ENERGY AND POWER CONVERSION FACTORS*

| ENERGY = FORCE × DISTANCE OR PRESSURE × VOLUME | | | | |
|---|---|---|---|---|
| ENERGY UNIT | ERG | JOULE | THERMOCHEMICAL CALORIE | BTU |
| CGS: 1 erg = dyne cm = | 1 | $10^{-7}$ | $2.38901 \times 10^{-8}$ | $9.4782 \times 10^{-11}$ |
| S.I.: 1 joule = newton meter = | $10^7$ | 1 | 0.23901 | $9.4782 \times 10^{-4}$ |
| 1 calorie = | $4.194 \times 10^7$ | 4.1840 | 1 | $3.9657 \times 10^{-3}$ |
| 1 Btu = British thermal unit = | $1.055056 \times 10^{10}$ | 1.055.056 | 252.164 | 1 |
| 1 kilowatt hour = | $3.600 \times 10^{13}$ | $3.600 \times 10^7$ | $8.6042 \times 10^5$ | 3,412.1 |
| POWER = ENERGY PER UNIT TIME | | | | |
| POWER UNIT | WATT | BTU HOUR$^{-1}$ | HORSEPOWER (ELECTRICAL) | TON (REFRIGERATION) |
| 1 watt = joule s$^{-1}$ = | 1 | 3.41443 | $1.340 \times 10^{-3}$ | $2.8435 \times 10^{-4}$ |
| Btu hour = | 0.292875 | 1 | $3.9259 \times 10^{-4}$ | $8.3278 \times 10^{-5}$ |
| Horsepower (electrical) = | 746.00 | 2,547.2 | 1 | 0.21212 |
| Ton (refrigeration)† = | 3,516.9 | 12,000 | 4.7143 | 1 |

| System | Force Unit |
|---|---|
| CGS | dyne = force to accelerate a mass of 1 g by 1 cm s$^{-2}$ |
| S.I. (MKS) | newton = force to accelerate a mass of 1 kg by 1 m s$^{-2}$ |

Force = mass × acceleration

*The most widely used systems of measurements are S.I. and CGS units. Given in this table are conversion factors between the two systems and some other common energy and power units.

†Ton of refrigeration is approximately the energy removal rate that will freeze 2,000 lb of water at 0°C in 24 hours.

warm a room in a house with a radiator. Hot air can be convected from one room to another by a forced-air heating unit. On the earth natural convection transfers heat by air masses in the atmosphere and seawater in the ocean from warmer equatorial regions towards cooler polar regions.

The third kind of heat transfer is by *conduction*. In conduction the transfer of energy from a hotter object to a cooler one is by collisions of neighboring atoms and free electron exchange. The free electron exchange explains why metals heat up much more rapidly than silicate rocks. Metals have clouds of electrons that transfer the heat. In contrast to convection in conduction the energy is transferred but there is no net motion of matter. Heat from a warm room can be conducted outside through a wall given enough time. In the earth heat from the interior is conducted outward to the surface through rocks. Understanding the mechanisms of heat transfer allows better use of solar energy, particularly for low quality purposes.

### Solar Water Heating

Heating water by sunlight is a widely used technology, particularly in China with 40% of the world's capacity; it is also heavily used in Europe, Japan, and India. It is growing at almost 20% per year as an increasing number of countries, states, and cities are mandating solar hot water systems in new building construction. Most of this is for domestic usage but solar hot water also has industrial applications. Over 70 million households worldwide used solar hot water systems in 2011 with 27 million rooftop solar water heaters in operation in China alone. This supplies the energy equivalency of over 100 large nuclear power plants. In Rizhao and Dezhou, China, 99% of households in the city center use solar hot water heaters. In 2011, Worldwatch Institute estimated that 2 million Germans are living in homes where both water and interior space are heated by rooftop solar collectors.

To heat water with sunlight a solar collector is installed in a location with as much of an unobstructed view of the sun as possible, typically on a roof. It can be a thin box or set of tubes filled with water that are painted black for maximum energy absorption. The hot water is stored in a tank if hot water is used directly or run through pipes to a heat exchanger for space heating (**Figure 4.3**).

## Turbines and Electrical Generators

Energy can be transported by electrons under a voltage. The force of water, steam, or air can be used to generate this voltage. What is required is a *turbine* and electrical *generator*. A turbine uses a stream of gas or liquid generated from heated water, water stored behind a dam, or wind to push against blades to rotate a shaft. Inside the generator electrical wire has been wrapped around the shaft. Rotating these wires in a magnetic field produces electricity. An electrical generator then converts the mechanical energy of the turning shaft into voltage that allows electrical energy to flow in a conductor as shown for a steam turbine and generator in **Figure 4.4**. This occurs because if one moves a wire through a magnetic field an electric current is produced. It can be thought of as a "reverse" motor as the generator converts the rotation of a shaft into electricity.

## Solar Power

The possibilities for solar power production are different depending on where you are. For instance

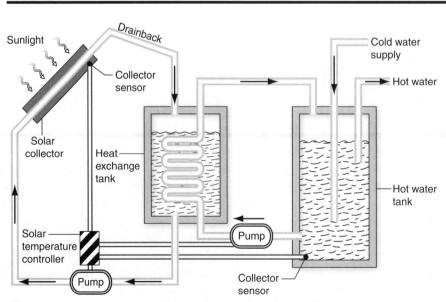

*Figure 4.3* Domestic solar water heating system supplying hot water for heat and for direct use.

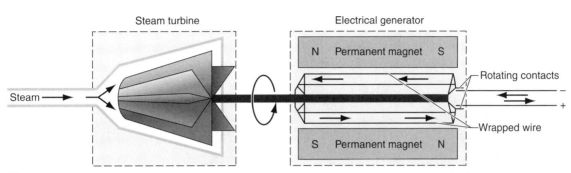

Steam turbine

Electrical generator

N Permanent magnet S

Rotating contacts

Steam →

+

Wrapped wire

S Permanent magnet N

*Figure 4.4* Steam turbine and electrical generator where the energy of the moving steam rotates a shaft with wires attached in a magnetic field to produce electricity.

the southwestern U.S. gets about twice the radiation from the sun of the eastern U.S. There are three ways to use the power. One way is to focus light from the sun with mirrors and lenses to heat a solar furnace. The heat can then be used to produce steam to power a turbine that runs a generator and produce electricity. A second way is to focus sunlight to provide the energy needed to run a chemical reaction; for instance, the splitting of water into $H_2$ and $O_2$. The $H_2$ produced can then be used as a fuel, similar to natural gas. Another use of this type is to generate heat from focused sunlight to drive industrial chemical reactions that require elevated temperatures. The third type of solar power usage is to produce an electric current in a *photovoltaic* (PV) *cell* as outlined below.

### Solar Furnace Electricity

One type of solar furnace is a solar power tower, also known as a "central tower" power plant or *heliostat* power plant. It uses a tower to receive focused sunlight and an array of flat, movable mirrors called heliostats to focus the sun's rays on the collector tower to heat water or another liquid. The heated liquid is transformed to a gas to power a turbine and run a generator. Recent designs have used molten salt in place of water. Molten salt has a higher ability to store heat, that is, heat capacity. This heat is then transferred to water to produce the needed steam. Because heat can be stored in molten salt, electricity can be generated at night.

The THEMIS Solar Power Tower Plant in France started operation in 1983. In 2004 the plant was modernized to 201 heliostats of 11,800 $m^2$ to produce 1 megawatt (MW) of electricity (**Figure 4.5**). In Seville, Spain the 11 MW PS10 solar power tower and 20 MW PS20 solar power tower have been recently completed. A 15 MW molten salt Solar Tres Power Tower is scheduled to be built in Andalusia, Spain consisting of 298,000 $m^2$ of solar collection surface on 2,590 heliostats. A 100 MW molten-salt-type central receiver solar power plant is planned with 4,000 to 5,000 heliostat mirrors for a site near Upington, South Africa.

The Odeillo-Font-Romeau Solar Furnace in Odeillo, France has 63 flat mirrors that automatically track the sun and concentrate light on a reflector (**Figure 4.6**). The reflector then focuses the sunlight to produce 1 megawatt of power and a temperature of up to 3,500°C in a solar oven. This oven is used to melt iron ore to produce steel.

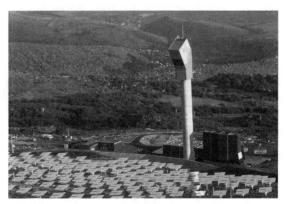

*Figure 4.5* THEMIS Solar Tower Power plant near Targassonne in the Pyrénées-Orientales of France.

*Figure* 4.6 Odeillo-Font-Romeau Solar Furnace in Odeillo, France.

## *Parabolic Trough Collector*

Another type of solar furnace is a *parabolic trough collector*. It is constructed as a set of long parabolic mirrors shaped like half-pipes, with polished aluminum or silver coated surfaces. A Dewar tube runs along its length at the focal point to which the sunlight is reflected (**Figure** 4.7). The Dewar tube is constructed of two silvered wall layers with a vacuum between them. This design reduces heat transfer by radiation (prevented by the silvering) and conduction (prevented by the vacuum). This is similar to a Dewar flask that keeps cold liquids cold or hot liquids hot. Inside the Dewar tube a fluid, such as synthetic oil, is heated from the sunlight to over 400°C. The trough is usually aligned on a north-south axis, and rotated to track the sun as it moves across the sky each day. Maintenance on this system includes an automated washing mechanism that periodically cleans the mirrors.

The Solar Energy Generating Systems (SEGS) site is the largest solar energy generating facility in the world. It consists of nine separate parabolic trough collector systems located in California's Mojave Desert with a 354 MW installed capacity (**Figure** 4.8). The facility uses natural gas along with the parabolic trough collector to generate electricity with 90% of the electricity produced

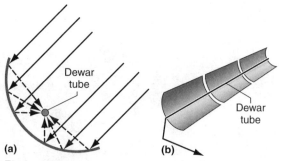

*Figure* 4.7 (a) Cross-section through a parabolic trough collector showing the reflection of energy to a Dewar tube at its focal point. (b) An array of parabolic trough collectors connected together with a continuous Dewar tube.

*Figure* 4.8 Parabolic mirror arrays of Solar Energy Generating Systems (SEGS) facility in the U.S. Mojave Desert.

by sunlight. Natural gas is only used when the solar power is insufficient to meet the demand.

## *Photovoltaic cells*

Photovoltaic effects that cause sunlight to produce a voltage can occur in semiconductors. A *semiconductor* is a substance that can conduct electricity under some conditions but not others, making it a good medium for the control of electrical current. Most commercial solar cells are made of crystalline or amorphous silicon semiconductors. When sunlight strikes N-type silicon, part of the light spectrum imparts enough energy to excite outer electrons in the silicon atoms. N-type silicon is fabricated by adding an impurity, an element of a valence of five such as phosphorus, to four valence silicon. This increases the number of outer free charges. The sunlight excited electron has enough energy to break free from the silicon and join others to produce a current of electricity that moves along elevated energy orbitals for electrons in the N-type silicon. A "hole" is also created where the electron has left. A potential barrier for electrons in the cell is set up using a junction with P-type silicon. P-type silicon is produced by adding an impurity of a three valence element such as boron to the silicon. This impurity takes away, that is accepts, weakly bound outer electrons from the four valence silicon creating a "hole." Therefore, electrons are favored in the N-type silicon and "holes" are favored in the P-type silicon. As an electron jumps from one silicon atom to another the electrons are observed to move to the N-type silicon and the holes to move to the P-type silicon. This causes an electric current with a voltage of about 0.5 volt to be produced between the N-type and P-type silicon (**Figure** 4.9). The amount of electrical current, or amperes, produced depends on the area of exposure of the solar cell and the intensity of sunlight. Remember that amperes × volts = watts. Photovoltaic cells are connected in series to increase current and in parallel to increase voltage.

Current energy efficiencies for capture of the sun's energy are 12% to 15% for single crystal silicon cells and 4% to 6% for cheaper amorphous silicon cells. The cost of photovoltaic cells has dropped from about $1,000 per watt in the 1950s to under $5 per watt at the present time. They can produce electricity for as little as 25 to 30 cents per kilowatt-hour.

The direct current produced by the cells is converted to alternating current (AC) with a power inverter. The inverter switches the + and − terminals 60 times a second creating a 60 cycles per second square wave, which is conditioned to produce the required sine wave. With AC, transformers can be used to change voltages, which allows efficient electrical transmission at high voltages. Nearly all commercial electricity is AC.

While many rooftop systems are installed worldwide the largest PV power plants in the world are in Olmedilla de Alarcón, Spain (60 MW), Puertollano, Spain (50 MW), Moura, Portugal (46 MW), and Waldpolenz, Germany (40 MW) (**Figure** 4.10). Either a fixed photovoltaic panel or a tracking system where the panel follows the sun across the sky is used.

Photovoltaic cells can be manufactured into flexible thin films although these are less efficient, but cheaper to make. Many of these are installed on roofs of buildings as well as incorporated into material for building facades in what is termed building-integrated photovoltaics (BIPV). Additionally, semitransparent thin film solar cells that use infrared light to create electricity have been developed. These can be applied to window glass or skylights to tint them to decrease the amount of sunlight that enters a room and also produce electricity.

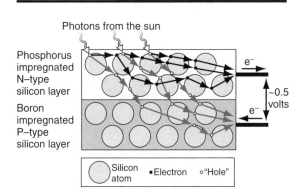

*Figure 4.9* Schematic diagram of a photovoltaic cell. Black arrows show the movement of electrons and grey arrows the movement of holes.

*Figure 4.10* Photovoltaic power plant at Serpa, Portugal that includes a tracking system to follow the sun's path across the sky.

# Fuel Cells for Energy and Hydrogen Gas

A *fuel cell* produces electricity in an electrochemical reaction similar to a battery. It, however, does not lose its charge with time because it runs on externally supplied fuel. In a hydrogen fuel cell with a proton exchange membrane (PEM) the fuel is hydrogen gas. In the reaction $H_2$ gives off electrons and is reduced to $H^+$ at an anode (negative electrode). The oxidant $O_2$ at a cathode (positive electrode) takes the electrons together with the $H^+$ and combines them to produce water (**Figure** 4.11). The maximum potential difference given in volts (V) is 1.23 V at 25°C but decreases to 1.18 V at 80°C. In real world situations at 80°C voltages near 0.7 V are obtained. The rest of the energy dissipates as heat.

Most electric motors to power vehicles operate at 200 to 300 V. This means a large number of fuel cells are connected in series to obtain the required voltage. The anode and cathode of a fuel cell are separated by a solid electrolyte. An *electrolyte* is a substance that has a high concentration of positive and negative charges mixed together, but is electrically neutral. A concentrated aqueous solution of NaCl is an electrolyte

solution as NaCl is present as $Na^+$ and $Cl^-$. The electrolyte used in PEM cells is a solid membrane of dissociated poly-perfluorosulfonic acid containing $H^+$ and negatively charged membrane sites. The half-reaction at the anode of the PEM cell is:

$$H_2 \rightarrow 2H^+ + 2e^- \qquad [4.1]$$

where $e^-$ stands for an electron.

In order to increase the reaction rate so it can supply enough power to run a vehicle, a platinum (Pt) catalyst is used. With this catalyst much more rapid reactions than reaction [4.1] occur consisting of

$$H_2 + 2Pt \rightarrow 2Pt\text{–}H \qquad [4.2]$$

and

$$2Pt\text{–}H \rightarrow 2Pt + 2H^+ + 2e^-. \qquad [4.3]$$

Problems with using a Pt catalyst include its high cost and the fact that it is easily contaminated and, therefore, requires a relatively pure source of $H_2$ gas. At the cathode the reaction

$$4e^- + O_2 + 4H^+ \rightarrow 2H_2O \qquad [4.4]$$

takes place. Note that the PEM cell produces only $H_2O$ with no $CO_2$ emissions. If the reactions [4.2], [4.3], and [4.4] are summed the overall reaction is

$$2H_2 + O_2 \rightarrow 2H_2O. \qquad [4.5]$$

Instead of using a fuel cell with its electrochemical reaction, the hydrogen gas can be burned in air that is also given by reaction [4.5]. The heat that is produced could be used to boil water and run a turbine to rotate a generator to produce electricity. The advantage of the fuel cell is that much more electricity can be produced for a given oxidation of $H_2$ as little waste heat is produced and a turbine and generator are not required.

Many kinds of fuel cell reactions that produce electricity are possible depending on the application. Given in **Table** 4.2 is a list of some common fuel cells and their characteristics. The PEM fuel cell is used in vehicles because it starts up quickly and has a low operating temperature. Vehicles run with a fuel cell would not have

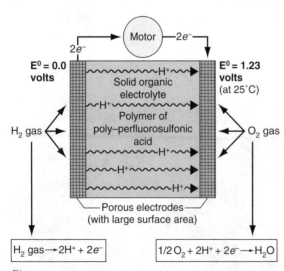

*Figure 4.11* PEM fuel cell showing how $H_2$ and $O_2$ are reacted to produce electrons.

Table 4.2 SOME COMMON FUEL CELL TYPES.

| FUEL CELL NAME | ELECTROLYTE | OPERATING TEMP, C° | ELECTROCHEMICAL REACTIONS | APPLICATIONS | ADVANTAGES |
|---|---|---|---|---|---|
| PEM (Polymer Exchange Membrane) | Solid organic polymer of poly (perfluorosulfonic) acid | 60–100 | An: $H_2 \rightarrow 2H^+ + 2e^-$ <br> Cat: $0.5O_2 + 2H^+ + 2e^- \rightarrow H_2O$ <br> Cell: $H_2 + 0.5O_2 \rightarrow H_2O$ | Electric utility Portable power Transportation | Solid electrolyte reduces corrosion Low temperature Quick start up |
| AFC (Alkaline) | Aqueous solution of KOH in a matrix | 90–100 | An: $H_2 + 2OH^- \rightarrow 2H_2O + 2e^-$ <br> Cat: $0.5O_2 + H_2O + 2e^- \rightarrow 2OH^-$ <br> Cell: $H_2 + 0.5O_2 \rightarrow H_2O$ | Military space | Cathode reaction faster in alkaline electrolyte—so high performance |
| PAFC (Phosphoric acid) | Liquid phosphoric acid soaked in a matrix | 175–200 | An: $H_2 \rightarrow 2H^+ + 2e^-$ <br> Cat: $0.5O_2 + 2H^+ + 2e^- \rightarrow H_2O$ <br> Cell: $H_2 + 0.5O_2 \rightarrow H_2O$ | Electric utility Transportation | Up to 85% efficiency in cogeneration of electricity and heat Can use impure $H_2$ as fuel |
| MCFC (Molten carbonate) | Liquid solution of Li, Na, and/or K carbonate soaked in a matrix | 600–1000 | An: $H_2 + CO_3^{2-} \rightarrow H_2O + CO_2 + 2e^-$ <br> Cat: $0.5O_2 + CO_2 + 2e^- \rightarrow CO_3^{2-}$ <br> Cell: $H_2 + 0.5O_2 + CO_2 \rightarrow H_2O + CO_2$ | Electric utility | High temperature advantages of higher efficiency, inexpensive catalysts and more flexibility in fuel types |
| SOFC (Solid oxide) | Solid Zr oxide with a small amount of yttria, $(Y_2O_3)$ | 600–1000 | An: $H_2 + O_2 \rightarrow H_2O + 2e^-$ <br> Cat: $0.5O_2 + 2e^- \rightarrow O_2$ <br> Cell: $H_2 + 0.5O_2 \rightarrow H_2O$ | Electric utility | High temperature advantages (see MCFC) Solid electrolyte advantages |

*Reproduced from Fuel cells - Green power: written by Sharon Thomas and Marcia Zalbowitz at Los Alamos National Laboratory.*

an internal combustion engine but an electric motor instead. The difference between a fuel cell-powered electric vehicle (EV) and a battery-powered EV is that the fuel cell vehicle requires a supply of hydrogen gas and the battery-powered vehicle requires recharging batteries when their charge is low.

## Water Power

Water wheels have been used at least back to 240 BC to grind grain and probably much earlier to help control water flow in crop irrigation. The first documented industrial use of hydropower to generate electricity occurred in 1880 when direct electric current was supplied to light a chair factory in Grand Rapids, Michigan. Water power uses the kinetic energy of the movement of water. This can be the movement of ocean water as well as water transported in rivers.

### Ocean Power

Electrical power is produced with the aid of ocean tides. Tidal power then converts the potential energy of water with changing tides into the kinetic energy of moving water to make electricity. In terms of renewable energy sources, tides are much more predicable than wind or solar energy. Tidal barricades can be used to control differences in height between high and low tides. The barricades are dams built across the entire width of an arm of the ocean that extends inland such as in a bay at the mouth of a river. Shown in **Figure 4.12** is the Rance Tidal Power Station located on the Rance River in France.

Another way to produce tidal power is to use the kinetic energy of moving water in a current in a *tidal estuary* as tides rise and fall to power a turbine under the water (**Figure 4.13**). This is similar to windmills using moving air. Because of the lower construction cost and lower ecological impact compared to a tidal barricade this method is gaining in popularity.

Wave power is another type of ocean power where energy from surface waves is employed to generate electricity or pump water. The energy

*Figure 4.12* The Rance Tidal Power Station produces electricity with 24 turbines and has a peak electrical power output of 240 MW.

*Figure 4.13* Proposed design for electrical power generation from tidal currents.

of the rising and falling motion of waves can be captured by a buoy or other floating object. This kinetic energy is applied to drive a turbine to produce electricity. The world's first commercial wave farm is located 5 km off the Portugal coast at the Aguçadoura Wave Park. It can produce 2.25 MW of electricity and plans to increase its capacity to 21 MW.

It has also been shown that the difference in higher temperature surface and cooler deep-water can be used to obtain energy in what is termed Ocean Thermal Energy Conversion (OTEC). Warm seawater is placed in a low-pressure container where it vaporizes until it cools to the lower temperature. The expanding vapor is used to drive a low-pressure turbine to produce electricity. In 1993 a power plant at Keahole Point, Hawaii, produced 50 kW of electricity during an OTEC experiment.

## Hydroelectric Power

Modern hydroelectric power plants convert the potential energy of dammed water in reservoirs on rivers into electricity. The kinetic energy of the falling water drives a water turbine, which powers an electric generator. In 2010 there was 860 gigawatts (GW) of generating capacity worldwide in hydroelectric power plants producing 3,400 TWh of electricity. One TWh (terawatt hour) equals $10^{12}$ watt-hours. This makes hydroelectric power plants the largest source of renewable energy accounting for ~16% of the world's electricity usage. The world's 10 largest hydroelectric dams are listed in **Table 4.3**. With the completion of the Three Gorges Dam China produces the most of any country. China is also in the process of building a significant number of new large hydroelectric dams which will have a combined generating capacity of over 50 GW. Problems with constructing dams for power production are outlined when dams and water resources are discussed in a following chapter.

*Table 4.3* TEN LARGEST HYDROELECTRIC POWER PLANTS.

| NAME | COUNTRY | YEAR OF COMPLETION | RIVER DAMMED | TOTAL CAPACITY (GW) | ANNUAL ELECTRICITY (TW HRS) | AREA FLOODED (KM²) |
|---|---|---|---|---|---|---|
| Three Gorges Dam | China | 2009 | Yangtze | 22.5 | 80.8 | 632 |
| Itaipu | Brazil/ Paraguay | 1984 | Paraná | 14 | 94.7 | 1,350 |
| Guri (Simón Bolívar) | Venezuela | 1986 | Caroni | 10.2 | 46 | 4,250 |
| Tucuruí | Brazil | 1984 | Tocantins | 8.37 | 21 | 3,014 |
| Sayano Shushenskaya | Russia | 1985/1989 | Yenisei | 6.4 | 26.8 | 621 |
| Krasnoyarskaya | Russia | 1972 | Yenisei | 6 | 20.4 | 2,000 |
| Grand Coulee | U.S. | 1942/1980 | Columbia | 6.809 | 22.6 | 324 |
| Robert-Bourassa | Canada | 1981 | La Grande | 5.616 | | 2,835 |
| Churchill Falls | Canada | 1971 | Churchill | 5.429 | 35 | 6,988 |
| Longtan Dam | China | 2009 | Hongshui | 6.3 | 18.7 | |

Shown in **Figure** 4.14 are the amounts of hydroelectric power production in million metric tons of oil equivalent, Mtoe, for 2011 for the countries that produce the most hydroelectric power. One can compare these values to the total world energy usage in Figure **PT1.2** in the introduction of Part One for 2011 where about 13.5 thousand Mtoe were used.

**Figure** 4.15 displays a cross-section through a hydroelectric dam. The conversion of potential energy to electrical energy is 80% to 90% efficient, as opposed to 35% to 40% for burning fossil fuels. The greater efficiency occurs because no steam with its heat loss needs to be produced to run the turbine. Also hydroelectric generators can be powered up quite rapidly, making it an excellent backup system when additional electricity is needed in a power grid.

Because of the efficiency and rapid response, in some locations pumped-water storage facilities have been constructed. When there is low electrical demand, the additional generation capacity is used to power a water pump that pumps water into a higher elevation reservoir. When electrical demand is higher than average, water falling through a turbine from the high elevation reservoir drives

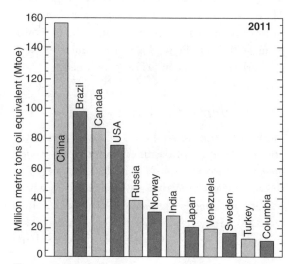

*Figure 4.14* Hydroelectric power production in million metric tons of oil equivalent (Mtoe) displayed for the countries that produce the most hydroelectric power. (Data from: *BP Statistical Review of World Energy*, June 2010.)

an additional electrical generator. About 70% to 85% of the electricity used to pump the water to the higher reservoir is recovered when the falling water is used to power an electric generator. Such an energy storage facility is shown in **Figure** 4.16.

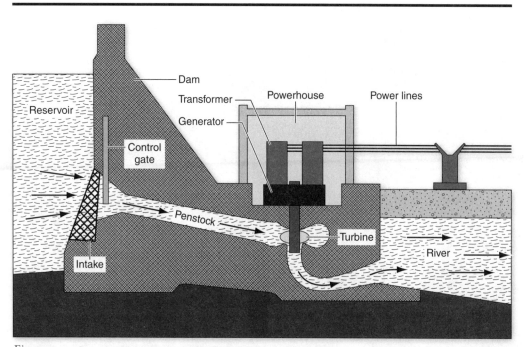

*Figure 4.15* Cross-section through a hydroelectric dam.

*Figure 4.16* Koepchenwerk pumped-storage facility with a net storage of 1.5 million m3 located along the Hengsteysee reservoir on the Ruhr River, near Dortmund, Germany.

# Electrical Power Grid

An *electrical power grid* is a set of power plants and energy storage facilities, transmission circuits, substations, and connected electrical users (**Figure** 4.17). A problem with the power grid is the difficulty in storing significant quantities of electrical energy in the system. A system of controls is required to match supply and demand to avoid brownouts and blackouts. *Brownouts* occur when the power supplied drops below the demand. *Blackouts* occur when there is no power at all transferred to the end user.

A voltage step-up transformer, which lowers the current, is used before the electric power from a power plant is put into the transmission lines to increase transmission efficiency. The high voltage transmission lines are uninsulated aluminum. Aluminum is used rather than copper because of its lower weight and cost. Transmitting electricity thousands of kilometers is reasonably cheap and efficient. Costs are on the order of US$ 0.005 to 0.02 per kilowatt hour (kWh). It is estimated that in the United States about 6.5% of the energy is lost during transmission and distribution judging from the difference in energy produced by power plants and that metered by end customers.

An average U.S. household uses about 10,650 kWh of electricity each year. In mid 2009, the average cost was 12¢/kWh, varying from a 7¢/kWh low average in North Dakota to 26¢/kWh high average in Hawaii. Therefore, transmitting costs are a small part of total costs. A problem with connecting a far away power source is the construction of transmission lines is expensive.

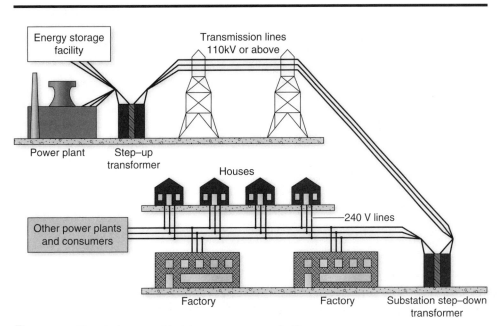

*Figure 4.17* Electrical power grid with an energy storage facility.

# Wind Power

The conversion of the kinetic energy of the wind into a useful form of energy is termed wind power. Historically this energy has been harnessed by sails on ships, windmills to move blades or grinding stones to "mill" grain, and windmills on ranches and farms to pump water from underground wells. Recently the use of wind turbines to produce electricity has been growing rapidly. At the end of 2011, worldwide capacity of wind-powered electrical generators was 238 GW, about 1.6% of worldwide electricity usage (**Figure** 4.18). The electrical wind power production by country for 2011 is given in **Table** 4.4.

Wind is produced by differential heating of the earth. The earth's poles receive less energy from the sun than the equator. Also, the continents heat up and cool down more rapidly than the ocean does. The differential heating drives a global atmospheric convection system. Most of the energy in these wind movements can be found at high altitudes where continuous wind speeds of over 160 km/h occur.

Wind-produced electricity is generated from turbines that often are as tall as a 20-story building

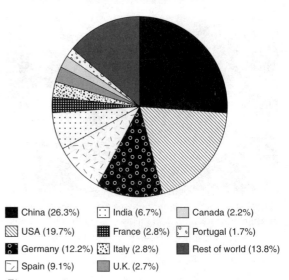

China (26.3%)   India (6.7%)   Canada (2.2%)
USA (19.7%)   France (2.8%)   Portugal (1.7%)
Germany (12.2%)   Italy (2.8%)   Rest of world (13.8%)
Spain (9.1%)   U.K. (2.7%)

*Figure 4.18* Countries with the largest installed wind turbine capacity of the 238.35 gigawatt total capacity as of 2011. (Data from: Global Wind Energy Commission, *GWEC Global Wind Statistics 2011*.)

and have two or three long blades. The wind spins the blades, which turns a shaft connected to a generator that produces electricity. Most common are wind turbines that rotate blades about a horizontal

*Table 4.4* 2011 WIND POWER ELECTRICAL CAPACITY AND WORLD PERCENT OF CAPACITY.

| NATION | GW CAPACITY | PERCENT OF WORLD TOTAL |
| --- | --- | --- |
| China | 62.73 | 26.3 |
| USA | 46.92 | 19.7 |
| Germany | 29.06 | 12.2 |
| Spain | 21.67 | 9.1 |
| India | 16.08 | 6.7 |
| France | 6.80 | 2.8 |
| Italy | 6.75 | 2.8 |
| U.K. | 6.54 | 2.7 |
| Canada | 5.27 | 2.2 |
| Portugal | 4.08 | 1.7 |
| Total | 238.35 | 100% |

*Data from: BP Statistical Review of World Energy, June 2009.*

axis but vertical axis wind turbines are also used and look like giant eggbeaters (**Figure 4.19**).

Shown in **Figure 4.20** is the internal working of a wind turbine. The wind rotates blades, which turn a low-speed shaft that is geared to rotate a high-speed shaft inside a generator. A device, termed an anemometer, measures wind speed and a wind vane determines its direction. This information is sent to a controller and drive to adjust the direction and maximize the efficiency of the turbine.

In order to evaluate the wind resources in a particular area it is important to know that the energy available is proportional to the cube of wind speed. Therefore, twice the wind speed increases the available energy by a factor of eight. The wind is seldom present in a steady, consistent flow. It can change with the time of day, season, height above ground, and type of terrain. Typically annual average wind speeds of 5 m/s (11 miles per hour) or greater are required for electrical grid connection.

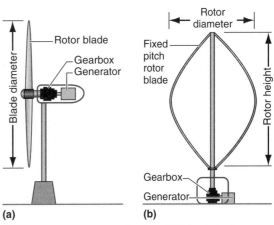

*Figure 4.19* (**a**) Horizontal axis and (**b**) vertical axis (egg beater) wind turbine.

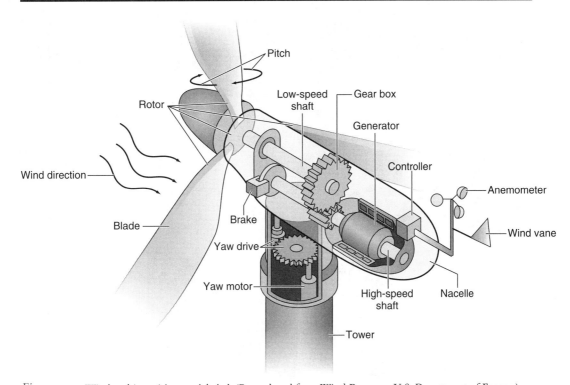

*Figure 4.20* Wind turbine with parts labeled. (Reproduced from Wind Program, U.S. Department of Energy.)

Power density is a useful measure of wind energy. This is energy per time per unit area of wind capture. Because air has mass it has kinetic energy equal to ½ $m v^2$ where $m$ is mass and $v$ is its velocity. Remembering that mass = density × volume, to determine wind power, $P$, the relationship is

$$P = \tfrac{1}{2}\,\rho\,A\,v^3 \qquad\qquad [4.6]$$

where $\rho$ stands for air density (~1.225 kg m$^{-3}$ at sea level and less at higher elevations) and $A$ denotes the area of wind captured perpendicular to its velocity. Wind speed and therefore wind power increases with height. At a particular location

these are typically considered at standard heights of 10 and 50 m. As given by the wind power classes in **Figure** 4.21, sites with a rating of 4 or higher are preferred for large-scale wind electrical generation plants.

The wind blows faster at higher altitudes because there is less drag from the earth's surface. Tower heights approximately two to three times the blade length have been found to balance material costs of the tower against better utilization of the expensive magnets in the electrical generators. Typical modern wind turbines have diameters of 40 to 90 meters and are rated between 500 kW and 2 MW although a few are rated as

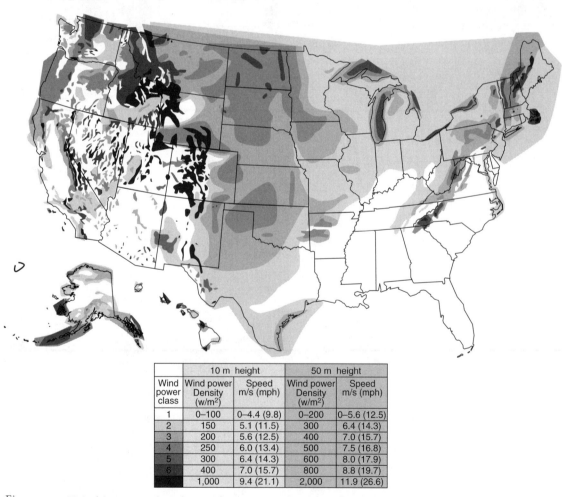

| Wind power class | 10 m height | | 50 m height | |
|---|---|---|---|---|
| | Wind power Density (w/m²) | Speed m/s (mph) | Wind power Density (w/m²) | Speed m/s (mph) |
| 1 | 0–100 | 0–4.4 (9.8) | 0–200 | 0–5.6 (12.5) |
| 2 | 150 | 5.1 (11.5) | 300 | 6.4 (14.3) |
| 3 | 200 | 5.6 (12.5) | 400 | 7.0 (15.7) |
| 4 | 250 | 6.0 (13.4) | 500 | 7.5 (16.8) |
| 5 | 300 | 6.4 (14.3) | 600 | 8.0 (17.9) |
| 6 | 400 | 7.0 (15.7) | 800 | 8.8 (19.7) |
| | 1,000 | 9.4 (21.1) | 2,000 | 11.9 (26.6) |

*Figure* 4.21 United States annual wind power density. (Reproduced from Wind Energy Atlas of the United States/NREL.)

high as 6 MW. Turbine installations in hilly or mountainous regions tend to be on ridgelines. This is done to exploit the topographic acceleration as the wind accelerates over a ridge.

Because of the strong winds produced by the differential heating of the land and ocean surface each day, many turbine installations are on land within 3 kilometers of a shoreline or on the water within 10 kilometers of shore. Offshore wind turbines are less obtrusive than turbines on land, as their apparent size and noise is mitigated by distance. Also water has less surface roughness than land and the average wind speed is, therefore, usually higher over open water.

## Concerns in Developing Wind Power

There are some environmental concerns with the use of wind turbines. These include bird fatalities by flying into the rotating blades. Statistical analysis appears to indicate bird fatalities are minimal when compared to other electrical power sources. There is, however, a significant problem with bat fatalities. Bats that use trees as roosts appear to be attracted to the wind turbines. This is a particular problem in migration and mating seasons.

Noise pollution is another concern. Wind turbine blades produce a low-frequency buzzing, swishing sound when contacting the air. The turbine's internal gears also make mechanical noise. Both types of noise have caused irritation to some individuals that are close to a wind turbine. Such concerns need to be addressed when siting wind turbines.

Concerns have also been expressed about the availability of rare earth elements going forward if wind power expands dramatically. The latest wind electrical generating technology uses special magnets made from neodymium and other rare earth elements that dramatically increase energy conversion efficiency. There is some worry about supply as the People's Republic of China dominates rare earth mining (see the chapter on specialty metals). They have been known to attempt to manipulate the market.

# Biofuels

Biofuels are a wide range of alternative energy sources derived from recently living organisms. This includes biomass used for heating and to produce electricity as well as biodiesel and ethanol. Biomass heating and electrical production uses forest wood, agricultural residues, and urban as well as industrial waste as the energy source.

Forest wood is burned in fireplaces, ovens, and furnaces. Worldwide it makes up almost 10% of the total primary energy supply. More than 2 billion people use wood energy for cooking and/or heating, particularly in developing countries. For many it is the only affordable and locally available source of energy.

In developed countries urban waste trash is burned to produce both heat for industrial operations and steam for electricity generation. There were 87 trash-burning power plants in the United States as of 2011. Other developed countries have many more per capita. There are about 200 across Europe.

Another biofuel is methane released from organic mater decay in waste dumps and landfills by the reaction:

$$2CH_2O \rightarrow CH_4 + CO_2. \qquad [4.7]$$
Organic      methane
matter

Most landfill gas occurs in developed countries, where the levels of waste production tend to be highest. This gas is being collected in many locales and the methane is separated from the $CO_2$ and used as an energy source.

Using biomass diverts wood waste from landfills and conserves fossil fuels but increases carbon addition to the atmosphere as biomass has lower energy content per carbon combusted than fossil fuels and is no longer serving as a sink for atmospheric carbon. Crops grown for their energy content include hybrid poplar, willow trees, and switchgrass. If biomass is produced for its energy it takes agricultural land out of food production.

Biodiesel is a renewable transportation energy source that can be used in standard diesel car and truck engines. Nearly 85% of the world's biodiesel is supplied by European Union countries. Biodiesel is generated from vegetable oil, mainly olive, rapeseed and soybean oils, and animal fat. Soybean oil accounts for almost 90% of production in the U.S., but waste vegetable oils after food frying, waste animal lard and grease, and chicken fat are also processed into biodiesel.

With higher energy costs a number of companies have looked into algae production as a biofuel energy source. The attractiveness is that waste water or ocean water rather than freshwater can be used to grow the algae, unlike most other biofuel sources. Also microalgae grow faster than terrestrial crops grown for their energy. Carbohydrate in the algae can be fermented into ethanol and oil present in the algae can be processed to produce biodiesel.

## Ethanol

Ethanol ($C_2H_5OH$) is a short, straight carbon chain liquid also called ethyl alcohol. It is the intoxicating ingredient present in alcoholic beverages. The largest single use of ethanol is as an additive making up about 10% of petroleum motor fuel in over 20 countries. This fuel can be burned in vehicles designed to run on 100% petroleum. This use of ethanol is generally motivated by a desire to conserve petroleum.

Vehicle engines with specially designed engines that burn nearly pure ethanol fuel are found in Brazil as well as the United States. In Brazil the fuel used is 96% ethanol and 4% water. The 4% water is a product of the distillation process. In the U.S. E85 is used, a fuel containing 85% ethanol and 15% gasoline. The 15% gasoline is added to increase the vapor pressure of ethanol. Pure ethanol has problems igniting in cold weather because the vapor pressure in the combustion chamber is so low. Ethanol when burned releases about 21.2 megajoules (MJ) per liter of energy while regular gasoline releases 44.4 MJ per liter. However, because higher, more efficient compression can be used in an E85 engine the decrease in mileage is only about 25% over regular gasoline.

Ethanol can be produced by the hydration of ethylene, $C_2H_4$, during petroleum refining by the reaction:

$$C_2H_4 + H_2O \rightarrow C_2H_5OH. \qquad [4.8]$$

However, most ethanol is produced commercially by grinding grain kernels into flour. The starch in the flour is converted to sugar with the help of enzymes. The sugars are then fermented with the aid of yeast to create carbon dioxide and ethanol by the reaction:

$$C_6H_{12}O_6 \rightarrow 2C_2H_5OH + 2CO_2. \qquad [4.9]$$
glucose

The 12% to 15% ethanol produced can be separated from the mixture by fractional distillation due to their differences in boiling temperatures. A mixture of 96% ethanol and 4% water is produced at 78.2°C. Dehydration of this ethanol by a molecular sieve that absorbs water but allows ethanol to pass is typically used to remove the water in the U.S.

The economics of reactions [4.8] and [4.9] depends on the relative prices of petroleum and grain. For the recent past, fermenting grain sugar has been the dominant production method with less than 5% produced from petroleum.

Ethanol production in 2009 was equal to 38 million metric tons of oil equivalent (Mtoe) (**Table 4.5**) or 20 billion gallons. The United States produced a little over half of this from corn kernels with Brazil producing another 1/3 of the world total mainly from sugar cane. A problem with pure ethanol distribution is that it easily absorbs water. Therefore, it cannot be efficiently shipped over long distances through underground pipelines as is done with liquid hydrocarbons.

Rather than using starches and sugars ethanol can also be produced from the cellulosic nonedible parts of plants composed of cellulose and hemicellulose. They, together with lignin, are the dominant components of woods and grasses. Ethanol from cellulosic plant material holds great promise due to the widespread availability and relatively low cost. However, cellulosic plant material requires a greater amount of processing

*Table 4.5* COUNTRIES WITH LARGEST 2011
ETHANOL PRODUCTION OF THE WORLD'S TOTAL
OF ABOUT 22.4 BILLION U.S. GALLONS.

| COUNTRY/ REGION | FUEL ETHANOL |
|---|---|
| United States | 13,900 |
| Brazil | 5,573 |
| European Union | 1,199 |
| China | 555 |
| Canada | 462 |

In millions of U.S. gallons.

than is needed with starches and sugars. Enzymes are first used to break down the complex cellulose structure into simple sugars before fermentation is undertaken. Although a number of fermentation processes are technically feasible, to date cost-effective processes have been difficult to achieve because of the complexity of pretreatment and fermentation to alcohol required.

# Geothermal Energy

The temperature of the earth increases downward. Therefore heat energy can be transferred from depth in the earth to the surface where it can be used as an energy source, termed geothermal energy. Geothermal energy is nearly a totally renewable energy resource as the extraction of heat in water is replaced by more heat from the rocks below and natural radioactive decay. It is not totally renewable as in some cases the extraction of heat occurs more rapidly than it can be replaced. However, geothermal energy can provide continuous base-load power with low environmental impact. *Base-load power* means producing power at a dependable rate 24 hours a day and 365 days a year except for routine maintenance. The main problem with tapping energy from the interior of the earth is the high capital costs because of the need to drill wells. Geothermal areas can produce electricity quite inexpensively once the plant is constructed because no external energy is required to produce steam to power a turbine.

About 10.7 GW of geothermal electric capacity was installed around the world as of the end of 2009, generating 0.3% of global electricity demand. The U.S. has 28.8% of the world's total installed geothermal power capacity followed by 17.8% for the Philippines and 11.2% for Indonesia. There is an additional 28 GW of direct heat geothermal resources installed worldwide for district heating, space heating, spas, industrial processes, desalination, and agricultural applications.

The heat that is available in the earth is different depending on the location. In areas that have experienced recent volcanic activity there are generally hot rocks at shallow depths. Circulating groundwaters are heated to high temperatures by these rocks and brought naturally to the surface in some areas. Surface expressions are hot springs, mud pots, geysers, and steam vents. These can be excellent places to construct a geothermal power plant because the hot water can be used to produce electricity. Given in **Figure 4.22** are the potential geothermal resources of the United States. Note the western U.S. has the high heat fluxes needed as a result of tectonic activity in the Sierra Nevada Mountains, Great Basin, and Rocky Mountains, which brought hot rocks and magmas closer to the surface.

## *Geothermal Power Plants*

There are three geothermal power plant technologies being used to convert hot water in the earth to electricity. The conversion technologies are *dry steam*, *flash steam*, and *binary cycle boiling* depending on the temperature of the hot water.

Dry steam power plants are the most common form of geothermal power plant. They use water at depths with temperatures greater than 182°C. The hot water is pumped to the surface from the geothermal production zone under a high enough pressure that it remains liquid. Therefore, the pressure needs to be greater than the water's boiling pressure. When it reaches the surface the pressure

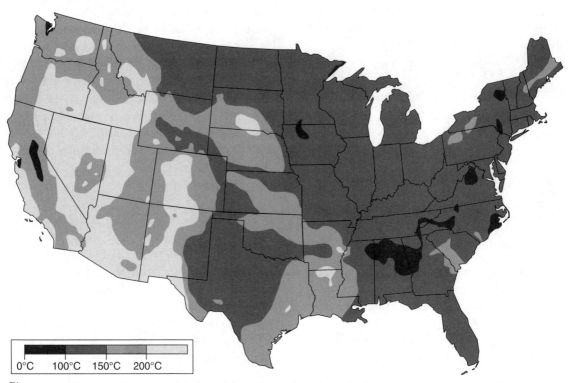

*Figure* 4.22 Geothermal resources of the United States indicating estimated subterranean temperatures at a depth of 6 kilometers. (Adapted from NREL, U.S. Department of Energy.)

is reduced to near 1 atmosphere (atm). As a result of the high temperature all the water is converted to steam, termed dry-steam. After the steam is run through a turbine to generate electricity it is cooled so it is transformed back to liquid water and returned to the reservoir to be heated again by the hot rocks in the production zone as shown in **Figure** 4.23. This is what is done in the Geysers area of California, the world's largest dry-steam geothermal production field where 1500 MW of capacity has been installed.

If the water is at lower temperatures and therefore contains less heat, a flash-steam geothermal plant can be constructed. Flashing is the process of rapidly dropping the pressure on water to suddenly produce steam by opening the hot water to a low pressure space. In this type of geothermal power plant the flashing of the water from the production zone produces some steam but a brine solution is also produced. It is brine because the

hot water typically has many dissolved constituents, which become concentrated with the loss of steam. The steam is separated from the waste brine and run through a turbine. This steam is cooled and injected together with the brine back into the production zone as shown in **Figure** 4.24. Liquid-dominated systems, such as at Wairakei, New Zealand, produce a mixture of steam and hot water on flashing.

In areas where groundwater is heated to below 175°C it can't be flashed to produce significant steam, but the heat can be extracted by using a heat exchanger in a binary flash system. Heat is transferred from the geothermal fluid to a working fluid. The working fluid is often butane or pentane, which have the low boiling temperatures at 1 atm required of −0.5°C and 36°C, respectively. When this working fluid is vaporized by boiling, the vapor can be directed to a turbine to power an electric generator. After it is run through the

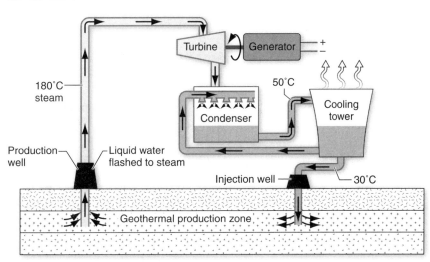

*Figure 4.23* Dry-steam geothermal power plant.

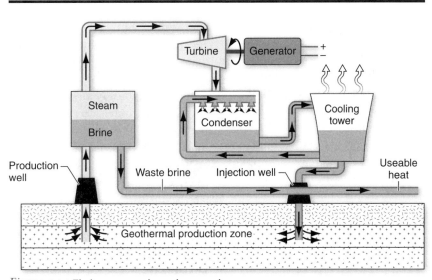

*Figure 4.24* Flash-steam geothermal power plant.

turbine the vapor is condensed to a liquid by cold air or water and fed back into the heat exchanger as shown in **Figure 4.25**. Binary cycle geothermal plants are in operation at Mammoth Lakes, California, Hilo, Hawaii, and Steamboat Springs, Nevada.

There is an enormous untapped geothermal energy resource base associated with oil and natural gas operations. In many of these wells hot water of less than 150°C is extracted along with the oil and natural gas. While not being utilized presently, this hot water can produce electricity by running it through a binary flash system (McKenna et al., 2005). The resource potential is estimated to be 985 to 5,300 MW using the water-rich fluids currently being produced during production of oil and natural gas in seven U.S. Gulf Coast states.

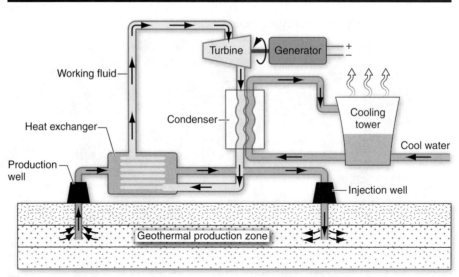

*Figure 4.25* Binary cycle geothermal power plant.

## Concerns in Developing Geothermal Power

The biggest issues with geothermal power revolve around water usage. The hot water withdrawn from underground can contain toxic elements such as mercury. Care must be taken so that after its use it can be safely injected underground and does not contaminate any drinking water supplies. Also, significant quantities of freshwater are used to cool used steam as is true of other power plants that use steam. In areas where freshwater is not plentiful, developing geothermal power decreases the amount of water available for farming and residential use.

At some geothermal production plants land subsidence has occurred due to loss of the extracted fluid. The change in the fluid regime underground has also been known to cause small earthquakes. Of most concern is with the development of enhanced geothermal systems.

## Enhanced Geothermal Systems

Hot rocks at depth are often nonporous and do not allow the circulation of water. A geothermal resource can be developed in these rocks by fracturing them to allow fluid flow. Recently crystalized granites with elevated geothermal gradients have been targeted. Cold water is pumped down into the granite at high pressure through a borehole typically drilled to a depth of 3 to 4 km. As the water warms it expands and stresses the granite. The granite then factures, increasing the rock's permeability. A second borehole is drilled nearby to recover the heated water from the fractured rock, and this water is used to make electricity.

In Cooper Basin, South Australia, Geodynamics Limited is in the process of developing a 25 MW enhanced geothermal power system (EGS) by December 2013 and is targeting production of more than 500 MW by 2018 by fracturing hot granites at depth.

## Geopressured Geothermal Energy

The U.S. Gulf Coast region is underlain by many deeply buried, low salinity sandstone reservoirs containing water in excess of hydrostatic pressure, that is, *geopressured*, and it often approaches the lithostatic pressure produced by the weight of the rocks themselves. Because the geothermal gradient is high in the region the fluid in these reservoirs is at significantly elevated temperatures. It has also been demonstrated that at least some of this water contains small amounts of dissolved methane gas. This geothermal resource is located near the petroleum refining plants along the Texas and Louisiana Gulf Coast, one of the major energy consuming areas in the United States. Dorfman and Kehle (1974) estimate the energy contained within the geopressured aquifers of Texas may be as great as 20,000 MW centuries excluding the natural gas production that could occur.

## Direct Use Geothermal

Warm water that is not hot enough to produce electricity can be tapped for space heating of homes and greenhouses. In Iceland, which sits on recently formed basaltic rocks of the mid-Atlantic ridge, groundwater of temperatures between 60° to 80°C is used to heat homes. Besides heating homes, warm water in shallow reservoirs through-out the world is used to heat schools, offices, health spas, greenhouses, and fish farms.

## Geothermal Heat Pumps

The heating and cooling needed for buildings can be done by extracting heat from and injecting heat into the earth with a *geothermal heat pump*. Geothermal heat pumps do not need high temperature water near the earth's surface and can therefore be used in areas with normal geothermal gradients. By pumping fluid through loops of pipe buried underground next to a building, these systems take advantage of the relatively constant temperature, 7° to 13°C of the earth's surface below that is heated by the sun. Heat can be transferred into buildings in winter and out of buildings in summer. There are about 300,000 heat pump installations in the United States. Switzerland and several other countries are implementing large heat pump programs. The geothermal heat pump shown in **Figure 4.26** only uses a small amount of energy to move water from one location to another with a fluid pump while controlling the temperature inside the building.

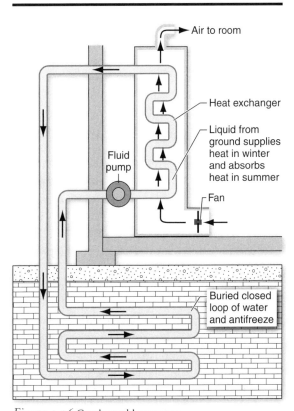

*Figure 4.26* Geothermal heat pump.

A related system is used at Cornell University and Ithaca High School, New York. Here, a lake source cooling system transfers 4°C water from the bottom of nearby Cayuga Lake to a heat exchanger to cool water in the campuses-chilled water cooling systems. It then returns the water to the lake's surface. This replaces using 51 MW of electricity and conventional refrigerants to cool buildings. A similar system operates in Toronto, Canada using Lake Ontario water.

# Total Cost of Electrical Production in the U.S. by Power Source

Total cost for the generation of electricity includes both power plant construction costs as well as electricity production costs. Each type of power plant is unique in the distribution of these costs. There are also regional differences in construction for the same power source including the external costs of required pollution abatement as well as load-balancing concerns that are not considered in **Figure 4.27** where power source costs are compared. This means the lower value of the intermittent production from wind and solar is not factored into these costs. Decommissioning costs are included only for nuclear power. Nuclear power, however, does not include the cost to dispose of spent fuel. This is presently a U.S. government cost. Hydroelectric power total cost does not consider building the required dam, just the electrical generating facility. Also, costs in the figure are for 2009 and some adjustment likely needs to be made for any changes in the price of coal, natural gas, and uranium with time.

For hydroelectric power 2/3 of total costs is for construction and 1/3 production costs, whereas for solar it is almost entirely construction costs. A natural gas power plant is the least expensive type of power plant to construct per kwh produced with most costs from the purchase of natural gas. Nuclear power plants are expensive to construct but produce more useable energy than an average sized coal or natural gas power plant for the same cost of fuel.

What Figure 4.27 shows is that relative to the fossil fuels, coal and natural gas, electrical power from geothermal, hydroelectric, and nuclear are competitive. However, going forward both geothermal and hydroelectric are limited in their ability to

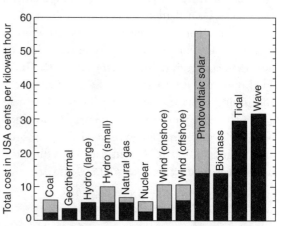

*Figure 4.27* Electrical power production costs in U.S. cents per Kwh in 2009 for the indicated energy source. Lighter grey area indicates range of costs. (Data from: *U.S. Energy Information Administration's Annual Energy Outlook 2009*; biomass, tidal, and wind data from the U.K. Department of Trade & Industry, May 2007, *Energy white paper: Meeting the energy challenge*, recalculated to 2009 values.)

expand in the future in industrialized nations as most prime locations have been exploited. As will be outlined in the next chapter nuclear power has significant safety concerns. The path to increased electricity production into the future is still not clear, but many options with their pluses and minuses exist. The preferred generating technology at a particular location will depend on the specific circumstances of the situation.

## SUMMARY

Solar energy comes from a whole spectrum of different wavelengths. Some we can see (visible light), some are too short to see (ultraviolet light), and some are too long to see but we feel as heat (infrared). Humankind uses low-quality solar energy for heating and illumination. This energy is transferred by radiation, convection, and conduction. Solar furnaces focus sunlight and photovoltaic cells absorb sunlight to produce electricity.

Fuel cells produce electricity in an electrochemical reaction. A proton exchange membrane (PEM) fuel cell runs on hydrogen and produces only water as a reaction product. PEM cells are used to power motors in electric vehicles.

Electrical power can be produced by tides, ocean waves, and differences in surface and deep-water temperatures. Hydroelectric dams convert the potential energy of water into electrical energy and are quite efficient in the process. Electrical power grids produce and distribute electricity. The supply must be matched to demand.

Wind power converts the kinetic energy of air particles into useful energy. The turbines that are used are constructed and located to take advantage of maximum sustained winds. Because the temperature in the earth increases downward, the heat energy at depth can be brought to the surface by water in what is called geothermal energy. Different types of geothermal power plants are constructed depending on the temperature of the water, dry-steam, flash-steam, and binary-cycle boiling. Heat pumps can move heat from a higher temperature heat source to a lower temperature heat sink.

When total costs are considered a geothermal power plant is the least expensive way to produce electricity if water is hot enough to produce dry steam. Photovoltaic solar, tidal, and wave energy are the most expense ways.

## KEY TERMS

| | |
|---|---|
| base-load power | geothermal heat pump |
| binary cycle boiling | greenhouse effect |
| blackout | heliostat |
| brownout | infrared radiation |
| conduction | parabolic trough collector |
| dry steam | photovoltaic cell |
| electrical power grid | radiation |
| electrolyte | semiconductor |
| energy convection | solar power |
| flash steam | tidal estuary |
| fuel cell | turbine |
| generator | ultraviolet radiation |
| geopressured | |

## PROBLEMS

1. a. How much electrical energy can be generated by all the water in a lake 2,000 meters wide by 8,000 meters long by 50 meters deep if all the water falls through a vertical distance of 650 meters? Assume that the electrical generator is 80% efficient and the density of water is 1.0 g cm$^{-3}$. Give your answer in joules.

   b. What would be the average electric power output in watts if this lake were drained over a period of one year?

2. If the wind speed increases by 60%, an ideal windmill will produce how much more power?

3. To heat an average house for one year in Provo, Utah takes about 10$^7$ kcal. Say you could get this heat by cooling off the rock under your yard by 10°C. If your yard is 30 m by 20 m, how deep a layer of rock (h) must you cool to extract this much heat? The heat capacity of rock is 0.21 kcal/kg/°C and the density of rock is ~2,700 kg/m$^3$.

## References

Dorfman, M. and Kehle, R. O. 1974. Potential geothermal resources of Texas: University of Texas, Austin. *Bur Econ Geology Geol Circ* 74–4:26–28.

McKenna, J. K., Blackwell, D. D., Moyes, C., and Patterson, P. D. 2005. Geothermal electric power supply possible from Gulf Coast, Midcontinent oil field waters. *Oil & Gas Journal* 103(33):34–40.

Sawin, J. L. et al. 2011. *Renewables 2011 global status report*. Paris: Renewable Energy Policy Network for the 21st Century (REN21). Available online from www.ren21.net.

Tester, J. W., et al. 2006. *The future of geothermal energy: Impact of enhanced geothermal systems (EGS) on the United States in the 21st century*, 2006 Report from the Massachusetts Institute of Technology. Idaho Falls: Idaho National Laboratory.

Thomas, S., Zalbowitz, M., and Gill, D. 1999. *Fuel cells: Green power* (pp. 1–33). Los Alamos, NM: Los Alamos National Laboratory. Document LA-UR-99-3231 available online from www.lanl.gov.

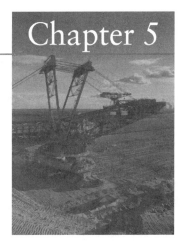

# Chapter 5

# Nuclear Power

Nuclear energy as a resource concerns the generation of electrical power from the energy released when splitting apart a heavy atom's nucleus or potentially fusing the nuclei of two light atoms together. Heavy atoms are those with a large numbers of protons plus neutrons in their nuclei while light atoms have a small number of protons plus neutrons in their nuclei.

## Binding Energy

A large amount of energy, the *binding energy*, is released if light elements are fused together, in a process called *nuclear fusion*. This is because there is a small decrease in mass when a new larger nucleus is formed from smaller ones. This mass is converted to energy given by Einstein's famous formula:

$$E = \Delta mc^2 \qquad [5.1]$$

where $E$ stands for energy, $\Delta m$ represents the mass change, and $c$ denotes the speed of light. While the mass change may be small, the speed of light is a large number and it is squared so the amount of energy produced by the mass change is large. The same is true of heavy elements with a large number of protons plus neutrons if they are split apart, that is, undergo *nuclear fission*. In this process mass is also lost and converted to energy. This occurs because the energy that holds protons and neutrons together, the binding energy, has a maximum at $^{56}$Fe (See **Figure 5.1**).

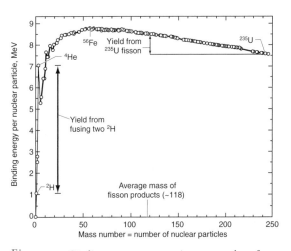

*Figure 5.1* Binding energy per atomic mass number of isotopes. Because the binding energy is the amount of energy lost on binding, higher binding energies are more stable. Also shown is the energy released (yield) when two hydrogen atoms undergo fusion and $^{235}$U experiences fission.

$^{56}$Fe stands for iron with an atomic mass number of 56. The atomic mass number is equal to the number of protons plus neutrons in the nucleus. Because iron has 26 protons in its nucleus, $^{56}$Fe is iron with 30 neutrons. There is also 2.1% of iron that is $^{57}$Fe and therefore has 31 neutrons in its nucleus. One refers to the two irons, $^{56}$Fe and $^{57}$Fe, with different atomic masses as different *isotopes* of iron.

## Nuclear Fission

All the isotopes of uranium and thorium undergo a natural fission process called radioactive decay at a very slow rate. Radioactive decay occurs when energy from repulsion of the positively charged protons in the nucleus becomes greater than the binding energy between the nuclear particles and an *alpha particle* made up of two protons plus two neutrons is ejected from the nucleus. This positively charged particle becomes a neutrally charged helium atom by attracting two electrons from the electron clouds surrounding nearby atoms.

Natural uranium contains 0.72% $^{235}U$ and 99.27% $^{238}U$. $^{235}U$ and $^{238}U$ nuclei can undergo fission, releasing binding energy by bombarding them with neutrons. With $^{235}U$ a fission reaction occurs if it absorbs a neutron to become $^{236}U$. This occurs with greater probability with slow neutrons. Unstable $^{236}U$ then fissions releasing neutrons and can produce over 20 different reaction products depending on how this occurs. One reaction is

$$^{235}U + n \rightarrow\ ^{236}U \rightarrow\ ^{141}Ba + ^{92}Kr + 3n + \text{energy}. \qquad [5.2]$$

The energy given off is $3.2 \times 10^{-11}$ joules per atom. If the $^{235}U$ is concentrated enough in the uranium above what is called the *critical mass*, the neutrons released in reactions like reaction [5.2] can fission other $^{235}U$ nuclei in a self-sustaining chain reaction that continues to release binding energy.

The amount of energy released is large. The density of $^{235}U$ is ~19 g/cm$^3$, and one mole of $^{235}U$ has a mass of 235 grams. Therefore, the volume of a mole of $^{235}U$ is 12.4 cm$^3$ or a cube less than an inch on each side. From reaction [5.2] when a mole of $^{235}U$ nuclei undergoes fission the release of energy is

$$3.2 \times 10^{-11} \text{ joules/atom} \times 6.02 \times 10^{23} \text{ atom/mole}$$

$$= 1.9 \times 10^{13} \text{ joules/mole}. \qquad [5.3]$$

Because burning one barrel of oil releases about $6.1 \times 10^9$ joules of energy, a mole of $^{235}U$ that has undergone nuclear fission is equivalent to 3,160 barrels of oil or a little over 600 metric tons of average coal.

With $^{238}U$ the fission reactions require a fast moving neutron and this fission reaction is not sustainable because slower moving neutrons are given off by the $^{238}U$ fission reactions. An atom of $^{238}U$ can, however, absorb a slow neutron to become $^{239}U$, which undergoes beta decay,

$$^{238}U + n \rightarrow\ ^{239}U \rightarrow\ ^{239}Np + \beta^-. \qquad [5.4]$$

*Beta particles* ($\beta^-$) are high-speed electrons emitted from a nucleus during radioactive decay when a neutron ($n$) becomes unstable and is transformed into a proton ($p^+$) by a reaction that can be written as

$$n \rightarrow p^+ + \beta^-. \qquad [5.5]$$

The 239-neptunium ($^{239}Np$) produced in reaction [5.4] undergoes further $\beta^-$ decay by the following reaction:

$$^{239}Np \rightarrow\ ^{239}Pu + \beta^-. \qquad [5.6]$$

The $^{239}Pu$ produced like $^{235}U$ can undergo fission by a slow neutron producing more neutrons in the decay products. These neutrons can fission other $^{239}Pu$ or $^{235}U$ is a self-sustaining reaction if these two isotopes are concentrated enough.

$^{239}Pu$ can be extracted from used nuclear reactor fuel and processed into a new fuel assembly to be used in a nuclear reactor. Alternatively, separating the plutonium (Pu) from the other elements present can produce a plutonium nuclear bomb. Unlike $^{235}U$ this can be done by ordinary chemical procedures as separating of isotopes of Pu is not required as all the plutonium produced is $^{239}Pu$.

# Nuclear Fuel Cycle

Nearly all nuclear reactors, both civilian and military, are fueled by $^{235}U$ and reactor-produced $^{239}Pu$. Because uranium is first extracted and finally buried back in the ground with its reaction products, the whole process is termed the nuclear fuel cycle as shown in **Figure 5.2**.

## Uranium (U)

Little uranium is present in the earth's mantle. Uranium is, however, reasonably common in the earth's continental crust, some 40 times more abundant than silver, as shown in Figure 9.1 in the chapter on specialty metals. When magma is

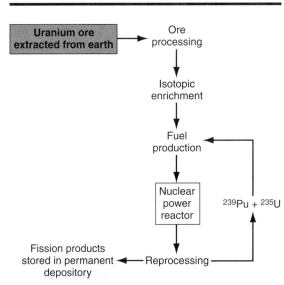

*Figure 5.2* The nuclear fuel cycle showing the path of uranium extraction to permanent disposal of its reaction products.

produced from partial melting of the continental crust it will have a concentration of U greater than the average of the rock that was melted because U is partitioned from the rock into the magma. As the magma starts to crystallize, U continues to be concentrated in residual magma as minerals without any U are crystallized first. U then becomes incorporated in accessory minerals during the final stages of crystallization where concentrations of 100 parts per million (ppm) or more can be reached. Accessory minerals are those that are present in small amount and are not essential constituents of the rock, like the mineral zircon ($ZrSiO_4$).

U in crystallized igneous rocks occurs in the reduced $U^{4+}$ oxidation state, which is very insoluble in aqueous fluids. U is, however, reasonably soluble in the oxidized $U^{6+}$ oxidation state as an uranyl ion, $UO_2^{2+}$, an uranyl carbonate complex, $UO_2(CO_3)_2^{2-}$, or in an uranyl chloride complex, $UO_2Cl^+$. These oxidized U species form in oxygenated near-surface groundwater from the breakdown of U-rich accessory minerals. The oxidized species have the ability to carry U to a site where it reacts with reduced material such as organic matter and precipitates insoluble, reduced $U^{4+}$ in a mineral. In this way a significant deposit of U can be formed.

Typically the sites of precipitation are in or at the margins of sandstone layers because they can have high permeability that allows the flow of substantial U containing fluid. Also sandstones can incorporate organic matter or other reduced phases to reduce $U^{6+}$ to $U^{4+}$ so it will precipitate in a mineral. The most important uranium ore mineral is *uraninite* ($UO_2$). A common impure variety that is black and massive is called *pitchblende*.

## Radium and Radon

Pitchblende is also the main source for radium (Ra) and radon gas (Rn) as Ra and Rn are decay products of uranium radioactive decay. They are recovered as a byproduct of uranium processing. The most stable isotope of radioactive radium is $^{226}Ra$ with a half-life of 1,602 years. The *half-life* is the time it takes for Ra to decay to half its initial value. Radium is usually sold as radium chloride ($RaCl_2$) or radium bromide ($RaBr_2$) and not as pure radium so it can easily be put into solution. $^{226}Ra$ decays to $^{222}Rn$ by emitting an alpha particle consisting of two protons and two neutrons. $^{222}Rn$ is a radioactive inert gas and is radon's most stable isotope with a half-life of 3.8 days.

### Use

The radiation given off by Ra and Rn can destroy living cells. This property of Ra and Rn is used in medicine to treat certain types of malignant growths such as cancer. The use of radon for radiation therapy is, however, being replaced by artificially produced radioisotopes made in particle accelerators and nuclear reactors. Radiation from Ra is also used to obtain X-ray images of material to examine them for flaws. Ra was once used to make luminous paint for dials of watches and clocks. However, given the harmful effects of the radiation given off this is not currently done. Other radioisotopes are now used because they are safer by being less radioactive and are less expensive.

## Uranium Ores

The minimum grade of uranium ores is typically 1 wt% $U_3O_8$ unless it is produced as a byproduct of mining a different ore. Generally, all uranium ore concentrations are reported in terms of the oxide $U_3O_8$. The International Atomic Energy Agency (IAEA) characterizes 15 different types of uranium deposits according to their geological settings. The four most important in descending order are: unconformity-related deposits, sandstone deposits, quartz-pebble conglomerate deposits, and vein deposits. With the large variety of different deposits there is a broad distribution of uranium ore resources on all continents.

## Unconformity-Related Deposits

Unconformity-related uranium deposits form at or near the contact of unconformable sedimentary beds. This contact represents a period of erosion or nondeposition of sediments. In the case of uranium deposits the contact presents an unconformity between older igneous and metamorphic rocks and younger sandstone beds. Investigations suggest the mineralizing fluid is a Cl-rich basinal brine that is oxidized, slightly acidic, and hot, up to 240°C in temperature. Basinal brines are warm, salt-rich fluids produced at depth from pore fluids during burial of sediments in developing sedimentary basins. Some of these sediments are salt-rich. This is the origin of the salt in the fluids. (See discussion of Figure 8.15 in the chapter on base metals).

The $U^{6+}$ in the fluid is likely transported through the sandstone in water by the aqueous species $UO_2Cl^+$. At the unconformity a strong oxidation gradient occurs due either to organic matter $CH_2O$ or pyrite ($FeS_2$). Reaction with this reduced material causes the ore mineral uraninite to precipitate by a reaction like

$$O_2 + 2UO_2Cl^+ + 4OH^- + 2CH_2O \rightarrow 2UO_2 + 2Cl^- + 2CO_2$$

Uranium     Organic     Ore     + $4H_2O$.  [5.7]
species     matter

Unconformity-related uranium deposits are some of the largest and richest making up about 1/3 of uranium resources. Large deposits occur in the Chu-Sarysu district of south central Kazakhstan, the Athabasca Basin and Thelon Basin of Canada,

as well as in the Pine Creek and Rudall River area of Australia.

## Sandstone Deposits

Sandstone uranium deposits occur within rocks formed from sand deposited in bars in meandering rivers or marine beach and sand dune environments. They make up ~18% of the world's uranium resources. Large sandstone uranium deposits occur in Niger, Kazakhstan, Uzbekistan, Gabon, and South Africa in the Karoo Basin. In these deposits $U^{6+}$ in solution precipitates when encountering reduced organic material such as plant debris, humic substances, or marine algae. Ore precipitation can also occur if solutions encounter pyrite or petroleum in the sandstone.

The most common type of U-rich sandstone deposit is what is called a roll-front sandstone-hosted deposit. These are arcuate, crescent-shaped bodies of ore in sandstones (**Figure 5.3**). Roll-front sandstone-hosted uranium deposits occur in the Powder River Basin in Wyoming, South Dakota, the Colorado Plateau, and along the U.S. Gulf Coast in south Texas. Oxidized waters containing $U^{6+}$ precipitates $UO_2$ along a reaction front when it encounters a reducing agent such as $CH_2O$ by a reaction like

$$1/2\ O_2 + UO_2(CO_3)_2^{2-} + Fe^{2+} + CH_2O \rightarrow UO_2 + FeCO_3 + 2CO_2$$

Uranium           Organic     Ore     + $H_2O$.  [5.8]
species           matter

## Quartz-Pebble Conglomerate Deposits

Uraninite occurs in some 3.1 to 2.2 billion-year-old quartz-pebble conglomerates formed in paleostream environments. Pebble conglomerates consist of pebble-sized fragments of preexisting rock set in a finer-grained mass of clay and silt. The pebbles are carried in streams during times of floods with its higher water velocities. Uraninite in these deposits was likely weathered from uraniferous veins produced during the last stages of granite magma crystallization. Uraninite-bearing granites were exposed in mountains and weathered. The anoxic atmosphere at this time in early earth history kept uraninite crystals weathered out

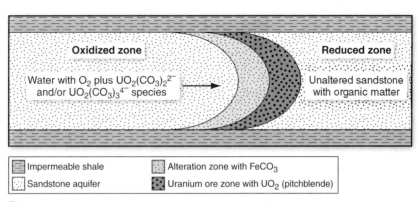

*Figure 5.3* Roll-front uranium deposit forming in a sandstone aquifer.

of the rock from being altered or dissolved as no $U^{6+}$ could form without oxygen in the atmosphere. Fluvial transport of the reduced uraninite crystals could therefore occur. Because of the high density of uraninite, $\sim$10.9 $g/cm^3$, the crystals became concentrated by washing out of lighter particles in stream pebble-rich deposits.

Deposits are generally low grade, below 0.15 wt% $U_3O_8$, but have large tonnages and make up approximately 13% of the world's uranium resources. These include the Elliot Lake deposits in Canada and the Witwatersrand gold-uranium deposits in South Africa. In these latter deposits uranium is recovered as a byproduct of gold mining, as the U grade is often only 0.01 wt% $U_3O_8$.

### Vein Deposits

In vein deposits water-rich fluids from the late stages of the crystallization of granites deposit pitchblende within cavities. These can be in the granitic rocks themselves (Central Massif, France), in veins in metasedimentary rocks produced from fluids that have exited granites (Erzgebirge Mountains, Germany/Czech Republic), or in mineralized fault and shear zones near granitic rocks (central Africa). Together they constitute 9% of the world's uranium resources.

## Uranium Prices

The uranium market has a history of volatility, moving with the standard forces of supply and demand as well as with geopolitical considerations

(**Figure 5.4**). The only significant commercial use for uranium is to fuel nuclear reactors for the generation of electricity. However, uranium fuel costs for nuclear plants are a small proportion of total generating costs. There are large capital costs for construction that are greater than those for coal-fired plants and much greater than those for gas-fired plants that need to be recovered over the lifetime of the nuclear power plant.

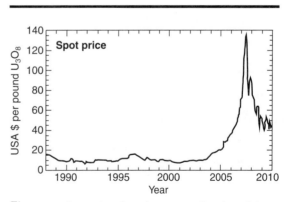

*Figure 5.4* Spot price of uranium ore as a function of time.

Uranium prices reached an all-time low of US $7 per lb of $U_3O_8$ in 2001. This was followed by a period of gradual rise until a price spike to $\sim$ US $137 per lb in 2007. This was the highest price adjusted for inflation in 25 years. A possible cause for the spike was the flooding of the newly developed Cigar Lake Mine in Canada, which is a very large high-grade uranium ore deposit. It is an unconformity-related deposit with an average grade reported to be 17.4 wt% $U_3O_8$. The higher price spurred new prospecting and the reopening of old mines bringing the price down from its historic high to US $50 per lb.

## Uranium Reserves

Given in **Table** 5.1 are the known recoverable resources of uranium to US $130 per kg of $U_3O_8$. This is US $59 per pound. The world total is near 5.5 million metric tons of $U_3O_8$. The United States resources at this price are 342,000 metric tons of $U_3O_8$. If one increases the cost to US $100 per pound the recoverable $U_3O_8$ becomes 558,000 metric tons. If one lowers the cost to recover to US$ 50 per pound the recoverable resource becomes 245,000 metric tons.

In 2009, 50,600 metric tons of $U_3O_8$ was mined with 27% from Kazakhstan, 20% from Canada, 16% from Australia, 9% from Nambia, 7% from Russia, and 6% from Niger. The United States produced 3% of the world's supply. The locations of uranium resources in the U.S. are shown in **Figure** 5.5. The grade of most of these resources is from 0.09% to 0.14% $U_3O_8$ so uranium in the U.S. is mined primarily as a byproduct.

*Table 5.1* KNOWN RECOVERABLE RESOURCES* OF URANIUM ($U_3O_8$).

| COUNTRY | METRIC TONS U[†] | WORLD RESOURCES |
|---|---|---|
| Australia | 1,243 | 23% |
| Kazakhstan | 817 | 15% |
| Russia | 546 | 10% |
| South Africa | 435 | 8% |
| Canada | 423 | 8% |
| USA | 342 | 6% |
| Brazil | 278 | 5% |
| Namibia | 275 | 5% |
| Niger | 274 | 5% |
| Ukraine | 200 | 4% |
| Jordan | 112 | 2% |
| Uzbekistan | 111 | 2% |
| India | 73 | 1% |
| China | 68 | 1% |
| Mongolia | 62 | 1% |
| Other | 210 | 4% |
| World totals | 5,469 | 100% |

* Reasonably Assured Resources plus Inferred Resources, to USA$ 130/kg U, 1/1/07, from Organisation for Economic Co-operation and Development (OECD), Nuclear Energy Agency, and International Atomic Energy Agency, *Uranium 2007: Resources, Production, and Demand* ("Red Book").

† In thousands.

*Data from*: OECD/International Atomic Energy Agency (IAEA) (2009), *Uranium 2009*: Resources, Production and Demand, OECD Publishing.

*Figure 5.5* Location of uranium resources in the United States. (Adapted from U. S. Department of Energy, Grand Junction Project Office. *National Uranium Resources evaluation. Interim report,* (June 1979) Figure 3.2; and GJPO data files.)

### Electricity Production

Global electricity use is likely to increase from the 13 trillion kilowatt-hours (kwh) used in 1999 to 22 trillion kwh in 2020, increasing by 2/3 during this period. One metric ton of natural uranium can produce ~40 million kwh of electricity. This number is variable, depending on the degree of isotopic separation done (see below) and the type of reactor used to produce the electricity. However, using this number it would take $13 \times 10^{12}$ kwh/$40 \times 10^6$ kwh per metric ton = 0.325 million metric tons to produce all the electricity produced in 1999 and 0.55 million metric tons would be needed in 2020. The uranium reserves at theses rates of use would last between 5.5 million metric tons/0.325 million metric tons per year = 17 years and 5.5 million metric tons/0.5 million tons per year = 10 years, respectively. If one only wants to produce the current percentage of electricity from nuclear reactors, ~14%, these numbers would need to be multiplied by 1/0.14 or a little over 7.

Decommissioning nuclear weapons and using their Pu as well as their highly enriched uranium would increase these times. However, if the world embraces nuclear power this means at sometime in the not too distant future either thorium fuel or breeder reactors would need to be employed.

## Uranium Production

After mining, the uranium ore undergoes extensive beneficiation by mechanical and chemical means. It is first crushed to a fine powder. The fine powder is reacted in concentrated acid, base, or peroxide solution to leach out uranium. After the leachate is dried and filtered, it produces *yellowcake*, a substance whose uranium concentration is 70% to 90% $U_3O_8$. The name comes from the color and texture of the processed ore obtained during earlier used processing techniques. The uranium in yellowcake is, however, only 0.72% $^{235}U$ (**Table 5.2**) and needs to be enriched to ~4% to be fuel for nuclear reactors. At 4% $^{235}U$ a self-sustaining chain reaction can occur in common nuclear reactors because atoms of $^{235}U$ are close enough together that the neutrons given off by the decay of one $^{235}U$ nucleus can be absorbed by another $^{235}U$ nucleus.

*Table 5.2* CONCENTRATION AND HALF-LIVES OF ISOTOPES IN NATURAL URANIUM.

| ISOTOPE | TOTAL U% | HALF-LIFE* |
|---------|----------|------------|
| $^{234}U$ | 0.0054 | 0.2455 |
| $^{235}U$ | 0.7204 | 703.8 |
| $^{238}U$ | 99.2742 | 4,468 |

* In million of years.

# Uranium Isotope Separation and Nuclear Fuel

To separate the isotopes of uranium in order to enrich the $^{235}U$ to nuclear fuel grade, a process that depends on the difference in the mass of the isotopes such as diffusion is required. The heavier $^{238}U$ isotope diffuses more slowly. With the diffusion technique $U_3O_8$ is first converted into gaseous $UF_6$. The rate of diffusion of a gas is proportional to the square root of its mass. Under this constraint $^{235}UF_6$ can be separated from $^{238}UF_6$.

The U.S. Energy Policy Act of 1992 created the United States Enrichment Corporation (USEC) to privatize uranium enrichment for civilian use. It contracts with the U.S. Department of Energy to produce enriched uranium for use in commercial nuclear power reactors. USEC has a gas diffusion isotope enrichment plant in Paducah, Kentucky. A large number of rotating cylinders (centrifuges), in series are used. This enhances the diffusion process of the U isotope in $UF_6$ as each cylinder's rotation creates a centrifugal force so that the heavier gas molecules containing $^{238}U$ moves outward and the lighter gas molecules rich in $^{235}U$ collect closer to the center of rotation. In 2010 an additional enrichment facility owned by the Uranium Enrichment Company (URENCO), a nuclear fuel company that operates a number of uranium enrichment plants in Europe, opened a uranium enrichment plant east of Eunice, New Mexico. This was undertaken because the old USEC plant in Paducah, Kentucky is scheduled to be closed in the near future. At present, Russia supplies about 40% of the enriched uranium needed for U.S. commercial reactors.

The separation work to separate isotopes is given in separation work units (SWU) per kilogram. This is not energy, as the same amount of SWU will require different amounts of energy depending on the efficiency of the separation technology. For example, to produce 1 kg of nuclear fuel of 3.6% $^{235}U$ requires ~8 kg of natural 0.72% $^{235}U$ and 4.5 SWU of enrichment with a waste product of uranium which has 0.3% $^{235}U$. However, if the waste product's $^{235}U$ is lowered to 0.2% $^{235}U$ then only 6.7 kg of 0.72% $^{235}U$ uranium is required but almost 5.7 SWU of enrichment is required. As these quantities move in opposite directions the changing costs of the natural 0.72% $^{235}U$ uranium changes the optimum separation cost efficiency.

When the $UF_6$ gas is enriched to 3.6% to 4% $^{235}UF_6$ it is converted into $UO_2$ powder to produce pellets. Pellets are loaded into 5-meter long fuel rods of about 1 cm in diameter. 220 ± 40 fuel rods are placed in a bundle and inserted into a nuclear reactor. There are about 160 ± 40 bundles in a reactor core. When fuel rods are close enough together they start a self-sustaining fission process. Typically fuel rods undergo sufficient reaction for three to five years before they need to be replaced because there is not a high enough concentration of $^{235}U$ remaining to sustain the reaction.

## Zirconium (Zr) and Hafnium (Hf)

Fuel rods are clad with a zirconium (Zr) alloy, the major use of Zr. This is because Zr has a low neutron capture cross-section meaning it will not absorb neutrons released during the fission reaction and has a high resistance to chemical corrosion. Zr is also used in jewelry as cubic zirconia, synthesized crystalline Zr dioxide, $ZrO_2$. It has a likeness to diamonds. Synthesized beads of $ZrSiO_4$ are used in milling and grinding of rocks as it has a Mohs hardness of 7.5 (see the introductory chapter).

Zr is obtained from zircon, $ZrSiO_4$, found weathered from igneous rocks in some sand deposits. Zircon with a density of 4.6 to 4.7 g cm$^{-3}$ is concentrated in these deposits by the winnowing of less dense minerals in stream gravels. Hafnium (Hf) substitutes for Zr in a ratio of 1 to 50 in most zircons. Hf is produced as a byproduct during the purification process to produce pure Zr. Hf is added to reactor *control rods* (see below) because it has a good absorption cross-section for thermal neutrons nearly 600 times that of Zr.

# Nuclear Reactor Types

Nuclear reactors are based on the fission of $^{235}U$ and reactor produced $^{239}Pu$. They can be divided into two types, *thermal* and *fast neutron reactors*, depending on the energy of the neutrons that sustain the fission. When a $^{235}$U nucleus fissions the free neutrons produced exit the nucleus at a high velocity, as fast neutrons. In thermal reactors fast neutrons are slowed by a *moderator* material. By slowing the neutrons they have a greater probability to be absorbed by $^{235}$U and $^{239}$Pu leading to their fission and a lower probability to fission $^{238}$U. This means less enrichment of $^{235}$U is needed in the fuel rods.

Fast neutron reactors use the fast neutrons produced to fission the $^{235}$U. They, therefore, do not use a neutron moderator. To maintain a chain reaction requires a higher concentration of $^{235}$U of $\geq 20\%$ due to the lower probability to fission. The advantage of fast neutron reactors is that the fast neutrons can fission all the actinides produced so there is less transuranic waste. Actinides are the radioactive elements with atomic numbers ranging from actinium with atomic number 89 to lawrencium with atomic number 103. However, fast neutron reactors are more difficult to build and more expensive to operate because of the extra costs of isotope enrichment. Therefore, thermal reactors dominate electrical power production.

Nuclear reactors for power generation generally use water as the coolant that transfers the heat from the fission reactions. One of the advantages of water is that it is also a neutron moderator that slows the neutrons produced. A disadvantage is that $H_2O$ absorbs some of the neutrons to produce HDO. This means they are not available to fission $^{235}$U. D denotes the hydrogen isotope *deuterium*, a hydrogen atom with a neutron in its nucleus.

Water-cooled reactors can be separated into pressurized water reactors (PWR) and boiling water reactors (BWR). Either of these can use light water ($H_2O$) or *heavy water* ($D_2O$). The extra neutrons make the water "heavy." The advantage of using $D_2O$ is that unlike $H_2O$ it does not absorb neutrons and therefore fuel near the natural concentration of 0.72% $^{235}$U can be used to achieve a chain reaction. The problem with heavy water reactors is that $D_2O$ is expensive to produce and the reactors are therefore more expensive to construct and operate.

A problem for all these reactor types in terms of nuclear weapon proliferation is that during the fission reactions some uranium is converted to $^{239}$Pu. $^{239}$Pu is easy to chemically separate from used fuel to produce a thermonuclear weapon. More plutonium is produced using heavy rather than light water as the coolant.

## Boiling Water Reactor

In a *boiling water reactor* (*BWR*) the same water is used as a moderator, coolant of the reactor core, and source for steam that drives the turbine blades (**Figure 5.6**). This reduces the complexity of the reactor and the construction costs. Water passing over the reactor core is heated and turns into steam. The steam produced is used to turn the turbine blades to make electricity. The used steam from the turbine is cycled through a condenser that cools it back into a liquid state. This water is reintroduced into the reactor core completing the loop. The BWR has a self-moderating design because an increase in fission causes increase steam production and therefore decreased moderators in the reactor. This slows fission.

## Pressurized Water Reactor

A *pressurized water reactor* (*PWR*) is a type of electric power reactor consisting of two separate loops for pressurized water and steam production (**Figure 5.7**). In the water loop, water is heated to ~300°C in the reactor but maintained by a pressurizer at a pressure of ~160 bars so it does not boil. The steam loop consists of steam produced from 1 bar liquid water that is used to run a turbine. The two loops are connected by a heat exchanger where heat from the high-pressure water converts the water in the steam loop to steam.

Because the water used in the high-pressure water loop is isolated from water in the steam loop no radioactive material is contained in the steam. This means the piping for the steam loop and turbine do not require special shielding materials. However the high pressures and temperatures of the high-pressure loop accelerate corrosion and this loop needs to be constructed of stainless steel.

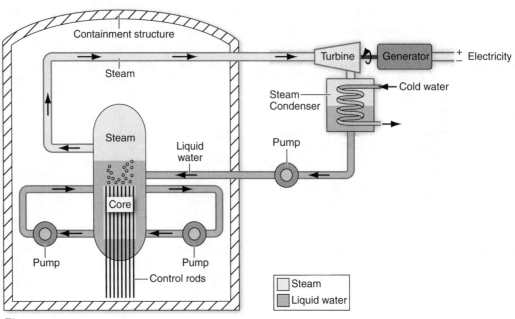

*Figure 5.6* Boiling water reactor.

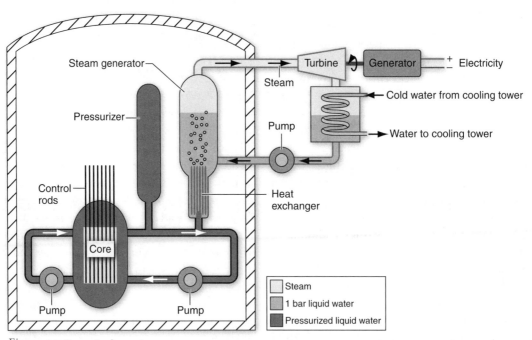

*Figure 5.7* Pressurized water reactor.

Because of the use of high water pressure a PWR cannot be refueled while operating and requires a reactor shutdown for 3 to 14 days to exchange fuel assemblies.

This decreases the availability of the reactor. Generally, about 1/3 of the fuel assemblies are replaced during each refueling on an 18- to 24-month shutdown cycle.

## Fast Neutron Reactor

A fast neutron reactor, also called a *breeder reactor* because it produces nuclear fuel as well as electricity, is shown in **Figure 5.8**. It uses fast neutrons to sustain the chain reaction. Because water is a neutron moderator, fast neutron reactors typically use liquid metal to transfer the heat. Fast neutrons are less efficient in $^{235}U$ fission so a higher degree of $^{235}U$ enrichment is required, generally 20% to 30%. The fast fission process produces more neutrons per fission than with slow neutron fission. A blanket of nonfissionable $^{238}U$ is wrapped around the core and the following reactions occur:

$$^{238}U + n \rightarrow {}^{239}U \qquad [5.9]$$

$$^{239}U \rightarrow \beta^- + {}^{239}Np \qquad [5.10]$$

$$^{239}Np \rightarrow \beta^- + {}^{239}Pu. \qquad [5.11]$$

The $^{239}Pu$ produced can be used as reactor fuel or to build a nuclear bomb. The time required for a breeder reactor to produce enough $^{239}Pu$ to fuel a second reactor is called its doubling time. This is on the order of 10 years. The extra neutrons can also be used to transmute a blanket of actinide waste to less troublesome isotopes.

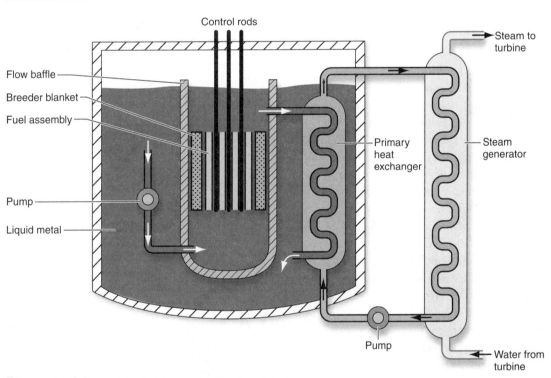

*Figure 5.8* Liquid metal fast breeder reactor of the "pool" design.

## Pebble Bed Reactor

Given the basic physics involved, new nuclear reactor designs are continuing to be considered. One is the pebble bed reactor where thousands of micro fuel particles coated with moderating graphite are encased in ceramic to hold them together in spherical, nearly tennis ball sized pebbles. The pebbles are stacked in a reactor and helium gas is circulated through the space between the fuel pebbles. The nuclear reactions in the fuel particles heat the helium to very high temperatures, 700° to 950°C. The helium can then be run through a turbine to power an electrical generator. Because the reactor is gas cooled, the extensive piping for water cooling is not required and the gas is much less likely to become radioactive. Despite more than 50 years of development, no commercial-scale design has been developed although the Chinese are trying. A design variation termed the Next Generation Nuclear Plant (NGNP), a demonstration project that uses 1 mm-sized graphite and silicon carbide coated enriched-uranium pebbles, is slated for construction by 2021 once reactor development work has been completed. It is to be located at the Idaho National Laboratory, Idaho Falls, Idaho.

# Reactor Operation

Nuclear reactor fuel rods can be UOX, $UO_2$ mixed with an organic binder and pressed into pellets, or MOX, a mixed-oxide blend of $PuO_2$ and $UO_2$. Reactors operate under conditions where the fission chain reaction occurs critically. Moderators are used to slow neutrons released in the fission process to make them more likely to be captured and fission other $^{235}U$ and $^{239}Pu$ nuclei. Moderators can be graphite, deuterium as deuterium oxide, heavy water, or normal light water.

Control rods are used to control the rate of reaction. These are stainless steel rods, shaped like the letter "x," filled with boron carbide or hafnium. The boron and hafnium are able to absorb many neutrons without undergoing fission themselves and thus slow the number of fission reactions. These rods are inserted between the fuel assemblies to control or stop the nuclear chain reaction.

Electrical nuclear power plants are generally located near a lake, river, or the ocean because of the need to cool the output steam from turbines. The steam is either sent to cooling towers where it is emitted as water droplets, literally a cloud, that undergoes evaporative cooling or the hot water is discharged into a lake, river, or the ocean.

# Nuclear Reactor Safety

Nuclear reactors are designed to operate safely even when systems fail. One possibility is a failure to be able to insert control rods. Reactors are typically equipped with ancillary control rods to slow a runaway reactor. These are inserted into the reactor core during an automatic reactor shutdown termed a

*scram*. The operator can also inject sodium polyborate or gadolinium nitrate into the reactor coolant water. These are neutron poisons, meaning they can absorb neutrons without undergoing fission.

If a loss of coolant fluid occurs temperatures in the core rise to extremely high values. If water is the coolant/moderator, it may boil, bursting out of its pipes. Therefore, pressure-operated relief valves are used and a backup supply of water under pressure is available to inject into the reactor. Also reactors are enclosed in containment structures to prevent radioactive material escape. Diesel generators and batteries are available and can be engaged if a loss of electrical power for reactor control occurs.

## Three Mile Island Reactor Accident

The most serous nuclear power plant accident in the United States occurred at the *Three Mile Island Nuclear Generating Station (TMI)* in Pennsylvania (**Figure 5.9**). It consists of two pressurized water reactors each inside their own containment building with each connected to two cooling towers. On March 28, 1979 the TMI-2 reactor lost its main water feed. It went into an automatic scram in the 15 seconds before backup pumps activated to restore water to the reactor. A pressure release value opened to release pressure as the temperature increased, but valves connecting the reserve water pumps were closed so large amounts of water escaped and were not replaced. The operators

*Figure 5.9* Three Mile Island Nuclear Power Plant showing steam release from two of the four cooling towers.

thought there was too much water in the reactor rather than too little. The core lay uncovered without water for several hours. The core heated up and about half of it melted. A small amount of radioactive gases including $^{131}$I was released into the air from the containment structure. The incident undermined confidence in the nuclear power industry in the U.S.

## Chernobyl Nuclear Accident

The world's worst nuclear power plant accident occurred on April 26, 1986 at the *Chernobyl Nuclear Power Plant* near Kiev, Ukraine during an experiment to test the ability to safely cool the core during an emergency. The Chernobyl scientists running the experiment disabled many security features, believing that a major incident would not occur. Among the systems that were disabled were an emergency core cooling system, an automatic control system, and the emergency power reduction system.

Power surged during the test of the backup system. The system overheated and a scram occurred. The steam pressure in the core increased rapidly and ruptured cooling water pipes. Without coolant the fuel rods began to melt. A steam explosion occurred blowing the 2,000 metric ton containment lid off of the reactor. About three seconds later, a second, more powerful explosion occurred throwing red-hot lumps of graphite moderator pieces and uranium-oxide fuel over the immediate area. Ten percent of the core was dispersed as particles in the atmosphere (**Figure 5.10**).

At the power plant 20 to 30 workers died immediately. Uncontrolled fires lasted for 10 days. After the accident 135,000 people were evacuated from the area because of the high radiation levels. In the year following the accident radiation released caused a 16% rise in the death rate in northern Ukraine.

The destroyed reactor was encased in a large concrete tomb to allow continuing operation of the other reactors at the power plant. These operated until December 2000. The containment structure is degrading because it contains 200 metric tons of highly radioactive material. A new better confinement structure is scheduled to be built around the reactor by sometime after 2015.

*Figure 5.10* Reactor 4 at the Chernobyl Nuclear Power Plant after the April 26, 1986 accident.

## Fukushima Nuclear Accident

Being in an earthquake prone area, the six boiling water reactors at the Fukushima Nuclear Power Plant in Fukushima, Japan were built to withstand an 8.2 magnitude earthquake. The 8.9 magnitude earthquake that occurred in March 2011 was five times larger. When the earthquake hit, the nuclear reactors instituted a scram. Within seconds after the earthquake started, the control rods were inserted into the cores and the nuclear chain reactions stopped. However, the cores still needed to be cooled. A large amount of heat was still being produced in the cores of the nuclear reactors due to the natural decay of the fission products created before shutdown. The earthquake knocked out the external electric supply to the nuclear reactors. For the first hour after the earthquake, the emergency diesel-powered generators provided the electricity needed for the cooling system pumps.

The earthquake produced a tsunami wave. When it arrived at Fukushima it was 10 m high and flooded the diesel-powered generators causing them to stop working. With the loss of the diesel generators the reactor operators switched to emergency battery power. The batteries provided power for pumping the water to cool the core for another eight hours.

After eight hours and loss of a supply of cooling water the heat production from the fission product decay in the cores caused the water present to heat significantly, producing steam and increasing steam and water pressure in the reactors. To control the pressure, steam and other gases along with some radioactive material were vented to the atmosphere. During this time, mobile generators were transported to the site and some power was restored. However, the water level dropped below the top of the fuel rods and a partial core meltdown occurred in reactors 1, 2, and 3. When the temperature exceeded 1200°C the Zr that clad the fuel assemblies reacted with water in the core, producing hydrogen gas. When this $H_2$ gas was vented to the atmosphere it reacted explosively with oxygen in the air causing damage to the containment structure.

Operators injected seawater mixed with boric acid, a neutron absorber into the reactors for a number of days. This ensured the reactor rods remained covered with water and reduced the core temperature. The temperature of the fuel rods finally decreased to a safe level as the short-lived isotopic material, whose decay accounted for most of the heat produced, was depleted. The pressure in the reactors then stabilized and venting of the gas in the reactor containment structure was no longer required.

Measurements in areas of northern Japan to a distance of 30 to 50 km from the reactors showed high levels of radioactive $^{137}$Cs. Food grown in this area was banned for sale. An exclusion zone for individuals of 20 km was established around the Fukushima plant as well as a 30 km voluntary evacuation zone. Officials in Tokyo recommended that for a limited time tap water should not be used to prepare food for infants. Measurements undertaken worldwide detected $^{131}$I and $^{137}$Cs in the atmosphere from the Fukushima reactors.

It is estimated that radioactive elements released from the Fukushima reactors are of the same order of magnitude as the releases from the Chernobyl power plant disaster. It is likely to take decades to clean up the area. The Fukushima nuclear plant will be decommissioned once the crisis is over. As a result of the accident Japan is considering curtailing the further use of nuclear power in the country.

The severity of the Fukushima nuclear accident is rated level 7 on the International Nuclear Event Scale. This scale considers an abnormal situation with no safety consequences as a level 0 event and increases to a maximum of level 7, indicating a major accident causing widespread contamination. The Chernobyl nuclear plant disaster is the only other level 7 event that has occurred. The Three Mile Island core meltdown was one of five accidents rated as level 5, while the disaster near Kyshtym, Russia in 1957 involving the explosion of stored radioactive waste at a nuclear reprocessing plant is the only level 6 event on the scale.

# Permanent Disposal of Concentrated Nuclear Waste

Concentrated nuclear waste in the United States is from two sources, Department of Defense military waste and Department of Energy civilian reactor waste. The concentrated military nuclear waste is from development and manufacturing of nuclear weapons and is termed *transuranic waste*. It contains significant amounts of transuranic elements, those elements with atomic numbers greater than the 92 of uranium. The civilian waste is primarily from electric nuclear power plants and is termed *high-level waste*. It consists of nuclear power plant used reactor fuel rods after they can no longer sustain nuclear fission and are ready for disposal.

Both types of waste have their own separate permanent disposal sites. The Waste Isolation Pilot Plant (WIPP) in New Mexico stores defense-related waste. In June 2008 the Department of Energy submitted a license application to authorize the construction of a high-level radioactive waste disposal site for civilian waste at Yucca Mountain, Nevada.

## Waste Isolation Pilot Plant

The *Waste Isolation Pilot Plant (WIPP)* stores U.S. defense related waste in a remote desert area ~42 km east of Carlsbad, New Mexico. Disposal operations began in 1999, but it was not licensed to take high-level wastes. Whether it will ever be able to take high-level wastes is unclear. It is the world's third licensed underground repository after two earlier repositories in Germany. At WIPP the waste is contained inside a 610 m thick bed of salt 260 m below the surface. The salt was deposited 250 million years ago in an ancient Permian sea that covered the area at that time. Salt was chosen to enclose the waste because of its self-sealing plastic behavior and the difficulty of transporting water and therefore nuclear waste through salt.

## Civilian High-Level Waste

Fuel assemblies used in civilian power plants last about 880 days. A large nuclear reactor produces about 3 cubic meters of spent fuel rods each year. Its general composition is given in **Table 5.3**. It is primarily composed of $^{238}U$ as well as significant quantities of $^{239}Pu$. The *actinides* uranium (atomic number 92), plutonium (atomic number 94), and curium (atomic number 96) are responsible for the bulk of the long-term radioactivity. Minor actinides are actinides other than uranium and plutonium. Actinides are highly toxic, long-lived radioactive elements. As mentioned they have atomic numbers from 89 to 103. The fission products include every element from zinc (atomic number 30) through lutetium (atomic number 71), with much of the fission yield concentrated in two peaks, one starting at atomic number 40 (zirconium, molybdenum, technetium, ruthenium, rhodium, palladium, and silver) while the other starts at atomic number 53 (iodine, xenon, cesium, barium, lanthanum, cerium, and neodymium).

Many of the fission products are either nonradioactive or short-lived radioactive isotopes. However, used fuel contains some radioactive material with medium to long half-lives such as 90-strontium, 137-cesium, 99-technetium, and 129-iodine. The radioactive isotopes in spent nuclear fuel of most concern are the artificial isotopes 237-neptunium, 241-americium, 243-americium, 242-curium through 248-curium, and 249-californium through 252-californium because of their significant half-lives (**Table 5.4**).

In order to encourage the development of nuclear power the United States passed the Nuclear Waste Policy Act of 1982. It made the federal government responsible for the burial of high-level waste from commercial nuclear power plants. The Act established a timetable and procedure for constructing a permanent, underground

---

*Table 5.3* COMPOSITION OF 4.2% ENRICHED NUCLEAR FUEL BEFORE AND AFTER 40,000 MW-DAYS PER METRIC TON* OF ENERGY HAS BEEN RELEASED.

| CONSTITUENT | NEW | USED |
|---|---|---|
| $^{238}U$ | 95.8% | 93.4% |
| $^{235}U$ | 4.2% | 0.71% |
| $^{239}Pu$ | 0.0% | 1.27% |
| Minor Actinides† | 0.0% | 0.14% |
| Fission products | 0.0% | 5.15% |

\* 1 MW-day = 24,000 kilowatt hours.

† Minor actinides include neptunium, americium, and curium.

*Table 5.4* HALF-LIVES OF SOME RADIOACTIVE
ISOTOPES PRESENT IN SPENT REACTOR FUEL.

| ISOTOPE | HALF-LIFE (YEARS) |
|---------|-------------------|
| $^{235}$U | 704 million |
| $^{239}$Pu | 24,000 |
| $^{90}$Sr | 28.8 |
| $^{137}$Cs | 30.23 |
| $^{99}$Tc | 211,100 |
| $^{129}$I | 15.7 million |
| $^{237}$Np | 2.14 million |
| $^{241}$Am | 432.2 |
| $^{243}$Am | 7370 |
| $^{242}$Cm | 152.5 |
| $^{248}$Cm | 340,000 |
| $^{249}$Cf | 351 |
| $^{252}$Cf | 2.645 |

repository to be completed by the mid-1990s. After considering sites in bedded salt, granite, and basalt the federal government settled on a repository site in volcanic rocks in the desert at *Yucca Mountain, Nevada*. Yucca Mountain is part of the old underground nuclear weapon testing facility north of Las Vegas, Nevada.

Recent estimates indicate the Yucca Mountain site could open in 2017 with a capacity for 70,000 metric tons of radioactive waste. As of 2009, $11 billion had already been spent to prepare the site. However, the Obama Administration rejected use of the site in the 2009 U.S. federal budget proposal, which eliminated all funding except that needed to answer inquiries from the Nuclear Regulatory Commission. On March 5, 2009, Energy Secretary Steven Chu told a Senate hearing "The Yucca Mountain site no longer was viewed as an option for storing reactor waste."

As of 2009, the United States had accumulated more than 64,000 metric tons of spent nuclear fuel from commercial nuclear reactors. Used fuel assemblies are presently kept on reactor sites in pools of cooling $H_2O$ waiting permanent storage or burial. The U.S. government pays nuclear power plant operators about $1 billion a year to cover the storage cost. After 10,000 years of radioactive decay, according to the U.S. Environmental Protection Agency (EPA), the spent nuclear fuel will no longer pose a threat to public health and safety.

## Reprocessing Spent Nuclear Fuel

Spent nuclear fuel can be reprocessed by dissolving it in nitric acid and allowing uranium and plutonium to be separated from the other waste. This material can then be incorporated into a mixed oxide (MOX) fuel pellet. This removes much of the long life radioactivity from the waste. Reprocessing of used civilian nuclear power reactor waste is currently done in Britain, France, Japan, and Russia and soon will likely be done in China and India. The United States stopped civilian waste reprocessing as part of their nuclear nonproliferation policy. This addressed concerns that plutonium from reprocessed material could be used in nuclear weapons. In 2006, a U.S. initiative, the Global Nuclear Energy Partnership, recommended reprocessing spent nuclear fuel in a manner that makes nuclear proliferation unfeasible. However, in the United States all spent nuclear fuel is currently treated as waste and is currently being stored at nuclear reactor sites so it could be easily reprocessed.

# Thorium (Th)

Thorium (Th) can be used to make nuclear fuel or a nuclear weapon. The dominant isotope of thorium is $^{232}$Th. $^{232}$Th can absorb a slow neutron (*n*) and produce $^{233}$U by the reactions:

$$n + {}^{232}\text{Th} \rightarrow {}^{233}\text{Th} \xrightarrow{\beta^-} {}^{233}\text{Pa} \xrightarrow{\beta^-} {}^{233}\text{U}. \qquad [5.12]$$

When $^{233}$U absorbs a neutron it usually undergoes fission in a reaction like

$$n + {}^{233}\text{U} \rightarrow 3n + \text{fission products} + \textit{energy}. \qquad [5.13]$$

The neutrons released by this fission reaction can then be absorbed by another $^{232}$Th present to start a chain reaction between reactions [5.12] and [5.13]. In order for the chain reaction to occur $^{233}U$, $^{235}U$, or $^{239}Pu$ must be present in a critical amount to supply the initial neutrons to start the reaction chain.

Thorium is about four times more abundant in the earth's crust than uranium and all but a trace amount of natural thorium is the needed $^{232}Th$ isotope. The world's thorium is obtained as a by-product of obtaining rare earths from the mineral, monazite (See Rare Earth Elements in the chapter on specialty metals) and is presently cheaper than uranium. A further advantage of thorium reactors is that the amount of transuranic waste is significantly decreased compared to reactors that use only uranium or plutonium. However, strong gamma radiation is released from the $^{233}U$ decay chain making its safe handling more difficult than with uranium or plutonium nuclear fuels.

# Nuclear Power Production

In 1954, the former USSR's Obninsk Nuclear Power Plant became the world's first nuclear power plant to generate electricity for a power grid. The first commercial nuclear plant in the United States became operational in 1956. After the initial development, nuclear power plant construction occurred relatively rapidly. Worldwide capacity increased from < 1 gigawatt (GW) in 1960 to a 100 GW capacity in the late 1970s. France and Japan took the lead because of the 1973 Arab oil embargo and their concerns about being dependent on foreign-derived fossil fuels. Nuclear power plant capacity reached 300 GW in the late 1980s, but construction of new nuclear power plants has since slowed. As of 2012, 435 nuclear power plants are in operation worldwide with an electric production capacity of about 375 GW. There are 65 plants with a capacity of 63 GW under construction with 27 of these in the People's Republic of China. **Figure 5.11** gives the nuclear electrical production in 2011 for the 10 largest producer countries. Note that because of the Fukushima nuclear accident in Japan its production in 2011 was significantly decreased.

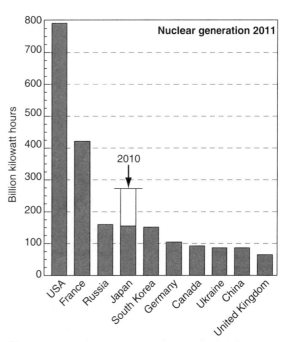

*Figure 5.11* Nuclear power electrical production for the top ten producing countries in 2011 (Modified from U.S. Energy Information Administration, *Energy in Brief*.)

Three factors contributed to the slowdown in nuclear power plant construction: lower fossil fuel prices over their initially projected costs; rising power plant construction and operating costs due in large part to increased regulations and litigation from antinuclear groups; and decreased enthusiasm for nuclear power by the general public. Much of the increased regulations and litigation was due to the 1979 accident at the Three Mile Island nuclear reactor and the 1986 accident at the Chernobyl nuclear reactor. Besides the reactor accidents the enthusiasm for nuclear power has been dampened by a general concern about exposure to radiation. There is a fear that plutonium produced in nuclear reactors could be used by terrorists to produce nuclear bombs, and apprehension about the ability to safely dispose of nuclear waste.

In 2011, 14% of the world's electricity was produced in nuclear power plants. See **Figure 5.12** for their locations and **Figure 5.13** for the breakdown by country. A large nuclear power plant generates about 1 million kilowatts, 1 GW, of electrical power. The United States produces the

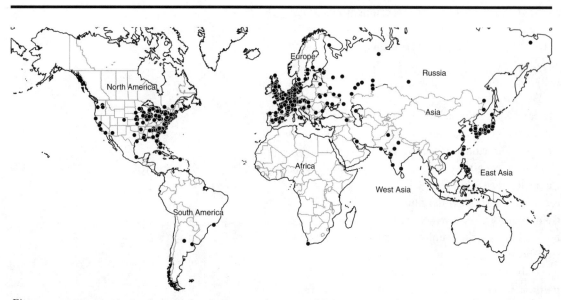

*Figure 5.12* Worldwide distribution of nuclear power plants to make electricity are shown as open circles. (Adapted from International Nuclear Safety Center at ANL, Aug 2005.)

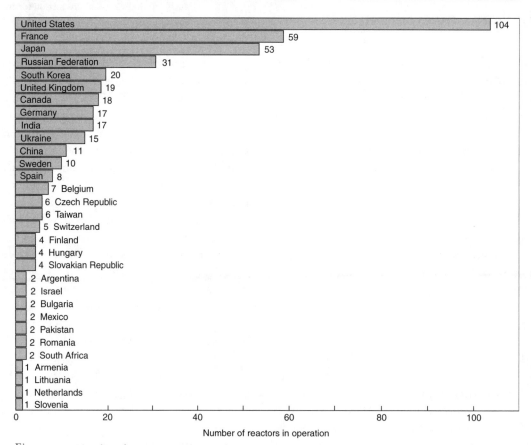

*Figure 5.13* Number of operating civilian nuclear power plants in the indicated country as of 2009.

most electricity from nuclear reactors, accounting for 20% of world production. However, France produces the greatest percentage of its electricity, 75%, from nuclear reactors as of 2011. In the European Union as a whole, nuclear energy provides 30% of the electricity.

The expansion of nuclear power in the future depends on political concerns influenced by reactor mishaps, and the politics are different in each country. Whether a particular country expands nuclear power generation or not depends on how the politics change. The politics are heavily involved in the availability of other forms of energy, reactor safety, and the perceived effects of global warming on the country.

The age of commercial reactors worldwide is given in **Figure 5.14**. Most of the older ones are in the U.S. There have been no new nuclear power plants built in the U.S. since 1978. Many in the U.S. were built in the 1960s and are approaching their permitted 40- to 50- year lifetimes. Presently new permits are being used to extend use of many of these older power plants by 20 more years. Permits for new reactors at current sites are also being submitted. Two double-loop pressurized water reactors were approved to be built at a preexisting nuclear plant, Vogtle, near Waynesboro, Georgia to come on line by 2020.

# Nuclear Radiation

Radioactive decay can be characterized in terms of its rate of decay termed activity, the energy imparted into an absorbing material termed absorbed dose, and the damage done to biological material termed dose equivalent. Depending on which radioactive isotope is decaying it releases particles of a particular energy that can cause damage to living tissue. The S.I. unit of decay activity is the *becquerel* (*Bq*), defined as one disintegration per second. Another older activity unit that is often used because large numbers of disintegration are typically considered is the *curie* (*Ci*). One *Ci* equals $3.7 \times 10^{10}$ *Bq*. The absorbed dose in S.I. units is the *gray* (*Gy*). One *Gy* is equal to 1 joule of energy per kg of material. An older unit that is encountered is the *radiation adsorbed dose* (*RAD*), where 1 *Gy* = 100 *RAD*.

The dose equivalent in S.I. units is the *sievert* (*Sv*). This unit takes into account the fact that a given absorbed dose of radiation of one type and/or

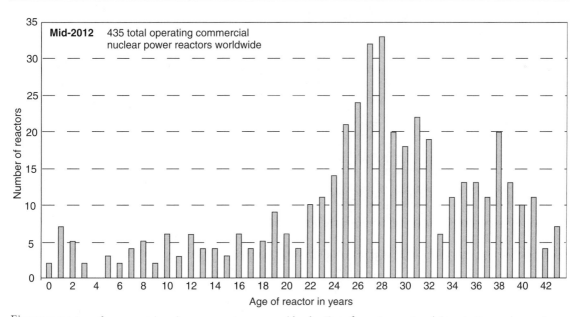

*Figure 5.14* Age of commercial nuclear reactors in years worldwide. (Data from: International Atomic Energy Agency.)

energy may give rise to a different biological effect than a radiation dose of another type and/or energy. It is equal to the absorbed dose times a quality factor ($Q$) so that $Sv = Gy \times Q$. $Q = 1$ for gamma rays, X-ray, and beta particles, but 10 for alpha particles because of their greater mass. Because of $Q$, an $Sv$ is somewhat imprecise. An older unit in the literature is the *roentgen equivalent man* (*REM*), where 1 $Sv$ is about 100 *REM*.

## Radiation Exposure and Risks

It is difficult to estimate risks for biological damage from radiation. Information from medical uses and from the survivors of the atomic bombs detonated during World War II in Japan have been used to set limits on exposure. The Biological Effects of Ionizing Radiation Committee of the U.S. Nuclear Regulatory Commission in 1990 placed the risk of cancer death at 0.08% per *REM* for doses received rapidly (acute) and is likely two to four times less, 0.04% per *REM*, for doses received over an extended period of time (chronic). There is, however, large uncertainty in the estimate. These estimates did not distinguish ages, sex, and types of cancer produced.

Natural doses received by the general population appear to be about 200 to 350 milli*REM* per year. No adverse health effects have been discerned from doses arising from these levels of radiation exposure. Radon gas released from the ground accounts for 2/3 of this and most of the variability. Cosmic rays from sunlight and radioactivity in the atmosphere account for the remainder of the exposure. In the United States this is highest in mountain states and lowest in Florida. This difference is due to an increase in cosmic rays at high elevations that are mediated by a thicker atmosphere at sea level. Also granitic rocks exposed in mountains contain more radioactive elements, in particular $^{40}K$ and $^{235}U$ whose decay releases radon gas.

Radiation contributed presently by nuclear power plants and earlier above ground nuclear weapon tests is estimated to be about 4 milli*REM* per year. About the same as that received from $^{14}C$

in fossil fuel burning of 3 milli*REM* per year. With this in mind the recommended exposure limit for the general public is set by the U.S. Nuclear Regulatory Commission at 500 milli*REM* per year. The maximum allowable by workers using nuclear material is 5 *REM* per year. It has been estimated that 500 *REM* at any one time is lethal.

### Radon Gas

As mentioned, radon (Rn) forms as one of the decay chain products when uranium or thorium naturally decays to a stable lead isotope. It is an odorless, colorless, radioactive gas that is released into the atmosphere. Rn undergoes alpha decay to polonium (Po). Radioactive Po can attach to airborne particles, such as household dust. If Rn or Po is inhaled, the radioactive particles decay in the lungs. This damages lung tissue leading to an increased risk of lung cancer. High levels of naturally occurring radon are most likely found where there are significant amounts of uranium in the soil or rocks. Depending on your geographic location, the radon levels of the air you breathe outside of your home may be as high as 0.75 p*Ci* per liter. p*Ci* is a picocurie = $10^{-12}$ curie. The United States average of outside radon levels is 0.4 p*Ci* per liter. The U.S. EPA has set an action level of 4 p*Ci* per liter. At or above this level of radon, it recommends that you take corrective measures to reduce your exposure.

# Nuclear Fusion

Fusing light elements together releases energy. Fusion does not produce fissionable material or radioactive waste. It is, therefore, argued that fusion is the ultimate energy source. The problem is how to harness a fusion reaction. The most promising fusion reaction for energy production by humankind is fusing two deuterium (D) atoms to form a 3-helium ($^3$He) atom given by the reaction

$$D + D \rightarrow {}^3He + n + energy \text{ (5.1} \times 10^{-13} \text{ joules)} \qquad [5.14]$$

or to form tritium ($^3$H) by the reaction

$$D + D \rightarrow {}^3H + n + energy \text{ (6.4} \times 10^{-13} \text{ joules).} \qquad [5.15]$$

Tritium plus deuterium then reacts to form 4-helium ($^4$He) that can be represented by

$$D + {}^3H \rightarrow {}^4He + n + energy \ (2.8 \times 10^{-12} \text{ joules}). \quad [5.16]$$

The net total reaction is

$$5D \rightarrow {}^4He + {}^3He + {}^1H + 2n + energy \ (4.0 \times 10^{-12} \text{ joules}). \quad [5.17]$$

There are very large technical problems in sustaining these reactions. The reactions require temperatures of 100 million °C and must be contained in a reaction space. At these temperatures atoms become a charged plasma. A plasma is a gas-like state where the atoms carry a positive charge due to being ionized. Because of the charges on the particles it has been suggested to contain the reaction in magnetic fields. It has also been suggested to use high-intensity lasers to generate energy to contain the reaction by inertial confinement. A heated outer layer is produced that explodes creating an inward inertia for containment. Scientists have yet to solve the containment problem.

An experimental nuclear fusion reactor using tokamak, a toroidal magnetic confinement field, to contain the plasma, the Joint European Torus (JET), was built in the United Kingdom in 1983 and upgraded in 2004. To date it has produced a peak of 16 MW of fusion power for less than a second. Another international experimental fusion reactor, International Thermonuclear Experimental Reactor (ITER), is being built near Marseille, France. The reactor is scheduled to be operational in 2018. It is anticipated to cost 16 billion euro to build and is slated to produce 500 MW of fusion power for as much as 1,000 seconds. Already in the planning stages is the DEMOnstration Power Plant (DEMO), a proposed nuclear fusion power plant constructed once designs solving the many problems of the current generation fusion reactors are engineered. It is slated to actually produce electricity.

## Summary

The binding energy that holds protons and neutron together is released if light elements are fused together. The same is true of heavy elements with a large number of protons plus neutrons if they are split apart in what is called nuclear fission. If a $^{235}$U nucleus undergoes fission it releases nearly three neutrons in the reactions as well as producing energy. If the density of $^{235}$U nuclei is high enough, above a critical mass, the neutrons released can fission other $^{235}$U nuclei in a self-sustaining reaction. $^{239}$Pu is produced from $^{238}$U during the fission reaction. As a fissionable isotope that releases neutrons $^{239}$Pu can also be used as fuel in a nuclear reactor. However, it is easily separated from used fuel and can be fabricated into a nuclear bomb. The nuclear fuel cycle outlines the process where uranium ore is extracted, processed, used in a reactor, and finally put in a permanent storage location.

Uranium is incorporated as $U^{4+}$ into accessory minerals during the final stages of fractional crystallization of rocks from silica-rich magma. This oxidation state of U is not soluble. Uranium is carried in solution when it is oxidized to $U^{6+}$ in oxygen bearing surface waters. When $U^{6+}$ reacts with reduced organic-rich material it precipitates $U^{4+}$ and a significant U ore deposit can develop. The most important uranium ore mineral is uraninite ($UO_2$), which when black and massive is often called pitchblende. The minimum grade of U ores is typically 1% $U_3O_8$. Pitchblende is also the main source for obtaining radium (Ra) and radon (Rn).

Uranium deposits are found in many geological settings, including unconformity-related deposits, sandstones, quartz-pebble conglomerates, and veins. After beneficiation a 70% to 90% $U_3O_8$ concentrate termed yellowcake is produced which contains 0.72% $^{235}U$. Yellowcake needs to be enriched to ~4% $^{235}U$ by isotope separation to be used as nuclear fuel.

The cost of uranium to fuel nuclear power plants is small compared to capital costs for construction that are greater than those for coal-fired plants and much greater than those for gas-fired plants. Worldwide reserves are sufficient for the near future unless the world decides to produce most of its electrical energy from enriched uranium. In this case, only a couple of decades of uranium are currently available. Decommissioning nuclear weapons and using their Pu as well as their highly enriched U would increase these times. However, if the world embraces nuclear power this means at sometime in the not too distant future either thorium fuel or breeder reactors would need to be employed.

Fission reactors are either thermal or fast neutron reactors, depending on the energy of the neutrons that sustain the fission. By slowing the neutrons in the thermal reactor they have a greater ability to fission $^{235}U$ and $^{239}Pu$. Moderators can be graphite, deuterium, or normal light water. Fuel rods are clad with a zirconium (Zr) alloy, zirconium's major use. Hafnium is added to reactor control rods because it has a good absorption cross-section for thermal neutrons.

Water-cooled thermal reactors can be separated into pressurized water reactors (PWR) and boiling water reactors (BWR). A PWR consists of two separate loops for pressurized water and steam production. In a BWR the same water is used as a moderator, the coolant, and the source for steam that drives the turbine blades.

Fast neutron reactors, also called breeder reactors, require a higher concentration of $^{235}U$ ($\geq 20\%$) due to the lower probability of fission. Because water is a neutron moderator, fast neutron reactors typically use liquid metal to transfer the heat. A blanket of nonfissionable $^{238}U$ is wrapped around the core and the fast neutrons convert some of the $^{238}U$ to fissionable $^{239}Pu$. Fast neutron reactors are more difficult to build and more expensive to operate. Therefore, thermal reactors are more common in electrical energy production.

Nuclear reactors are designed to operate safely even when systems fail. Ancillary control rods are inserted into the reactor to shut down a runaway reactor in what is termed a scram. The most serious nuclear power plant accident in the United States occurred at Three Mile Island in Pennsylvania but the worst nuclear power plant accidents in the world were the ones at Chernobyl near Kiev, Ukraine and in Fukushima, Japan.

The Waste Isolation Pilot Plant (WIPP) in the U.S. has been built to store defense-related transuranic radioactive waste. Yucca Mountain in Nevada is being considered for disposal of the high-level radioactive civilian power plant fuel rod waste. Used fuel assemblies are presently kept on reactor sites in pools of $H_2O$ waiting permanent storage or burial. Spent nuclear fuel can be reprocessed by dissolving it in nitric acid allowing unreacted uranium and plutonium to be separated and reused. This is currently done in Britain, France, Japan, and Russia and soon will likely be done in China and India.

Thorium ($^{232}Th$) can be used to make nuclear fuel or a nuclear weapon. Thorium is about four times more abundant in the earth's crust than uranium and all but a trace amount of natural thorium is $^{232}Th$. However, strong gamma radiation is released from the decay making its safe handling more difficult than with uranium or plutonium nuclear fuels.

In 2010, 14% of the world's electricity was produced in nuclear power plants. A large nuclear power plant generates about a 1 GW of electrical power. As of 2012, 435 nuclear power plants are in operation worldwide with an electric production capacity of about 375 GW. There have been no new nuclear power plants built in the U.S. since 1978, but they are being built elsewhere.

Radioactive decay can be characterized in terms of its rate of decay (activity) of becquerels (*Bq*), the energy imparted onto an absorbing material (absorbed dose) of grays (*Gy*), and the damage done to biological material (dose equivalent) of sieverts (*Sv*). It is difficult to estimate risks for biological damage from radiation.

Fusing light elements releases energy and does not produce radioactive waste. The problem is how to harness a fusion reaction. There are very large technical problems in sustaining these reactions. Scientists have yet to solve the containment problem.

## KEY TERMS

| | |
|---|---|
| actinides | moderator |
| alpha particle | nuclear fission |
| becquerel (Bq) | nuclear fusion |
| beta particle, $\beta^-$ | pitchblende |
| binding energy | pressurized water reactor (PWR) |
| boiling water reactor (BWR) | radiation absorbed dose (RAD) |
| breeder reactor | roentgen equivalent man (REM) |
| Chernobyl Nuclear Power Plant | scram |
| control rod | sievert (Sv) |
| critical mass | thermal reactor |
| curie (Ci) | Three Mile Island Nuclear Generating Station (TMI) |
| deuterium | |
| fast neutron reactor | transuranic waste |
| gray (Gy) | uraninite |
| half-life | Waste Isolation Pilot Plant (WIPP) |
| heavy water | yellowcake |
| high-level waste | Yucca Mountain, Nevada |
| isotope | |

## Problems

1.  By using the index at the U.S. Energy Information Administration website: www.eia.doe.gov/fueloverview.html, please find the appropriate page and answer the following questions:

    a.  How many and what are the types of reactors used to produce electrical power in the U.S.?

    b.  How many reactors are in the state where you reside? Where are they located?

    c.  Draw a graph of the percent of total electricity from nuclear power used in the U.S. from 1973 to 2010.

2.  Using the notation $^{\text{Mass number}}_{\text{Atomic number}}E$, where $E$ is an element symbol or neutron ($n$), fill in the blanks in the following reactions:

    a.  $^{235}_{92}U + ^{1}_{0}n \rightarrow ^{99}_{42}Mo + \underline{\quad\quad} + 2^{1}_{0}n$

    b.  $^{239}_{94}Pu + ^{1}_{0}n \rightarrow ^{90}_{38}Sr + \underline{\quad\quad} + 3^{1}_{0}n$

## References

Ferguson, C. D. 2011. *Nuclear energy: What everyone needs to know.* Oxford, UK: Oxford University Press.

Henderson, H. 2000. *Nuclear power: A reference handbook.* Santa Barbara, CA: ABC-CLIO.

Deutch, J. and Moniz, E. (co-chairs), Ansolabehere, S., Driscoll, M., Gray, P., Holdren, J., Joskow, P., Lester, R., and Todreas, N. 2003 and 2009 update. *The future of nuclear power.* Cambridge: Massachusetts Institute of Technology. Available online at web.mit.edu/nuclearpower/.

U.S. Department of Energy. 1979. *National Uranium Resource Evaluation*, Interim Report, June 1979. Open-file Report GJO-m(79), Washington, DC: U.S. Department of Energy.

# Part Two

## Metals

Metals account for 1/4 of the earth's mass and nearly 2/3 of all the known elements. Given in **Figure PT2.1** are some of the more common metals. A *metal* is a chemical element whose atoms readily lose electrons to form a positive ion, termed a cation, and that forms metallic bonds with other metals in a chemical compound. Metallic bonds occur from the interaction of delocalized sharing of outer electrons of the two bonded metal atoms. Because the outer electrons can move freely, metals are good conductors of heat and electricity. This also gives them the ability to deform without breaking making them malleable and ductile.

The availability of any metal depends on how it forms an ore deposit. An *ore deposit* of a metal is a rock that contains the metal and can be extracted at a profit. These occur in limited abundance and are localized within the earth's crust. The factors that determine whether or not a volume of rock is an ore deposit are those that determine how costly it will be to extract the metal. The quantity or profitability of ore in a particular property is rarely known with accuracy. The profit that can be produced depends on the *grade* (concentration) and quantity of extractable ore as well as the costs of mining, processing, and selling the mineral product.

Grade is determined by assaying an ore. Assaying uses physical and chemical techniques to determine the content of valuable constituents in the rock. Typically the grade is not constant in an ore deposit. The cutoff grade is the lowest grade of ore that is extracted for processing. Ore of various grades or physical properties can be blended before processing to produce a uniform ore for production. Head grade or mill head grade is the grade of the ore fed into the mill.

Not only is the grade important, but also the kind of mineral, its size, and the texture of the rock that the metal is found in are important. Other factors of significance include the extent of the deposit, its depth, and geographical location. Also to be considered are predictions of the market value of the metal and whether there are any profitable *byproduct*s that can be obtained during extraction of the primary metal.

This is a secondary or incidental product easily produced when mining and processing the primary metal. Byproducts, such as a small amount of gold in a copper deposit, often produce the revenues that make a mining operation profitable.

New mines typically bring prosperity to an area, but often have unwanted environmental impact. Mines have finite lifetimes but the environmental impact can last many generations. When a mine is in operation the safety of workers as well as dust, noise, and release of toxic substances from the mine are concerns that must be addressed. After the mine shuts down and the prosperity brought to the region disappears, changes to the landscape and release of acidic water due to mine wastes can still be a problem.

Ores can be classified based on the transport medium responsible for their development. These include:

1. **Magmatic**: Due to a crystallizing magma, which is molten rock. These include deposits of nickel, chromium, and vanadium found concentrated in low silica igneous intrusions (see Chapters 6 and 7).

2. **Surface aqueous fluid**: A deposit formed by interaction of rocks with rainwater, groundwater, or seawater or when ore minerals precipitate directly out of seawater. These include nickel-rich soils as well as banded iron formations precipitated from seawater (see Chapters 6 and 7).

3. **Hydrothermal solutions**: Deposits formed by fluid-rock interaction with an aqueous fluid at elevated temperatures derived either from the last stages of crystallization of a magma or from heated groundwater. These include porphyry and skarn ore deposits of copper, lead, zinc, and tungsten. Porphyry deposits are formed by aqueous fluids associated with intrusive igneous rocks of intermediate silica content produced at convergent plate boundaries. Skarn deposits are formed from reactions of hot fluids at the contact of crystalizing magma and carbonate rocks like limestone, $CaCO_3$. The reactions that occur cause a loss of $CO_2$ and enrichment of silica at the mineralized contact zone (see Chapter 8).

In some deposits a combination of more than one of these fluids is responsible for the final ore deposit. For instance, an ore deposit produced by a magma or hydrothermal fluid can be exposed at the earth's surface. Weathering of the ore with the aid of a surface aqueous fluid can then increase the grade of the ore body. This is termed *supergene enrichment*.

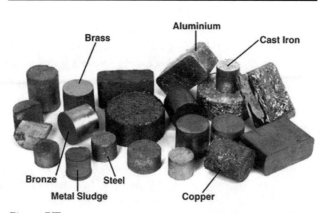

*Figure PT2.1* Some common metals and metal products used by humankind.

## Chapter 6

# Abundant Metals

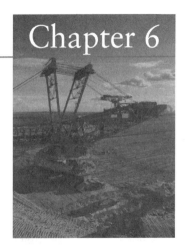

The most abundant elements in the earth's crust are given in **Table 6.1**. All the elements in the table are metals except oxygen, hydrogen, and silicon. While silicon (Si) is sometimes considered a metal it is best to consider it a metalloid. A *metalloid* is an element with some metal and some nonmetal properties. Silicon and the true metals in this table will be considered in this chapter except for sodium, potassium, and calcium. These elements will be considered when fertilizers and minerals

*Table 6.1* RELATIVE ABUNDANCE OF MOST ABUNDANT ELEMENTS IN EARTH'S CONTINENTAL CRUST AND COMMON ORE MINERALS OF MOST ABUNDANT METALS.

| ELEMENT | wt% | ORE MINERALS |
|---|---|---|
| Oxygen (O) | 45.20 | |
| Silicon (Si) | 27.20 | Quartz ($SiO_2$) |
| Aluminum (Al) | 8.00 | Gibbsite [$Al(OH)_3$], Diaspore [$AlO(OH)$], Boehmite [$AlO(OH)$], Kaolinite [$Al_2Si_2O_5(OH)_4$], Anorthite ($CaAl_2Si_2O_8$) |
| Iron (Fe) | 5.8 | Hematite ($Fe_2O_3$), Magnetite ($Fe_3O_4$), Goethite [$FeO(OH)$], Siderite ($FeCO_3$) |
| Calcium (Ca) | 5.06 | Calcite ($CaCO_3$), Dolomite [$CaMg(CO_3)_2$] |
| Magnesium (Mg) | 2.77 | Magnesite ($MgCO_3$), Dolomite [$CaMg(CO_3)_2$] |
| Sodium (Na) | 2.37 | Halite ($NaCl$), Trona [$Na_3H(CO_3)_2 \bullet 2H_2O$], Mirabilite ($Na_2SO_4 \bullet 10H_2O$) |
| Potassium (K) | 1.68 | Sylvite ($KCl$), Langbeinite ($2MgSO_4 \bullet K_2SO_4$), Kainite ($KCl \bullet MgSO_4 \bullet 3H_2O$), Carnallite ($KMgCl_3 \bullet 6H_2O$) |
| Titanium (Ti) | 0.86 | Rutile ($TiO_2$), Ilmenite ($FeTiO_3$) |
| Hydrogen (H) | 0.14 | |
| Manganese (Mn) | 0.10 | Pyrolusite ($MnO_2$), Psilomelane ($BaMn_9O_{18} \bullet 2H_2O$), Rhodochrosite ($MnCO_3$) |
| Total | 99.13 | |

*Data from:* WebElements periodic table on the web. Available online from www.webelements.com/.

for chemical use are discussed in another section. Also given in Table 6.1 are the most common ore minerals of Si and the metals listed. Note that these ore minerals are oxides, hydroxides, or carbonates of the metals and not sulfides and generally not silicates, except for Si from quartz and to a lesser extent aluminum (Al) from anorthite. This is because the abundant metals form stable metal-oxygen and metal-carbonate bonds leading to their prevalence. The abundant metals are not obtained from silicate minerals because the costs to separate them from the other constituents in the mineral are higher for silicates than with oxides, hydroxides, and carbonates, which dissolve in acid solutions relatively easily.

# Iron (Fe)

Iron (Fe) is the second most abundant metal in the earth being nearly as prevalent as Si. However, most of the Fe is located in the earth's core.

## Use

While iron is an important element required by the human body, Fe's major use by humankind is to produce steel. Steel is dominantly Fe combined with smaller amounts of other elements, including carbon, chromium, nickel, silicon, and/or molybdenum. Given in **Figure 6.1** are the uses of steel in the United States. Note the largest application of steel is for support beams and plates in the construction of buildings and in automobile frames and engines. However, steel in automobiles is being replaced by lighter aluminum to conserve weight and therefore less steel is being employed in automobiles.

Before the development of the process to create steel nearly pure iron was fabricated for nails, chains, bolts, and cutlery as well as to build railways and ironclad warships. As an inexpensive metal nearly pure iron still has some uses in the manufacture of fencing, outdoor furniture, and cookware. Fe is also used for making permanent magnets as well as electromagnets because metallic iron can be magnetized. Significant amounts of Fe metal are also used to make a number of different dyes, paints, and pigments.

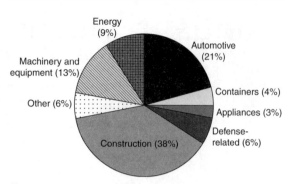

*Figure 6.1* Uses of steel in the United States.

# Oxidation State of Iron in the Earth

Fe on earth exists in three oxidation states. Fe metal of zero *valence*, that is charge, occurs in the earth's core. Of the accessible iron in the mantle and crust, iron is found in the $Fe^{2+}$ valence state, termed reduced iron, and in the $Fe^{3+}$ valence state of oxidized iron. Olivines contain a fayalite ($Fe_2SiO_4$) component where iron is in the reduced state. Iron is present in the $Fe^{3+}$ oxidized valence state in minerals like hematite ($Fe_2O_3$) and goethite [FeO(OH)]. These different valences of Fe are related through oxygen by what is termed an *oxidation-reduction reaction*:

$$Fe_2SiO_4 + 0.5O_2 \leftrightarrow SiO_2 + Fe_2O_3. \qquad [6.1]$$

As this reaction goes forward one speaks of the Fe on the left, $Fe^{2+}$, becoming oxidized, that is reacting with oxygen, to the state of $Fe^{+3}$ in the mineral on the right. Iron on the left is then reduced relative to the oxidized Fe on the right.

Minerals from the mantle typically contain reduced iron in the $Fe^{2+}$ state. Besides Fe in the iron-rich olivine component fayalite ($Fe_2SiO_4$) Fe in the iron-rich pyroxene component, ferrosilite ($FeSiO_3$) is also in the $Fe^{2+}$ state. In the mineral magnetite ($Fe_3O_4$), also crystalized in magma from the mantle, there are two $Fe^{3+}$ and one $Fe^{2+}$ atoms in its structure. Because magnetite contains $Fe^{3+}$ a reaction can be written between magnetite and the fayalite component in olivine to help define the oxidation state of the mantle:

$$2Fe_3O_4 + 3SiO_2 \leftrightarrow 3Fe_2SiO_4 + O_2. \qquad [6.2]$$

The equilibrium constant **K** of reaction [6.2] is

$$\mathbf{K} = f_{O_2} = P_{O_2} = 10^{-18} \text{ to } 10^{-14} \text{ bar} \qquad [6.3]$$

at mantle conditions depending on pressure and temperature. This indicates that the fugacity of oxygen $(f_{O_2})$ that is the partial pressure of $O_2$ $(P_{O_2})$ in the mantle is extremely small, $<10^{-14}$ bar.

Because air is 21% $O_2$ the partial pressure of $O_2$ in air is about 0.21 bar. With this high concentration of $O_2$ the stable oxidation state of iron in equilibrium with the earth's atmosphere is oxidized, $Fe^{3+}$. This is why reduced iron metal and reduced iron in minerals like Fe-rich olivine oxidize to a +3 state on weathering on the earth's surface as manifested by their turning rusty red.

Why is the earth's atmosphere so oxidizing when the earth's mantle is so reduced? The answer is photosynthesis. The reaction for photosynthesis can be written in a simple form as follows:

$$CO_2 + H_2O + \text{sunlight (\textit{energy})} \rightarrow CH_2O_{organic} + O_2. \quad [6.4]$$

Note that for each mole of organic matter ($CH_2O_{organic}$) produced, a mole of $O_2$ is also produced. Given the large amount of organic matter in coals and other sedimentary rocks, it is not surprising there is so much $O_2$ in the atmosphere. If photosynthesis stopped on the earth, with time $O_2$ in the atmosphere would react with $CH_2O_{organic}$, $Fe^{2+}$, and other reduced species stripping the atmosphere of $O_2$.

## *Resource Location*

Fe solubility in water is reasonably high as the $Fe^{2+}$ ion but very low as the $Fe^{3+}$ ion. This means to transport Fe in solution to be concentrated in an ore deposit a low oxygen environment is required. To reduce $Fe^{3+}$ at surface conditions and make it mobile in aqueous solutions an isolated microenvironment, where the amount of $O_2$ is lowered, is needed. This would be an environment rich in organic matter because the reaction

$$CH_2O + O_2 \rightarrow CO_2 + H_2O, \qquad [6.5]$$
$$\text{Organic}$$
$$\text{matter}$$

being the reverse of reaction [6.4] would consume any free $O_2$. Therefore, to understand the movement of Fe at the earth's surface, an understanding of oxidation-reduction reactions in solutions is needed.

### *Oxidation-Reduction Reactions*

Oxidation is an electron transfer reaction where electrons are lost from a species. As an example, consider the oxidation of $Fe^{2+}_{aq}$ in solution to $Fe^{3+}$ in the mineral goethite given by

$$4Fe^{2+}_{aq} + 6H_2O + O_2 \rightarrow 4FeO(OH) + 8H^+. \qquad [6.6]$$
$$\text{goethite}$$

Where do the electrons lost by iron in reaction [6.6] go? Note that oxygen has been reduced, that is, gained electrons. It has a valance of zero in $O_2$ but $-2$ in goethite. Anytime a species is oxidized in a reaction a species must also be reduced to conserve electrons. Oxidation-reduction reactions are often termed *redox* reactions.

The energy of these reactions can be quantified as a potential to move electrons. This is measured in volts relative to a *standard hydrogen electrode*. This voltage is specified as *Eh* as *E* is the standard symbol for voltage and *h* signifies it is the voltage relative to the standard hydrogen electrode. The standard hydrogen electrode has the ability to oxidize, that is, cause the loss of electrons, from $H_2$ gas to produce $H^+$. It has *Eh* = 0 at standard conditions on this scale. A more oxidized state would have a positive *Eh* and any more reduced state a negative *Eh*. As given in equation [6.6] redox reactions typically also involve $H^+$. Therefore, besides *Eh* the reactions and environments they outline are also *pH* dependent.

Given in **Figure 6.2** is an *Eh* vs *pH* diagram for iron species stability at earth surface conditions. Note that as expected the species $Fe^{3+}_{aq}$ is stable at a higher *Eh* than $Fe^{2+}_{aq}$ and magnetite stability is more reduced than hematite. The iron minerals siderite, pyrite, and magnetite are present at *Eh* < 0 as they contain $Fe^{2+}$. Because 1 bar of $O_2$ in equilibrium with $H_2O$ lies along the upper dashed line, environments in equilibrium with the earth's atmosphere with 0.21 bar of $O_2$ would plot only slightly lower. Therefore, to crystallize siderite, pyrite, or magnetite requires a local reducing environment isolated from the earth's oxidizing atmosphere.

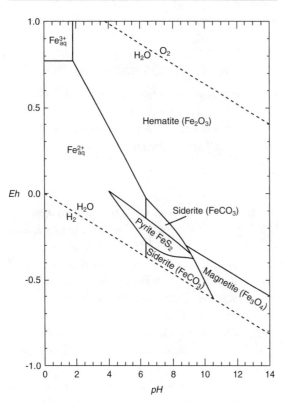

*Figure 6.2* Stability of iron oxides, sulfide, and carbonate in $H_2O$ at 25°C and 1 bar. Total dissolved carbonate is 1 molal and dissolved sulfur is $10^{-6}$ molal. Note the stability of pyrite with only a small concentration of dissolved sulfur in solution. (Adapted from *Solutions, Minerals, and Equilibria*, Robert M. Garrels and Charles L. Christ (1965) (2nd ed. Freeman Cooper Co, 1982 and revised ed. 1990) ISBN 0-86720-148-7 (1990 ed.))

## Iron Minerals and Ore Deposits

Because of its abundance in the earth's crust only about a nine-fold increase over the average concentration of iron in common rocks of about 5.8 wt% or ~ 50 wt% Fe is needed to form an ore deposit. There are a wide variety of geological processes that can produce this extent of Fe enrichment.

### Layered Mafic and Ultramafic Intrusions

Large amounts of Fe are transported from the mantle into the earth's crust in mafic and ultramafic magmas. Mafic magmas are those with $SiO_2$ between 45 and 52 wt%. Ultramafic magmas have $SiO_2$ < 45 wt%. These low silica magmas are the product of melting of rocks in the earth's mantle. A small percentage of melting of the mantle produces a mafic magma and a large percentage of melting an ultramafic magma because silica is preferentially released to the magma on melting. If the magmas do not reach the earth's surface the process of solidification on cooling in the crust produces differentiated layers of minerals in a magma chamber. During cooling these magmas saturate with and crystallize magnetite ($Fe_3O_4$). The magnetite crystals settle on the floor of the magma chamber forming a layer of almost pure magnetite. The largest of these bodies, with layers that extend for 370 km, is the mafic to ultramafic Bushveld Complex in the Republic of South Africa. The Bushveld Complex intruded into the earth's crust and crystallized in place producing layers of magnetite about 2 billion years ago. The magnetite in the Bushveld Complex is not currently mined for iron because other Fe deposits can be mined and the iron processed more economically. However, the magnetite in the Bushveld Complex is mined for the significant vanadium it contains. (see Chapter on ferro alloy metals).

### Skarns

Significant iron deposits occur in contact metamorphic zones between mafic magmas and carbonate host rock, termed *skarns*. These consist of both fluid-altered carbonates and igneous rocks. Generally, Fe skarns are too small for economic recovery of iron but there are Fe-skarn deposits large enough to mine in Daiquiri, Cuba; Shinyama, Japan; Kachar, Kazakhstan; and Sarbay, Russia. In the United States Fe-skarn deposits that have been mined in the past are located in Pennsylvania at Cornwall and Morgantown, in Utah at Iron Springs, and in Missouri at Pilot Knob. These skarns formed when acidic iron-rich fluids emitted by the cooling mafic intrusions react with limestones by a reaction like

$$CaCO_3 + H^+ \rightarrow Ca^{2+} + HCO_3^-. \qquad [6.7]$$

As limestone reacts with $H^+$ the *pH* of the solution increases and a skarn is formed. The solubility of iron decreases as *pH* increases. This can be deduced from Figure 6.2 where stabilities of Fe species in solution are replaced by mineral stabilities as *pH* increases. Magnetite and hematite are then precipitated by reaction [6.7] going forward.

Given in **Figure 6.3** is a cross-section through the rocks at the Cornwall mine in Pennsylvania where a contact metamorphic Fe skarn formed. Fluids from an intrusive diabase reacted with a limestone bed and magnetite was deposited, replacing the limestone. Diabase is fine-grain crystallized basaltic magma that has intruded and cooled in the earth's shallow crust. Its texture is intermediate between the glass found in rapidly cooling basalt rock and the large crystals found in more slowly crystallizing gabbro. Heat from the cooling diabase metamorphosed the limestone to marble. The dark area in the figure is the magnetite ore zone.

## Sedimentary Deposits

Iron concentrations can increase in the residual minerals produced in soil weathering profiles termed laterites. Laterites are hard, residual soils rich in Fe and Al and sometimes rich in other elements not mobile in oxidized aqueous solutions. Deposits are typically small and occur in the tropics with their somewhat elevated temperatures and abundant rainfall. Weathering destroys the primary minerals in the soils. Any $Fe^{2+}$ in these minerals is oxidized to $Fe^{3+}$ due to the oxygen in the air-charged water. Insoluble $Fe^{3+}$ and $Al^{3+}$ accumulate

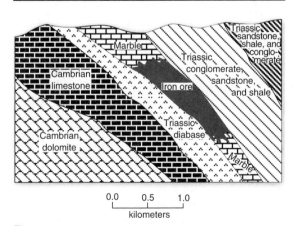

0.0    0.5    1.0
kilometers

*Figure 6.3* Cross-section through the Cornwall, Pennsylvania Fe-skarn ore deposit. (Adapted from Spencer, Arthur C., 1908, *Magnetite Deposits of the Cornwall Type in Pennsylvania*, USGS Bulletin 359, 102 p. Washington Government Printing Office.)

as other elements are leached out by the large flux of water through the soil. The primary iron ore minerals that form are the Fe-hydroxides, goethite [FeO(OH)], and limonite [FeO(OH)•$nH_2O$]. They make the soil hard and difficult to absorb moisture once they have formed. The weathering of iron gives the soils a brown, yellow, or red color. Laterites are an important resource base but not yet profitable to mine on a large scale because other deposits are more economic.

**Bog iron and ironstones.** Bog iron deposits are historically important Fe ore deposits but too small to be economically mined today. They form in bog environments produced by retreating glaciers and in coastal plain sediments during sea level fall. Bogs are wetlands that have no significant outflows of water. They develop acidic sphagnum mosses, which become peat. Organic matter in the bog reduces iron to $Fe^{2+}$. The reduced iron is carried in groundwater until it encounters an oxidizing environment at the edge of the bog. Iron is then precipitated in lenses as amorphous sedimentary Fe cements and limonite [FeO(OH)•$nH_2O$].

Another environment where iron ores are formed is in shallow, near-shore marine deposits where Fe-rich groundwater derived from organic-rich landward sediments precipitate and grow Fe minerals when they encounter oxygenated ocean water. In other situations Fe-rich laterites are thought to be weathered releasing Fe to solution in the active near-shore environment. The wave and tidal forces produce oolites in this environment from saturated Fe solutions. *Oolites* are spherical grains 0.25 to 2 mm in diameter composed of concentric growth layers. Some investigators believe the Fe-rich oolites are formed as a result of replacement of carbonate oolites occurring in Fe-rich solutions. The deposits contain a mixture of Fe oxide, silicate, and carbonates in oolitic form. These are termed *ironstones* of Minette-type in Europe and Clinton- or Wabana-type in eastern North America.

Ironstones are larger and more important than bog iron as they form in continuous sedimentary beds. However, they are of low grade compared to many banded iron formations and can't generally compete with them on the world market.

**Banded iron formations.**    Some of the oldest known rock formations on the earth include *banded iron formations* (*BIFs*) (**Figure 6.4**). BIFs occur exclusively in Precambrian (> 542 million years old) rocks and are common features in sediments for much of the earth's early history. Most of the world's iron ores are supply by BIFs. BIFs are less common in rocks that are younger than 1.8 billion years, although some are known from the late Precambrian as outlined in **Figure 6.5**. They consist of fine-grained magnetite ($Fe_3O_4$), hematite ($Fe_2O_3$), and/or siderite ($FeCO_3$) in finely laminated bands alternating with iron-poor shale, chert, or carbonate bands. Chert is a rock composed of fine-grained microcrystalline $SiO_2$. Each pair of bands is about 0.2 to 2.0 mm in thickness often incorporated within mesoscale bands of 10 to 50 mm in thickness. The total thickness of the BIF deposits are typically between 30 to 700 m and can extend over thousands of square kilometers. All have been metamorphosed to some extent. Some have been weathered which enriches the concentration of Fe.

The source of the iron and silica appears to be from hydrothermal solutions emanating from sea-floor igneous rocks. The ocean was oxygen-free, that is anaerobic, at this early time in the earth's history keeping Fe in solution as reduced $Fe^{2+}$. Vast amounts of iron were likely dissolved in the ocean. Eventually, photosynthetic cyanobacteria evolved in surface waters generating oxygen. The $Fe^{2+}$ in the earth's ocean surfaces was then oxidized to $Fe^{3+}$ and precipitated out as iron oxides. For magnetite the reaction would be

$$3Fe^{2+} + 1/2\,O_2 + 3H_2O \rightarrow Fe_3O_4 + 6H^+. \qquad [6.8]$$

The magnetite is thought to have settled on an anoxic silica-rich mud on the ocean floor forming

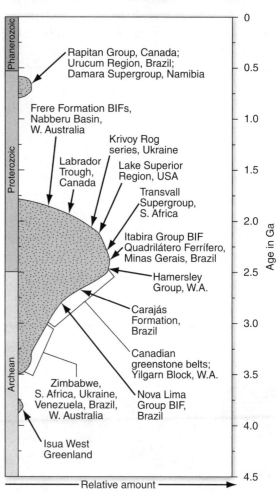

*Figure 6.5* Age distribution of relative amounts of banded iron formation in Ga = $10^9$ yr. (Adapted from Klein, Cornelis, 2005, Some Precambrian banded iron-formations (BIFs) from around the world: Their age, geologic setting, mineralogy, metamorphism, geochemistry, and origins, *American Mineralogist*; October 2005; v. 90; no. 10; p. 1473–1499; DOI: 10.2138/am.2005.1871 http://ammin. geoscienceworld.org/cgi/content/short/90/10/1473.)

*Figure 6.4* Banded iron formation showing dark bands of hematite varying in thickness from about 0.2 to 2.0 mm separated by light bands of chert.

alternating layers of magnetite and Fe-rich shale or chert. Because silica-secreting organisms had not yet evolved, the source of the silica was likely from precipitation of amorphous Si from a silica-saturated ocean. A cyclic process such as evaporate-supersaturation or coprecipitation with both iron oxide minerals and silica forming could produce the alternating Fe- and Si-rich layers. Therefore each set of layers is considered a "varve-like" feature in that it was due to cyclic variations in oxygen and silica supply. These could be seasonal oscillations or some other cyclic control until the ocean generally became permanently oxygenated about 1.8 billion years ago. Whether the iron precipitated chemically or by microbial processes is still a matter of debate. The total amount of oxygen locked up in BIFs by reactions like reaction [6.8] is estimated to be perhaps 20 times the amount of oxygen present in the modern atmosphere.

There are two types of BIF, *Algoma type* and *Lake Superior type*. The Algoma type occurs in Archean greenstone belts. *Greenstone* belts are primarily metamorphosed submarine basalt with minor interdispersed sedimentary rocks. These rocks are green because of the presence of chlorite and green amphiboles. They are thought to form in basins near oceanic spreading centers or volcanic arcs. Hydrothermal solutions rich in iron and silica mixed with cooler seawater saturating the water mass with Fe minerals and silica. Alternating bands of chert or quartz and magnetite or hematite were precipitated out of the seawater. Algoma-type BIFs can contain bands of a few millimeters in thickness representing the varve-like features outlined above. Because the deposition requires still water, they were likely deposited below the storm wave base of 200 m. They also have Al content < 1.8 wt%, which implies they contain only a minor detrital component and, therefore, formed significantly offshore. Deposits are commonly 30 to 100 m thick, and a few kilometers in length. Most are too small to be economic. Examples include the Vermillion Iron-Formation in Minnesota and the Temagami greenstone belt in Ontario, Canada.

The younger Lake Superior-type BIFs are commonly granular and oolitic. As shown in Figure 6.6 the name derives from the fact that these deposits are found around Lake Superior. Some of these magnetite-rich BIFs are mineable (~30 wt% Fe), but all high-grade ore bodies (60% to 68 wt% Fe) have been upgraded by secondary weathering processes and are hematite-rich. Lake Superior-type BIFs were likely deposited in a marine setting that was not as deep as the Algoma type as they do not contain microbands. Perhaps deep continental shelves or platforms as macrobands of a number of meters that can be correlated over 10,000 km² of area are present.

**Supergene enrichment of BIFs.** A banded iron formation typically has 20 to 30 wt% Fe. However, in numerous areas BIFs are exposed in equatorial tropical climates and are subjected to *supergene*, that is surface, enrichment of ore. Surface waters have percolated through the rocks and leached out Si relative to Fe. A hematite-goethite ore can develop with ~65 wt% Fe, a doubling on the iron content. Examples are the Cerro Bolivar deposits in Venezuela and the N4 ironstone deposits in the Carajás region of Brazil.

## Production and Reserves

Estimated iron ore production for 2011 and reserves in millions of metric tons are given in **Table 6.2**. Note that with total reserves of 87 billion metric tons and yearly total production of 2.8 billion metric tons a 29-year supply of Fe is available. However, the present world reserve base of Fe ore is 350 billion metric tons or over a 100-year supply. Remember the reserve base is the sum of measured reserves + indicated reserves + marginal reserves + a portion of subeconomic reserves that can be recovered at somewhat higher prices. Sufficient iron ore, therefore, exists to meet needs far into the future.

Given in **Figure 6.7** are annual total world and U.S. iron ore production and prices from 1995 to 2010. Note the world price and production was relatively constant to 2002 when China's consumption started to grow dramatically combined with slower demand growth by other countries producing a yearly 12% growth

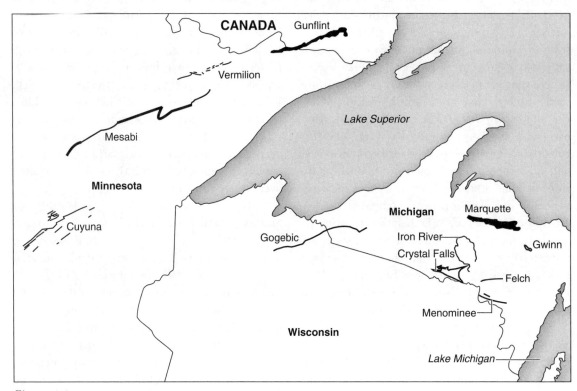

*Figure* 6.6 The major banded iron formations of the Lake Superior region. (Adapted from James, H.L., 1954, 'Sedimentary facies of iron formation', *Economic Geology*, 49, pp. 235–293.)

in production for the world as a whole. Much of this was due to increased industrialization in China and other countries of the industrializing world leading to greater demand for the uses of steel. Note that even with increased prices United States iron ore production has not increased as the increase demand has been supplied primarily by China.

### Steel Production

Shown in **Figure 6.8** is raw steel production worldwide and that for China and the United States since 1999. China is the largest producer accounting for nearly half the world's production while in 2011 the United States accounted for less than 5%. Note that raw steel production in China and the U.S. was nearly the same in 1999, however the Chinese production has increased along with the total world production while steel production in the U.S. has decreased

somewhat. This has occurred because production is driven by cost and older U.S. steel making furnaces and foundries have difficulty competing with newer more energy efficient facilities used elsewhere.

**Recycled steel.** Fe is one of the few magnetic metals present in most waste so steel is easy to separate in the waste stream. Steelmaking electric arc furnaces (EAFs) can handle 90% to 100% of recycled steel while basic oxygen steelmaking (BOS) furnaces can use up to 30% recycled material (see Iron and Steel Smelting section below). The use of recycled steel is currently limited by availability and its price and not processing capacity.

### Mining and Beneficiation

Open pit mines account for 85% of Fe ore production. The United States is at a disadvantage in this respect, as 5 to 6 metric tons of overburden must

*Table 6.2* IRON ORE PRODUCTION IN 2011 AND IRON RESERVES BY COUNTRY.*

| COUNTRY | PRODUCTION | RESERVES (FE) |
|---|---|---|
| China | 1,200 | 7,200 |
| Australia | 480 | 17,000 |
| Brazil | 390 | 16,000 |
| India | 240 | 4,500 |
| Russia | 100 | 14,000 |
| Ukraine | 80 | 2,100 |
| South Africa | 55 | 650 |
| United States | 54 | 2,100 |
| Canada | 37 | 2,300 |
| Iran | 30 | 1,400 |
| Sweden | 25 | 2,200 |
| Kazakhstan | 24 | 1,000 |
| Venezuela | 16 | 2,400 |
| Mexico | 14 | 400 |
| Mauritania | 11 | 700 |
| Others | 50 | 6,000 |
| Total worlds | 2,800 | 80,000 |

* In million metric tons.

*Data from:* U.S. Geological Survey, *Mineral Commodity Summaries*, January, 2012.

After ore is milled the iron oxide and carbonate powder is separated from the Fe-poor silicate mineral powder, termed gangue. *Gangue* is the unwanted material that surrounds a desired mineral in an ore deposit. Magnetite is typically separated with a magnetic separator. Low magnetic minerals such as hematite, limonite, or siderite are usually separated using a flotation and/or gravity separation process. The end product is a Fe oxide/carbonate concentrate. Powdered Fe oxide/carbonate concentrate is difficult to feed into a furnace so iron

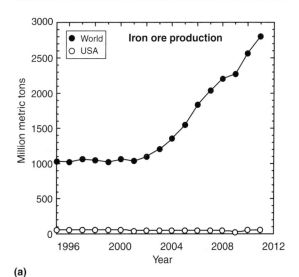

(a)

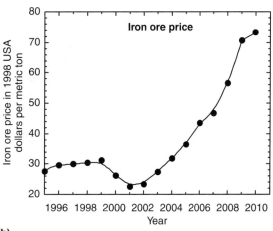

(b)

*Figure 6.7* (a) Worldwide and United States annual iron ore production from 1995 to 2011. (b) Iron ore price per metric ton in 1998 U.S. dollars. (Data from: U.S. Geological Survey, *Mineral Commodity Summaries*, January 2012.)

be removed for each metric ton of ore mined. In Brazil and Australia removal of 1.5 to 2 metric tons of overburden for each metric ton of ore is required.

Direct shipping ore is mined and used in blast furnaces with only simple preparation. It generally has > 60 % Fe and is made up almost entirely of magnetite, hematite, goethite, and siderite. Lower grade ores are upgraded by what is called beneficiation or mineral processing. Blocks of ore undergo comminution, that is, size reduction to less than a millimeter in diameter by first crushing the ore and then milling it. Milling is grinding the ore between two hard surfaces.

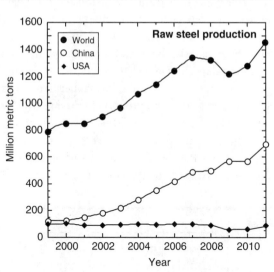

*Figure 6.8* Worldwide, China, and U.S. yearly raw steel production in millions of metric tons for the indicated year. (Data from: World Steel Association *World Statistical Yearbook 2012*.)

oxide together with limestone and bentonite clay are compressed into ~1 cm sized pellets which can then be easily dropped into a furnace.

### Iron and Steel Smelting

To separate pure metal from the iron oxide/carbonate it is smelted. *Smelting* is the process of producing metal from its oxide. In a typical Fe smelting operation 1.6 tons of iron oxide ore plus 0.2 tons of limestone in pellets together with 0.7 tons of *coke* are added to a *blast furnace* (**Figure 6.9**). Coke is a nearly pure solid porous carbon fuel produced from very soft bituminous coal. This fuel can generate 1600°C when burned because the high porosity allows easy access to $O_2$ and therefore fast reaction. The coke reacts with $O_2$ from oxygen-enriched air blown into the blast furnace to produce heat plus carbon monoxide gas (CO)

$$C + 1/2 \, O_2 \rightarrow CO + heat. \qquad [6.9]$$
$$\text{coke} \qquad\quad \text{air}$$

The CO is then reacted with molten iron oxide ore to produce iron metal

$$3CO + Fe_2O_3 \rightarrow 2Fe + 3CO_2. \qquad [6.10]$$
$$\text{iron ore} \quad \text{metal}$$

Limestone aids in formation of a *slag*, a calcium-alumina-silicate formed by reaction of limestone with $SiO_2 + Al_2O_3$ impurities in the molten iron. Being less dense than molten Fe, slag floats on the surface and forms a separate non-Fe containing silicate phase. This can be skimmed off the top of the molten iron. What is left is termed *pig iron*. Pig iron has a very high carbon content, typically 3.5 to 4.5 wt%, which makes it brittle and not of great practical use. Two-thirds of pig iron is used to make steel. Pig iron is typically poured directly

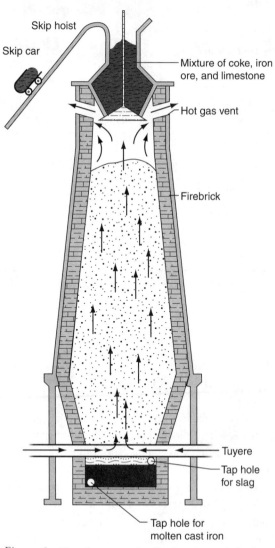

*Figure 6.9* Blast furnace used to produce iron metal, pig iron, from iron oxide ore.

out of the bottom of the blast furnace through a trough into a ladle car for transfer to the steelmaking furnace in a mostly liquid form. Therefore, it does not need to be melted again.

A steelmaking furnace takes the molten pig iron from the blast furnace as well as scrap steel and burns off excess carbon as well as sulfur and phosphorus impurities. The steel scrap helps lower the carbon content of the starting material. Alloying agents such as manganese, chromium, vanadium, and tungsten can be added depending on the product desired. *Carbon steel* is the most widely produced steel. It is iron with 0.20 to 2.14 wt% carbon as opposed to 3.5 to 4.5 wt% carbon in pig iron. In the U.S. 85% of all steel produced is carbon steel. Carbon acts as a hardening agent, preventing dislocations in the iron crystal lattice. Increasing the carbon content makes it less ductile and more difficult to weld.

The vast majority of steel manufactured in the world is produced using the *basic oxygen steelmaking (BOS) furnace* (**Figure 6.10**). A water-cooled oxygen lance is lowered into the molten iron present in a ladle. The lance is used to pump 99% pure oxygen at supersonic velocities into the melt. Carbon in the molten steel and iron is burned, that is oxidized, forming CO and $CO_2$ gas fumes and causing the temperature of the melt to rise to ~1700°C from the heat of reaction. Slag forms by oxidizing Si and other non-Fe metals in the melt that are removed from the BOS. The Fe stays in the reduced iron metal state. Sulfur impurities can be removed by adding magnesium. The magnesium reacts and forms magnesium sulfide (MgS) in an exothermic reaction. It then floats to the surface of the melt and is incorporated in the slag and can be raked off.

An *electric arc furnace (EAF)* is shown in **Figure 6.11**. It produces a melt by applying an AC electrical current to cast iron and steel scrap by means of graphite electrodes. Electrodes are lowered to a position just above the material and the current turned on. The electrodes then produce an electrical arc to the material and it begins to melt. The electrodes can then bore downward into the material producing a pool of liquid metal at the

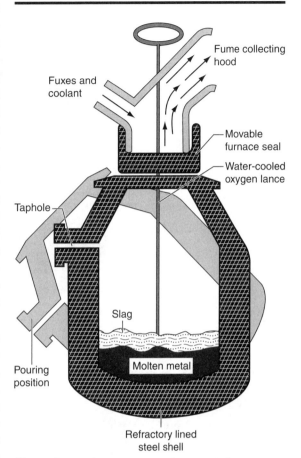

*Figure 6.10* Basic oxygen steelmaking (BOS) furnace.

bottom of the furnace. Eventually all the material in the furnace is melted. Oxygen is injected to oxidize the carbon in the steel to CO and $CO_2$ and these gases rise through the melt and help carry any nitrogen and hydrogen out of the melt. It requires about 500 kWh to produce 1 metric ton of steel.

Over the past two decades the use of electric arc furnaces has grown dramatically in the U.S. In 1975, EAFs accounted for only 20% of the steel production; by 1996 the number had risen to 39% and is presently 50%. There are two reasons for this trend: lower capital cost for an EAF than a blast furnace-BOS system and significantly less energy is required to produce the steel.

A number of different elements are alloyed with Fe in steel including Mn, P, S, Si, Ni, Cr, Mo, V, and Cu. However, carbon is the single most

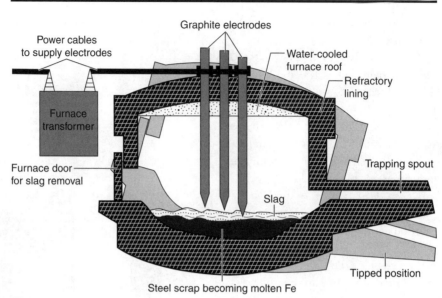

*Figure 6.11* Electric arc furnace (EAF).

common and important alloying element for steel. Given in **Table 6.3** are the types of carbon steels.

### Casting

The molten steel from the steelmaking furnace is typically continuously cast (CONCAST) into solid slabs, rounds, or bars. The primary shapes are then hot rolled, cold drawn, wire drawn, machined (e.g., drilling), and/or joined (e.g., welding). The steel can be heat-treated, that is tempered, or carbon added to the surface to harden it. Alternatively a protective coating such as Zn can be galvanized to the surface.

# Aluminum (Al)

Corrosion-resistant aluminum (Al) is a malleable and ductile metal that is easily machined and cast. It is about 33% of the weight of either copper or iron and readily conducts electricity and heat.

*Table 6.3* CARBON STEELS.

| TYPE | ULTRA LOW C | LOW C | MEDIUM C | HIGH C |
|---|---|---|---|---|
| **Carbon wt %** | < 0.01 | < 0.2 | 0.2–0.8 | 0.8–1.5 |
| **Strength** | High | Medium | High | Very high |
| **Toughness** | High | High | High | Low |
| **Fabrication ease** | Good | Excellent | Good | Fair |
| **Applications** | Sheets, autos, pipes | Structures, sheets | Rods, shafts, gears, rails | Dies, tools, springs |

Adapted from American Iron and Steel Institute.

## Use

About 85% of aluminum is used in products where cast bars and blocks of Al metal have been rolled, extruded, and/or drawn into the required final shape. Aluminum is often alloyed with Cu, Mg, Si, and/or Zn. Al-Mg alloys are particularly common as they have greater corrosion resistance and are lighter in weight than pure Al. They are widely used in aerospace applications, storage tanks, boat hulls, and superstructures.

**Table 6.4** gives the industries that use Al products along with the percentage of yearly production used in the United States. Examples of the products are also given. Note the large employment of Al in packaging and transportation. The packaging and container industry has expanded its production of Al beverage containers in recent years leading to an increased share of Al metal production. Today's motor vehicles use large amounts of Al metal in place of Fe despite its higher cost. Its lighter weight affords increased fuel mileage.

## Resource Location

Mining of aluminum ores is inexpensive because large Al-rich deposits, *bauxites*, occur in extensive layers near the surface and the ore can be obtained from strip mines. Strip mines remove the surface layer of soil and rocks and mine the ore in long strips using large-sized earth-moving equipment (see Figure 2.23b in the chapter on petroleum).

Bauxites are formed by intensive weathering of preexisting rocks and soil typically in tropical areas with abundant rainfall and elevated temperatures (**Figure 6.12**). These rocks include granite, basalt, and shale. The rain and elevated temperatures cause weathering of the Al-rich primary minerals such as feldspars, micas, and hornblende to Al-oxides and Al-hydroxides. Soluble ions like $Na^+$, $K^+$, $Mg^{2+}$, $Ca^{2+}$, and $H_4SiO_4$ are removed in the soil solution. The ore minerals gibbsite $[Al(OH)_3]$, boehmite $[AlO(OH)]$, and diaspore $[AlO(OH)]$ are formed. While most bauxites develop from weathering of silicate rocks, *terra rosa*, a type of Al-rich red clay soil forms when there is extensive limestone weathering leaving behind their small clay component. The clay then alters to Al-oxides and Al-hydroxides.

Greater than 90% of discovered bauxite deposits have formed in the last 60 million years with the largest deposits forming less than 25 million years ago. Because of glaciation and uplift with erosion of these surface deposits earlier formed bauxites have been destroyed over longer geological time periods.

Mines are typically in tropical areas that have neither abundant cheap electricity nor any markets for the product. Large amounts of electricity are required to process the Al-rich ore into Al metal. Therefore, ores are shipped to processing plants in countries with access to cheap electricity, typically near hydroelectric power plants.

*Table 6.4* Al USE BY INDUSTRY AND Al PRODUCTION % IN THE U.S. IN 2011.

| USAGE | PRODUCTION % | EXAMPLES |
|---|---|---|
| Transportation | 34 | 8.6% of an auto or > 200 kg |
| Packaging and containers | 27 | Beverage containers, foil |
| Construction products | 12 | Roofing material and panels |
| Electrical | 8 | High voltage cable |
| Machinery | 8 | Construction, agricultural |
| Consumer durables | 7 | Refrigerators, cooking utensils, furniture, boats |

*Source of production %:* U.S. Geological Survey, *Mineral Commodity Summaries*, January 2012.

*Figure 6.12* Bauxite, an ore of aluminum, showing characteristic pisolitic texture. The texture results from cyclic weathering processes that leach alkalis, alkaline earths, and silica from the parent rock, leaving aluminum oxides/hydroxide and some iron oxides/hydroxide residues.

## Production and Reserves

**Table 6.5** gives the yearly production and reserves for the largest Al ore producing countries for 2011. Note that over 80% of the total Al ore is mined in Australia, China, Brazil, India, and Guinea. All these countries have regions with tropical climates. With a worldwide ore production of 211 million metric tons per year there is over a 100-year supply in reserves. The U.S. does not mine significant Al oxides/hydroxides but imported 9 million metric tons to process to Al metal and Al metal alloys in 2011. Given in **Table 6.6** is the extent of production of Al metal from ores for the largest producing countries. Al metal production is dominated by China producing over 40% of the world's total of Al metal. There is also a large scrap market in Al fed by recycled cans, foil, wire, and engine parts. Over 41% of the aluminum produced in the U.S. in 2011 was from reprocessing scrap.

### Al Processing and Smelting

Bauxite ore is first crushed then mixed with sodium hydroxide (NaOH) before being heated

*Table 6.5* BAUXITE MINE PRODUCTION FOR 2011 AND RESERVES BY COUNTRY.*

| COUNTRY | PRODUCTION | RESERVES |
|---|---|---|
| Australia | 67.0 | 6,200 |
| China | 46.0 | 830 |
| Brazil | 31.0 | 3,600 |
| India | 20.0 | 900 |
| Guinea | 18.0 | 7,400 |
| Jamaica | 10.2 | 2,000 |
| Russia | 5.8 | 200 |
| Kazakhstan | 5.4 | 160 |
| Suriname | 5.0 | 580 |
| Venezuela | 4.5 | 320 |
| Greece | 2.1 | 600 |
| Guyana | 2.0 | 850 |
| Sierra Leone | 1.7 | 180 |
| Vietnam | 0.08 | 2,100 |
| United States | 0.0 | 20 |
| Other countries | 2.6 | 3,300 |
| **World totals** | **220** | **29,000** |

* In millions of metric tons.

*Data from:* U.S. Geological Survey, *Mineral Commodity Summaries*, January, 2012.

to produce liquid sodium aluminate and some undissolved mineral residues. The residues consist of Si-, Fe-, and Ti-containing minerals that sink to the bottom of the reaction vessel and are removed. Cooling the sodium aluminate produces crystals of hydrated $Al_2O_3$. These hydrated crystals then undergo calcination, a heating process that removes the water from the structure, to produce anhydrous $Al_2O_3$. This alumina is then smelted.

It takes considerable energy to produce Al metal, aluminum, from alumina because of the strength of the O-Al bond. Difficulty of separating

*Table 6.6* Al METAL PRODUCTION FOR 2011 BY COUNTRY.*

| COUNTRY | PRODUCTION |
|---------|------------|
| China | 18.00 |
| Russia | 4.00 |
| Canada | 2.97 |
| United States | 1.99 |
| Australia | 1.93 |
| United Arab Emirates | 1.80 |
| India | 1.70 |
| Brazil | 1.41 |
| Bahrain | 0.87 |
| Norway | 0.80 |
| South Africa | 0.80 |
| Iceland | 0.79 |
| Mozambique | 0.56 |
| Germany | 0.45 |
| Venezuela | 0.38 |
| Other countries | 5.23 |
| **World total** | **41.10** |

\* In millions of metric tons.

*Data from:* U.S. Geological Survey, *Mineral Commodity Summaries*, January, 2012.

Considerable heat energy is required for melting and electrical energy for electrolysis to produce Al metal. Electric currents as high as 200,000 amperes at 1,000 volts are used. The electrical energy required is over twice the theoretical efficiency due to loss of heat from the pots so that about 20 kWh per kg of Al is needed. With wholesale electric rates of $0.07 per kWh the electrical cost of processing would be $1.40 per kg. These energy costs are a significant part of the total cost of producing Al metal that in 2011 sold for somewhat over $2 per kg.

# Silicon (Si)

## Use

Humankind has a great demand for silicon (Si), both as an oxide, silica, and as pure silicon. The oxide $SiO_2$ is used largely in glasses, cements, and ceramics (see Chapter on building materials). The largest use of silicon is in silicon-steel alloys that contain up to 6.5 wt% Si. In steelmaking, molten steel is rid of oxygen by the addition of small amounts of silicon, typically as ferrosilicon, a solid

Al from oxygen in the oxide ores is overcome by the use of cryolite as a flux to dissolve the oxide minerals. Cryolite ($Na_3AlF_6$) melts below 900°C. Large bathtub-like pots containing liquid cryolite make up what is termed a pot line and are used to dissolve the Al oxides and produce $Al^{3+}$ ions in the melt. The $Al^{3+}$ is separated by electrolysis where electrodes under a voltage are inserted into the melt. $Al^{3+}$ moves towards the negatively charged carbon anode. Electrons from the anode combine with $Al^{3+}$ forming liquid Al metal. These droplets settle to the bottom of the reaction pot as shown in **Figure 6.13**.

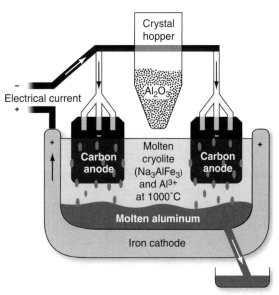

*Figure 6.13* Smelter for Al ore that uses electrical current to melt $Al_2O_3$ and turn dissolved $Al^{3+}$ into Al metal.

solution with Fe made up of 15% to 90 wt% Si. The ferrosilicon melts in the steel. The Si reacts with $O_2$ in the melt and forms the silicate $SiO_2$. The silicate being less dense than molten Fe floats to the surface and is removed. Si also has beneficial alloying properties in steel. Silicon lowers magnetic hysteresis, that is, remnant magnetism during the production of steel. Ordinary steel contains less than 0.03 wt% Si. Silicon steel, which typically has 1% to 20% Si is used for the cores of electrical transformers where induced currents of charge need to be avoided. Silicon is, also, at times alloyed with copper, brass, and bronze.

*Silanes* are silicon-hydrogen compounds with the general formula $Si_nH_{2n+2}$. They are used in silicone resins, caulks, lubricants, and water-repellent compounds. Silicones are a class of polymers with a chemical structure based on chains of alternate silicon and oxygen atoms, with organic groups attached to the silicon atoms. Typically heat-resistant, nonsticky, and rubberlike, silicones are commonly used in cookware, medical applications such as breast implants, sealants, adhesives, lubricants, and thermal insulation.

Silicon carbide (SiC), sometimes termed carborundum, is a widely used abrasive. Grains of silicon carbide can be bonded together by sintering to form very hard ceramics that are widely used in applications requiring high physical endurance, such as car brakes and ceramic plates in bulletproof vests. Sintering is a heating process below a substance's melting point that bonds solid particles in a powder together. Silicon carbide is also widely used to produce light-emitting diodes and detectors in high-temperature semiconductor electronics.

As a metalloid, Si conducts electricity only poorly, which allows electrical currents in Si to be controlled at the atomic level. Silicon has, therefore, found a use in microchips, both in ultra pure single crystals of silicon and by introducing small quantities of elements to change electrical properties. Silicon also forms the basis of solar cells to obtain electricity from sunlight and for photocells which act as sensors to detect light (see Chapter on alternative energy).

## Resource Location

There are virtually limitless supplies of silicon as it is the most abundant element in the earth's crust after oxygen. Si is obtained from quartz ($SiO_2$) with the rock quartzite generally the deposit of choice. Because of the abundance of quartzite, ore deposits are of high purity and generally mined near the plant that produces silicon.

## Production and Reserves

Given in **Table 6.7** is the world's production of silicon by country for 2011. Note the dominance of China in the world market. Large amounts of energy are required to break the Si to O bonds in $SiO_2$. Because of the low costs of mining high grade $SiO_2$ and the significant costs of electrical energy to produce silicon, countries with low electric power costs like Norway and Iceland, with their hydroelectric and geothermal power, respectively, will likely increase their share of the future world market.

Silicon (Si) is commercially produced by reacting high-purity $SiO_2$ with wood, charcoal, or coal in an electric arc furnace using carbon electrodes. At temperatures over 1,900°C the carbon reduces the silica to silicon by the reactions:

$$SiO_2 + C \rightarrow Si + CO_2 \qquad [6.11]$$

and

$$SiO_2 + 2C \rightarrow Si + 2CO. \qquad [6.12]$$

Liquid silicon collects in the bottom of the furnace, and is then drained and cooled. Using this method, silicon carbide (SiC) can also be produced by the reaction:

$$SiO_2 + 3C \rightarrow SiC + 2CO, \qquad [6.13]$$

if the concentration of C to Si is kept higher.

# Magnesium (Mg)

Magnesium is both abundant and widespread on the earth being the eighth most abundant element in the earth's crust.

*Table 6.7* SILICON PRODUCTION FOR 2011 BY COUNTRY.*

| COUNTRY | PRODUCTION |
| --- | --- |
| China | 5,400 |
| Russia | 670 |
| United States | 350 |
| Norway | 320 |
| Brazil | 230 |
| France | 140 |
| South Africa | 130 |
| Ukraine | 100 |
| Iceland | 75 |
| India | 68 |
| Venezuela | 62 |
| Canada | 52 |
| Other counties | 400 |
| **World total** | **8,000** |

* In thousands of metric tons.

*Data from:* U.S. Geological Survey, *Mineral Commodity Summaries*, January, 2012.

## Use

Magnesium is used principally as the oxide magnesia (MgO) and the silicate forsterite ($Mg_2SiO_4$), although there are a number of applications that use Mg metal. Both the Mg oxide and silicate are cut into refactories. A *refractory* is a solid with a high melting point used to line high temperature furnaces such as those used in the production of steel. In the United States 60% of the Mg consumed is used for refractories. Mg is also used in fertilizers and in animal feed as it is an essential nutrient. MgO and $MgCl_2$ are added to some cements in place of the portlandite $CaOH_2$ when making concrete as Mg-bearing cements have greater strength and can bond to cellulose.

Mg is combined with Al metal to produce a corrosion-resistant lightweight alloy used in beverage cans, aircraft, vehicles, and machinery, accounting for 41% of Mg metal use. Structural uses of Mg metal in castings and wrought products accounted for 32% of consumption. Mg metal used to desulfur iron and steel accounted for 13% of U.S. consumption. In desulfurization Mg metal is injected into molten Fe or steel and the following reaction occurs:

$$Mg + S \rightarrow MgS. \qquad [6.14]$$

The MgS particles are removed from the melt producing ultra-low concentrations of sulfur in iron or steel (< 0.0002 wt%).

## Production and Reserves

MgO is obtained from magnesite ($MgCO_3$) and dolomite [$CaMg(CO_3)_2$] by calcination. In the process heat is applied to the ore bringing about the thermal decomposition of magnesite and dolomite to its oxides by driving off $CO_2$ by a reaction like

$$MgCO_3 \rightarrow MgO + CO_2\uparrow. \qquad [6.15]$$

Given in **Table 6.8** are the worldwide production and reserves of magnesite by country. The MgO can be smelted to produce Mg metal as outlined above for iron in the section on Iron and Steel Smelting.

Mg metal is also obtained from normal seawater which contains 1,290 ppm Mg as well as from Mg-rich natural brines. The water is evaporated to precipitate magnesium hydroxide [$Mg(OH)_2$]. The magnesium hydroxide is reacted with hydrochloric acid (HCl) to produce magnesium chloride ($MgCl_2$):

$$Mg(OH)_2 + 2HCl \rightarrow MgCl_2 + 2H_2O. \qquad [6.16]$$

The magnesium chloride is decomposed to chlorine gas ($Cl_2$) and Mg metal in an electrolytic cell. In the cell $MgCl_2$ is dissolved into molten

*Table 6.8* Magnesite production for 2011, reserves, and reserve base by country.*

| COUNTRY | PRODUCTION | RESERVES | RESERVE BASE |
|---|---|---|---|
| China | 4,100 | 550,000 | 860,000 |
| Russia | 350 | 650,000 | 730,000 |
| Turkey | 300 | 49,000 | 160,000 |
| Austria | 200 | 15,000 | 20,000 |
| Slovakia | 190 | 35,000 | 320,000 |
| Spain | 130 | 10,000 | 30,000 |
| Brazil | 115 | 160,000 | 100,000 |
| India | 100 | 6,000 | 55,000 |
| Australia | 90 | 95,000 | 120,000 |
| Greece | 90 | 30,000 | 30,000 |
| Korea, North | 45 | 450,000 | 750,000 |
| United States | w | 10,000 | 15,000 |
| Other countries | 150 | 390,000 | 440,000 |
| **World totals** | **5,900** | **2,500,000** | **3,600,000** |

\* In thousands of metric tons.

w = Withheld to avoid disclosing company proprietary data.

*Data from:* U.S. Geological Survey, *Mineral Commodity Summaries*, January, 2012.

carnallite to produce $Mg^{2+}$ and $Cl^-$. Carnallite is a hydrated K-Mg chloride salt that forms in evaporates. Electricity is applied between two electrodes in the molten solution. The electrons supplied convert $Mg^{2+}$ to Mg metal while the $Cl^-$ gives up electrons to produce $Cl_2$ gas. This accounted for about 54% of U.S. Mg metal production in 2011. **Table 6.9** gives the Mg metal production by country.

Given humankind's usage, the resource base of Mg is nearly limitless. Resources from which magnesium compounds can be recovered are large to virtually unlimited and globally widespread. Identified world resources of magnesite total 12 billion tons, and for brucite [$Mg(OH)_2$], several million tons. Resources of dolomite, forsterite, magnesium-bearing evaporite minerals, as well as magnesia-bearing brines are estimated to constitute a resource of Mg in billions of tons.

*Table 6.9* Magnesium metal production for 2011 by country.*

| COUNTRY | PRODUCTION |
|---|---|
| China | 670 |
| Russia | 37 |
| Israel | 28 |
| Kazakhstan | 20 |
| Brazil | 16 |
| Ukraine | 2 |
| Serbia | 2 |
| United States | w |
| **World total** | **780** |

\* In thousands of metric tons.

w = Withheld to avoid disclosing company proprietary data.

*Data from:* U.S. Geological Survey, *Mineral Commodity Summaries*, January, 2012.

# Titanium (Ti)

Titanium (Ti) makes up 0.56 wt% of the earth's crust and is found mainly in the minerals rutile ($TiO_2$), ilmenite ($FeTiO_3$), and Ti-rich magnetite ($Fe_3O_4$ – $Fe_2TiO_4$).

## Use

Titanium's primary use, accounting for 95% of consumption, is as titanium oxide ($TiO_2$). Titanium oxide is a white, highly opaque pigment that does not fade. It is the principle white pigment used in paints, paper, plastics, rubbers, and toothpaste. Because of its corrosion resistance and because Ti has the highest strength-to-weight ratio of any metal it is also used in steel alloys. These Ti steel alloys are used in jet engines and airframes for aircraft. In 2011, aerospace applications used an estimated 75% of the titanium metal produced. Because of its strength Ti steel can be fabricated with thin walls that increases its ability to transfer heat. It is used in tubing in heat exchangers used to transfer heat between two fluids such as in electricity-generating plants.

## Production and Reserves

Ti minerals occur most commonly in mafic igneous rocks ranging from gabbro to anorthosite. Anorthosite is a type of intrusive igneous rock composed predominantly of calcium-rich plagioclase feldspar crystals that are large enough to be seen by the naked eye but can contain up to 10% ilmenite and Ti-rich magnetite. Anorthosite bodies are thought to form from fractional crystallization of high-Al basalt magma producing an accumulation of plagioclase crystals in a magma chamber. Rutile is found as a common accessory mineral in igneous rocks and high-grade metamorphic rocks.

Ilmenite supplies about 92% of the world's demand for titanium minerals with rutile making up most of the rest. Ilmenite and rutile are mined mainly in placer deposits where they concentrate because of their high density and low reactivity relative to other minerals (**Table 6.10**). These are typically sands, called black sands because of their color. Black sands also contain many other low reactivity metamorphic minerals such as kyanite, sillimanite, garnet, and tourmaline. Large Ti placer sand deposits occur in Quaternary shoreline strata located along the coastal areas of India, Australia, Mozambique, Canada, South Africa, Kenya, Sierra Leone, Madagascar, and Vietnam (**Figure 6.14**). The Quaternary is the geological time period from 2.6 million years ago to the present.

Ti placer deposits also occur along the shores of some large lakes such as in Malawi in southeast Africa. Deposits are often produced from Ti sand-sized minerals weathered from high-grade metamorphic terrains with the sand accumulating in shoreline deposits. Magnets are used to separate the Ti-rich minerals from the other sand particles.

Ilmenite is also mined from some anorthosite igneous intrusions in open pit mines. The largest

*Table 6.10* TITANIUM-CONTAINING MINERALS AND THEIR DENSITIES.

| MINERAL NAME | COMPOSITION | DENSITY (g/cm³) |
|---|---|---|
| Rutile | $TiO_2$ | 4.23–5.5 |
| Ilmenite | $FeTiO_3$ | 4.70–4.78 |
| Zircon | $ZrSiO_4$ (with Ti in place of Zr and Si) | 4.6–4.7 |
| Magnetite | $Fe_3O_4$–$Fe_2TiO_4$ | 4.78–5.2 |
| Ulvöspinel | $Fe_2TiO_4$ | 4.78–4.91 |

*Figure 6.14* These Quaternary beach sands are located near Morgan's Bay, Wild Coast on the Eastern Cape Province of South Africa. The shoreline sands contain high concentrations of titanium and are mined in some localities.

open pit ilmenite mine is the Tellnes mine located in Norway. An ilmenite-rich crystal mush was injected into surrounding Precambrian anorthosite magma. Given in **Table 6.11** are the 2011 production and reserves of ilmenite and rutile for the major producer countries. Note that together Australia and South Africa supply over 70% of the world's supply of rutile and over 1/3 of the ilmenite.

Ti metal is obtained by the Kroll process where ground rutile and ilmenite ore is heated to 1000°C in an oxygen-free atmosphere by burning *coking coal* and then reacting it with chlorine gas ($Cl_2$) to produce $TiCl_4$ gas. The $TiCl_4$ gas is then reacted with liquid Mg at 800° to 850°C to obtain Ti metal by the reaction:

$$2Mg + TiCl_4 \rightarrow 2MgCl_2 + Ti. \qquad [6.17]$$

*Table 6.11* ILMENITE AND RUTILE PRODUCTION FOR 2011 AND RESERVES BY COUNTRY.*

| COUNTRY | ILMENITE PRODUCTION | ILMENITE RESERVES | RUTILE PRODUCTION | RUTILE RESERVES |
|---|---|---|---|---|
| South Africa | 1,030 | 63,000 | 131 | 8,300 |
| Australia | 900 | 100,000 | 400 | 18,000 |
| Canada | 700 | 31,000 | 0 | ? |
| India | 550 | 85,000 | 24 | 7,400 |
| Mozambique | 510 | 16,000 | 6 | 480 |
| China | 500 | 200,000 | 0 | ? |
| Vietnam | 490 | 1,600 | 0 | ? |
| Norway | 300 | 37,000 | 0 | ? |
| Ukraine | 300 | 5,900 | 57 | 2,500 |
| Madagascar | 280 | 40,000 | 0 | ? |
| United States | 200 | 2,000 | a | 400 |
| Brazil | 45 | 43,000 | 3 | 1,200 |
| Sierra Leone | 0 | 0 | 60 | 3,800 |
| Other countries | 37 | 26,000 | 0 | 400 |
| World totals | 6,000 | 650,000 | 700 | 42,000 |

* In thousands of metric tons.

a = Included with ilmenite to avoid disclosing company proprietary data.

*Data from:* U.S. Geological Survey, *Mineral Commodity Summaries*, January, 2012.

# Manganese (Mn)

Manganese (Mn) makes up about 0.1 wt% of the earth's crust. It occurs in multiple common valence states: $Mn^{2+}$, $Mn^{3+}$, and $Mn^{4+}$. Similar to Fe, Mn is soluble in water in the $Mn^{2+}$ state, but quite insoluble when it is oxidized to the $Mn^{4+}$ state.

## Use

Manganese metal is used primarily in the production of steel. It improves the hardness, stiffness, and strength of iron and is also used in what are called superalloys. Superalloys are high-performance metal alloys that can withstand high temperatures, high stresses, and highly oxidizing atmospheres. They often have significant concentrations of Ni, Cr, Co, Mo, W, and/or Co. They typically do not contain Fe if oxidizing gases or liquids will be present.

Manganese dioxide ($MnO_2$) is used to make the heads of matches, fireworks, and dry-cell batteries. Manganese chloride ($MnCl_2$) is employed as a catalyst when organic compounds are chlorinated and as a nutrient in animal feed. Manganese sulfate ($MnSO_4$) is used as a fertilizer and as a livestock nutritional supplement. In glazes and varnishes $Mn^{2+}$ produces a pink color. As a strong oxidizing agent potassium permanganate ($KMnO_4$) is used for water purification and in waste-treatment plants.

## Resource Location

Manganese is obtained mainly from sedimentary deposits, generally laterites with > 35 wt% Mn concentrated by weathering. The high Mn content is a result of the presence of Mn-rich original rock. A large resource base, but no ores, occurs in the United States, Japan, and western Europe. Pyrolusite ($MnO_2$) is the most important ore mineral but other more exotic Mn minerals can also occur. High Mn concentrations are also found in *karst deposits* in the minerals rhodochrosite ($MnCO_3$) and rhodonite ($MnSiO_3$). Karst deposits develop in highly weathered carbonate rock terrains marked by caves and underground water drainage.

The largest Mn mine in the world occurs on the island of Groote Eylandt on the north coast of Australia. This open pit mine produces more than 3.8 million metric tons of Mn annually, about a quarter of the world's total. Pisolitic (> 2 mm) and smaller oolitic (0.25 to 2 mm) concretionary grains of manganese ores have > 40 wt% Mn. The ore developed from lateritic supergene enrichment of an original shallow-water marine clay.

Other important deposits occur north of the Black Sea in Oligocene (34 to 23 million years old) rocks in the Chiat'ura deposit of the country of Georgia and the Nikopol deposit in the Ukraine, in the Jurassic rocks of the Molango District of Mexico, and in the Precambrian rocks of the Kalahari Field of South Africa. All these deposits, like the Groote Eylandt deposit, are generally thought to have formed in non-volcanogenic shallow marine near-shore environments at locations starved of clastic sediments. They formed on the edges of anoxic basins during high sea-level stands. The Mn is likely supplied by the weathering of Mn-rich rocks surrounding the basin.

## Production and Reserves

When the world reserves of Mn of 630 million metric tons (**Table 6.12**) and the world reserve basis of 5,200 million metric tons are compared to total annual production of about 14 million metric tons, ample reserves are available. A potential source of manganese in ferromanganese nodules, or often just called *manganese nodules*, occurs on the deep sea floor. These nodules are complex mixtures of Fe and Mn oxides and hydroxides that grow about 1 mm per 10,000 years or slower in a series of concentric layers. They are often partly or completely buried in surface sediments (**Figure 6.15**). Manganese nodules contain 27% to 30 wt% manganese as well as ~0.2 wt% cobalt, ~1.5 wt% nickel, 1.0 to 1.4 wt% copper, and ~0.1 wt% zinc. There is an estimated 500 billion metric tons of manganese nodules on the sea floor. To date, a cost-effective way to harvest and process the metals in these nodules has not been developed.

*Table 6.12* MN PRODUCTION FOR 2011, RESERVES, AND RESERVE BASE BY COUNTRY.*

| COUNTRY | PRODUCTION | RESERVES | RESERVE BASE |
|---|---|---|---|
| South Africa | 3,400 | 150,000 | 4,000,000 |
| China | 2,800 | 44,000 | 100,000 |
| Australia | 2,400 | 93,000 | 160,000 |
| Gabon | 1,500 | 21,000 | 90,000 |
| India | 1,100 | 56,000 | 150,000 |
| Brazil | 1,000 | 110,000 | 150,000 |
| Ukraine | 340 | 140,000 | 520,000 |
| Mexico | 170 | 4,000 | 8,000 |
| Other countries | 1,400 | Small | Small |
| World totals | 14,000 | 630,000 | 5,200,000 |

\* In thousands of metric tons.

*Data from:* U.S. Geological Survey, *Mineral Commodity Summaries*, January, 2012.

*Figure 6.15* (a) Close-up of a 4 cm diameter manganese nodule, recovered from the ocean floor, which grows by adding rings of material to its outer layer. (b) Manganese nodules lying on the ocean floor partly buried in sediment.

## SUMMARY

Ore minerals of the abundant metals Fe, Al, Si, Ca, Mg, Na, K, Ti, and Mn are oxides, hydroxides, or carbonates of the metals. Unlike other metals they are not sulfides or sulfates and generally not silicates, except for Si from quartz ($SiO_2$) and to a lesser extent Al from the mineral anorthite ($CaAl_2Si_2O_8$).

Natural Fe generally exists in two oxidation states: reduced $Fe^{2+}$ and oxidized $Fe^{3+}$. Iron from the mantle is mostly $Fe^{2+}$ whereas minerals formed at the earth's surface have $Fe^{3+}$. If electrons are lost from a metal it is oxidized and if electrons are added it is in a reduced oxygen state. A reaction can be written that oxidizes reduced mantle Fe when it is introduced into the earth surface environment:

$$Fe_2SiO_4 + 0.5O_2 = SiO_2 + Fe_2O_3. \hspace{2cm} [6.18]$$

Mantle $Fe^{2+}$         Oxidized $Fe^{3+}$

    olivine             hematite

This extent of electron transfer is quantified by determining a partial pressure of oxygen for the environment or its *Eh*. This is a voltage change (E) relative to the standard hydrogen (h) electrode. In the $Fe^{2+}$ state Fe has reasonable solubility in water. Therefore, to transport Fe to make an ore deposit a reduced-oxygen, or low *Eh*, environment must be present.

Types of Fe deposits include magnetite layers produced during differential crystallization on cooling in large igneous intrusions of basalt and ultramafic rocks from the mantle. Metamorphic zones between mafic magmas and carbonate host rocks, termed skarns, can concentrate Fe. Laterites and ironstones can also have large amounts of Fe. However, banded iron formations (BIFs) and weathered BIFs are the major suppliers of Fe. They formed in the Precambrian when the oxidation state of the oceans was very low leading to high concentrations of $Fe^{2+}$ in seawater. Reserves are sufficient to meet needs far into the future.

Iron ore is smelted in a blast furnace to separate Fe from O. Coke, formed from bituminous coal, is burned together with Fe ore and limestone to produce pig iron. A basic oxygen steelmaking or electric arc furnace takes the molten pig iron as well as scrap steel to produce carbon steel.

Aluminum with 1/3 the weight of Cu or Fe is used in vehicles where lower weights are at a premium. Al is also used in beverage cans and foils because it does not rust. Electrical wires are often made from Al because it is lightweight and less expensive than Cu. Ore deposits are formed by intensive weathering of Al-rich rocks to produce Al-rich soils, termed bauxites. Aluminum is extensively recycled, producing a secondary market of Al. Production of Al metal occurs where the required electricity needed for production is plentiful.

Silicon's largest use is to make silicon-steel alloys. Silanes, silicon-hydrogen caulks, and resins are also in significant demand. Silicon carbide is a widely used abrasive. Because Si is a semiconductor of electricity, wafers of Si are fabricated to make microchips and solar cells. Resources are virtually limitless with China controlling much of the world production. Pure quartzite is typically the mineral deposit of choice.

Magnesium oxides are used as refactories to build high temperature furnaces. $Mg^{2+}$ is incorporated into fertilizers and animal feed. Mg metal is alloyed with Al for the packaging and transportation industries and used to desulfur steel. It is obtained from Mg carbonates by calcination and from seawater and seawater brines.

Titanium is employed primarily as $TiO_2$. It is a white highly opaque pigment used as the white color in paints, paper, plastics, rubbers, and toothpaste. Ti metal in steel increases its corrosion resistance and has the highest strength-to-weight ratio of any metal. Ti occurs in mafic igneous rocks;

ilmenite supplies about 92% of the world's demand. Ilmenite and other Ti containing minerals is mined from black sand placer deposits developed along ancient coastlines.

Manganese is incorporated into steel and superalloys. Similar to Fe, $Mn^{2+}$ is soluble in water but quite insoluble when it is oxidized to the $Mn^{4+}$ state. It is mined from laterite soils with > 35 wt% Mn. Ample reserves are available. A potential source of Mn occurs in ferromanganese nodules on the ocean floor. These are complex mixtures of Fe and Mn oxides and hydroxides that have grown slowly over time in a series of concentric layers.

## KEY TERMS

Algoma-type BIF
banded iron formation (BIF)
basic oxygen steelmaking (BOS)
  furnace
bauxite
blast furnace
byproduct
carbon steel
coke or coking coal
*Eh*
electric arc furnace (EAF)
gangue
grade
greenstone
ironstones
karst deposits
Lake Superior-type BIF

manganese nodules
metal
metalloid
oolites
ore deposit
oxidation-reduction reaction
pig iron
redox
refactory
silane
skarn
slag
smelting
standard hydrogen electrode
supergene enrichment
terra rosa
valence

## PROBLEMS

1.  Of the abundant metals, which can undergo oxidation/reduction reactions? What are two possible oxidation states of each of these?
2.  What is ferrous as opposed to ferric iron? Which is more soluble in water?
3.  Which of the following are redox reactions?
    a.  $Cu_{(solid)} + 2Ag^+_{(aq)} \rightarrow Cu^{2+}_{(aq)} + 2Ag_{(solid)}$
    b.  $SiO_2 + CaCO_3 \rightarrow CaSiO_3 + CO_2$
    c.  $C_6H_{12}O_{6(aq)} + 6O_{2(gas)} \rightarrow 6CO_{2(gas)} + 6H_2O_{(liq)}$
    d.  $SiO_2 + KAl_3Si_3O_{10}(OH)_2 \rightarrow KAlSi_3O_8 + Al_2SiO_5 + H_2O$
    e.  $SiO_2 + Fe_3O_4 \rightarrow Fe_2SiO_4 + O_2$
    f.  $Fe_3O_4 + FeTiO_3 \rightarrow Fe_2TiO_4 + Fe_2O_3$
4.  Write a balanced reaction between pyrite ($FeS_2$), hematite ($Fe_2O_3$), sulfur dioxide gas ($SO_2$), and oxygen gas ($O_2$).

# REFERENCES

Garrels, R. M. and C. L. Christ, 1965, *Solutions, Minerals and Equilibria*, Harper & Row, New York, 450 pp.

James H. L., 1954, Sedimentary facies of iron formation, *Economic Geology*, v. 49, p. 235–293.

Klein, C., 2005, Some Precambrian banded iron-formations (BIFs) from around the world: Their age, geologic setting, mineralogy, metamorphism, geochemistry, and origins, *American Mineralogist*, v. 90, no. 10, p. 1473–1499.

Pohl, W. L. 2011. *Economic geology, principles and practice: Metals, minerals, coal and hydrocarbons—an introduction to formation and sustainable exploitation of mineral deposits.* Chichester, UK: Wiley-Blackwell.

Skinner, B. J. 1986. *Earth resources*, 3rd ed. Englewood Cliffs, NJ: Prentice Hall, Inc.

Spencer, A. C. 1908. Magnetite deposits of the Cornwall type in Pennsylvania. *U.S. Geol Surv* 359, Washington Government Printing Office.

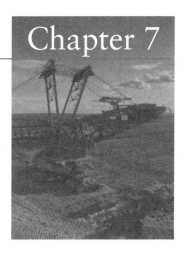

## Chapter 7

# Scarce Metals: Ferro-Alloy Metals

## Scarce Metals

The next three chapters discuss what are termed scarce metals. In general, when geologists refer to scarce metals they are considering those that occur at less than 0.1% on average in the earth's crust. In fact the sum of all scarce metals together with "fertilizer" elements and "chemical" elements account for less than 1% of the earth's crust. For the purposes of the treatment in this book the scarce metals are divided into four groups based on their usage: ferro-alloy metals, base metals, precious metals, and specialty metals. Ferro-alloy metals are used as alloys with Fe to give greater strength and resistance to corrosion of steel products. They often exist in ores in the +3 oxidation state. Base metals are a group of common inexpensive metals that easily oxidize and corrode, but do not have a natural +3 oxidation state of their atoms. Precious metals are metals of high economic value, but are not radioactive. Specialty metals are a collection of scarce metals that have particular properties that make them useful in specific industrial applications.

Scarce metals have two different modes of occurrence in the earth's crust. They can be dispersed at very low concentration in common silicate minerals by atomic substitution to form a solid solution. A *solid solution* is a homogeneous mixture of the elements considered in a single mineral. For instance, nickel ($Ni^{2+}$) can substitute for $Mg^{2+}$ in forsterite to produce a nickel-rich forsterite, $(Mg^{2+}, Ni^{2+})_2SiO_4$. The other mode of occurrence of scarce metals is in a number of unusual non-silicate ore minerals.

## Common Ore Minerals

It is helpful to have some understanding of the composition of common ore minerals before specific metal resources are considered. This then allows one to place these minerals into the context of ore deposits of a particular element of interest for the type of rock they occur in.

Given in **Table** 7.1 are the average concentrations of the scarce metals in the crust and the minerals mined to obtain them. As discussed previously, the abundant metals Si, Al, Fe, Ti, and Mn form oxide or hydroxide ore minerals. For Ca and Mg, carbonates are the dominant ore minerals. For the scarce minerals, except for beryllium (Be) obtained from the uncommon silicate mineral beryl ($Be_3Al_2Si_6O_{18}$), and including rare earth metals obtained from the phosphate monazite ($CeYPO_4$), and a few obtained from simple oxides or found in their native state, most are obtained from *sulfides*.

## Sulfides

A sulfide mineral contains sulfur in a negative *oxidation state*, $S^{-2}$ or $S_2^{-2}$. Sulfur is then the major anion in sulfides. Sulfide minerals have compositions that range from two metals per sulfur to one metal per two sulfurs. **Table** 7.2

*Table 7.1* SCARCE ELEMENTS, REPORTED AVERAGE CONCENTRATIONS BY WEIGHT IN EARTH'S UPPER CONTINENTAL CRUST*, AND THEIR ORE MINERALS.

| FERRO-ALLOY METALS | AVERAGE UPPER CRUST CONCENTRATION (PPM) | ORE MINERALS |
|---|---|---|
| Chromium (Cr) | 35–115 | Chromite ($FeCr_2O_6$) |
| Vanadium (V) | 50–110 | Solid solution in magnetite ($Fe_3O_4$) |
| Nickel (Ni) | 19–60 | Pentlandite [$(Ni,Fe)_9S_8$] |
| Molybdenum (Mo) | 0.8–1.5 | Molybdenite ($MoS_2$) |
| Cobalt (Co) | 12–18 | Linnaeite ($Co_3S_4$), Co-pyrite [$(Fe,Co)S_2$] |
| Tungsten (W) | 0.9–3.3 | Wolframite ($FeWO_4$), Scheelite ($CaWO_4$) |
| **BASE METALS** | **AVERAGE UPPER CRUST CONCENTRATION (PPM)** | **ORE MINERALS** |
| Copper (Cu) | 14–32 | Chalcopyrite ($CuFeS_2$), Chalcocite ($Cu_2S$) |
| Lead (Pb) | 17–18 | Galena (PbS) |
| Zinc (Zn) | 52–70 | Sphalerite (ZnS) |
| Tin (Sn) | 1.7–5.5 | Cassiterite ($SnO_2$) |
| Mercury (Hg) | 0.01–0.06 | Cinnabar (HgS) |
| Cadmium (Cd) | 0.075–0.102 | Greenockite (CdS), Hawleyite (CdS) |
| **PRECIOUS METALS** | **AVERAGE UPPER CRUST CONCENTRATION (PPM)** | **ORE MINERALS** |
| Gold (Au) | 0.0012–0.0018 | Native gold |
| Silver (Ag) | 0.050–0.055 | Native silver, Argentite ($Ag_2S$) |
| *PLATINUM GROUP ELEMENTS* | | |
| Platinum (Pt) | 0.0005 | |
| Palladium (Pd) | 0.0005 | |
| Rhodium (Rh) | 0.00007 | Occur in sperrylite ($PtAs_2$), cooperite (PtS) and as native elements |
| Iridium (Ir) | 0.00002 | |
| Osmium (Os) | 0.00003 | |
| Ruthenium (Ru) | 0.0003 | |
| **SPECIAL METALS** | **AVERAGE UPPER CRUST CONCENTRATION (PPM)** | **ORE MINERALS** |
| Niobium (Nb) | 12 | Columbite [$(Fe,Mn)Nb_2O_6$] |
| Tantalum (Ta) | 1.7 | Tantalite [$(Fe,Mn)Ta_2O_6$] |
| Arsenic (As) | 4.8 | Native arsenic, Arsenopyrite (FeAsS) |

| | | | |
|---|---|---|---|
| Antimony (Sb) | 0.4 | Stibnite ($Sb_2S_3$) | |
| Bismuth (Bi) | 0.16 | Bismuthinite ($Bi_2S_3$), Bismite ($Bi_2O_3$) | |
| Germanium (Ge) | 1.4 | Sphalerite (ZnS) with substitution of Ge for Zn | |
| Gallium (Ga) | 18 | Sphalerite (ZnS) with substitution of Ga for Zn | |
| Indium (In) | 0.06 | Sphalerite (ZnS) with substitution of In for Zn | |
| Beryllium (Be) | 2 | Beryl ($Be_3Al_2Si_6O_{18}$), Bertrandite [$Be_4Si_2O_7(OH)_2$] | |
| Rare Earth Elements (REEs) | 0.3 (Tm) to 64 (Ce) | Monazite ($CeYPO_4$), Bastnäsite ($CeFCO_3$) | |

\* Average upper crust concentrations from a variety of sources.

*Table 7.2* COMMON SULFIDE MINERALS.

| NAME | FORMULA | COLOR | MOHS HARDNESS* | DENSITY (g/cm³) |
|---|---|---|---|---|
| Argentite | AuS | Black lead-grey | 2–2.5 | 7.2–7.4 |
| Bornite | $Cu_5FeS_4$ | Copper red, bronze brown, purple | 3–3.25 | 4.9–5.3 |
| Chalcocite | $Cu_2S$ | Dark grey to black | 2.5–3 | 5.5–5.8 |
| Chalcopyrite | $CuFeS_2$ | Brassy yellow | 3.5–4 | 4.1–4.3 |
| Cinnabar | HgS | Brownish-red | 2–2.5 | 8.2 |
| Covellite | CuS | Indigo blue | 1.5–2 | 4.6–4.8 |
| Galena | PbS | Silvery lead-grey | 2.5–2.75 | 7.2–7.6 |
| Pentlandite | $(Fe,Ni)_9S_8$ | Yellowish bronze | 3.5–4 | 4.6–5.0 |
| Pyrite | $FeS_2$ | Pale brass yellow | 6–6.5 | 4.9–5.1 |
| Sphalerite | ZnS | Brown, yellow, red, green, black | 3.5–4 | 3.9–4.2 |

\* Mohs hardness is a scale of resistance to scratching from 1 for talc (which can be scratched with a fingernail) to 10 for diamond (see Table 7.3).

gives a list of 10 sulfide minerals of particular importance. Note sulfides have densities much greater than common silicate rock of ~2.7 g/cm³ and except for pyrite have a lower Mohs hardness. The *Mohs hardness* scale characterizes the relative scratch resistance of minerals with values from 1 for soft talc to 10 for extremely hard diamond as given in **Table 7.3**.

*Table 7.3* MOHS HARDNESS SCALE.

| MOHS HARDNESS | MINERAL |
| --- | --- |
| 1 | Talc |
| 2 | Gypsum |
| 2.5 | Fingernail |
| 3 | Calcite |
| 3.5 | Copper Coin |
| 4 | Fluorite |
| 5 | Apatite |
| 5.1 | Knife blade |
| 6 | Orthoclase |
| 6.5 | Steel Tool |
| 7 | Quartz |
| 8 | Topaz |
| 9 | Corundum |
| 10 | Diamond |

## Concentration of Scarce Metal Ores

Because most of these scarce metals appear at very low concentrations in silicate minerals, it is likely their concentration in silicates is distributed according to a normal probability distribution throughout rocks in the crust. It is also likely that in their other mode of occurrence, as nonsilicate ore minerals, they are also normally distributed throughout the crust. Because silicates so dominate the minerals in the crust they will dominate the distribution of a scarce metal's concentrations in the crust as shown in **Figure 7.1** but with a smaller normal distribution peak located at higher concentrations for the nonsilicate ore minerals where they are found in abundance.

The average concentration of a scarce metal in the crust will be close to the peak in the normal distribution for the silicates shown in **Figure 7.1**. For instance, molybdenum (Mo) has an average continental crustal concentration near 1 ppm. This would be near the higher maximum in the figure. Rocks with only 10 ppb (0.01 ppm) Mo would occur much

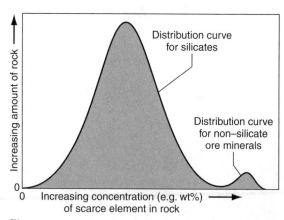

*Figure 7.1* Distribution of the concentration of a scarce element in crustal rocks distributed as a solid solution in typical silicate rocks and concentrated in non-silicate ores.

less frequently. Rocks with somewhat more than the average say 10 ppm Mo would also occur much less frequently. However, the rocks that are mined for Mo occur at the concentration on the smaller nonsilicate ore mineral probability peak where molybdenite ($MoS_2$) occurs with a concentration of ~0.1% (1,000 ppm) molybdenum in the rock.

This is 1,000 times greater than the crustal average. Only when the scarce metal occurs at concentrations greater than 25 to 1,000 times that found in average crustal rocks are they mined. These ore deposits then have high concentrations of these unusual nonsilicate minerals.

Typically, the most concentrated ores are most profitable and are mined first. The implication of **Figure** 7.1 is that with time as ores with lower concentrations are exploited there comes a time in the future when a maximum in abundance is reached on the smaller nonsilicate ore peak. When this occurs humankind will have decreased abundances of the scarce metal to exploit even though it is obtained from lower grade deposits. Similar to peak oil and the other energy peaks discussed earlier, peaks in scarce metals can also occur in the future. A particular concern of the scarce metal peaks is that unlike energy sources, some scarce metals have unique properties for which substitutions are difficult or impossible.

## Underground Mining

Most of the scarce metals are concentrated and therefore mined underground. Underground mining is done when ore is too far below the surface to make surface mining, such as in an open pit,

practical. In order to obtain the ore, a shaft is sunk into the earth and the ore is mined at various levels in what are called *stopes* as shown in **Figure** 7.2. Once blocks of ore are blasted from a working face in the mine the ore is loaded in cars and transported along a tunnel to be loaded into a *skip* and brought to the surface. Also shown in **Figure** 7.2 is an earlier now abandoned open pit mine that was used to obtain ore near the surface and an *adit* used to transport ore and waste in the original underground mine with the waste left outside the adit portal.

## Metal Production from Sulfide and Oxide Minerals

After being mined, ore blocks containing sulfide or oxide minerals are typically crushed to diameters of 0.1 mm or less. By doing this each grain is generally either ore or gangue. Mechanical techniques such as density sorting or magnetic separation can then be used to separate the ore minerals of interest.

If the ore mineral is a sulfide, the mineral powder is typically put in a flotation tank with water, 1% pine oil, and organosulfur compounds to conglomerate the sulfide particles in the ore. The tank is then agitated creating a slurry. Air is bubbled through the slurry causing the conglomerated sulfide particles to form an oily froth on the top of the tank while

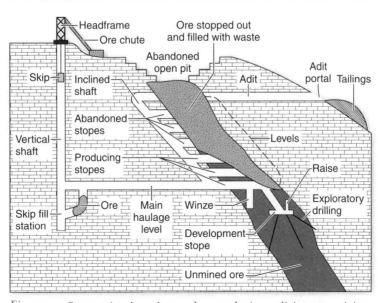

*Figure* 7.2 Cross section through an underground mine outlining some mining terms used.

silicates stay in the slurry. The sulfide froth can be skimmed from the top of the slurry.

The separated gangue from density sorting and flotation processes is typically stored in a large pile of what are called *tailings*. The separated sulfide ore is then roasted. Roasting is heating in the presence of oxygen to remove the sulfur in the ore as $SO_2$ gas and produces an oxide of the metals by the reaction:

$$\text{Metal-sulfide} + O_2 \rightarrow \text{Metal-O} + SO_2. \qquad [7.1]$$

The $SO_2$ is a byproduct that is converted to sulfuric acid ($H_2SO_4$) through the reaction:

$$2SO_2 + 2H_2O + O_2 \rightarrow 2H_2SO_4 \qquad [7.2]$$

and sold or used in subsequent processing.

The metal oxide from reaction (7.1) or the oxide ore mineral separated from the gangue is combined with coking coal in a blast furnace. The coking coal is burned to produce very high temperatures and CO gas. The CO gas reacts with the metal oxide giving rise to metal plus $CO_2$ by the reaction:

$$\text{CO} + \text{Metal-O} \rightarrow \text{Metal} + CO_2. \qquad [7.3]$$

This is similar to the reduction of iron described in the chapter on abundant metals. If the liquid metal from the blast furnace is lowered to 330°C, any copper in the liquid metal crystallizes and rises to the surface as a scum that can be skimmed off. Gold and silver can be removed from the liquid metal by adding small quantities of Zn. The liquid metal is then cooled slightly, which produces a scum of precipitated Zn containing the Au and Ag that is also skimmed off and processed for its metal content.

If a hydrometallurgical process is used to process the metal oxides the Metal-O can be leached with sulfuric acid producing positively charged Metal$^{2+}$ species in solution:

$$H_2SO_4 + \text{Metal-O} \rightarrow \text{Metal}^{2+} + SO_4^{2-} + H_2O. \qquad [7.4]$$

An electrical charge is put on the metal solution that causes Metal$^{2+}$ species to obtain electrons from the negative anode and precipitate as pure metal. Solvent extraction techniques introduce an organic solvent in which a particular metal is more soluble. The metal is then separated from the solvent with the final products typically one or a number of nearly pure metals.

## Biomining

While tailings are mine waste consisting of the uneconomic part of the ore, this gangue contains a small background concentration of ore minerals. The gangue typically exists as a fine-grained powder produced as a result of the crushing process. Aqueous solutions containing microorganisms, generally bacteria, can be passed through piles of tailings as well as piles of fine-grained low-grade ores in what is called biomining. Gold, copper, iron, and uranium have been recovered by bacteria that obtain energy from oxidizing these powders. Employing bacteria to obtain metals is becoming more common as high-grade ores become scarcer. Because biomining does not need to roast or smelt ore it can be a very cost-effective way to extract low-grade ore metals from rocks.

Biomining relies on the fact that most sulfide ore deposits contain large amounts of pyrite. Sulfur-oxidizing archaea bacteria obtain energy and produce heat and acid from reactions like

$$FeS_2 + 7/2O_2 + H_2O \rightarrow Fe^{2+} + 2SO_4^{2-} + 2H^+. \qquad [7.5]$$

Iron-oxidizing archaea bacteria obtain energy from oxidation of $Fe^{2+}$ by the reaction:

$$Fe^{2+} \rightarrow Fe^{3+} + e^-. \qquad [7.6]$$

The low pH and heat produced by reactions [7.5] and [7.6] increase the solubility and rate of reaction of $Fe^{3+}$ with any copper sulfide minerals present, such as in the reaction:

$$CuS + 8Fe^{3+} + 4H_2O \rightarrow Cu^{2+} + 8Fe^{2+} + SO_4^{2-} + 8H^+. \qquad [7.7]$$

The $Cu^{2+}$-rich solutions produced exit the tailings or low-grade ore and are precipitated by reacting the solution with solid $Fe^0$ by the reaction:

$$Fe^0 + Cu^{2+} \rightarrow Cu^0 + Fe^{2+} \qquad [7.8]$$

to obtain solid copper. Note these reactions could happen without the bacteria, but they would proceed much too slowly.

# Ferro-Alloys

The ferro-alloy metals are those that are combined with iron and a small amount of carbon to make various kinds of *steel*. When combined with iron the metals lower the melting temperature of the mixture and harden the final product. They include chromium, vanadium, nickel, tungsten, titanium, and molybdenum. Ferro-alloy metals are unique in that they are unsuitable by themselves for most applications but have important alloying properties when combined with iron and other metals.

# Chromium (Cr)

## Use

It was discovered in the early 1900s that steel could be made highly resistant to corrosion and discoloration by adding greater than 10% chromium (Cr) to produce "stainless" steel. Over 75% of the world's chromium is utilized in the production of stainless steel. Another major use of Cr is in chrome plating. A thin layer of chromium is electroplated on the surfaces of other metals.

In *electroplating* a negative charge is put on the object by connecting it to an external battery and immersing the object and the positive battery electrode into a solution containing $Cr^{3+}$. The positively charged $Cr^{3+}$ ions in the solution are attracted to the negatively charged object to be plated. Here electrons from the anode reduce the $Cr^{3+}$ to a zero charged with Cr metal plating on the object's surface.

Various ions of Cr are added to inks, paints, and dyes to produce specific colors: $Cr(OH)_6^{3-}$ (green), $CrO_4^{2-}$ (yellow), and $Cr_2O_7^{2-}$ (orange). An additional utilization of Cr is in leather tanning solutions. Chromium sulfate ($CrSO_4$) bonds to amino acids in animal skins making them softer and stretchier.

The aircraft and other Al processing industries employ chromic acid ($H_2CrO_4$) to anodize aluminum. Anodizing produces a porous $Al_2O_3$ layer on the surface of Al metal. The surface $Al_2O_3$ has increased resistance to corrosion, surface hardness, and adhesion for the application of color dyes. Chromic acid is the acid of choice because it can dissolve a high concentration of Al in solution.

The use of the Cr-containing mineral chromite for its mineral properties rather than the Cr it contains is discussed later where industrial minerals are considered.

## Resource Location

The average upper continental crust concentration of Cr has been estimated to be anywhere from 35 to 115 ppm. About 3,000 ppm Cr on average is in the earth's mantle. It is, therefore, rocks from the mantle that are emplaced into the upper crust that have the necessary enrichments to form a Cr ore deposit.

Chromium occurs in the mineral chromite ($FeCr_2O_4$). Chromite is found crystalized in small amounts in most mafic to ultramafic rocks from the mantle. *Podiform* masses consisting almost entirely of chromite crystals can occur in ultramafic rocks. These are irregular lens of chromite enclosed in dunites, rock made up almost entirely of olivine, or enclosed in hydrated olivine rocks termed serpentinites. Podiform Cr deposits form in mantle rocks that have been squeezed up during plate tectonic continent-continent collisions. The tectonic stresses congregate the chromite in pods. They are characteristic of deformed ultramafic sections of *ophiolites*, that is, ocean crust plus some underlying upper mantle emplaced on continents.

The other occurrence of chromite is in *stratiform deposits* of large mantle-derived layered mafic to ultramafic igneous magma intrusions into the crust. The chromite crystallizes from the magma in a cooling magma chamber. It sinks rapidly to the floor of the magma chamber because of its elevated density, 4.5 to 4.8 g cm$^{-3}$ forming a layer of chromite. Layers up to 1 m thick and extending for kilometers are found in some large ultramafic intrusions (**Figure 7.3**). A current theory is that chromites precipitate as a result of introduction and mixing of chemically primitive ultramafic magma with a more evolved magma in the magma chamber. This leads to supersaturation of chromite in the mixture.

Much of the production and reserves of chromium are located in the very large mafic to ultramafic igneous intrusion of the Bushveld Complex of South Africa. Lesser amounts occur in the mafic

*Figure 7.3* Dark chromite and light anorthosite layers in the Critical Zone rocks of the Bushveld Complex, Mononono River outcrop, near Steelpoort, South Africa.

to ultramafic Great Dyke of Zimbabwe. Some very large podiform deposits are mined in Kazakhstan. Podiform plus layered chromite intruded into the mafic to ultramafic Iron Ore Supergroup of the Sukinda Valley of India are also mined for chromium. In the U.S. seams of stratiform chromite of the mafic to ultramafic Stillwater Complex in Montana have been mined in the past but are not presently exploited.

## Bushveld Complex, South Africa

An estimated 12.5 km diameter meteorite impacted north of Pretoria, South Africa 2,060 million years ago. This impact produced the Vredefort Ring and the three lobes of the *Bushveld Complex* (**Figure 7.4**). The rocks of the Bushveld

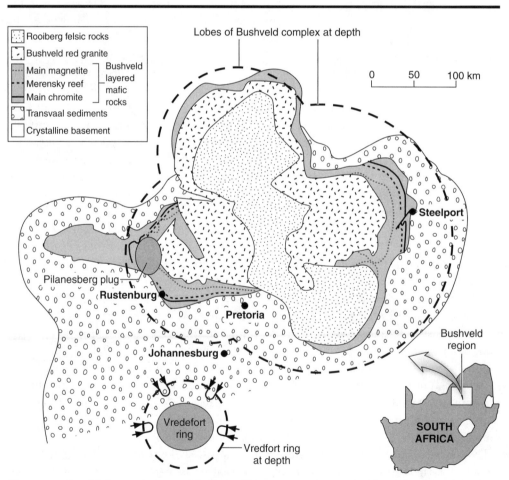

*Figure 7.4* Simplified geological map of the Bushveld igneous complex, South Africa. The dashed lines show the subsurface extent of the three lobes of the Bushveld complex and that of the Vredefort Ring. The inset gives the location of the Bushveld in South Africa (Adapted from Philpotts, A. R. and Ague, J. J., 2009, *Principles of Igneous and Metamorphic Petrology*, 2nd ed., Cambridge University Press, New York, Fig 15.17).

Complex crystalized from a greater than 350 km in diameter sheet-like mafic to ultramafic magma of up to 9 km thickness produced by the impact. The Bushveld is presently tilted about 25° to the south. Recent work suggests a shallow, stratified magma chamber was produced that was repeatedly filled at different locations (**Figure 7.5**).

The Bushveld Complex hosts the largest known deposits of Cr, V, and platinum group elements (PGE) on the earth (Willemse, 1969). As the Bushveld Complex cooled, chromite accumulated into layers on the floor of the magma chamber, the thickest of which are about 1 m. There are 12 such layers, but only three are of adequate thickness and exposure to be economically mined. The main zone magma produced the Merensky Reef and Platreef. Together with the underlying Upper Group 2 (UG2) Reef, produced by lower zone magma, these three reefs contain 70% of the world's known reserves of chromium as well as the PGEs of platinum, palladium, rhodium, ruthenium, iridium, and osmium. In mining terminology a reef is the name given a lode or vein of ore.

### Great Dyke, Zimbabwe

The *Great Dyke of Zimbabwe* (**Figure 7.6**) is not really a dike, as it was not intruded as a sheet cutting across preexisting rock. The 500 km long and 5 km wide Great Dyke is a lopolith formed by a number of magmas intruding into a narrow trough between two parallel faults. A *lopolith* is a large lenticular igneous body emplaced concordantly by a feeder dike between the surrounding rocks. Therefore, the intrusion is inappropriately named. It was emplaced 2,580 million years ago according to recent age dating.

The Great Dyke is made up of north and south chambers that have funnel-shaped cross-sections with feeder zones below. Each chamber consists of three subchambers composed of lower ultramafic layers of banded peridotites, pyroxenites, and norites capped by gabbro layers. *Peridotites* are rocks composed dominantly of the minerals olivine and pyroxene, pyroxenites are rocks consisting dominantly of pyroxene, and norite is rock made up dominantly of Ca-rich plagioclase, Mg + Fe-rich pyroxene, and olivine. Much of the gabbro has been removed by erosion. Chromite occurs in lenses and pods 5 to 100 cm thick near the base of the ultramafic sequence. They are mined throughout the lopolith. It is estimated that the Great Dyke contains 100 million metric tons of chromite.

## Production and Reserves

Approximately 24 million metric tons of chromite ore were produced in 2011, worldwide.

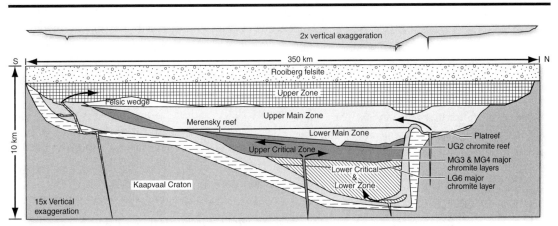

*Figure 7.5* North-south cross-section through the Bushveld Complex. Arrows show pathway of filling magmas. Note the large vertical exaggeration (Adapted from Kruger, F. Johan, 2005, Filling the Bushveld Complex magma chamber: lateral expansion, roof and floor interaction, magmatic unconformities, and the formation of giant chromitite, PGE and TiV-magnetitite deposits, *Mineralium Deposita*, v 40, p. 451–472).

*Figure 7.6* The Great Dyke in Zimbabwe. The lighter color in the dike shows the location of the platinum deposits.

The largest producers are South Africa (46%), Kazakhstan (16%), and India (16%). Mineable reserves are estimated to be 480 million metric tons of chromite and the resource base is greater than 12 billion metric tons. While reserves are somewhat low the resource base is sufficient to meet demand far into the future. About 95% of the world's Cr resources are present in Kazakhstan and South Africa.

In the U.S., Cr is considered a strategic metal. A strategic metal is one essential for industry and national security, but for which the nation has little or no domestic supply. The U.S. Defense National Stockpile Center (DNSC) has ferrochromium (FeCr) stocks to last the military for two years and chromium metal for three years at the present rate of use. It is estimated that the Stillwater Complex in Montana has 600,000 metric tons of mineable chromite.

## Health Concerns

There are several different valence states of chromium that differ in their effects upon organisms. $Cr^{3+}$ occurs naturally in the environment and is an essential nutrient for humans. Cr in stainless steel can be considered to have zero valence, which when corroded releases $Cr^{3+}$ to the environment and is not a health hazard.

$Cr^{6+}$, which is produced industrially, is a recognized carcinogen. $Cr^{6+}$ at elevated levels of 100 to 1,000 times higher than those found in normal air occur in industrial air, particularly in steel and textile plants. Long-term exposure to elevated $Cr^{6+}$ has been correlated with increased lung cancer and death. Besides industrial workers' exposure there is also a concern about the hazards of the release of $Cr^{6+}$ into surface and groundwater from old industrial sites that contain large amounts of Cr-bearing wastes. Also, $Cr^{6+}$ in leather products has been shown to cause skin rashes in some people.

# Vanadium (V)

With an estimated average abundance anywhere between 50 to 110 ppm in the earth's upper continental crust vanadium (V) is one of the most common geochemically scarce metals. Vanadium is unusual in that it has four common oxidation states each producing a different color in solution: $V^{+2}$ (lilac), $V^{+3}$ (green), $V^{+4}$ (blue), and $V^{+5}$ (yellow).

## Use

Vanadium is mainly used in steels and to produce vanadium carbide (VC). Adding as little as 0.2 wt% V to ordinary carbon steel considerably increases its *tensile strength*, hardness, and fatigue resistance. Vanadium carbide is an extremely hard ceramic. It is used to produce machine tool bits, knives, and other types of cutters. Vanadium is also mixed with Al in titanium alloys employed in jet engines and high-speed airframes to increase strength. Vanadium pentoxide ($V_2O_5$) is an oxidation catalyst used in the production of sulfuric acid from $SO_2$ gas. $V_2O_5$ is also employed for this purpose in automobile catalytic converters and organic compound synthesis.

## Resource Location

$V^{+3}$ occurs as a solid solution with concentrations up to 5 mole% along with titanium substituting for iron in magnetite [(Fe,V,Ti)$_3$O$_4$]. Concentrations of this magnetite are found in ultramafic and mafic rocks. Three sources account for 98% of vanadium production. These are layers of titaniferous magnetite in the Bushveld Complex in South Africa, in a dunite-pyroxenite intrusion in Kachkanar, Russia, and in a gabbroic, layered intrusion in Panzhihua, China. The V containing titaniferous magnetite is thought to have precipitated in the magma chambers of these bodies by episodic increases in the magmas oxidation state, that is its fugacity of oxygen ($f_{O2}$) by a reaction like

$$6\ FeSiO_3\ +\ 2\ Fe_3O_4\ \rightarrow\ 6\ Fe_2SiO_4\ +\ O_2 \qquad [7.9]$$
*in* pyroxene *in* magnetite       *in* olivine *in* magma

going forward. The names under the formula indicate where the components in the igneous system reside. In this reaction olivine becomes more Fe-rich at the expense of Fe in pyroxene and magnetite. The V and Ti replace the Fe lost from the magnetite.

Vanadium is also present in bauxite at up to 2,600 ppm derived from the weathering of mafic and ultramafic rocks containing V-rich magnetite. Remember that bauxite is a rock with high concentrations of aluminum plus oxygen and hydrogen produced from intensive weathering. V$^{3+}$ in magnetite becomes oxidized during the weathering forming a V$^{5+}$ aqueous species such as VO$_2^+$.

VO$_2^+$ is reasonably soluble and can precipitate on concentration from evaporation of water in arid climates or reduction to V$^{3+}$ leading to the precipitation of some rather rare V-containing minerals.

Vanadium concentrations up to 1,200 ppm are also found in reduced carbon-containing deposits such as crude oil, coal, oil shale, and tar sands. The Canadian Athabasca Tar Sands has 250 ppm V in places while some Venezuelan Orinoco Tar Sands contain 500 ppm V. Carbon in these deposits reduces mobile V$^{5+}$ in solution to insoluble V$^{3+}$ that then precipitates. This has also occurred in uranium-rich sandstones present on the Colorado Plateau of the western U.S. Most of these carbon-rich sources of V are not presently being utilized but are considered for the V reserve base particularly because vanadium, like other heavy metals, is concentrated in the fly ash when these fossil fuels are burned. At the present time about 5% of the world's production of V comes from fly ash recovery and reprocessing spent catalytic converters from motor vehicles.

## Production and Reserves

As given in **Table 7.4** the total world production of V was 60,000 metric tons and reserves were estimated to be 14 million metric tons in 2011.

*Table 7.4* VANADIUM PRODUCTION FOR 2011, RESERVES, AND RESERVE BASE BY COUNTRY.*

| COUNTRY | PRODUCTION | RESERVES | RESERVE BASE |
|---|---|---|---|
| China | 23 | 5,100 | 14,000 |
| South Africa | 20 | 3,500 | 12,000 |
| Russia | 15 | 5,000 | 7,000 |
| United States | 0 | 45 | 4,000 |
| Other countries | 1.5 | NA | 1,000 |
| World totals | 56 | 13,600 | 38,000 |

* In thousands of metric tons.

NA = Not available.

*Data from:* U.S. Geological Survey, *Mineral Commodity Summaries*, January, 2012.

Therefore, reserves are sufficient for over 100 years even with increased usage into the future. However, because vanadium is usually recovered as a byproduct or coproduct of mining of other metals its price can fluctuate widely.

## Health Concerns

The main health concern for vanadium is inhalation of vanadium-containing dust. About 110,000 metric tons of vanadium generally as $V_2O_5$ per year is released into the atmosphere mainly by burning fossil fuels. Exposure to vanadium dusts can lead to severe eye, nose, and throat irritation and causes bronchitis and pneumonia. Dusts containing vanadium at harmful levels are produced in the mining of vanadium bearing ores. Significant exposure to vanadium dust also occurs during the cleaning of oil and coal ash from oil and coal burners.

# Nickel (Ni)

Average nickel (Ni) concentrations in the earth's upper continental crust are estimated to be 19 to 60 ppm. It is not evenly distributed as ultramafic rocks have up to 3,000 ppm mainly in olivine and sulfide minerals.

## Use

Nickel is a malleable metal that can be hammered into a desired shape, for protective nickel covers and plating. However, nickel-steel alloys consume 60% of the Ni mined each year. Nickel adds strength, ductility, and resistance to corrosion and heat damage to iron. This is dominantly in the production of stainless steel, whose composition can vary but is iron with about 18% chromium and 8% to 10% nickel. High-strength structural steels have a Ni-content from 5% to 9%.

Fourteen percent of Ni is used to produce nickel-copper alloys including MONEL®, an alloy of 63% nickel, 28% to 34% copper, with a maximum of 2% manganese and 2.5% iron. MONEL is widely used in oil refining and marine applications where long corrosion-free service is required.

Another common nickel-copper alloy is a silver-white colored alloy termed nickel silver. It has no silver content but is composed of about 60 wt% copper, 20 wt% nickel, and 20 wt% zinc. Nickel silver is used to make zippers and wind musical instruments. The 5-cent coin in the U.S. and Canada, called the nickel, is a nickel-copper alloy made from 25% nickel and 75% copper.

Ni is an important component of superalloys like INCONEL® and HASTELLOY® accounting for 15% of Ni production. A *superalloy* is a high-performance metal alloy typically containing Ni and Co with superior mechanical strength at high temperatures. About 3% of Ni production is used for nickel-plated cast iron such as used in cookware. Another 3% is used for heat and electric resistance alloys, such as Nichrome wire. A powdered nickel-aluminum catalyst is used for hydrogenation of vegetable oils and organic syntheses. Nickel is also added to glass to produce a green color.

## Resource Location

Most nickel is mined from two different types of ore deposits. One is accumulations of the Ni-rich magmatic sulfide pentlandite $[(Ni,Fe)_9S_8]$ in what is termed a magmatic massive sulfide deposit. *Magmatic massive sulfides* are large accumulations nearly totally made up of sulfide minerals found in mafic and ultramafic rocks. They can contain significant quantities of Ni + Cu + Co. The other type of deposit where Ni ores are found is in laterites. *Laterites* are residual soils rich in oxides of iron and aluminum formed by extensive weathering of silicate minerals in tropical and subtropical regions. If they are an ore of Al they are often referred to as lateritic bauxites.

Most mafic magmas from the mantle that intrude into the earth's crust are likely undersaturated with sulfides. They become saturated as the magma cools and crystallizes silicates, enriching the residual magma in sulfide. As the magma becomes sulfur saturated, sulfide minerals begin to precipitate along with silicate minerals. To form the large concentration of sulfides without silicates, such as found in a magmatic massive sulfide, another process must also be important.

Mafic magmas do not have enough heat from crystallization to melt a significant concentration of country rock. Therefore, most investigators believe that heat released on crystallization of the magma produces a sulfur-rich magma phase from melting of only sulfates and sulfides in the country rock. Deposits are then formed by *liquid immiscibility*. A separate silica-rich magma and a sulfide-rich magma are produced on cooling of the magma as droplets of iron sulfide liquid form in the silica-rich magma. These contain high concentrations of Ni, Cu, Co, Pt, etc., partitioned into the sulfur-rich magma phase as it forms. The sulfide liquid droplets are denser than the silicate melt and sink. A magmatic massive sulfide ore deposit forms at the bottom of the magma chamber with up to 3% to 4% Ni. Besides pentlandite, pyrrhotite ($Fe_{1-x}S$) and chalcopyrite ($CuFeS_2$) are also crystallized from the separated iron sulfide liquid.

### Norilsk-Talnakh Massive Sulfide Deposit

Norilsk, Russia and Talnakh, a town merged into Norilsk, lay above the Arctic Circle in Northern Siberia. The magmatic massive sulfide deposits there contain 1/3 of the earth's known reserves of nickel and accounts for 22% of world production. The deposits also contain 2/5 of the earth's platinum (Pt) reserves and accounts for 9% of world Pt production. Additionally, it is one of the world's largest producers of copper and palladium at 3% and 38%, respectively. Recovery of the metals from this deposit requires working in weather with extremely low temperatures. Average temperature is $-10°C$ and temperatures as low as $-58°C$ have been recorded.

The deposit was formed during the eruption of the Siberian Traps magma about 250 million years ago. The *Siberian Traps* are a large outpouring of flood basalts that cover about 2 million km$^2$ on the earth's surface on which 1 to 4 million km$^3$ of lava was erupted (**Figure 7.7**). Most investigators believe a mantle plume "hot spot" is responsible for the basalt but some believe a large asteroid impact caused the eruption.

The deposits at Norilsk are mined at a depth of between 0.5 km and 1.5 km beneath a series of flood basalts and sediments as shown in **Figure 7.8**. The magmatic massive sulfide ore bodies occur within the poorly mineralized Talnakh intrusive gabbro

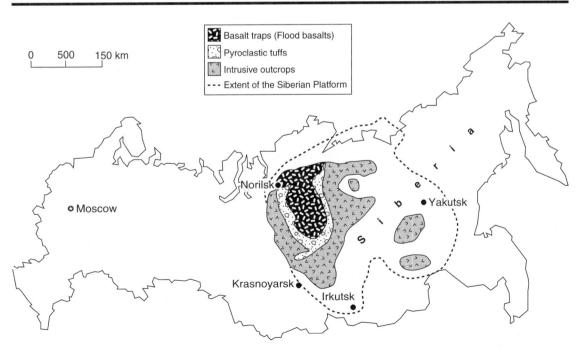

*Figure 7.7* Map giving the extent of the Siberian Traps igneous rocks. (Adapted from Environmental News Service).

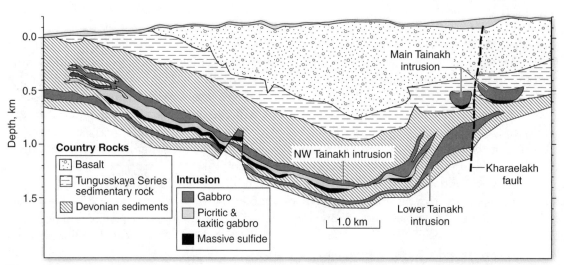

*Figure 7.8* Geologic cross-section through the Norilsk-Talnakh massive sulfide deposits with 2 × vertical exaggeration. (Modified from Li, C., Ripley, E. M. and Naldrett, A. J., 2003, Compositional variations of olivine and sulfur isotopes in the Noril'sk and Talnakh intrusions, Siberia: Implications for ore-forming processes in dynamic magma conduits, *Economic Geology*, v. 98, p. 69–86).

complex produced at the same time as the flood basalts. The ore is thought to have formed when the magma became saturated in sulfur, forming globules that crystallized pyrrhotite ($Fe_{(1-x)}S$), with x = 0 to 0.2, pentlandite [$(Fe,Ni)_9S_8$], and chalcopyrite ($CuFeS_2$). Most investigators believe much of the sulfur was derived from the gypsum ($CaSO_4 \cdot 2H_2O$) beds through which the flood basalts were erupted. It has been argued that the magmatic massive sulfide deposits are contained in exit conduits for the flood basalts in country rock with the sulfides precipitating from the first flow of a gypsum-contaminated basalt.

### Sudbury, Canada Massive Sulfide Deposit

The Sudbury massive sulfide deposit is generally considered to have formed by a 10 km in diameter meteorite impact event 1.85 billion years ago creating a 250 km long crater. The crater filled with basaltic magma produced from impact melted crustal rocks that contained a high concentration of sulfides. On cooling, to produce a gabbro, a separate iron sulfide liquid phase that had high concentrations of nickel, copper, platinum, palladium, and gold formed in the magma. Since the metal sulfides were heavier then the basaltic melt they settled to the bottom of the magma chamber.

The meteorite crater was deformed into its current smaller oval shape during later metamorphism termed the Grenville orogeny, occurring between 1.1 and 1.0 billion years ago (**Figure 7.9**). Most of the mining is around the rim of the crater where the massive sulfide at the bottom of the original crater is exposed.

### Ni Laterites

Ni is concentrated in laterite soils produced in tropical areas during weathering of mafic to ultramafic rocks and can contain about 0.2 to 0.3 wt% Ni (**Figure 7.10**) A Ni-rich hydrated iron hydroxide mineral termed limonite [$(Fe,Ni)O(OH) \cdot nH_2O$], where *n* can be variable, and the hydrous mixed Ni-silicate ore termed garnierite are produced during weathering. The abundant rainfall percolates through the rock dissolving and removing Mg and Si, which leads to a residual concentration of Fe and Al and 1 to 2 wt% Ni in limonite and with up to 5 wt% Ni in garnierite. Laterite Ni deposits occur in a variety of geological settings spanning the Precambrian to the Tertiary. About 70% of the earth's nickel resources are contained in laterites. They currently account for about 40% of the world's nickel production.

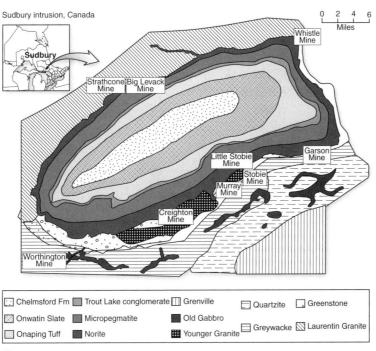

*Figure 7.9* Geological map of the Sudbury nickel intrusion. Inset gives its location in Canada (Adapted from Naldrett, A. J., 2005, A history of our understanding of magmatic Ni-Cu sulfide deposits, *The Canadian Mineralogist*, v 43, p. 2069–2098).

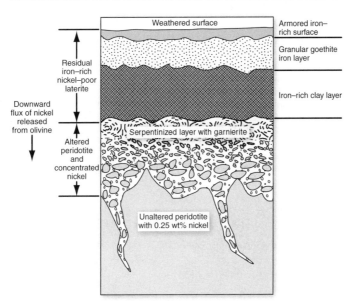

*Figure 7.10* Cross-section through a nickel laterite deposit in New Caledonia, one of the largest Ni deposits ever discovered Nickel is concentrated in laterite soils produced in tropical areas during weathering of mafic to ultramafic rocks. (Adapted from De Chetelat, E., 1947, *La genese et l'evolution des gisement de nickel de la Nouvelle-Caledonie. Société géologique France Bulletin*, Ser. 5, v 17, p. 105–160).

## Production and Reserves

The Norilsk deposit in Siberia, Russia produces about 17% of the world's supply of Ni. The Sudbury massive sulfide deposit in Canada produces another 15% (**Table 7.5**). Other major deposits of nickel are found in laterites in Indonesia (13% of world production), Australia (11% of world production), and New Caledonia (6% of world production). Recently, a nickel laterite deposit in western Turkey has been exploited, with this location being especially convenient for European smelters, steelmakers, and factories. Given the world production and the reserves plus

reserve base outlined in **Table 7.5** supplies should be sufficient for the next 100 years.

The one locality in the U.S. where nickel was commercially mined is Riddle, Oregon, where a number of km$^2$ of nickel-bearing garnierite surface deposits are located. The mine closed in 1987. The U.S. has no high-grade Ni deposits. Low-grade deposits occur in the Duluth Gabbro, a 16 km thick mafic igneous intrusion in Minnesota on the north shore of Lake Superior (0.21 wt% Ni). The *Duluth Gabbro* formed 1.1 billion years ago when the North American continent split apart and basaltic magma from the mantle rose through the earth's crust in what is called the Midcontinent Rift.

*Table 7.5* NICKEL PRODUCTION FOR 2011, RESERVES, AND RESERVE BASE BY COUNTRY.*

| COUNTRY | PRODUCTION | RESERVES | RESERVE BASE |
|---|---|---|---|
| Russia | 280 | 6,000 | 9,200 |
| Indonesia | 230 | 3,900 | 13,000 |
| Philippines | 230 | 1,100 | 5,200 |
| Canada | 200 | 3,300 | 15,000 |
| Australia | 180 | 24,000 | 29,000 |
| New Caledonia | 140 | 12,000 | 15,000 |
| Brazil | 83 | 8,700 | 8,300 |
| China | 80 | 3,000 | 7,600 |
| Cuba | 74 | 5,500 | 23,000 |
| Colombia | 72 | 720 | 2,700 |
| South Africa | 42 | 3,700 | 12,000 |
| Botswana | 32 | 490 | 920 |
| Madagascar | 25 | 1,600 | NA |
| Dominican Republic | 14 | 1,000 | 1,000 |
| Other countries | 100 | 4,600 | 6,100 |
| United States | 0 | 0 | 150 |
| World totals | 1,800 | 80,000 | 150,000 |

* In thousands of metric tons.

NA = Not available.

*Data from:* U.S. Geological Survey, *Mineral Commodity Summaries*, January, 2012.

# Molybdenum (Mo)

Molybdenum (Mo) abundance in the upper continental crust is estimated to be 0.8 to 1.5 ppm. Molybdenum is a refractory metal, meaning it has extraordinary resistance to wear at high temperature and has the sixth highest melting point of any element. For a metal, molybdenum is, however, soft and ductile with a high elastic modulus. Elastic modulus is a measure of its ability to deform elastically, that is, not permanently.

## *Use*

Molybdenum readily forms hard, stable carbides generally of $Mo_2C$, and for this reason it is often used in high-strength steel alloys and superalloys for cutting tools and armaments. The various uses of Mo are outlined in **Figure 7.11**.

## *Resource Location*

Ores of molybdenum contain the mineral molybdenite ($MoS_2$). It occurs in porphyry-type igneous intrusions that are described when copper is discussed in a later section. Porphyry is a term used to denote an igneous rock that has large conspicuous crystals, especially of feldspar, set in a matrix of fine-grained crystals. Mo is usually mined as a byproduct of copper production. However, molybdenite porphyries occur together with tin and tungsten but without copper at the Climax, Henderson, and Urad mines in Colorado and the Questa mine in New Mexico. The Climax mine for many years supplied 3/4 of the world's molybdenum. Theses mines are currently inactive because of the low price of Mo when produced as a byproduct of mining of copper. Therefore, the production of Mo is currently from porphyry copper deposits.

## *Production and Reserves*

Reserves appear adequate for years to come as shown by the production and reserves given in **Table 7.6**. China is the largest producer of Mo. These are mainly from ores associated with granitic porphyry deposits formed by collision of the North China and Siberian tectonic plates during the Early to Middle Triassic.

China's high level of steel production and consumption continues to generate strong internal consumption of Mo. This consumption, coupled with reduced Chinese exports in 2007 and 2008 owing to export quotas and duties imposed in July 2007, continued to support historically high Mo prices despite the worldwide economic crisis which started in late 2007 and lasted into 2009 (**Figure 7.12**). This also happened in late 2004 and during 2005. In contrast, the market price of nickel is not strongly tied to Chinese exports and surged throughout 2006 and the early months of 2007 due to market speculation, but the economic crises brought the price down rapidly. The Mo market was not affected by the crises until late 2008.

# Cobalt (Co)

Cobalt (Co) is a ferromagnetic metal, meaning it can be permanently magnetized. Most of the earth's cobalt is thought to be in the earth's core. Cobalt

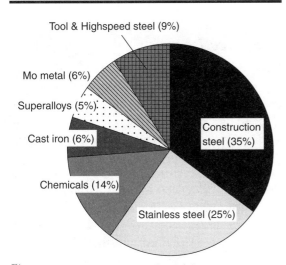

*Figure 7.11* The various uses of molybdenum.

*Table 7.6* MOLYBDENUM PRODUCTION FOR 2011, RESERVES, AND RESERVE BASE BY COUNTRY.*

| COUNTRY | PRODUCTION | RESERVES | RESERVE BASE |
|---|---|---|---|
| China | 94 | 4,300 | 8,300 |
| United States | 64 | 2,700 | 5,400 |
| Chile | 38 | 1,200 | 2,500 |
| Peru | 18 | 450 | 230 |
| Mexico | 12 | 130 | 230 |
| Canada | 8.3 | 220 | 910 |
| Armenia | 4.2 | 200 | 400 |
| Russia | 3.8 | 250 | 360 |
| Iran | 3.7 | 50 | 140 |
| Mongolia | 2.0 | 160 | 200 |
| Uzbekistan | 0.55 | 60 | 150 |
| Kazakhstan | 0.36 | 130 | 200 |
| Kyrgyzstan | 0.25 | 100 | 180 |
| World totals | 250 | 10,000 | 19,000 |

* In thousands of metric tons.

*Data from:* U.S. Geological Survey, *Mineral Commodity Summaries*, January, 2012.

occurs at an average concentration of only 12 to 18 ppm in the earth's upper continental crust.

## Use

As outlined in **Figure 7.13** cobalt's major uses are in metal alloys, including corrosion-resistant metal alloys such as superalloys used for parts in gas turbines and aircraft engines. Because of its ferromagnetic properties Co is also used in magnets and magnetic recording media. The remainder of the cobalt is used in various other metallic and chemical applications such as catalysts for the petroleum and chemical industries and battery production. Lithium cobalt oxide ($LiCoO_2$) is commonly used as the positive electrode of lithium ion batteries. Cobalt-based colors and pigments have been used since ancient times for jewelry and paints. Cobalt salts can impart blue and green colors in glass and ceramics. Radioactive $^{60}Co$ is used in the treatment of cancer.

## Resource Location

Cobalt is obtained as a byproduct of mining Cu, Ni, and Ag when the minerals linnaeite ($Co_3S_4$) and cobalt pyrite [$(Fe,Co)S_2$] are present. The U.S. is the world's largest consumer of cobalt. It imports 100% as it does not mine or refine cobalt at present. Congo and Zambia produce 2/3 of the world's Co in sediment-hosted Cu + Co stratiform ores in the central African copper belt. Stratiform ores are concordant ore bodies that are parallel to bedding. These ores were deposited in the Katangan basin, which has been deformed into the Katangan fold belt as shown in **Figure 7.14**. It should be noted that cobalt is also produced as byproduct from the

Sudbury mine in Canada and the Norilsk deposits in Siberia.

## African Copper Belt

The African copper belt is estimated to contain up to 140 million metric tons of native Cu and Cu in sulfide minerals plus 6 million metric tons of Co in sulfide minerals in Late Precambrian sedimentary deposits. The genesis of these deposits is much debated. It is argued that mineralizing brines developed in hypersaline lagoons formed during deposition of intertidal and supratidal sediments. The sulfur isotopes of the ores lead some to argue the sulfur in the sulfides comes from bacterial reduction of seawater sulfates. Others consider inorganic changes in oxygen fugacity, pH, and/or salinity are responsible for the sulfide deposition (Cailteux and others, 2005).

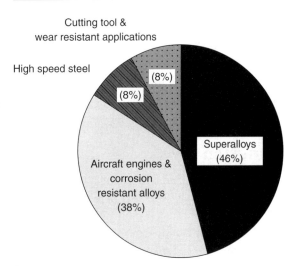

*Figure 7.13* The major uses of cobalt.

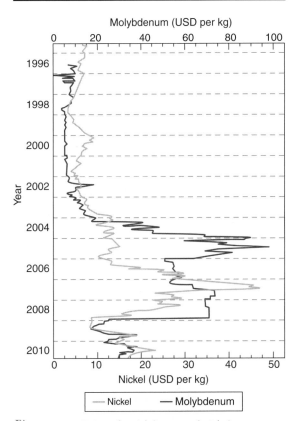

*Figure 7.12* Prices of molybdenum and nickel as a function of time. Note the difference in the effect of the 2007–2009 worldwide economic crises on molybdenum as opposed to nickel prices.

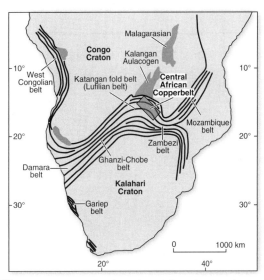

*Figure 7.14* Central and southern Africa showing the Katangan fold belt which contains the Central African Copperbelt between the Kalahari and Congo cratons. A fold belt is a large region where all the bedrocks have been subjected to similar folding and generally a mountainous terrain is produced. A craton is a significant portion of a continental landmass that has been relatively undisturbed since the Precambrian era (Adapted from Cailteux, J. L. H., Kampunzu, A. B., Lerouge, C., Kaputo, A. K., and Milesi, J. P., 2005, Genesis of sediment-hosted stratiform copper-cobalt deposits, central African Copperbelt, *Jour. African Earth Sciences*, v. 42, Issue 1–5, July–September, p. 134–158).

## Production and Reserves

Co reserves appear to be no problem at present as outlined in **Table** 7.7. At today's production rates over 80 years of supply exists in reserves that can be extended for an additional 100 years with the available reserve base.

# Tungsten (W)

Upper continental crustal average concentration of tungsten (W) is reported to be from 0.9 to 3.3 ppm. The most common oxidation state of tungsten is +6, but it can exhibit in oxidation states as low as −2. Tungsten has the highest melting point, 3,422°C, of any metal and at temperatures above 1,650°C the highest tensile strength. Tungsten also has the lowest coefficient of thermal expansion of any pure metal. Most of the unusual properties of W are due to the half-filled 5d electron shell with a very high binding energy of the tungsten metal lattice.

## Use

Alloying small quantities of tungsten with steel greatly increases its toughness. Tungsten is also a component of many superalloys. Tungsten carbide, WC and $W_2C$, has a hardness approaching diamond. Because of its high-tensile strength tungsten carbide is used for cutting tools and drill bits. About half of the tungsten used goes into tungsten carbide and tungsten alloys for use in machine tools and oil and gas well drilling equipment. Due to its electrical conductivity and high melting point W is also used in electrodes and filaments for incandescent light bulbs and X-ray tubes as shown in **Figure** 7.15. Because of these properties, tungsten, tungsten alloys and some

*Table 7.7* COBALT PRODUCTION FOR 2011, RESERVES, AND RESERVE BASE BY COUNTRY.\*

| COUNTRY | PRODUCTION | RESERVES | RESERVE BASE |
|---|---|---|---|
| Congo (Kinshasa) | 52 | 3,400 | 4,700 |
| Canada | 7.2 | 130 | 350 |
| China | 6.5 | 80 | 470 |
| Russia | 6.3 | 250 | 350 |
| Zambia | 5.7 | 270 | 680 |
| Australia | 4.0 | 1,400 | 1,800 |
| Cuba | 3.6 | 500 | 1,800 |
| Morocco | 2.5 | 20 | NA |
| New Caledonia | 2.0 | 370 | 860 |
| Brazil | 1.7 | 89 | 40 |
| Other countries | 7.0 | 740 | 1,100 |
| United States | 0.0 | 33 | NA |
| World totals | 98 | 7,500 | 13,000 |

\* In thousands of metric tons.

NA = Not available.

*Data from:* U.S. Geological Survey, *Mineral Commodity Summaries*, January, 2012.

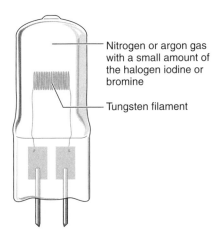

*Figure 7.15* Incandescent tungsten halogen lamp used in automotive headlamps. Tungsten is employed because of its high melting point. This lamp can be operated at a higher temperature than a similar standard gas-filled lamp. The higher operating temperature results in light of a higher color temperature, which gives it a higher luminosity.

tungsten compounds cannot be substituted for in many important applications.

## Resource Location

Tungsten occurs in veins in hydrothermal ore deposits in the minerals scheelite ($CaWO_4$) and wolframite [$(Fe,Mn)WO_4$] in granitic rocks, often associated with tin. The source of the metals is from a magmatic aqueous fluid. During cooling of a granitic magma, fractional crystallization of anhydrous minerals occurs saturating the magma with water. A separate aqueous metal-rich fluid is produced. Scheelite and wolframite are generally found in veins precipitated from these aqueous fluids in rocks in the earth's upper crust. In these deposits hot aqueous fluids have penetrated cracks or reacted with limestones to produce skarns. Most of the tungsten

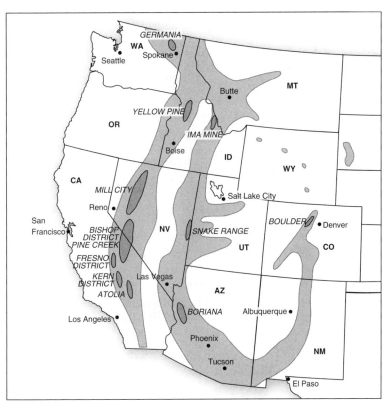

*Figure 7.16* Western U.S. tungsten belt is shown in light grey. Most productive districts are labeled and shown in darker shade. (Modified from Kerr, P. F., 1946, *Tungsten Mineralization in the United States*, Waverly Press, Baltimore, 241 p.).

*Table 7.8* TUNGSTEN PRODUCTION FOR 2011, RESERVES, AND RESERVE BASE BY COUNTRY.*

| COUNTRY | PRODUCTION | RESERVES | RESERVE BASE |
|---|---|---|---|
| China | 60 | 1,900 | 4,200 |
| Russia | 3.1 | 250 | 420 |
| Canada | 2.0 | 120 | 490 |
| Portugal | 1.3 | 4.2 | 62 |
| Bolivia | 1.2 | 53 | 100 |
| Austria | 1.1 | 10 | 15 |
| Other countries | 3.4 | 600 | 750 |
| United States | 0.0 | 140 | 200 |
| World totals | 72 | 3,100 | 6,300 |

* In thousands of metric tons.

*Data from:* U.S. Geological Survey, *Mineral Commodity Summaries, January*, 2012.

deposits are found in younger mountain belts, that is, the Alps, the Himalayas, and the circum-Pacific mountain belts. **Figure 7.16** shows the location of deposits in the U.S., none of which are mined at present. The concentration of W in economic ores is usually between 0.25 and 1.0 wt% $WO_3$. This is greater than most of the deposits shown in **Figure 7.16**. Mount Morgan tungsten mines in the Bishop tungsten district in California, produced over 2,000 metric tons W per day, but the mines have been in caretaker status since 1992.

## Production and Reserves

The U.S. imports needed tungsten as ores and concentrates, as well as waste and as scrap. China is responsible for over 75% of total world production, with most of the remaining coming from Austria, Bolivia, Canada, Portugal, and Russia. Sufficient reserves exist far into the twenty-first century (**Table 7.8**). China and Russia have eight of the world's 10 largest deposits. These eight deposits contain about half of the world's known resources of tungsten.

## SUMMARY

Scarce metals have two different types of occurrences. They can be dispersed at very low concentration in common silicate minerals by atomic substitution to form a solid solution. They can also be concentrated in a number of unusual nonsilicate minerals, generally simple oxides or sulfides. It is the latter occurrence that produces ore deposits. Scarce minerals are mined both in open pit surface mines and underground.

Ferro-alloy metals are alloyed with Fe to give greater strength and resistance to corrosion to steel products. They include chromium (Cr), vanadium (V), nickel (Ni), tungsten (W), titanium (Ti), and molybdenum (Mo).

Chromium is often electroplated onto steel object surfaces to protect them from rusting and produce a silvery metallic finish that can be highly polished.

Stainless steel contains > 10% Cr. Chromium is concentrated in mantle rocks. When these rocks are emplaced in the upper crust a Cr ore body can form. Podiform chromite ($FeCr_2O_4$) bodies are squeezed up from the mantle during plate tectonic collisions. Stratiform chromite deposits crystallize from large mafic to ultramafic magma bodies injected from the mantle into the crust. These include the Bushveld Complex of South Africa with 70% of world reserves.

Vanadium is added to steel to increase its strength and to produce high-speed V-carbide machine tools. Vanadium, like Cr, is higher in concentration in mantle rocks than crustal rocks and is found in titaniferous magnetite in mafic and ultramafic intrusions. It is also present in bauxites and fossil fuel deposits.

Nickel is a common steel-alloying agent and used as a plating agent on steel. Stainless steel contains 8% to 10% nickel. Ni is concentrated in mantle rocks and found as a sulfide, pentlandite $[(Ni,Fe)_9S_8]$, in massive sulfide deposits. It is also mined from laterite soils developed on ultramafic rocks.

Molybdenum forms hard carbides of $Mo_2C$ and is used in superalloys, for cutting tools, and for armaments. Ores of Mo contain molybdenite ($MoS_2$), which occurs in porphyry-type igneous intrusions. Reserves appear adequate for years to come with large productions in China, the U.S., and Chile.

Cobalt is used in many alloys and superalloys. Cobalt is obtained as a byproduct of mining Cu, Ni, and Ag from the minerals linnaeite ($Co_3S_4$) and cobalt pyrite $[(Fe,Co)S_2]$. The Congo and Zambia account for 2/3 of worldwide production of Co from sediment-hosted Cu + Co stratiform ores.

Tungsten is used to make superalloys and tungsten carbide, WC and $W_2C$, which has a hardness approaching diamond. It is also used in electrodes and filaments for incandescent light bulbs. Tungsten occurs in veins in hydrothermal vein and skarn deposits. Ore minerals are scheelite ($CaWO_4$) and wolframite $[(Fe,Mn)WO_4]$, found in granitic rocks, often with tin. China produces over 75% of the total world demand.

## KEY TERMS

| | |
|---|---|
| adit | peridotites |
| Bushveld Complex | podiform |
| Duluth Gabbro | Siberian Traps |
| electroplating | skip |
| Great Dyke of Zimbabwe | solid solution |
| laterites | steel |
| liquid immiscibility | stopes |
| lopolith | stratiform deposits |
| magmatic massive sulfides | sulfide |
| Mohs hardness | superalloy |
| ophiolites | tailings |
| oxidation state | tensile strength |

## PROBLEMS

1.  Consider the ferro-alloy metals in **Table 7.1**. Using the lowest ppm by weight value of the reported average upper crust concentration, which metal has the highest concentration by number of atoms? The lowest?

2.  At the present rate of consumption calculate the length of time it will take worldwide reserves of each of the ferro-alloys to be used up.

3.  If a pentlandite has one atom of Ni for every 19 atoms of Fe what is the wt% of Ni in this sulfide?

4.  What is the number of ppm by weight of tungsten in the mineral scheelite?

## REFERENCES

Cailteux, J. L. H., Kampunzu, A. B., Lerouge, C., Kaputo, A. K., and Milesi, J. P. 2005. Genesis of sediment-hosted stratiform copper-cobalt deposits, central African copper belt. *Jour African Earth Sciences* 42(1–5):134–158.

De Chetelat, E. 1947. La genese et l'evolution des gisement de nickel de la Nouvelle-Caledonie. *Bulletin de la Société Géologique de France* 5(17):105–160.

Kerr, P. F. 1946. *Tungsten mineralization in the United States*. Baltimore: Waverly Press.

Kruger, F. J. 2005. Filling the Bushveld Complex magma chamber: Lateral expansion, roof and floor interaction, magmatic unconformities, and the formation of giant chromitite, PGE and TiV-magnetitite deposits. *Mineralium Deposita* 40:451–472.

Li, C., Ripley, E. M., and Naldrett, A. J. 2003. Compositional variations of olivine and sulfur isotopes in the Norilsk and Talnakh intrusions, Siberia: Implications for ore-forming processes in dynamic magma conduits. *Economic Geology* 98:69–86.

Naldrett, A. J. 2005. A history of our understanding of magmatic Ni-Cu sulfide deposits. *The Canadian Mineralogist* 43:2069–2098.

Philpotts, A. R. and Ague, J. J. 2009. *Principles of igneous and metamorphic petrology*, 2nd ed. (Figure 15.17). New York: Cambridge University Press.

Willemse, J. 1969. The geology of the Bushveld Complex, the largest repository of magmatic ore deposits in the world. *Economic Geology Monograph* 4:1–22.

Zolotukhin, V. V. and Al'Mukhamedov, V. V. 1988. Traps of the Siberian Platform. *In*: J. D. Macdougall (Ed.), *Continental flood basalts* (p. 273–310). Dordrecht, Germany: Kluwer Academic Publishers.

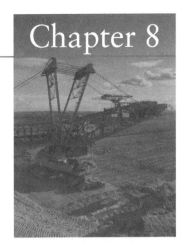

# Chapter 8

# Scarce Metals: Base Metals

In this text and when considering mining in general, base metals typically refer to the scarce nonprecious, nonferric (i.e., not commonly occurring in the +3 valence state) metals of copper, lead, zinc, tin, mercury, and cadmium. Base metals oxidize or corrode relatively easily making them nonprecious. At the end of the chapter the nonmetal fluorine will be discussed because it is often found in deposits with base metals.

# Copper (Cu)

A base metal of particular importance is copper (Cu). Cu is estimated to have an average abundance in the earth's upper continental crust of 14 to 32 ppm by weight.

Native copper is easily distinguished from silicate rocks and therefore became an important ore used by humankind before recorded history, at least 10,000 years ago. Pure Cu is rather soft and malleable and therefore easily fashioned into jewelry or tools. Some of these are preserved at archeological sites. Archeologists refer to the 1,000-year period between the Neolithic and Bronze Age, roughly from 4500 to 3500 BC, when copper objects are found in abundance as the Copper Age.

In the sixth through third centuries BC Romans used copper lumps as money. In the first century BC Julius Caesar had coins with his likeness made from bronze, a copper-zinc alloy. The first emperor of China used circular copper coins with a square hole in the middle, starting about 221 BC. Both England and the U.S. have long used copper for their most common coin, the penny.

## Use

Copper has found a great many uses in modern society. Besides being ductile it has very high thermal and electrical conductivity. Its major use is in electrical wires, but substantial amounts of Cu are also used in plumbing pipes, coins, and in the production of brass, a Cu plus zinc alloy. Constructing electric power grid distribution wires in and between homes and businesses in communities across the globe has required large amounts of Cu for wire for the last 100 years. Additionally, nearly all electric motors are made with wound copper wire.

Significant amounts of copper are used in motor vehicles. An average car contains 20 to 45 kg of copper in wires, the radiator, and as an alloying element with aluminum in cylinder heads and engine blocks. Copper compounds such as copper sulfate ($CuSO_4$) are used to control fungal diseases and as snail bait while copper acetate [$Cu(C_2H_3O_2)_2$] is produced for insecticides and fungicides.

## Resource Location

Besides native copper, copper is found in a number of sulfide minerals. The important ones include chalcopyrite ($CuFeS_2$), digenite ($Cu_9S_5$), chalcocite ($Cu_2S$), bornite ($Cu_5FeS_4$), enargite ($CuAsS_4$), and tetrahedrite ($Cu_{12}Sb_4S_{13}$). Copper is concentrated in some magmatic segregation deposits along with nickel. (see chapter on ferro-alloy metals). The following are the most important types of copper deposits:

1. Porphyry copper deposits with their associated skarns,
2. Volcanic massive sulfide (VMS) deposits, and
3. Stratabound or stratiform sediment-hosted Cu (SSC) deposits.

## Porphyry Copper Deposits

*Porphyry copper deposits* (*PCDs*) contain concentrations of copper sulfide minerals within altered felsic to intermediate composition porphyritic intrusive rocks. *Porphyritic intrusive rocks* are igneous rocks whose magmas have intruded into the rocks above and solidified with a number of large crystals set in a fine-grained mineral groundmass. The magma from which the porphyry rocks crystallized is thought to form in the mantle wedge above subducting oceanic lithosphere that produces a volcanic arc above. It is called the mantle wedge because the mantle above the subducting slab is wedge shaped in cross-section.

Below these volcanic arcs, where magmas have risen to a depth of from 1 to 3 km of the surface and crystallized in place as a stock, a porphyry intrusive can develop. A *stock* is an intrusive igneous rock that covers less than 100 square kilometers in area and has steep contacts with the surrounding rocks. Many stocks are considered to be crystallized magma chambers at depth below a volcanic edifice on the surface. Multiple intrusive events producing stocks are common in areas where porphyry Cu mineralization has occurred. The Cu mineralized intrusion is typically the youngest and most chemically differentiated stock that contains the highest silica content.

Slow cooling of magma, in the magma chamber that will become the stock of interest, crystallizes large anhydrous minerals. The remaining magma becomes saturated with water. A separate water-rich phase containing abundant silica, HCl, $H_2S$, KCL, NaCl, and dissolved metals, such as Cu and Mo, develops. Because the volume of the aqueous phase plus anhydrous crystals is greater than the hydrous magma from which it formed, the pressure in the magma increases with increasing crystallization.

The elevated pressure causes country rocks to fracture and the aqueous fluid escapes from the magma chamber. The magma expands into the space previously occupied by the aqueous fluid, immediately decreasing in pressure and causing it to cool rapidly. This causes a sudden crystallization of the magma resulting in the fine-grained groundmass of the porphyritic texture. Small fractures that have developed in the country rock from rock failure at the top of the chamber are filled with the water-rich phase and precipitate quartz plus sulfide minerals in veinlets. At later times heat from the solid intrusion causes groundwater circulation through the fractures in the stock and the surrounding country rocks setting up a fluid convection system as shown schematically in **Figure 8.1**. The circulating fluid scavenges Cu from the surrounding rocks and precipitates Cu ore minerals in veins along the sides and the top of the stock. Thus, two separate water-mediated ore forming events, driven by magmatism and its heat, are superimposed in a porphyry ore deposit.

There are about 200 important porphyry copper and porphyry molybdenum deposits distributed from the Aleutian Islands southward through Canada, the U.S., and Mexico to Chile and Peru along the current and earlier volcanic arc terrains. In the U.S., the volcanic arc is represented by the intrusive rocks of the Sierra Nevada Mountains. Note in **Figure 8.2** that most of the porphyry copper deposits that have not been uplifted and eroded by the rise of the Sierra Nevada occur on their extension southeastward through Arizona and northern Mexico.

Gases, such as HCl and $H_2S$ dissolved in the aqueous phase, produce acid on cooling by reactions like

$$HCl_{(gas)} \rightarrow H^+ + Cl^- \qquad [8.1]$$

and

$$2H_2S_{(gas)} \rightarrow H^+ + HS^-. \qquad [8.2]$$

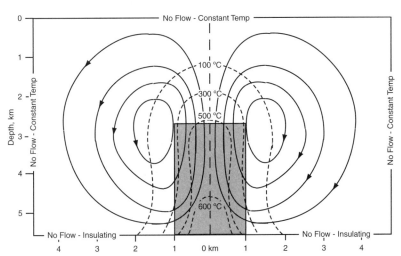

*Figure 8.1* Model of H$_2$O circulation and temperature distribution around an intruded igneous stock (shaded) with initial temperature of 750°C after 20,000 years of cooling. The solid lines with arrows are fluid flow paths and the dashed lines are isotherms of the indicated temperature. (Adapted from Norton, D. and Cathles, L. M., 1979, Thermal aspects of ore deposition. In Barnes, H. (ed.) *Geochemistry of Hydrothermal Ore Deposits*. John Wiley & Sons, New York, pp. 611–631.)

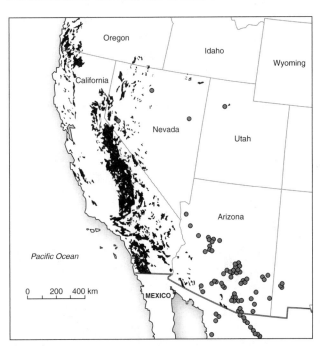

*Figure 8.2* Distribution of Mesozoic (250 million years ago to 65 million years ago) intrusive rocks shown in black and the locations of major porphyry copper deposits identified by a circle in the southwestern U.S. (Adapted from Theodore, T. G., 1977, Selected copper-bearing skarns and epizonal granitic intrusions in the southwestern United States, *Geol. Soc. Malaysia*, Bull 9, p. 31–50.)

This acid alters the rock, which increases the pH of the fluid through reactions like

$$2H^+ + 3KAlSi_3O_8 \rightarrow KAl_3Si_3O_{10}(OH)_2$$
K-feldspar          muscovite
$$+ 6SiO_2 + 2K^+. \qquad [8.3]$$
aqueous

On cooling the increased $SiO_2$ from reaction [8.3] precipitates out as quartz in veins along with copper sulfides like chalcopyrite by a reaction like

$$Cu^{2+} + Fe^{2+} + 2HS^- \rightarrow CuFeS_2 + 2H^+ \qquad [8.4]$$
chalcopyrite

that is driven to the right by the increase in pH. The alteration zones in the deposits and the location of ore minerals are shown schematically in **Figure 8.3**.

Porphyry Cu deposits can have some large veins of very high grade Cu, but most of the Cu tends to be in tiny veinlets dispersed throughout the ore zone with grades from 0.25 to 2.0 wt% Cu. Because this ore zone is generally near the surface and of large volume, open pit mining techniques are typically used to obtain the ore. The largest open pit mine in the world excavates porphyry copper at Bingham, Utah. The pit is about 4 km across and almost 1 km deep. Cu reserves at Bingham are estimated to be 1,700 million metric tons (down to 0.71 wt% Cu) with the ore containing the sulfides chalcopyrite ($CuFeS_2$), bornite ($Cu_5FeS_4$), pyrite ($FeS_2$), and molybdenite ($MoS_2$).

**Secondary enrichment.** When porphyry Cu deposits are exposed at the earth's surface secondary or supergene enrichment of ores can occur by oxygenated surface water reacting with the sulfides. The abundant pyrite in most sulfide deposits reacts with groundwater and produces ferric hydroxides, *limonite*, and acid by the reaction:

$$5H_2O + 2FeS_2 + 15/2\ O_2 \rightarrow 2FeO(OH)$$
pyrite                    limonite
$$+ 8H^+ + 4SO_4^{2-}. \qquad [8.5]$$

The limonite produces a red-stained gossan because of its oxidized $Fe^{3+}$. Highly weathered rock that occurs above sulfide ore deposits is known as *gossan*. It can be weathered to such an extent that only iron oxides and quartz remain. Weather-resistant quartz and iron oxides in the gossan often causes it to be somewhat elevated above the surrounding land surface. Historically, gossans have been important guides for prospectors to find buried sulfide ore deposits.

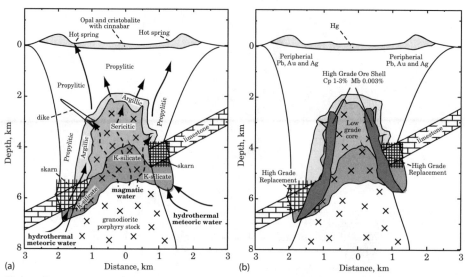

*Figure 8.3* Schematic diagram showing the development of a porphyry copper ore deposit (a) Alteration zones; (b) Location of ore deposition.

The acid produced by reaction [8.5] can mobilize Cu in Cu sulfides like chalcopyrite by dissolving them in solution:

$$2H^+ + CuFeS_2 + 4O_2 \rightarrow Cu^{2+} + Fe^{2+} + 2HSO_4^-. \quad [8.6]$$

There solutions can precipitate hydrated copper sulfate, or chalcanthite, at greater depth in an oxidized ore as shown in **Figure 8.4** by the reaction:

$$5H_2O + Cu^{2+} + HSO_4^- \rightarrow CuSO_4 \bullet 5H_2O + H^+. \quad [8.7]$$
$$\text{chalcanthite}$$

In the water below the water table Cu can react with pyrite decreasing the pH of the solution by the reaction:

$$28H_2O + 8FeS_2 + 15Cu^{2+} \rightarrow 15CuS + 8Fe(OH)_3$$
$$\text{pyrite} \qquad \text{covellite} \quad \text{limonite}$$
$$+ SO_4^{2-} + 32H^+ \qquad [8.8]$$

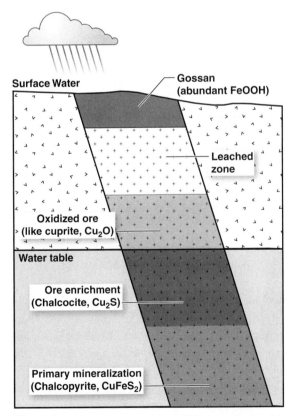

*Figure 8.4* Cross-section through a sulfide mineral vein with supergene enrichment showing the zones that form. (Adapted from Bateman, A. M., 1950, *Economic mineral deposits*, Wiley, New York, 916 p.)

to produce covellite and limonite. This enriches the rock with additional Cu. Alternatively the reaction:

$$10H_2O + 2FeS_2 + 6Cu^{2+} \rightarrow 3Cu_2S + 2Fe(OH)_3$$
$$\text{pyrite} \qquad \text{chalcocite} \quad \text{limonite}$$
$$+ SO_4^{2-} + 14H^+ \qquad [8.9]$$

that produces chalcocite and limonite can occur.

### Skarns

Skarns form at the contact of igneous intrusions and country rocks, most commonly granites intruding into carbonate rocks. An alteration zone of calc-silicate rocks is produced at the contact. When dolomite is present diopside and andradite garnet are produced and with calcite as the carbonate wollastonite is formed (**Figure 8.5**).

Skarns are divided into exoskarn indicating the altered sedimentary rock part and endoskarn to denote the altered igneous rock part of the skarn. Zonation of most skarns reflects the geometry of the igneous contact and the extent and composition of the metasomatic fluid given off by the igneous intrusion as outlined for a typical Cu skarn zonation around a porphyry igneous intrusion in **Figure 8.5**.

Depending on the skarn, economic concentrations of Au, Cu, Fe, Mo, Pb, Pt, REEs, Sn, U, W, and/or Zn can be present. The largest skarn deposits are iron skarns, mined for their magnetite content as shown in **Figure 8.6**. Examples of economic skarn deposits include the Pine Creek tungsten skarn in California, Ok Tedi gold and copper skarn deposit in New Guinea, zinc and lead skarn in Groundhog, New Mexico, and the tin mines in Moina, Tasmania.

### Volcanic Massive Sulfide Deposits

*Volcanogenic massive sulfide (VMS) deposits* contain a significant amount of the world's known ore of Cu, Zn, Pb, Au, and Ag. VMS deposits form from hydrothermal solutions produced by crystallizing magma and leached from country rocks that erupted on or intruded into sediments on the seafloor. They are deposits of massive concentrations of sulfide minerals of which ~90% is pyrite but

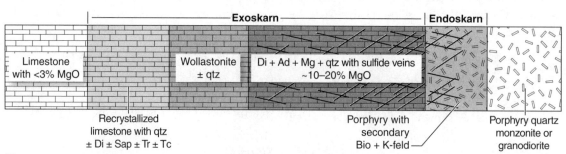

*Figure 8.5* Schematic cross-section of zonation in a mineralized porphyry skarn. Straight lines represent pyrite + chalcopyrite veins. Ad = andradite, Bio = biotite, Di = diopside, Mg = magnetite, K-feld = K-feldspar, qtz = quartz, Sap = saponite, Tc = talc, and Tr = tremolite. (Adapted from Theodore, T. G., 1977, Selected copper-bearing skarns and epizonal granitic intrusions in the southwestern United States, *Geol. Soc. Malaysia*, Bull 9, p. 31–50.)

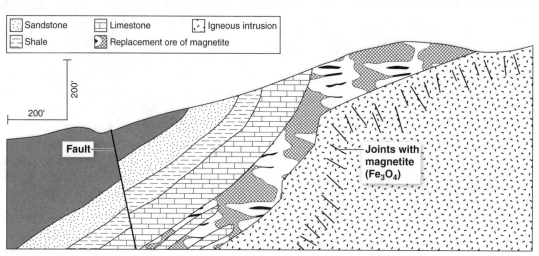

*Figure 8.6* Cross-section through the Fe-skarn deposit at Iron Spring, Utah. (Adapted from Gilluly, J, Waters, A.C., and Woodford, A.O., 1968, *Principles of geology*, 3rd ed,, Freeman, San Francisco, 687 p. and Mackin, J. H., 1947, Some structural features of the intrusions in the Iron Springs district [Utah]: Utah Geological Society, *Guidebook to the geology of Utah*, no. 2, 62 p.)

sometimes pyrrhotite ($Fe_{1-x}S$ with x = 0 to 0.2) is present together with the ore sulfides.

The deposits occur conformably with bedrock in lenses or sheets between volcanic layers or at volcanic-sedimentary interfaces of all geologic ages. Feeder zones that contain vein and disseminated ore occur in a pipe-like structure under the massive sulfide layer. VMS deposits are classified by their host rock type and on the basis of ore composition. These include:

> *Cyprus-type* (**Figure 8.7**) mafic volcanic hosted (30% of deposits),
>
> *Kuroko-type* (**Figure 8.8**) felsic volcanic hosted (50% of deposits), and

> *Besshi-type* volcanic-sedimentary boundary hosted deposits (20% of deposits).

Cyprus-type massive sulfides tend to be small and are thought to form in an environment similar to the black smokers presently observed at ocean crust spreading centers. Black smokers are hot springs on the seafloor that precipitate iron sulfides forming a dark cloud in the venting solution. Kuroko-type deposits apparently form in underwater silicic calderas of convergent oceanic plate boundaries. A caldera is a volcanic cauldron feature formed by collapse of the volcanic surface after an eruption. Examples of these environments occur along the submerged Izu-Bonin-Mariana

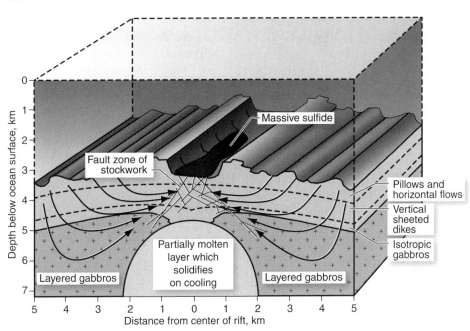

*Figure 8.7* Massive sulfide deposit of the Cyprus type forming at a mid-oceanic rift system showing pathways of circulation of fluid.

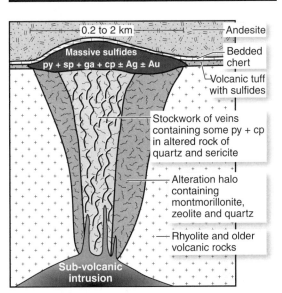

*Figure 8.8* Cross-section through a Kuroko-type zinc-lead-copper massive sulfide deposit. py = pyrite, sp = sphalerite, ga = galena, cp = chalcopyrite.

andesitic arc between Tokyo, Japan, and Guam where the Pacific plate is undergoing subduction beneath the Philippine Sea plate.

Besshi-type deposits occur in mixed volcanic-sedimentary environments. They often contain identifiable turbidites. These are sediments deposited by a turbidity current, that is an underwater avalanche of sediments. They are thought by many to be deposited in back-arc oceanic basins associated with sediments derived from island arcs.

Examples of VMS deposits include the following:

1.  Cyprus-type: Skouriotissa, Cyprus; Betts Cove, Newfoundland; Turner-Albright, Oregon; Big Mike, Nevada;
2.  Kuroko-type: Kuroko, Japan; Kidd Creek, Ontario; Iron King and Penn Mine, California;
3.  Besshi-type: Besshi, Japan; Windy Craggy, British Columbia.

### Stratabound or Stratiform Sediment-Hosted Cu Deposits

*Stratiform sediment-hosted copper (SSC) deposits* are found in sedimentary layers similar to MVT deposits and are thought by many to have a similar origin. 70% of the time the deposits occur in organic-rich calcareous shale layers and 30% of the time in sandstones. Deposits contain Cu sulfides together with some native Cu. Researchers are still unclear as to how the sulfides in these deposits formed. Are they *syngenetic*, having obtained their metals and formed the sulfides during the rock formation process, or *epigenetic*, enriched in the desired metal sulfides after the rock was first formed?

**Kupferschiefer deposit.** The Kupferschiefer SSC deposit of northern Europe occurs in 250- to 260-million-year-old sedimentary rocks. The Kupferschiefer sediments were deposited in the Zechstein *epicontinental sea* as given in **Figure 8.9**, shortly after the sea was formed by a rapid marine transgression. Epicontinental seas are produced when seawater extends over the interior of a continental land mass. The Kupferschiefer is a thin (< 5.5 m) bed of marine bituminous and calcium carbonate-rich mudstone containing sulfides. It rests on the Rotliegend sandstone and is overlain by

at least five depositional cycles of evaporite sediments of the Zechstein formation. Ore in the overlying Zechstein, as well as the underlying Rotliegend sandstone, is associated with the Rote Fäule facies of postdepositional oxidation. This suggests an epigenetic ore-forming process (**Figure 8.10**). However, the bulk of the evidence including sulfur isotope data points to fixation of metals as sulfides by bacteriogenic, syngenetic processes when the Kupferschiefer was being deposited or when sediments were starting to solidify. Most investigators argue the Rote Fäule alteration was restricted to sediments deposited on paleotopographic highs in the Kupferschiefer stage of deposition.

The Kupferschiefer has greater than 0.3 wt% Cu over 5% of its entire exposed area with some locations averaging 1.5 wt% Cu. The mine at Lubin, Poland has reserves of 2,600 million metric tons of 2 wt% Cu as well as 30 to 80 g Ag and 0.1 g Au per metric ton. The more mineralized regions in Germany and Poland have been mined systematically since medieval times.

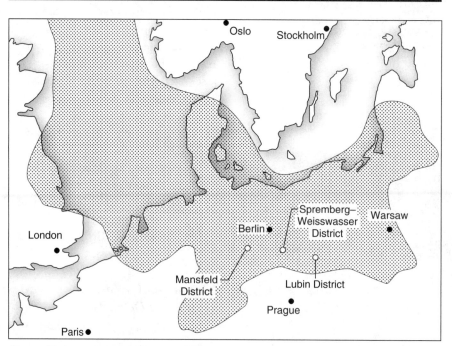

*Figure 8.9* Extent of the epicontinental Upper Permian Zechstein Sea in northern Europe and current mining districts in Germany and Poland. (Adapted from Evans, A. M., 1993, *Ore Geology and Industrial Minerals—An Introduction*, 3rd ed., Blackwell Scientific, Oxford, 400 p.)

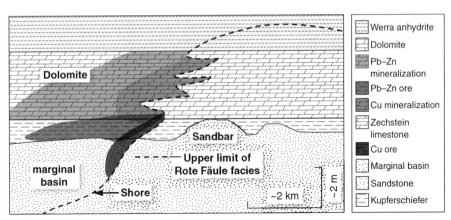

*Figure 8.10* Typical stratigraphic section of the lowermost Zechstein with Rote Fäule facies alteration. Copper ore bodies are next to the Rote Fäule with Pb-Zn mineralization further away. (Adapted from Brown, A. C., 1978, Stratiform copper deposits – Evidence for their post-sedimentary origin, *Minerals Sci. Engng.*, v. 10, n. 3, p 172–181.)

Next to porphyry Cu deposits SSC deposits are the most important source of Cu with grades from 1 to 5 wt% Cu. Co and Ag are often significant byproducts. These deposits include the Kupferschiefer of northern Europe, Zambian Copper Belt, White Pine deposits in Michigan, Udokan and Dzhezkazgan in Kazakhstan, Corocoro in Bolivia, and Dongchuan of China (Misra, 2000).

### Hydrothermal Vein Deposits

There are three sources of water that can carry the material deposited in hydrothermal vein deposits. First, a cooling magma can produce a separate water-rich phase during the final stages of its crystallization as outlined above for porphyry copper deposits. Secondly, fluid can be released from dehydration reactions taking place during metamorphism when water-rich minerals, such as micas and amphiboles, are buried and become unstable. This water can scavenge and transport metals to a fracture where ore minerals are precipitated. Thirdly, near the earth's surface, *meteoric* groundwater can be heated and scavenge, transport, and deposit metals. A combination of the three sources is possible. Observations made in some hydrothermal ore deposits suggest meteoric water has mixed with magmatic or metamorphic waters to produce the ore forming fluid.

**Hydrothermal veins at Butte, Montana.** Veins containing ore minerals were mined at Butte beginning in 1864 and mining operations terminated in 1983 when the open pit mines were closed (**Figure 8.11**). Because of the extensiveness of mineralization in the Butte district it was called "the richest hill on earth." The mines produced more than \$300 billion worth of metal in its lifetime. There are two vein types: pre-main-stage veins and main-stage veins. The host rock for the veins is a quartz monzonite occurring in the southwest corner of the Boulder Batholith, emplaced between 68 and 78 million years ago. A quartz monzonite is an intrusive igneous rock that contains 5% to 20% quartz, somewhat less than a true granite. A batholith is the term used to denote a very large mass of igneous rock, often a granite, solidified at depth and has least 100 km$^2$ of exposure on the earth's surface by erosion.

The pre-main stage mineralization consists of small quartz veins containing chalcopyrite ($CuFeS_2$) that evolved in time to contain molybdenite ($MoS_2$), found in a deep central mineralized zone. These veins have alteration halos of K-feldspar, biotite and sericite, a fine-grained muscovite. This is typical porphyry copper type alteration mineralization. The main stage mineralization occurred millions of years later than the pre-main stage mineralization. It likely leached

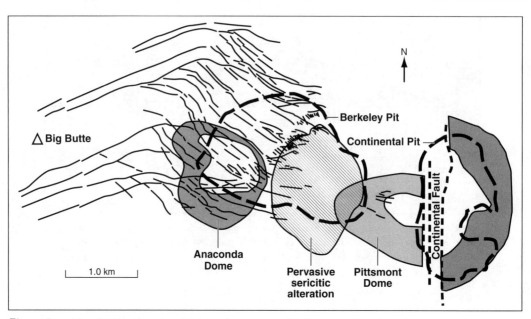

*Figure 8.11* Map showing the subsurface Main stage veins as black lines and the disseminated pre-Main stage porphyry Cu mineralization of the Anaconda and Pittsmont Domes in dark grey at Butte, Montana. Also shown in a lighter gray is a zone of pervasive sericitic alteration separating the two Cu mineralization domes. The outline of the Berkeley and Continental open pits are shown with dashed lines as well as the location of the Continental Fault zone that brought the lower Pittsmont Dome to the surface. (Modified from Rusk, B., Reed, M.H., and Dilles, J.H., 2008, Fluid inclusion evidence for magmatic-hydrothermal fluid evolution in the porphyry copper-molybdenum deposit, Butte, Montana: *Economic Geology*, v. 103, p. 307–334.)

the earlier mineralization and produced a set of large veins termed the Anaconda and Blue veins. All these veins exhibit similar mineralization with a central zone of copper mineralization containing chalcocite ($Cu_2S$) and enargite ($Cu_3AsS_4$), that changes outwardly to an intermediate zone of chalcopyrite ($CuFeS_2$) with minor sphalerite (ZnS). An outermost peripheral zone contains sphalerite, rhodochrosite, $MnCO_3$, and Ag mineralization (**Figure 8.12**).

## Production and Reserves

**Table 8.1** gives 2011 Cu production and reserves. Worldwide 60% of copper is produced from porphyry Cu and associated skarn deposits, 20% from sediment-hosted stratiform deposits, and 12% from volcanic-hosted massive sulfides. The land-based Cu reserve base is estimated to be 1 billion metric tons. *Deep-sea manganese nodules* could supply an additional 700 million tons of Cu at significantly

increased costs (See discussion of Figure 6.15). Note that Chile produces 1/3 of the world's copper. Given the reserves, the reserve base, and increasing yearly consumption of Cu, reserves appear adequate for 50 years but for less than 100 years.

# Lead (Pb)

Lead (Pb) is estimated to be present on average at the 17 to 18 ppm level in the earth's upper continental crust. It is considered a toxic heavy metal. Heavy metals are metallic elements that have high density and molecular weight.

## Use

Pb became the dominant base metal used into the early 1900s as it was employed in weights, solders, ceramic glazes, and glassware. Lead has a combination of properties that make it valuable. In particular, it is dense, soft, easily worked, and

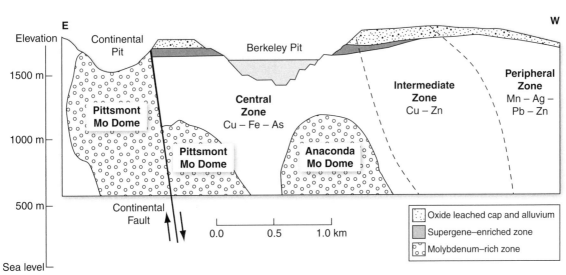

*Figure 8.12* Simplified East-West schematic cross-section through the Butte, Montana mining district showing the mineralization that has occurred and the vertical extent of the Anaconda and Pittsmont domes. (Adapted from Gammons, C.H., Metesh, J.J., and Duaime, T.E., 2006, An overview of the mining history and geology of Butte, Montana, Technical Communications, Special Publication, *Mine Water and the Environment*, 25(2): 70–75.)

*Table 8.1* COPPER PRODUCTION FOR 2011, RESERVES, AND RESERVE BASE BY COUNTRY. *

| COUNTRY | PRODUCTION | RESERVES | RESERVE BASE |
|---|---|---|---|
| Chile | 5,420 | 190,000 | 360,000 |
| Peru | 1,220 | 90,000 | 120,000 |
| China | 1,190 | 30,000 | 63,000 |
| United States | 1,120 | 35,000 | 70,000 |
| Australia | 940 | 86,000 | 100,000 |
| Zambia | 715 | 20,000 | 35,000 |
| Russia | 710 | 30,000 | 30,000 |
| Indonesia | 625 | 28,000 | 38,000 |
| Canada | 550 | 7,000 | 20,000 |
| Poland | 425 | 26,000 | 48,000 |
| Mexico | 365 | 38,000 | 40,000 |
| Kazakhstan | 360 | 7,000 | 22,000 |
| Other countries | 2,000 | 80,000 | 110,000 |
| World totals | 16,100 | 690,000 | 1,000,000 |

* In thousands of metric tons of Cu.

*Data from*: U.S. Geological Survey, *Mineral Commodity Summaries*, January, 2012.

corrosion resistant. Due to its toxicity it has been used as an antifouling agent in paints on boats to stop the growth of algae and barnacles. Pb can also be used as an antiknock compound in petroleum fuel as it increases petroleum performance by increasing its octane rating.

At the present time lead-acid battery (**Figure 8.13**) production for automobiles and trucks is the principal use of lead, accounting for 87% of U.S. consumption for 2011. About 10% of lead is incorporated into ammunition, weights, pipes, sheets (including radiation shielding), cable covers, caulk, solder and oxides for ceramics glazes, glass, and pigments. Many of these are being phased out because of the toxicity of lead to humans.

## Resource Location

Lead, along with zinc, occurs in stratabound or stratiform sediment-hosted exhalative (SEDEX) deposits and in *Mississippi Valley-type (MVT) deposits*. Given in **Figure 8.14** are the tonnage and grade of known MVT and SEDEX deposits. Note the impressive size and grade of the SEDEX deposits. In these deposits lead occurs in the mineral galena (PbS), while zinc is in sphalerite (ZnS).

### Sedimentary Exhalative Deposits

Sedimentary exhalative (SEDEX) deposits consist of finely layered ores present in shales, siltstones, and carbonates. They are formed during sediment deposition and early rock-forming processes from reactions with hydrothermal solutions. Most occur in black shales. Black shales are organic-rich shales deposited under anaerobic conditions. SEDEX deposits are considered to form by release of ore-bearing hydrothermal fluids into the ocean or a lake resulting in the precipitation of ore minerals in the depositing sediments before the sediments are lithified. Water depths of 50 to 800 m in a continental rift environment have been postulated for most (Sangster, 1990).

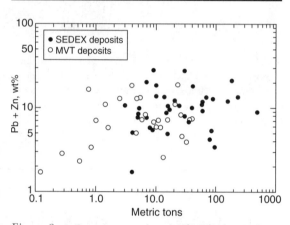

*Figure 8.14* Ore tonnage and grade of Mississippi Valley-type (MVT) and sedimentary exhalative (SEDEX) deposits. (Adapted from Sangster, D. F., 1990, Mississippi Valley-type and SEDEX lead-zinc deposits: a comparative examination: Institution of Mining and Metallurgy, *Transactions, Section B: Applied Earth Science*, v. 99, p. 21–42.)

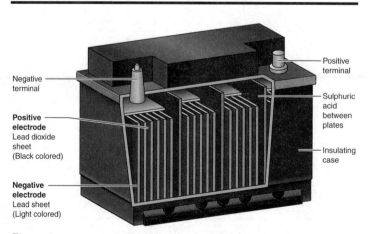

*Figure 8.13* Lead-acid automotive battery with separators between the + and − plates to keep them from short-circuiting.

Important deposits include: Broken Hill, McArthur River, and Mount Isa, Australia; Howard's Pass and Sullivan, Canada; Jiashengpan, China; Silvermines, Navan, Ireland; and Red Dog, Alaska.

## Mississippi Valley-Type Deposits

Mississippi Valley-type (MVT) sulfide ore deposits are either carbonate-hosted where Zn > Pb >> Cu or sandstone-hosted with Pb > Zn >> Cu. There is a reasonable consensus that topographically driven connate basinal fluids have traveled hundreds of kilometers in stable continental rocks and precipitated Zn-Pb ores where Zn is dominant in carbonate sediments with Pb-rich ores found in sandstones (**Figure 8.15**). Connate fluids are those that were trapped in the pores of a sedimentary rock when it was formed. Most deposits occur near the tops or along the flanks of regional structural highs. In the U.S. this would be the Ozark dome and Cincinnati Arch as shown in **Figure 8.16**. These deposits are termed Mississippi Valley-type because they are described from the watershed of the Mississippi River in Missouri, Oklahoma, Kansas, Illinois, and Tennessee. MVT deposits also occur at Pine Point and Polaris in Canada, the Silesian district of Poland, the Mechernich of Austria, the Pering of South Africa, and the Lennard Shelf, Sorby Hills, and Coxco of Australia.

Fluids that transported the Zn, Pb, and Cu to the site of deposition appear to have greater than 15 wt% NaCl equivalent and contain $SO_4^{2-}$. Temperatures of the brines are determined to be between 100° and 150°C. The Pb and Zn were likely carried from the basins to their site of deposition by Pb-Cl and Zn-Cl aqueous species. Most investigators believe the needed sulfur was carried as sulfate in solution and the reduction of this sulfur to sulfides occurred by reaction with organic matter or hydrocarbons at the site of ore deposition.

The sulfate in solution can be reduced by a reaction like

$$2H^+ + SO_4^{2-} + C_8H_{18} \rightarrow H_2S + C_8H_{10} + 4H_2O \qquad [8.10]$$
$$\text{octane} \qquad \text{ethylbenzene}$$

if the reducing agent is octane. If a natural gas petroleum cap is encountered then the reaction is likely

$$2H^+ + SO_4^{2-} + CH_4 \rightarrow H_2S + CO_2 + 2H_2O. \qquad [8.11]$$
$$\text{natural gas}$$

These reactions imply the mixing of a reduced and more oxidized fluid leads to the production of $H_2S$ and that this causes supersaturation of the

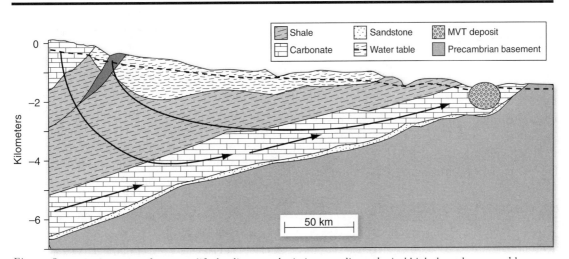

*Figure 8.15* Fluid transport from an uplifted sedimentary basin into an adjacent basinal high through a permeable carbonate unit. (Adapted from Garven, G., and Freeze, R. A., 1984, Theoretical analysis of the role of groundwater flow in the genesis of stratabound ore deposits 1—Mathematical and numerical model: *American Journal of Science*, v. 284, p. 1085–1124.)

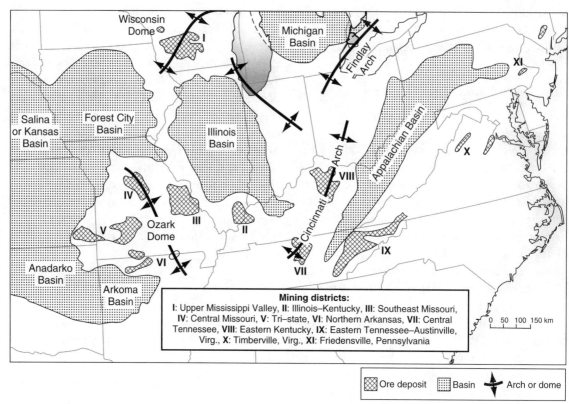

*Figure 8.16* Map showing the relationship of MVT deposits to basins, arches, and domes in the mid-U.S. mining districts: **I**-Upper Mississippi Valley; **II**-Illinois-Kentucky; **III**-Southeast Missouri; **IV**-Central Missouri; **V**-Tri-State; **VI**-Northern Arkansas; **VII**-Central Tennessee; **IX**-Eastern Tennessee—Austinville, Virginia; **X**-Timberville, Virginia; **XI**-Friedensville, Pennsylvania. (Adapted from Heyl, A. V., 1969, Some aspects of genesis of zinc-lead-barite-fluorite deposits in the Mississippi Valley, U.S. A.: *Inst. Mining Metallurgy Trans., sec. B*, v. 78, p. 148–160.)

mixed fluid with galena and sphalerite. Therefore, these minerals precipitate by reactions like

$$PbCl_2 + H_2S \rightarrow PbS + 2H^+ + 2Cl^- \qquad [8.12]$$

and

$$ZnCl_2 + H_2S \rightarrow ZnS + 2H^+ + 2Cl^-. \qquad [8.13]$$

It is argued the reason most of these deposits are near the tops or along the flanks of regional structural highs is this is where the needed reducing fluid exists in the form of petroleum accumulations.

## Production and Reserves

Given in **Table 8.2** are mine production and reserves of Pb for the indicated country. Worldwide yearly production in 2011 was about 8 million metric tons with about half of it produced from recycled scrap.

Production has been increasing worldwide. China's mines supply over 40% of the world's added Pb. At current use rates, the word's supply of lead is estimated to run out in 2050 (Cohen, 2007).

## Health Effects

If lead is inhaled or ingested, it interferes with the production of red blood cells in the body. Lead poisoning causes damage to nerve connections, especially in young children, as well as blood and brain disorders. More than 90% of the lead intake accumulates in the bones, where it is stored. Lead in bones is released into the blood, re-exposing organ systems long after the original exposure. Serious, permanent organ damage occurs at levels above 800 µg/L in the blood. The mean level of Pb in U.S. adults is 30 µg/L.

*Table 8.2* Lead mine production for 2011, reserves, and reserve base by country.*

| COUNTRY | PRODUCTION | RESERVES | RESERVE BASE |
|---|---|---|---|
| China | 2,200 | 14,000 | 36,000 |
| Australia | 560 | 29,000 | 59,000 |
| United States | 345 | 6,100 | 19,000 |
| Peru | 240 | 7,900 | 10,000 |
| Mexico | 225 | 5,600 | 10,000 |
| India | 120 | 2,600 | NA |
| Russia | 115 | 9,200 | NA |
| Bolivia | 85 | 1,600 | NA |
| Canada | 75 | 450 | 5,000 |
| Sweden | 70 | 1,100 | 2,000 |
| South Africa | 55 | 300 | 700 |
| Ireland | 50 | 600 | NA |
| Poland | 40 | 1,700 | 5,400 |
| Other countries | 340 | 5,000 | 30,000 |
| World totals | 4,500 | 85,000 | 170,000 |

* In thousands of metric tons.
NA = Not available.

*Data from*: U.S. Geological Survey, *Mineral Commodity Summaries*, January, 2012.

Given the hazardous nature of Pb many uses are being discontinued. Since the early 1970s lead-based paint for most uses and the use of lead solder in food cans has been eliminated. By the end of 2000, 42 countries including the U.S. had phased out lead in gasoline and in 2002 leaded gasoline was banned in the European Union. As of 2011 all countries except Afghanistan, Myanmar, North Korea, Algeria, Yemen, and Iraq have banned lead in gasoline.

# Zinc (Zn)

Zinc is the fourth most widely used metal after Fe, Al, and Cu. The average abundance of Zn in the earth's upper continental crust is estimated to be between 52 and 70 ppm, making it one of the most abundant of the scarce metals and the most abundant base metal. Only Cr and V possibly are more abundant scarce metals in the earth's crust.

## Use

The uses of zinc are outlined in **Figure 8.17**. More than half of the Zn mined is used for galvanizing. Galvanizing is a process where a thin coat of Zn is electro-deposited on steel to keep it from rusting. The second largest use is in metal alloys that are diecast to make metal parts. With diecasting, molten metal under high pressure is forced into a mold. This produces an object with good surface finish and dimensional consistency. About 10% of Zn is used to make brass, a copper plus zinc alloy. Sheets of zinc are used as roofing material. Other uses include ZnO in paint pigments, ointments, lotions, and creams to prevent sunburn.

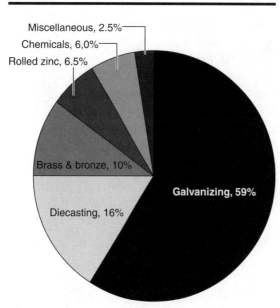

*Figure 8.17* Applications for the use of zinc.

## Resource Location

Zinc occurs in the mineral sphalerite (ZnS). Sphalerite is precipitated together with other sulfide minerals such as lead, copper, gold, and/or silver sulfides in VMS, SEDEX, and MVT-type deposits. Zn mines co-produce, in order of decreasing tonnage, Pb, $H_2SO_4$, Cd, Ag, and Au. The largest known concentration of Zn in the world is the SEDEX deposit at Red Dog, Alaska. A Zn-rich solution was likely injected from below into a 330-million-year-old organic and sulfur-rich mud on the sea floor. It precipitated sphalerite on cooling.

## Production and Reserves

Given in **Table 8.3** are the estimated production and reserves for Zn for the indicated country. Note that China accounts for over 30% of world production while the U.S. production is a little over 6%. Both these numbers are greater than each country's share of the world population. At the current rate of consumption, worldwide Zn reserves and reserve base are estimated to last until 2053 (Cohen, 2007). However, a large resource base exists with identified zinc resources estimated to be 1.9 billion metric tons.

*Table 8.3* ZINC PRODUCTION FOR 2011, RESERVES, AND RESERVE BASE BY COUNTRY.*

| COUNTRY | PRODUCTION | RESERVES | RESERVE BASE |
|---|---|---|---|
| China | 3,900 | 43,000 | 92,000 |
| Peru | 1,400 | 19,000 | 23,000 |
| Australia | 1,400 | 56,000 | 100,000 |
| India | 790 | 12,000 | NA |
| United States | 760 | 12,000 | 90,000 |
| Canada | 660 | 4,200 | 30,000 |
| Mexico | 630 | 17,000 | 25,000 |
| Kazakhstan | 500 | 12,000 | 35,000 |
| Bolivia | 430 | 5,000 | NA |
| Ireland | 350 | 1,800 | NA |
| Other countries | 1,600 | 68,000 | 87,000 |
| World totals | 12,400 | 250,000 | 480,000 |

* In thousands of metric tons.

NA = Not available.

*Data from*: U.S. Geological Survey, *Mineral Commodity Summaries,* January, 2012.

# Tin (Sn)

Tin (Sn) is present in the earth's upper crust on average at the 2 to 6 ppm level. Since the late fourth millennium BC bronze tools and weapons were produced by adding 4% to 5% tin to copper to harden it. This discovery led to what is called the Bronze Age. This age grew out of the Stone Age but was before the Iron Age where advanced metalworking was widespread. Tin foil was once a common material used for wrapping food and other small items. It was replaced in the early twentieth century by cheaper aluminum foil that is still sometimes referred to as "tin foil."

## Use

As given in **Figure 8.18** a little over half of the tin produced is combined with another element to lower its melting temperature and produce solder. Most of this is a combination with lead to produce tin-lead solder. This is used extensively for electrical as well as plumbing connections. Lead-free tin-silver-copper solder is used extensively in the electronics industry.

Presently about 16% of Sn is used as tinplate, in tin cans, and in other containers but this is decreasing. Al and plastic-lined containers now dominate the "tin can" market. Chemical uses of tin compounds include stannous fluoride ($SnF_2$), which is added to toothpaste as a source of fluoride to prevent tooth decay. However, the main chemical use of Sn is in polyvinyl chloride (PVC) to prevent heat damage due to HCl loss. Window glass is currently produced by floating molten glass on a bed of molten tin. Liquid tin is used because it is immiscible with the molten glass and has a high density so glass floats on it.

## Resource Location

Tin is obtained from the mineral cassiterite ($SnO_2$). Cassiterite is precipitated from aqueous fluids in pegmatites associated with silica-rich igneous rocks such as granites, in porphyry tin deposits, and in skarns. Pegmatites are coarse-grained granites often rich in rare elements. They will be discussed later.

Given in **Figure 8.19** is a model of a porphyry tin deposit. These are thought to occur in and under andesitic-rhyolitic volcanic cones. They differ from Cu porphyry deposits in occurring in more silica-rich rocks (see **Figure 8.3**). Similar to Cu porphyries, Sn porphyries also produce their porphyritic texture by creating a $H_2O$-rich fluid phase exsolved from a silica magma phase that fractures the rocks above and causes rapid cooling as pressure drops in the magma. The fractures also allow groundwater circulation and therefore alteration to occur around the stock. Late stage veins of $SnO_2$ + native Ag typically cut across the altered rocks.

Large Sn placer deposits occur in alluvium derived from pegmatite containing granite intrusives in the Kochiu deposits of China. *Alluvium* is material left by flowing water in a river or stream. This includes riverbed, flood plain, and delta deposits. Deposits in the pegmatites themselves are mined in the Andes Mountains. Bolivia has major tin deposits associated with porphyry stocks.

Cassiterite ($SnO_2$) has a high density of 7.0 g/cm³. It is also chemically unreactive. Therefore, cassiterite is a detrital mineral found in *placer deposits*. Placers are alluvial deposits that concentrate a

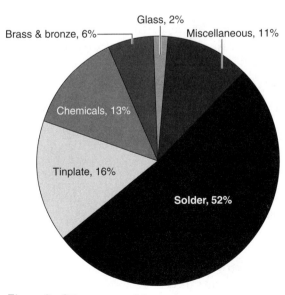

*Figure 8.18* Present uses of tin.

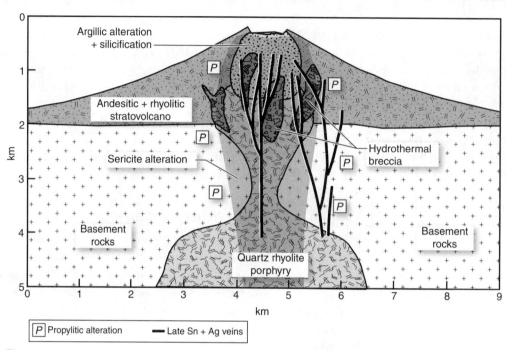

*Figure 8.19* Cross-section through a typical porphyry tin deposit showing a hydrothermally altered quartz rhyolite porphyry with late-stage veins of Sn + Ag. Quartz rhyolite is a rock that contains large crystals of quartz set in a rhyolitic groundmass.

mineral of interest by density differences. These include unconsolidated alluvial deposits located in gravel beds along stream channels close to highly weathered granite intrusions. Dredges are often used to extract the Sn-rich gravel. Significant placer deposits occur in the southeast Asian countries of China, Thailand, Myanmar, Malaysia, and Indonesia. The deposits typically only contain 0.015 to 0.025 wt% Sn. However, over 80% of the world's tin is extracted from these low-grade gravel deposits. Secondary, or scrap tin is also an important source in the supply of tin.

## Production and Reserves

Given in **Table 8.4** is the estimated production and reserves of Sn. Note that China and Indonesia are the major producers with no production in the U.S. Given the current production and the available reserve base, sufficient Sn is available for less than 50 years.

# Mercury (Hg)

Mercury (Hg) is also referred to by the old-fashioned word, quicksilver, as it is a liquid metal with a silver color. Hg is a rare element in the earth's upper crust, having an estimated average abundance by mass of only 0.01 to 0.06 ppm.

## Use

One-third of mercury is manufactured into cathodes for mercury-zinc batteries. These batteries have a long shelf life, up to 10 years, and produce a steady voltage. Button-type batteries have been used in watches and calculators, although these are being replaced because of health concerns about Hg.

Another third of Hg is used in the production of chlorine gas ($Cl_2$) and sodium hydroxide (NaOH). An electric current is passed through a solution of NaCl brine in the presence of Hg. Na forms

*Table 8.4* Tin production for 2011, reserves, and reserve base by country.*

| COUNTRY | PRODUCTION | RESERVES | RESERVE BASE |
|---|---|---|---|
| China | 110 | 1,500 | 3,500 |
| Indonesia | 51 | 800 | 900 |
| Peru | 34.6 | 710 | 1,000 |
| Bolivia | 20.7 | 400 | 900 |
| Australia | 19.5 | 180 | 300 |
| Brazil | 12 | 590 | 2,500 |
| Vietnam | 6.0 | NA | NA |
| Congo (Kinshasa) | 5.7 | NA | NA |
| Malaysia | 2.0 | 250 | 600 |
| Russia | 1.0 | 350 | 350 |
| Portugal | 0.1 | 70 | 80 |
| Thailand | 0.1 | 170 | 200 |
| Other countries | 2.0 | 180 | 200 |
| United States | 0.0 | 0.0 | 40 |
| World totals | 270 | 1,800 | 11,000 |

* In thousands of metric tons of Sn.
NA = Not available.

*Data from*: U.S. Geological Survey, *Mineral Commodity Summaries*, January, 2012.

an amalgam with the mercury at the cathode and chlorine gas is released. Amalgam is a term used to denote an Hg-containing alloy. Mercury forms alloys, that is, amalgams with most other metals. The Na in the amalgam is reacted with water to produce NaOH.

A final third of Hg is used in lamps, measuring devices for temperature and pressure, and in amalgams that do not contain Na. Amalgams of Ag are used as tooth fillings. Both compact and traditional fluorescent lamps often use amalgams and all contain mercury vapor.

Given the high price of gold a large number of small, informal mining operation have been started to obtain gold from low-grade gold-rich sediments and rocks. The time-honored technique to separate flakes of gold from the sediments or from crushed rocks, going back to the time of the Romans, is to add mercury. An amalgam of gold-mercury is formed which is larger, heavier, and easier to separate from the sediments or pulverized rock. In a slurry of silicate mineral that contains the amalgam, the amalgam will sink more rapidly leading to its separation.

An estimated 10 to 15 million miners worldwide, including millions in every province and municipality except Shanghai in China, are employed in small informal gold mining operations that use mercury. Other locations in the world where large numbers of informal gold mining operations with mercury are in operation include Central Kalimantan, Indonesia and Madre de Dios, Peru. After separation the gold is purified by heating the amalgam to release the mercury as vapor. It has been estimated that informal mining releases between 650 and 1,000 metric tons of mercury a year into the air and water.

## Resource Location

Hg occurs in cinnabar (HgS) found in deposits produced by hot springs at the tops of hydrothermal deposits and also in silica-carbonate type deposits. Small amounts of liquid mercury as well as cinnabar can be present within these rocks. Silica-carbonate deposits obtain Hg, $H_2S$, and $CH_4$ gasses from organic-rich sedimentary rocks in regions of elevated heat flow. These gasses rise, are trapped, and react to form silica-carbonate rocks above. Often these are ophiolite ocean crust sections that contain the mineral serpentine. The released methane reacts with serpentine and oxygenated water by a reaction like

$$Mg_3Si_2O_5(OH)_4 + 4CH_4 + 7O_2 \rightarrow 3MgCO_3 + 2SiO_2$$
$$\text{serpentine} \qquad\qquad \text{carbonate silica}$$
$$+ CO_2 + 8H_2O. \quad [8.14]$$

Oxygenated water in these rocks reacts with the Hg and $H_2S$ gasses to produce cinnabar:

$$2Hg + 2H_2S + O_2 \rightarrow 2HgS + 2H_2O. \quad [8.15]$$

The Almadén mines in Spain opened in 0 AD but closed in 2000 having produced the most mercury of any mine in the world, about 250,000 metric tons. At these mines cinnabar is found in quartzite. One model of formation of this deposit is that deltaic sand was deposited over mercury containing black shales. Hg was released from the black shales and deposited when oxidized sulfur containing seawater was encountered in the sands precipitating cinnabar.

## Production and Reserves

Presently the world's three largest producers of Hg in descending order are China, Kyrgyzstan, and Peru. Most of the Chinese mercury deposits are carbonate dominant silica-carbonate deposits while those in Kyrgyzstan and Peru are associated with hydrothermal hot spring deposits. No mercury is produced and no reserves exist in the U.S.

Mercury is recovered from recycled products that include batteries, thermostats, compact and traditional fluorescent lamps, dental amalgam, and other medical devices. A significant amount of mercury is also obtained from cleanup of mercury-contaminated soils.

The price of mercury has been highly volatile over the years and in 2011 it averaged about US $540 per 76-pound (34.46 kg) flask. One metric ton (1,000 kilograms) is a little over 29 flasks. Given in **Figure 8.20** is the price of mercury and its worldwide production back to 1970. Note the

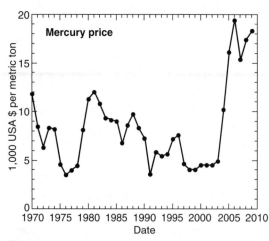
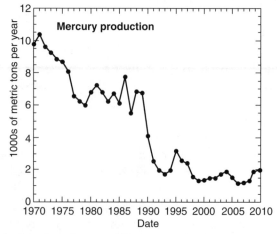

*Figure 8.20* Average annual mercury price and production by date. (Data from: Mercury Statistics U.S. Geological Survey, October, 2010.)

*Table 8.5* MERCURY PRODUCTION FOR 2011, RESERVES, AND RESERVE BASE BY COUNTRY.*

| COUNTRY | PRODUCTION | RESERVES | RESERVE BASE |
|---|---|---|---|
| China | 1,400 | 21,000 | NA |
| Kyrgyzstan | 250 | 7,500 | 13,000 |
| Peru | 35 | NA | NA |
| Other countries | 130 | 38,000 | 61,000 |
| Spain | NA | NA | 90,000 |
| United States | 0.0 | 0.0 | 7,000 |
| World totals | 1,930 | 93,000 | 240,000 |

\* In metric tons.
NA = Not available.

*Data from*: U.S. Geological Survey, *Mineral Commodity Summaries*, January, 2012.

production has been low for the last 10 years, but the price has tripled in the last 6 years.

As given in **Table 8.5** total world mercury production is almost 2,000 metric tons per year. There are reserves of Hg of 67,000 metric tons and a reserve base of 240,000 metric tons. These observations, combined with the fact the estimated worldwide resources of mercury are 600,000 metric tons, indicates the mercury supply, despite its low concentration in the upper continental crust, should not be a concern for over a hundred years.

## Health Concerns

Mercury is highly toxic to humans with many debilitating effects, generally from attacks on the nervous system. Exposure can be from inhalation, ingestion, or absorption through the skin. The largest source of mercury pollution in the world is from burning of coal. An estimated 3,000 metric tons per year is released as Hg vapor into air. Atmospheric concentrations of Hg are two to three times what they were 150 years ago. Much of this Hg is incorporated into rain and finds its way to the ocean from rivers and groundwater as well as direct precipitation from the atmosphere.

Mining and processing mercury ore can expose workers to mercury vapor as well as to direct contact with the skin. Historically Hg has been used to help in the gravity separation of fine-grained gold from silicate minerals as outlined under Uses. The gold is separated from mercury by boiling away the Hg. This was discontinued by most large-scale mining companies because of the harmful effects of the Hg vapor. However, most small-scale gold production operations still use mercury to extract the gold because it is not difficult to use and cheap. China is the largest user of mercury in gold mining. Mercury exposure is also a problem in the production of $Cl_2$ and NaOH.

Outside of industrial settings, most exposure to Hg comes from the ingestion of fish into which mercury has accumulated. Hg bioaccumulates in an organism. *Bioaccumulation* is the progressive increase in the amount of a chemical in an organism because the rate of intake exceeds the organism's ability to remove it. Of particular concern is the eating of sharks, swordfish, and king mackerel as they are large consumers of other fish that can contain mercury that has already bioaccumulated in their bodies with time from seawater. Although highly variable on average these fish contain about 1 ppm Hg while fresh water fish contain only about 0.02 ppm Hg.

The maximum contaminant level set by the U.S. for inorganic mercury in drinking water is 2 ppb and mercury in seafood is 1 ppm. Children under age 6 and women who are pregnant are most vulnerable to mercury's harmful effects. The U.S. Environmental Protection Agency (EPA) has banned the use of Hg for many applications because of the concerns about environmental contamination.

# Cadmium (Cd)

Cadmium (Cd) is an extremely toxic high-tech metal. It is rather rare in the earth's upper continental crust with an estimated abundance between 0.075 and 0.102 ppm. It also has a low concentration in seawater of ~0.1 µg/L.

## Use

Three-fourths of Cd is used in NiCd batteries. It is also added as a pigment in paints making very bright yellows, reds, blues, and greens. Cadmium telluride is employed in constructing thin-film solar cells. Cd-electroplating protects Fe metals from oxidation and these coatings reduce wear by lowering surface friction and are easy to solder.

## Resource Location

Cd most often occurs as a replacement of Zn in sphalerite (ZnS), where between 0.25 and 0.5 wt% of the Zn is often replaced. Occasionally greenockite (CdS) is found. This is the only Cd mineral of importance. Most Cd is obtained as a byproduct of smelting sphalerite ores.

## Production and Reserves

While production of Cd is tied to production of Zn, significant Cd is also recovered from spent consumer and industrial NiCd batteries. Given in **Table 8.6** is Cd production by country and their reserves. Note that the largest producer of Cd is China, but the Republic of Korea, Kazakhstan, and Japan are also significant producers. Global use of Cd is expected to remain constant into the near future. Increased applications are balanced by

*Table 8.6* CADMIUM PRODUCTION FOR 2011, RESERVES, AND RESERVE BASE BY COUNTRY.*

| COUNTRY | PRODUCTION | RESERVES | RESERVE BASE |
|---|---|---|---|
| China | 7,500 | 92,000 | 280,000 |
| Republic of Korea | 2,500 | NA | NA |
| Japan | 2,000 | NA | NA |
| Kazakhstan | 1,800 | 35,000 | 89,000 |
| Canada | 1,300 | 18,000 | 84,000 |
| Mexico | 1,500 | 48,000 | 39,000 |
| India | 660 | 130,000 | 49,000 |
| United States | 600 | 39,000 | 67,000 |
| Netherlands | 580 | NA | NA |
| Poland | 550 | 16,000 | NA |
| Germany | 400 | NA | 8,000 |
| Peru | 400 | 45,000 | 87,000 |
| Australia | 380 | 61,000 | 280,000 |
| Russia | NA | 21,000 | 37,000 |
| Other counties | 1,300 | 130,000 | 2000,000 |
| World totals | 21,500 | 640,000 | 1,200,000 |

* In metric tons.
NA = Not available.

*Data from*: U.S. Geological Survey, *Mineral Commodity Summaries*, January, 2012.

lower NiCd battery use as NiCd batteries are being replaced by Li-ion batteries. The reserve base for the future appears to be adequate for over 50 years.

## Health Concerns

Cd doesn't have a constructive purpose in the human body. Instead Cd is quite toxic even in low concentrations as it bioaccumulates in human tissue. Cd poisoning is a hazard associated with industrial processes that use Cd such as metal plating and the production of nickel-cadmium batteries, pigments, plastics, and other synthetics. The primary route of poisoning in industrial settings is by inhalation. In some industrial and mining areas that handle Cd, cadmium levels have increased in the water and soil. In Japan a large community of people consumed rice that was grown in water that was contaminated with Cd from mine tailings. This lead to itai-itai disease, also termed "ouch-ouch sickness," a softening of the bones and kidney failure. Cd in metal painted toys can lead to poisoning by sucking or chewing on the toys. Environmental Cd levels peaked in the 1960s. Since then, these levels have decreased due to better production technologies and Cd disposal methods.

# Fluorine (F)

Fluorine (F) is a highly reactive nonmetal present in fluorite, $CaF_2$. It is commonly found in Pb and Zn deposits and exists in the earth's crust at an average concentration reported to be between 290 and 585 ppm.

## Use

The principal use of fluorine depends on the grade of fluorite ore from which it is obtained. The lowest grade fluorite ore is used as a flux in steel and aluminum production and fluoridation of municipal water supplies. In steel and aluminum production fluorine lowers the melting point and increases fluidity of the melt. It also helps in the removal of sulfur and phosphorous from the metal into the slag. In municipal water supplies fluorine is added to drinking water to prevent cavities from forming in people's teeth. Teeth are primarily the minerals hydroxyapatite

$[Ca_5(PO_4)_3(OH)]$ and carbonated hydroxyapatite $[Ca_5(PO_4)_3(CO_3)_{x/2}(OH)_{1-x}]$. Fluorine in water promotes the formation of fluorapatite $[Ca_5(PO_4)_3F]$, which is harder and less prone to decay.

Intermediate-grade fluorite ore is used to produce opalescent glass from clear glass and reduce the viscosity and melting point of enamels so they better fuse on metals. The highest-grade fluorite ore, with 97 wt% or more of $CaF_2$, is used to make hydrofluoric acid (HF). Hydrofluoric acid is the primary source of fluorine in the production of virtually every organic and inorganic fluorine-containing compound including Teflon® and Gore-Tex®. HF is also used to etch glass and for the preparation of uranium hexafluoride ($UF_6$), utilized in the gaseous diffusion process of separating $^{235}U$ from $^{238}U$ for nuclear reactor fuel and nuclear bombs.

## Resource Location

Fluorine is present in fluorite ($CaF_2$), also termed fluorspar. Fluorite is a mineral formed from low temperature hydrothermal fluids where it often occurs in veins with the sulfide minerals galena and sphalerite or with barite and quartz (**Figure 8.21**). Significant deposits are widespread and often associated with skarns produced by the intrusion of granitic rocks. It has been argued that fluorine is carried in magma as a $SiF_4$ species that is deposited as fluorite ($CaF_2$) when fluids from the magma react with limestone during the formation of a skarn.

*Figure 8.2*1 Fluorite ($CaF_2$) crystals from the Hilton mine, Scordale, Cumbria, England. The mine was originally worked for galena (PbS) and later for witherite ($BaCO_3$) and barite ($BaSO_4$), but abandoned in 1963.

*Table 8.7* FLUORINE PRODUCTION FOR 2011 AND RESERVES BY COUNTRY.*

| COUNTRY | PRODUCTION | RESERVES |
|---|---|---|
| United States | NA | NA |
| Brazil | 65 | NA |
| China | 3,300 | 24,000 |
| Kazakhstan | 70 | NA |
| Kenya | 115 | 2,000 |
| Mexico | 1,080 | 32,000 |
| Mongolia | 430 | 22,000 |
| Morocco | 90 | NA |
| Namibia | 100 | 3,000 |
| Russia | 250 | NA |
| South Africa | 270 | 41,000 |
| Spain | 140 | 6,000 |
| Other countries | 300 | 110,000 |
| World totals (rounded) | 6,200 | 240,000 |

* In thousand metric tons.
NA = Not available.

*Data from*: U.S. Geological Survey, *Mineral Commodity Summaries*, January, 2012.

However, the mineral is deposited in many other types of host rock and at a considerable distance from an igneous intrusion. Significant amounts of fluoride are present in Mississippi Valley-type (MVT) lead–zinc deposits.

## Production and Reserves

The largest fluorite producing country is China with the important Shuangjiangkou-Jiangjunmiao vein deposits of Hunan Province in south China (**Table 8.7**). These are associated with early Yenshanian granites. The second largest producing country is Mexico where the San Martin skarn deposit occurs in Zacatecas. In the U.S. some production occurs as a byproduct of Pb-Zn production from Mississippi Valley-type deposits. With world fluorite reserves at 230 million metric tons and resources estimated to be approximately 500 million metric tons, supplies appear ample for at least the next 40 years.

## SUMMARY

Base metals refer to the scarce nonprecious metals of copper, lead, zinc, tin, mercury, and cadmium. They do not occur naturally in the +3 valence state.

Besides a small amount of native copper, most copper is found in a number of sulfide minerals. The most common one is chalcopyrite ($CuFeS$)$_2$. Porphyry copper deposits with their associated skarns, volcanic massive sulfide (VMS) deposits, and stratabound or stratiform sediment-hosted Cu (SSC) deposits are the most important types of ore deposits.

Lead, along with zinc occurs in stratabound or stratiform sediment-hosted exhalative (SEDEX) deposits and in Mississippi Valley-type (MVT) deposits. They are present in galena, (PbS) and sphalerite (ZnS), respectively. Tin is obtained from the mineral cassiterite ($SnO_2$). Sn occurs in pegmatites associated with granites and in their associated skarns in porphyry rock deposits. Cassiterite is also found in placer deposits.

Mercury occurs in cinnabar (HgS), found in veins above hydrothermal deposits. Significant mercury is also recovered from recycled products. Mercury is highly toxic to humans with many debilitating effects. Most cadmium is obtained as a byproduct of smelting Zn sulfide ores where it occurs when Cd replacements Zn in sphalerite (ZnS). Three-fourths of Cd is used in NiCd batteries. It is a toxic chemical that bioaccumulates in organisms. Fluorine is obtained from fluorite ($CaF_2$), often from granitic skarn deposits and Pb-Zn deposits. It is used as a flux in steel and Al production, the fluoridation of drinking water, and the production of hydrofluoric acid (HF).

## KEY TERMS

| | |
|---|---|
| alluvium | Mississippi Valley-type (MVT) deposit |
| Besshi-type massive sulfide deposit | placer deposit |
| bioaccumulation | porphyry copper deposit (PCD) |
| Cyprus-type massive sulfide deposit | porphyritic intrusive rock |
| deep-sea manganese nodule | stock |
| epicontinental sea | stratiform sediment-hosted copper (SSC) deposit |
| epigenetic | |
| gossan | stratiform sediment-hosted exhalative (SEDEX) deposit |
| Kuroko-type massive sulfide deposit | |
| limonite | syngenetic |
| meteoric | volcanic massive sulfide (VMS) deposit |

## PROBLEMS

1.  Calculate the lifetime availability of each of the base metals outlined in the text given its 2011 production and estimated current reserves.

2.  For the base metals plot the upper crustal abundance in ppm versus reserves in millions of metric tons. Which base metal falls off the trend and why?

3.  An ore deposit of Cu contains on average 0.5 wt% Cu. On the London Metal Exchange Cu metal sells for US $7,700 per metric ton. It is estimated to cost $13 per ton to mine the ore, $5 per ton to transport the ore to a processing plant, $16 per ton to process and smelt the ore to produce pure metal, and $7 per ton of metal to ship it to the buyer. How much profit does 1 ton of ore make when sold? If the start-up costs of purchasing the property mineral rights, setting up the mine, equipping it, and constructing a processing facility is $90 million, how much ore needs to be processed to recoup the start-up costs? How many ore truckloads of 150 tons need to be hauled to the processing plant to break even?

4. For each of the seven elements discussed in this chapter what percentage of world production is mined in China and what percentage in the U.S? For which element is the amount produced nearly the same?

## REFERENCES

Bateman, A. M. 1950. *Economic mineral deposits*. New York: John Wiley & Sons, Inc.

Brown, A. C. 1978. Stratiform copper deposits: Evidence for their post-sedimentary origin. *Minerals Sci Engng* 10(3):172–181.

Evans, A. M. 1993. *Ore geology and industrial minerals: An introduction*, 3rd ed. Oxford, UK: Blackwell Scientific Publishing.

Cohen, D. 2007. Earth audit. *New Scientist* 194(2605):34–41.

Gammons, C. H., Metesh, J. J., and Duaime, T.E. 2006. An overview of the mining history and geology of Butte, Montana, Technical Communications, Special Publication. *Mine Water and the Environment*, 25(2):70–75.

Garven, G. and Freeze, R. A. 1984. Theoretical analysis of the role of groundwater flow in the genesis of stratabound ore deposits 1: Mathematical and numerical model. *American Journal of Science* 284:1085–1124.

Gilluly, J., Waters, A. C., and Woodford, A. O. 1968. *Principles of geology*, 3rd ed. San Francisco: W. H. Freeman and Company.

Heyl, A. V. 1969. Some aspects of genesis of zinc-lead-barite-fluorite deposits in the Mississippi Valley, United States. *Transactions of the Institution of Mining and Metallurgy, Section B* 78:148–160.

Hyrsl, J. and Petrov, A. 2006. Famous mineral localities: Llallagua, Bolivia. *The Mineralogical Record* 37(2):117–162.

Mackin, J. H. 1947. Some structural features of the intrusions in the Iron Springs district [Utah]. Utah Geological Society, *Guidebook to the geology of Utah*, No. 2.

Misra, K. C. 2000. *Understanding mineral deposits*. Dordrecht, The Netherlands: Kluwer Academic Publishers.

Rusk, B., Reed, M. H., and Dilles, J. H. 2008. Fluid inclusion evidence for magmatic-hydrothermal fluid evolution in the porphyry copper-molybdenum deposit: Butte, Montana. *Economic Geology* 103:307–334.

Sangster, D. F. 1990. Mississippi Valley-type and SEDEX lead-zinc deposits: A comparative examination. *Transactions of the Institution of Mining and Metallurgy, Section B: Applied Earth Science* 99:21–42.

Sillitoe, R. H., Halls, C., and Grant, J. N. 1975. Porphyry tin deposits in Bolivia. *Economic Geology*, 70:913–927.

Theodore, T. G. 1977. Selected copper-bearing skarns and epizonal granitic intrusions in the southwestern United States. *Geol Soc Malaysia Bull* 9:31–50.

Turneaure, F. S. 1960. A comparative study of major ore deposits of central Bolivia, *Economic Geology* 55(Part 1):217–254; (Part II):574–606.

U.S. Geological Survey, 2010 Mercury Statistics, *in* Kelly, T.D., and Matos, G.R. comps. Historical statistics for mineral and material commodities in the United States: U.S. Geological Survey Data Series 140.

U.S. Geological Survey, 2011, *Mineral commodity summaries 2011*, Washington, DC: U.S. Government Printing Office.

Walther, J. V., 2009, *Essentials of Geochemistry*, 2nd ed., Jones and Bartlett Publishers, Boston, 797 pp.

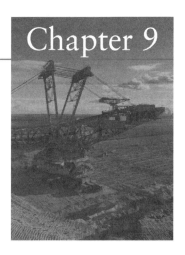

# Chapter 9

# Scarce Metals: Precious and Specialty Metals

A precious metal is a scarce metal with high economic value. Gold (Au), silver (Ag), and the six platinum group metals (PGMs) of platinum (Pt), palladium (Pd), rhodium (Rh), iridium (Ir), ruthenium (Ru), and osmium (Os) are generally designated as the precious metals. These chemical symbols are in shadow in **Figure 9.1**. Note that silver is more abundant in the earth's upper crust than tellurium (Te) and of similar abundance to mercury (Hg), selenium (Se), and indium (In), and yet is considered a precious metal. This is because of its high demand and

therefore its high economic value relative to these other elements of equal abundance in the earth's crust.

This chapter also discusses what are referred to as specialty metals. Specialty metals are a set of metals that play an important role in high technology manufacturing. Cesium (Cs), rhenium (Re), tellurium (Te), niobium (Nb), tantalum (Ta), arsenic (As), antimony (Sb), bismuth (Bi), germanium (Ge), gallium (Ga), indium (In), beryllium (Be), and the rare earth elements (REEs) are generally classified as specialty metals.

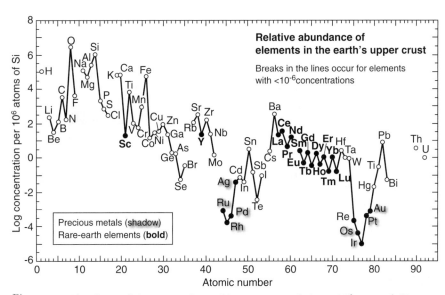

*Figure 9.1* Abundance of elements in the earth's upper crust relative to $10^6$ atoms of silica. Note the relative abundance of rare earth elements and precious metals. (Modified from Haxel, G. B., Hedrick, J. B., and Orris, G. J. (2002) Rare Earth Elements—Critical Resources for High Technology, U. S. Geological Survey Fact Sheet 087-02.)

*Table 9.1* RARE EARTH ELEMENTS.

| NAME | CHEMICAL SYMBOL | ATOMIC NUMBER | CONTINENTAL CRUSTAL ABUNDANCE (ppm BY WEIGHT) |
|---|---|---|---|
| Scandium | Sc | 21 | 26 |
| Yttrium | Y | 39 | 29 |
| Lanthanum | La | 57 | 34 |
| Cerium | Ce | 58 | 60 |
| Praseodymium | Pr | 59 | 8.7 |
| Neodymium | Nd | 60 | 33 |
| Promethium | Pm | 61 | – |
| Samarium | Sm | 62 | 6.0 |
| Europium | Eu | 63 | 1.8 |
| Gadolinium | Gd | 64 | 5.2 |
| Terbium | Tb | 65 | 0.94 |
| Dysprosium | Dy | 66 | 6.2 |
| Holmium | Ho | 67 | 1.2 |
| Erbium | Er | 68 | 3.0 |
| Thulium | Tm | 69 | 4.5 |
| Ytterbium | Yb | 70 | 2.8 |
| Lutetium | Lu | 71 | 0.56 |

Rare earth elements include scandium (Sc), yttrium (Y), and the 15 *lanthanides* given in **Table 9.1** and shown in bold in Figure 9.1. The term "rare earth" stems from the fact that they were first extracted from uncommon (rare) oxide-type minerals in the earth (earths) rather than from typical sulfide ore minerals. As shown in Figure 9.1 rare earth elements have reasonable concentrations in the earth's upper continental crust relative to many other elements.

## Precious Metals

Precious metals have lower chemical reactivity, higher melting points, and are softer or more ductile than most other metals. Coins and jewelry are made from precious metals because of their high luster. The demand for precious metals is driven by their industrial use, the fabrication of jewelry, and their role as investments in the accumulation of wealth.

## Gold (Au)

Gold (Au) is a very unreactive quite rare element in the earth's upper crust with an average concentration of just 1.2 to 1.8 ppb.

### Use

Gold has a pretty yellow color and metallic luster. Because it is soft and malleable but very resistant to chemical attack such as tarnishing, jewelry and the creative arts account for 80% of the demand for gold. Dentistry, coinage, and government stockpiles make up almost 12% of the total. Electrical and electronics uses account for 8%. Because of its high value and tendency not to corrode, gold is extensively recycled. This then helps supply demand. Over 15% of annual gold demand is satisfied from recycled Au. This implies much Au is recycled many times.

With its high value and very limited supply gold has been considered a safe investment in uncertain economic times. This accounts for the change in the price of gold through time as given in **Figure 9.2**. Until 1960 the price of gold in the U.S. was fixed at ~$35 per troy ounce. After 1960 the price of gold was allowed to fluctuate with market demand. At the middle of 2012 it was almost US $1,600 per troy ounce.

## Resource Location

Gold is most commonly found in a native elemental state, usually alloyed to some extent with silver, in what is called *electrum*. The rare ore minerals calaverite ($AuTe_2$) and sylvanite ($AuAgTe_4$) are sometimes found. Te is the element tellurium and these minerals are then tellurides.

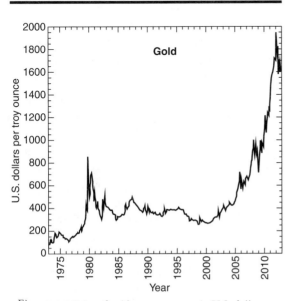

*Figure 9.2* Price of gold per troy ounce in U.S. dollars at the indicated time. (Data from: Gold Price.)

In aqueous solutions Au occurs in the $Au^+$ and $Au^{3+}$ oxidation states and forms the stable aqueous species: $AuCl_2^-$ and $Au(HS)_2^-$ in $Cl^-$ and $HS^-$ rich aqueous solutions, respectively. A reaction to precipitate gold from the chloride aqueous complex is

$$4AuCl_2^- + 2H_2O \rightarrow 4Au^\circ_{(solid)} + 8Cl^- + 4H^+ + O_2. \quad [9.1]$$

This reaction occurs with decreasing temperature and concentration of $Cl^-$ or $O_2$. It also occurs with increasing pH. Besides cooling on transport into lower temperature rocks, precipitation from an aqueous solution may occur with mixing of ore fluid with groundwater or boiling of the fluid. Boiling of the aqueous fluid both increases its pH and increases the concentration of Au in the remaining liquid phase. Because these Au solutions can be quartz or carbonate saturated, native gold metal often precipitates along with quartz or carbonate, producing gold-quartz, gold-carbonate, or gold-quartz-carbonate veins.

Gold-quartz veins are typically associated with granitic and granodioritic igneous rocks. Granodiorite is similar to granite but contains less quartz and more plagioclase than K-feldspar. The gold is thought to concentrate in a quartz saturated aqueous-rich fluid phase that is produced during final crystallization of these Si-rich magmas. As the hydrothermal solutions cool in the range from about 350° to 200°C gold plus quartz are deposited. These gold-quartz veins can be of high grade.

Gold-carbonate veins are typically found in greenstones. Greenstone is mafic to ultramafic lava that was originally extruded onto the ocean floor. It has since been metamorphosed as it became part of the continents, turning it somewhat green. The carbonate is present because the release of Ca and Mg to a hydrothermal fluid from these rocks combines with $CO_3^{2-}$ to precipitate calcite and dolomite. The source of the $CO_3^{2-}$ is thought to be from alkaline groundwater or produced from $CO_2$ derived from the gold-rich hydrothermal fluid. These reactions occur at lower temperatures (< 200°C) than the precipitation of gold-quartz veins as fluids contain little silica at these temperatures, limited by quartz solubility. However, because carbonates display *retrograde solubility* $Ca^{2+}$, $Mg^{2+}$, and $CO_3^{2-}$ will be high. Retrograde solubility occurs when the mineral becomes more soluble in water as temperature decreases. At intermediate temperatures gold-quartz-carbonate veins can form.

Once formed these veins can weather, yielding nuggets of gold and gold-containing gravels that are released to streams. Gold with a density of 17.3 g cm$^{-3}$ is denser than regular stream sands and gravels of ~2.7 g cm$^{-3}$. The nuggets collect in dead water depressions in a river while the moving water carries the less dense silicate sands and gravels away producing what is termed a placer gold deposit. However, most gold is currently mined from deposits in solid rock where it occurs as fine disseminated low-grade gold. Besides gold mines, gold is also produced where it is not the principal metal mined, but is a profitable byproduct.

### California Mother Lode

The California Mother Lode is a 195-km belt of gold-bearing quartz veins in metamorphic rocks of central California in the western foothills of the Sierra Nevada Mountains. The Sierra Nevada is a huge Cretaceous age (145 to 65 million years ago) batholith of granodiorite and related rocks formed at the edge of the continent from fluids given off by a subducting slab called the Farallon plate. The granodiorite stock shown in Figure 8.3 in the chapter on base metals is a smaller body and likely produced from a batholith at depth. The Mother Lode is not a continuous single lode, but rather a complex collection of veinlets that pinch, branch and terminate. They were likely initially deposited from fluid a km or more below the surface (**Figure 9.3**).

In January 1848 James Marshall, the construction boss for a sawmill being built on the south fork of the American River northeast of present-day Sacramento, California for John Sutter, discovered gold in the sands of a channelway of water under the mill. Within six months miners were working all the streams of the lower foothills of the Sierra Nevada Mountains for placer gold. The gold was first obtained from the river sand by panning. This involves submerging a pan in water and shaking it to remove the sand from the denser gold particles.

Operations then developed where gold-rich sediments were washed down an inclined wooden platform with an uneven bottom surface, in what is called a sluice box. Gold then collects in the dead water zone between bulges on the bottom of the box because of its greater density. Mercury (density ~13.6 g/cm$^3$) was often added to the box to improve the separation of the gold. Larger gold-mercury amalgam particles formed between gold and mercury particles aiding in the separation process. See discussion in the chapter on base metals on the uses of mercury.

By winter 1849 the number of miners had swelled to 80,000. During the California gold rush some 546 camps and digging sites were operating. Placer miners followed the gold-bearing sands upstream and discovered its source, the Mother Lode of gold and silver quartz veins in the upper foothills of the Sierra Nevada.

It has been argued that the mineralization of the Mother Lode reflects deposition from 250° to 300°C fluids derived from deep-seated magmas, now represented by the Sierra Nevada. The Mother Lode ore is also present at depth. In Amador County, California the mines are 1.7 km deep and the ore is of similar grade as that found on the surface.

### Witwatersrand Goldfield

The 2,800-million-year-old Witwatersrand Goldfield in South Africa was discovered in 1852. It has produced 40% of all the gold ever mined and still contains 30% of the world's present gold resources. The gold appears to have weathered from gold-bearing rocks and washed into the 36,000 km$^2$ Witwatersrand Basin and deposited as placers in the Central Rand Group of sediments. The arrows in **Figure 9.4** show the major transport directions

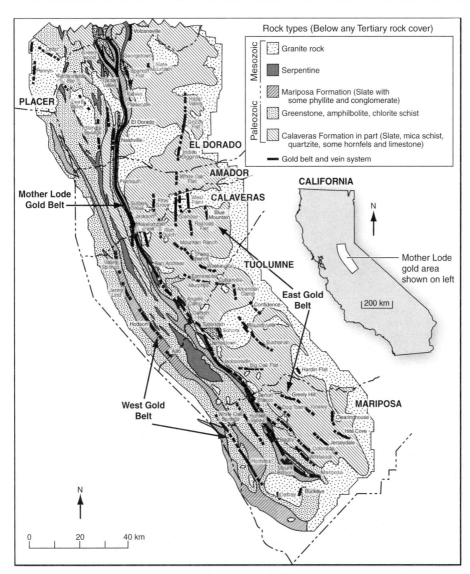

*Figure 9.3* Geological map showing the Mother Lode and east and west gold belts in the foothills of the Sierra Nevada Mountains in California. (Modified from Clark, W. B., 1998, *Gold Districts of California*, Bulletin 193, Sesquicentennial edition, California Gold Discovery to Statehood, California Division of Mines and Geology 1976, 199 p.)

of gold into the basin. Mining for gold has reached a depth of 4 km in some mines, the deepest mines in the world. Cost of mining is increasing with greater depth but is supported by the high price of gold. Apparently at these great depths some mines are starting to run out of ore.

The Basin is the site of an ancient lake or inland sea. The gold was transported by rivers and streams flowing from surrounding mountains and highlands and deposited into a number of alluvial fans of the Central Rand Group as shown in **Figure 9.5**. *Alluvial fans* are fan-shaped

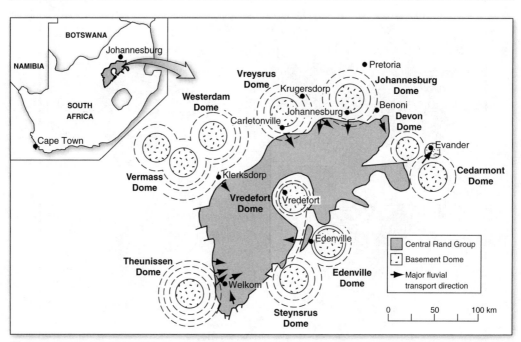

*Figure* 9.4 Witwatersrand gold deposits with inset giving location in South Africa. The Central Rand group contains the principal Au and U mineralization in conglomerates with the arrows giving the major fluvial transport directions for the mineralization into the basin. (Adapted from Evans, A. M., 1997, *An Introduction to Economic Geology and Its Environmental Impact, Blackwell Science*, Oxford, 364 p.)

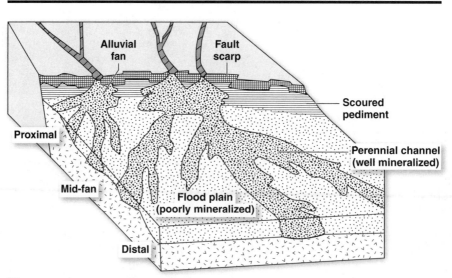

*Figure* 9.5 Depositional environment for gold in channelized reefs of alluvial fans showing proximal, mid-fan, and distal regions. (Adapted from Tucker, R. F. and Viljoen, R. P., 1986, The geology of the West Rand Goldfield. In: Annhaeusser, C. R. & Maske, S. (eds) *Mineral Deposits of South Africa*, p. 649–598, Geological Society of South Africa, Johannesburg.)

deposits of sediments from rivers typically produced when their velocity slows as they exit the mouth of a canyon. A gold-rich pebble conglomerate of about 5 to 10 g Au per metric ton developed. Grains of pyrite and uranium minerals occur with the gold gravel. Similar smaller Au deposits also occur in 1,900-million-year-old rocks in the Serra de Jacobina region of Brazil and in 2,500-million-year-old rocks of the Elliot Lake region of Canada.

### Homestake Gold Mine

The Homestake disseminated gold ore bodies occur near Lead, South Dakota in the northern Black Hills. Between 1877 and when mining stopped in 2002, 41 million troy ounces of gold were recovered. The ore zones are associated with about a 2-billion-year-old iron formation and interlayered metavolcanic rocks. The rocks have experienced many stages of metamorphism and are very tightly folded with the ore bodies present in secondary folds at nearly right angles to the primary folding. Gold ore mined at Homestake was from a very large deposit with a concentration of somewhat less than 1 troy ounce of Au per metric ton of rock.

### Low-Grade Carlin-Type Gold Deposits

The recent production of gold in the U.S. is mainly from 40 similar type deposits in Nevada and Utah. The type locality is the carbonate-hosted stratabound-disseminated gold deposits near Carlin, Nevada (**Figure 9.6**). These deposits are characterized by the dominance of "invisible gold." This is submicron-sized Au particles present in iron sulfide. They are thought to form from 150° to 250°C, low pH, and somewhat saline fluids derived from both meteoric and magmatic sources related to regional late Eocene (56 to 34 million years ago) magmatism in the region.

Low-grade Carlin-type gold, ~1.0 g Au per metric ton, was mined when bulk mining cyanide

techniques were developed and the price of gold continued to increase. After mining the ore undergoes comminution, where the ore is transformed to a powder. In this finely-sized powder the fine-grained Au can be separated from the silicate rock. A solution of sodium cyanide is sprayed on large piles of the powdered ore and the cyanide reacts with the gold producing a water-soluble Au-cyanide complex by the reaction:

$$4Au + 8Na(CN)_2 + O_2 + 2H_2O \rightarrow$$

$$\text{sodium}$$
$$\text{cyanide} \quad 4Au(CN)_2^- + 8Na^+ + 4OH^-. \quad [9.2]$$
$$\text{dicyanoaurate}$$
$$\text{complex}$$

If the pH in solution decreases the reaction

$$CN^- + H^+ \rightarrow HCN_{(g)} \quad\quad [9.3]$$

occurs. Because hydrogen cyanide ($HCN_{(g)}$) is a highly toxic gas, typically calcium or sodium hydroxide is added to the solution to keep the pH above 10.5. The Au-rich solutions are then subjected to electrowinning where a current is passed thorough the solution containing $Au(CN)_2^-$ by inserting a cathode and anode in it. Electrons from the cathode cause the following reaction:

$$Au(CN)_2^- + e^- \rightarrow Au + 2CN^- \quad\quad [9.4]$$

and Au particles precipitate on the cathode.

## Production and Reserves

Given in **Figure 9.7** is the world annual gold production since 1900. There has been a general rise in gold production from 1900 to 2001. Beginning in 2001 gold production has started to slow somewhat. Production has declined in all the major producer countries except China. Apart from China, no significant new sources have come into production and current mines are producing somewhat less with time. This is manifest in **Table 9.2**, which indicates that China has surpassed South Africa to become the world's largest producer of gold.

In 1980 South Africa produced 55% of all gold, but now it produces less than in the U.S.,

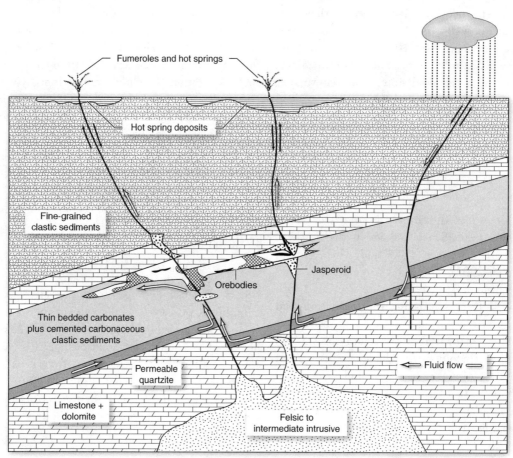

*Figure* 9.6 Carlin-type hot spring gold deposits showing the transport direction of the hydrothermal solutions along faults and permeable beds with open arrows. The solid arrows give the sense of motion on the faults. (Adapted from Evans, A. M., 1997, An Introduction to Economic Geology and Its Environmental Impact, Blackwell Science, Oxford, 364 p. and Nelson, C. E. and Giles, D. L. 1985, Hydrothermal eruption mechanisms and hot spring gold deposits. *Economic Geology*, v. 80, p. 1633–1639. )

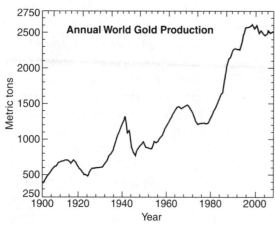

*Figure* 9.7 Annual world gold production from 1900 to 2011 in metric tons. (Data from: Gold Sheet Mining Directory.)

about 9% of world production. The price of gold has risen recently because production hasn't kept up with demand (Figure 9.2). The mining and processing of gold is both capital- and time-intensive. Supplying the demand has been a challenge throughout history.

Gold is mined in over 50 countries. However, as indicated in Table 9.2, China, U.S., South Africa, Australia, Peru, and Russia together account for 50% of production. Reserves appear adequate for 20 years. Judging from the reserve base it is not clear whether supplies will meet needs into the late twenty-first century except at increasing higher costs of extraction as lower grade deposits are brought on line.

*Table 9.2* GOLD PRODUCTION FOR 2011, RESERVES, AND RESERVE BASE BY COUNTRY.*

| COUNTRY | PRODUCTION | RESERVES | RESERVE BASE |
|---|---|---|---|
| China | 355 | 1,900 | 4,100 |
| Australia | 270 | 7,400 | 6,000 |
| United States | 237 | 3.000 | 5,500 |
| Russia | 200 | 5,000 | 7,000 |
| South Africa | 190 | 6,000 | 31,000 |
| Peru | 150 | 2,000 | 2,300 |
| Canada | 110 | 920 | 4,200 |
| Indonesia | 100 | 3,000 | 6,000 |
| Ghana | 100 | 1,400 | 2,700 |
| Uzbekistan | 90 | 1,700 | 1,900 |
| Mexico | 85 | 1,400 | 3,400 |
| Papua, New Guinea | 70 | 1,200 | 2,300 |
| Brazil | 55 | 2,400 | 2,500 |
| Chile | 45 | 3,400 | 3,400 |
| Other countries | 630 | 10,000 | 22,000 |
| World totals | 2,700 | 51,000 | 100,000 |

* In metric tons.

*Data from*: U.S. Geological Survey, *Mineral Commodity Summaries*, January, 2012.

# Silver (Ag)

## Use

For at least the last 5,000 years high-value objects have been made of silver (Ag) to reflect the status and wealth of the owner. Jewelry and silverware are traditionally made from sterling silver, an alloy of 92.5 wt% silver with 7.5 wt% copper. Because pure silver is very soft it is alloyed with copper to increase its strength. As the amount of silver in the alloy decreases, the amount of tarnishing, that is oxidation, of silver increases. Early coins were made from electrum, a naturally occurring alloy of Au and Ag. Electrum coins of 10% to 55% Ag with a pale to bright yellow color are known from as early as 700 BC.

Many industrial applications use silver because it has the highest electrical conductivity, that is, lowest resistance to the flow of electricity of any element. Other usages take advantage of the fact that silver has the best ability to conduct heat, that is, highest thermal conductivity of any metal. Silver is also a common coating for mirrors because it has the highest optical reflectivity of any element. Reflectivity is the ratio of the energy of a wave reflected from the surface to the energy absorbed by the surface.

Ag has found a use as a catalyst for some oxidation reactions. Due to its antimicrobial properties Ag is sometimes used as a disinfectant. Because some Ag compounds like silver iodide are very light sensitive, large amounts of Ag were used in photographic film. Much of this has been replaced, however, by digital photography techniques that do not require film. Dental fillings are often made of amalgams, Ag alloyed with mercury. Ag is also used in batteries, brazing and soldering compounds, for catalytic converters in automobiles, in inks, and some electroplating applications.

## Resource Location

Silver occurs in its native form, alloyed with gold in electrum, and is present in the minerals argentite ($Ag_2S$), chlorargyrite (AgCl), pyrargyrite ($Ag_3SbS_3$), and as a solid solution in tetrahedrite [$(Cu,Fe,Ag)_{12}Sb_4S_{13}$]. There are a number of mines where Ag minerals are the primary ore. However, Ag is dispersed in the matrices of minerals like chalcopyrte, galena, and sphalerite. Most Ag is currently obtained as a byproduct from ores mined primarily for their copper, copper + nickel, lead, or lead + zinc content.

### Epithermal Vein Ag Deposits

*Epithermal* vein deposits occur at a shallow depth at some distance from any magma source that provided the metals in the ore minerals. The water that precipitated the ore minerals ranged in temperature from about 50° to 300°C. Epithermal vein deposits develop in geothermal systems often with associated hot springs in and above intrusive silica-rich to moderate silica-bearing igneous rocks. The metals of interest in the aqueous fluids are derived from the last stages of crystallization of magma at depth or leached from country rocks by hydrothermal solutions circulated by the heat of the intrusion. These aqueous fluids are at elevated temperatures but under high enough pressure to remain in a liquid state. They rise toward the surface along fractures. At some depth, typically above 500 m depth, which depends on temperature and the salinity of the liquid, the aqueous fluid starts to boil. This occurs due to a pressure drop in the fluid often triggered by a rupturing fault. This is similar to decompression of a can of carbonated soda and the formation of $CO_2$ bubbles when the top is opened.

This boiling leads to the deposition of minerals out of the liquid part of the fluid as they become supersaturated with loss of $H_2O$ to the gas phase and increases in the liquid's pH. Metal-bearing minerals deposit first followed by quartz, calcite, and/or K-feldspar. The growth of the minerals seals the fracture. Fluid pressure can build again and the process repeated. **Figure 9.8** shows a cross-section through such a deposit that precipitates Ag and Au above the level of boiling and Pb, Zn, and Cu below. The low pH vapor produced during boiling spreads out laterally and upward. This causes alteration in the rocks as the $H^+$ in the vapor reacts with the rocks.

**Comstock Lode.** The Comstock district is a set of world-class Ag plus Au epithermal quartz vein deposits located 24 km southeast of Reno, Nevada. The Comstock's silver, discovered in 1859, was the first important silver-mining district in the U.S. **Figure 9.9** shows a cross-section along the Sutro Tunnel that was used to drain water and supply air to the working mines. Lodes are labeled where they have been intersected by the Sutro Tunnel before the Comstock fault is reached. A *lode* is a vein of ore deposited within a rock fissure. At the Comstock fault, which contains the Comstock Lode, the Sutro Tunnel goes north and south underneath the major mines. The Comstock district produced 7.3 million kg of silver and over 300,000 kg of gold from veins occurring 4.8 km along the strike of the Comstock fault from the surface to a depth of 1 km.

The Comstock Lode is present in 13.7-million-year-old andesite. Ore was determined to have been deposited from hydrothermal solutions of temperatures between 200° and 300°C with up to 6 wt% NaCl. The ore fluid is thought to be composed of both meteoric and magmatic water, depositing ore minerals where fluid boiling had occurred. This boiling is marked by the presence of the mineral kaolinite. Andesite rock adjacent to veins is altered to the quartz sericite (a fine-grained muscovite ± paragonite mica) and the clay montmorillonite. The ore minerals are primarily electrum (alloy of Au and Ag) and argentite ($Ag_2S$), which has been altered to acanthite, a low temperature modification of argentite.

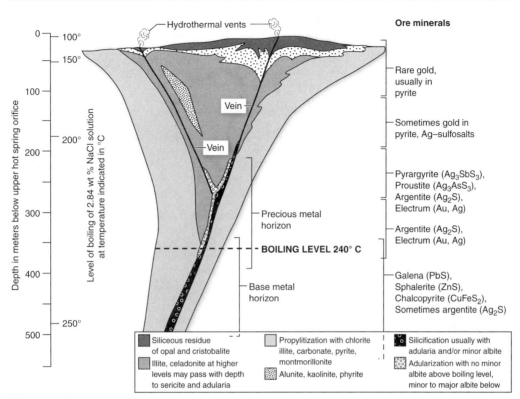

*Figure 9.8* Epithermal vein ore deposit. Alteration boundaries are more diffuse than shown. In some deposits the zones are telescoped together in others multiple ore stages have developed. In many locations the top of the deposit has been eroded leaving precious metal veins exposed near the surface. (Adapted from Buchanan, L. J. (1981) Precious metal deposits associated with volcanic environments in the Southwest; in Relations of Tectonics to Ore Deposits in the Southern Cordillera; *Arizona Geological Society Digest*, Volume 14, pages 237–262.)

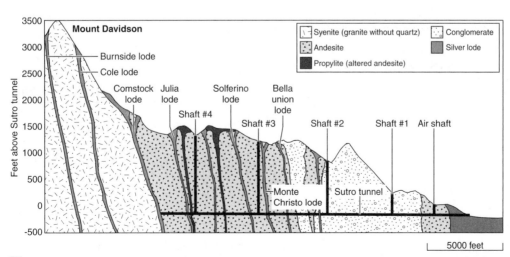

*Figure 9.9* Comstock and related lode veins along the Sutro tunnel towards Mount Davidson. (Modified from Brechin, G. and Kirshenbaum, N. W., 2004, Underground power generation at the Comstock lode and the Sutro tunnel, Nevada, *Bulletin of the Peak District Mines Historical Society*, v. 15, No. 4/5, p. 104–110.)

The discovery of the Comstock Lode stimulated prospecting for silver across the Great Basin, from east of the Sierra Nevada Mountains to west of the Rocky Mountains. This resulted in the opening of a number of silver mining districts in Nevada, including Austin in 1862, Eureka in 1864, Pioche in 1869, and finally Tonopah in 1900.

## Production and Reserves

Worldwide annual Ag production since mid-1998 is shown in **Figure 9.10**. The leading silver producing countries in descending order are Mexico, Peru, China, Australia, Chile, and Russia as given in **Table 9.3**. Note that the world production in 2011 was 23.8 thousand metric tons = 765 million troy ounces. This is almost 10 times the production of gold.

The U.S. imported 2/3 of the Ag it required in 2011. This was 3,840 metric tons with a little over 1/2 from Mexico and 1/4 from Canada. Domestically 1,600 metric tons of silver was recovered from scrap including 60 to 90 metric tons from photographic wastewater.

As a metal that is obtained principally as a byproduct of the mining of other metals, if demand increases it will be difficult to increase supply without considerably higher costs. Fortunately, some uses of silver, such as for photographic

*Table 9.3* SILVER PRODUCTION FOR 2011 AND RESERVES BY COUNTRY.*

| COUNTRY | PRODUCTION | RESERVES |
|---|---|---|
| Mexico | 4.5 | 37 |
| Peru | 4.0 | 120 |
| China | 4.0 | 43 |
| Australia | 1.9 | 69 |
| Chile | 1.4 | 70 |
| Russia | 1.4 | NA |
| Bolivia | 1.35 | 22 |
| Poland | 1.2 | 85 |
| United States | 1.16 | 25 |
| Canada | 0.7 | 7 |
| Other countries | 2.2 | 50 |
| World totals | 23.8 | 530 |

* In thousands of metric tons.
NA = Not available.

*Data from*: U.S. Geological Survey, *Mineral Commodity Summaries*, January, 2012.

applications, are decreasing as digital photography has replaced the need for silver in film.

### Silver Prices

The industrial demand for silver has increased since 2001 so, on average, has its price as shown in **Figure 9.11** (except for the economic downturn in 2008). Even excluding this event the price of silver has been quite volatile as a function of time. Silver like gold is used to store wealth. As a result its price tends to increase in poor economic times. There was a speculation spike in prices in 2011 due to a slowdown in production in the U.S. and reported slowdown in China. There was also a large spike in silver prices in early 1980. Beginning in 1973 the Americans Nelson and William Hunt, who made a fortune in oil, began accumulating large amounts of silver thinking they could corner the world market in silver and therefore control its price. In late 1979 it was

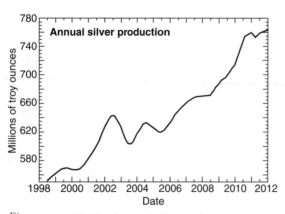

*Figure 9.10* Worldwide annual silver production as a function of time in troy ounces. (Adapted from Thomas Chaize, Energy and Mining Newsletter.)

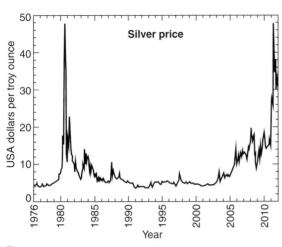

*Figure 9.11* Price of silver in U.S. dollars per ounce at the indicated time. (Data from: Silver Price.)

*Table 9.4* AVERAGE PLATINUM GROUP METAL CONCENTRATIONS IN THE EARTH'S CRUST.

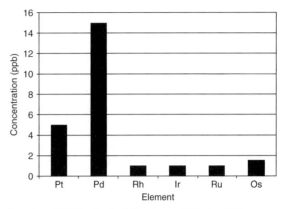

*Data from*: CRC Handbook of Chemistry and Physics, 2009.

estimated they owned or had future contacts for 100 million troy ounces. Silver prices rose from US $11 a troy ounce in September 1979 to nearly US $50 a troy ounce in January 1980. Before they could sell, silver prices collapsed to below US $11 a troy ounce in March 1980 leading to the bankruptcy of Nelson Hunt.

# Platinum Group Metals (Pt, Pd, Rh, Ir, Ru, Os)

The platinum group metals (PGMs), which have similar chemical and physical properties, include platinum (Pt), palladium (Pd), rhodium (Rh), iridium (Ir), ruthenium (Ru), and osmium (Os). They tend to occur together in the same ore deposits. As indicated in **Table 9.4** the platinum group minerals have low concentrations in the earth's crust with Pt and Pd being more abundant than the other PGMs.

## *Use*

PGMs are particularly resistant to oxidation and corrosion, have high melting points, and very good thermal and electrical conductivity. In addition they have outstanding catalytic properties. Being biocompatible, PGMs work particularly well in organic molecule synthesis. As precious

metals the PGMs, like gold and silver, are also used to store wealth.

## *Platinum (Pt)*

A little over half of the platinum (Pt) produced is fabricated for catalytic converters to reduce toxic emissions from internal combustion engines in motor vehicles. About 20% is exploited in fine jewelry because of its resistance to both wear and tarnish. The electronics industry uses 5% of Pt production to make electrical contacts and coatings. Another 5% is employed in the chemical industry for catalysts. For instance, platinum and platinum-rhodium alloys are fabricated as gauze to use as a catalyst for the oxidation of ammonia to produce nitric acid by the reaction:

$$NH_3 + 2O_2 \rightarrow HNO_3 + H_2O. \qquad [9.5]$$

Nitric acid is then used to make fertilizers and explosives. Pt catalysts are also employed extensively in the petrochemical industry during the refining and reformation of hydrocarbons. Other uses include crucibles to contain reactions at high temperature, thermocouples, and dental crowns and bridges.

## *Other PGMs*

Less dense and softer than platinum, palladium (Pd) is alloyed with platinum to optimize platinum's working characteristics. Presently the largest use of palladium is in catalytic converters for motor vehicles. Palladium-based alloys are also used for dental crowns and bridges.

Rhodium (Rh) has a higher melting point and lower density than Pt. It is typically used for its catalytic properties with platinum in catalytic converters. Rh is also used as an alloying agent with Pt and Pd for hardening and improving the corrosion resistance of these metals. These alloys are made into thermocouples and furnace windings as well as aircraft spark plugs.

Iridium (Ir) is the most corrosion-resistant metal, even at temperatures up to 2000°C. It is alloyed with other metals to increase hardness. For instance, iridium-titanium alloys are used in pipes laid on the seafloor. Ruthenium (Ru) has the ability to harden platinum and palladium so it is used in platinum and palladium alloys to produce wear-resistant electrical contacts. It has a lower cost but similar properties to rhodium and is often used as a substitute. This includes its use as plating material for electric contacts. Osmium (Os) is the densest natural element. Osmium is alloyed with other platinum group elements in applications where extreme hardness and durability are required such as in electrical contacts and fountain pen tips.

## Resource Location

The mantle is the principal reservoir on earth of PGMs. PGMs are obtained from mafic to ultramafic magmas from the mantle intruded into the earth's crust. Platinum is often found in its native state as well as being alloyed with iridium in platiniridium. Some uncommon sulfides (i.e., [(Pt,Pd)S]), tellurides (i.e., PtBiTe), antimonides (PdSb), and arsenides (i.e., $PtAs_2$) are PGM bearing and some native alloys occur with PGMs and nickel or copper.

The largest known reserves of PGMs are in the Bushveld Complex in South Africa where the PGMs occur with Ni and Cu. The Great Dyke in Zimbabwe also contains extensive PGM deposits (see Figure 7.6). Large amounts of PGMs are obtained from the copper–nickel deposits near Norilsk in Russia, and the Sudbury nickel deposits in Canada (see chapter on ferro-alloy metals). The Stillwater and East Boulder Mines in Montana are the only PGM producing mines in the U.S. These are in the Stillwater igneous complex, a large, layered mafic igneous intrusion similar in nature but smaller than the Bushveld Complex in South Africa.

PGMs are also found in alluvial sands, but production is presently small. Large alluvial PGM deposits were discovered in the Ural Mountains in 1823, and are still being worked although they are presently of low grade. Alluvial PGM deposits also occur in Alberta and Yukon, Canada; New South Wales, Australia; and Alaska, among other locations.

Significant quantities of ruthenium, rhodium, and palladium are formed as fission products in nuclear reactors. As the prices escalate and with increasing global demand, reactor-produced PGMs can be considered a potential source. Because some asteroids have PGM contents that exceed 50 ppm it has been suggested they may be able to supply future needs.

## Production and Reserves

Unlike gold, there is not a large supply of PGMs in stockpiles ready to use. The two major supply sources are the Bushveld Complex in South Africa (75%) and the Norilsk region in Russia (12%).

Given in **Table 9.5** is production and reserves estimates for countries with the largest production of Pt and Pd. Total annual PGM production is near 400 metric tons. With reserves of PGMs of about 66,000 metric tons over a 150-year supply is available. PGMs are also recovered from a variety of discarded consumer products with the largest source from recycling automotive catalytic converters. **Figure 9.12** outlines the price of Pt as a function of time. Note the rise in prices since 2002 with a precipitous drop during the economic downturn in 2008 as the number of automobiles manufactured and their Pt-containing catalytic converters decreased.

*Table 9.5* PLATINUM AND PALLADIUM PRODUCTION FOR 2011 AND RESERVES FOR ALL PLATINUM GROUP METALS BY COUNTRY.*

| COUNTRY | PT PRODUCTION | PD PRODUCTION | PGM RESERVES |
|---|---|---|---|
| South Africa | 139 | 78 | 63,000 |
| Russia | 26 | 85 | 1,100 |
| Canada | 10 | 18 | 310 |
| Zimbabwe | 9.4 | 7.4 | NA |
| United States | 3.7 | 12.5 | 900 |
| Colombia | 1.0 | NA | NA |
| Other countries | 2.5 | 6.1 | 800 |
| World totals | 192 | 207 | 66,000 |

* In thousands of kilograms.
NA = Not available.

*Data from*: U.S. Geological Survey, *Mineral Commodity Summaries*, January, 2012.

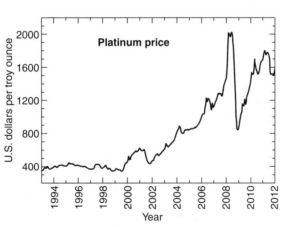

*Figure 9.12* Platinum price in U.S. dollar per troy ounce as a function of time. (Data from: Johnson Matthey.)

The world's largest production of Pd comes from the Norilsk-Talnakh nickel deposits in northernmost Siberia. Rh, Ru, and Os are generally obtained as a byproduct in the recovery of Pt and Pd from PGM ores.

Ir is found as an uncombined element or in natural alloys of iridium–osmium, osmiridium (Os rich), and iridosmium (Ir rich). It also occurs with other platinum group metals substituting for platinum in sulfides, tellurides, and arsenides. Ir annual production and consumption is only about 3 metric tons.

Meteorites contain an abundance of Ir that is much higher than the average crustal abundance. When Ir was found at high concentrations in clay layers deposited at the K-T (Cretaceous-Tertiary) boundary it lead to a theory that the extinction of many species, including dinosaurs 65.5 million years ago, was caused by the impact of a massive extraterrestrial object.

# Specialty Metals

Specialty metals are metals used by modern industries with none of them found or mined in large quantities. Each, however, has a unique set of properties that makes them very useful to modern industrial societies. Many specialty metals are found in pegmatite.

## Pegmatite

*Pegmatite* is very coarse-grained igneous rock with grain size ≥ 20 mm, generally of granitic composition. It is typically made up of large crystals of quartz, feldspar, and biotite, plus a wide range of accessory minerals that contain specialty metals. Generally associated with large granite bodies, pegmatites are distributed along their margins as irregular dikes, lenses, or veins as shown in **Figure 9.13**. They occur in rocks of all ages. Pegmatites form from water-rich fluids separated from the granitic magma during slow cooling. As the water-rich magmatic fluid cools it promotes low rates of nucleation of crystals. This coupled with high element diffusivity in the fluid promotes the growth of large-size crystals.

Pegmatite fluids are generally enriched in incompatible uncommon elements such as boron, cesium, lithium, uranium, tin, tungsten, and tantalum. Frequently they also contain rare earth elements. These uncommon elements promote the formation of a set of unusual accessory minerals like spodumene [$LiAl(SiO_3)_2$] and the hydrous boron silicate, tourmaline. The minerals topaz [$Al_2SiO_4(F,OH)_2$], beryl [$Be_3Al_2(SiO_3)_6$], and micas with unusual compositions can also be common.

## Cesium (Cs)

The largest use of nonradioactive cesium (Cs) is to make cesium formate ($CsCHO_2$) for drilling muds. (see chapter on petroleum). Cs metal is also used as a catalyst in the hydrogenation of organic compounds due to its ability to decrease high oxidation states of transition metals bonded to oxygen. Cs nitrate is used in making optical glass. The larger cesium replaces sodium and strengthens the glass. Atomic clocks use the transition between two electron energy levels of the $^{133}Cs$ atom as an oscillator to accurately regulate short intervals of time.

Cs is obtained from the zeolite pollucite [$(Cs,Na)_2Al_2Si_4O_{12} \cdot 2H_2O$], and to a lesser extent from a Cs-rich lepidolite mica [$(K,Cs)Li_2Al(Al,Si)_3O_{10}(F,OH)_2$]. A zeolite is a hydrated aluminosilicate mineral with an open three-dimensional crystal structure in which the water molecules in the chemical formula are held. The majority of the world's known reserves of cesium occur in pollucite and lepidolite from the Tango pegmatite located near Bernic Lake in Manitoba, Canada. The U.S. imports all the cesium it uses, mainly from Bernic Lake.

## Rhenium (Re)

The largest use of rhenium is the up to 6 wt% Re often added to nickel-based superalloys used in jet engines. The second largest use is as a catalyst for hydrocarbon hydrogenation and petroleum cracking.

With an average abundance of 7 ppb Re is quite rare in the earth's crust. It occurs, however, in amounts up to 0.2%, substituting for Mo in the mineral molybdenite ($MoS_2$). Re is produced as a byproduct of molybdenum production that is found in porphyry copper and porphyry molybdenite deposits. The main producers

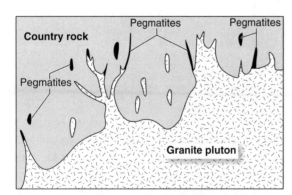

*Figure 9.13* Cross-section through a granite pluton showing the location of pegmatites on the fringes of the pluton, particularly where the country rock has been fractured. Pluton is the general term used for an igneous rock formed below the surface by cooling magma.

are Chile, U.S., and Peru (**Table 9.6**). With low availability relative to demand, Re is among the most expensive metals available. In 2011 the price exceeded US $6,000 per kg.

## Tellurium (Te)

Half of the tellurium consumed each year is used to improve the machinability of specialty iron and steel products. It can be alloyed with copper to improve copper ductility while not reducing conductivity for wire production. Te is alloyed with lead to prevent corrosion and improve resistance to vibration damage and fatigue.

With an average abundance of about 1 ppb, Te is quite rare in the earth's crust. It is most often found as a telluride of gold in the minerals calaverite ($AuTe_2$) and sylvanite [$(Au, Ag)Te_2$], sometime present in copper and gold deposits. The chief source of Te is as a byproduct of electrolytic copper refining where $Te^{2-}$ preferentially concentrates at the anode in concentrations of up to 5 wt%. Production of 1,000 metric tons of copper ore

typically yields 1 kilogram of Te. Although there are significant quantities of tellurium in some gold and lead deposits, this tellurium is not recovered. Te is also present in some coals but the cost of recovering the Te from this source is also presently too high to be economic.

Somewhat greater than 200 metric tons of Te metal are produced worldwide each year with about half coming from the U.S. Reserves are estimated to be 22,000 metric tons, mainly in Cu deposits. This means there is over a 100-year supply. While Te is only slightly more common than Au in the earth's upper crust (See Figure 9.1), it cost only about US $200 per kg in 2011.

## Niobium (Nb) and Tantalum (Ta)

Niobium and tantalum are transition metals and have many similar physical and chemical properties. Their average abundance in the earth's upper crust is about 8 ppm and 0.7 ppm respectively.

### Use

Ninety percent of Nb is used in high-temperature alloys to fabricate parts in such demanding environments as gas turbines and jet engines. Superconducting magnets employ Nb as it becomes superconductive at temperatures near absolute zero. Ta is employed to make capacitors, rectifiers, and high-temperature cutting tools. Ta together with $Ta_2O_5$ can achieve highly reliable high capacitance in a small volume. Ta capacitors are used on computer motherboards to remove interference and smooth power in such devices as DVD players and mobile phones as well as in automotive electronics. Ta capacitors have been estimated to account for more than 60% of Ta use.

### Resource Location

Niobium and tantalum are found together in the same mineral. When the mineral contains more Nb than Ta it is termed columbite with an endmember composition [$"(Fe,Mn)"Nb_2O_6$] and when it contains more Ta than Nb it is termed tantalite with an endmember composition [$"(Fe,Mn)"Ta_2O_6$]. Nb is also present in the mineral pyrochlore [$(Na,Ca)_2Nb_2O_6(OH,F)$]. These minerals are present

*Table 9.6* RHENIUM PRODUCTION FOR 2011 AND RESERVES BY COUNTRY.*

| COUNTRY | PRODUCTION | RESERVES |
|---------|-----------|----------|
| Chile | 26 | 1,300 |
| United States | 6.3 | 390 |
| Peru | 5.0 | 45 |
| Poland | 4.7 | NA |
| Kazakhstan | 3.0 | 190 |
| Canada | 1.2 | 32 |
| Russia | 0.5 | 310 |
| Armenia | 0.6 | 95 |
| Other countries | 1.5 | 91 |
| World totals | 49 | 2,500 |

* In metric tons.
NA = Not available.

*Data from*: U.S. Geological Survey, *Mineral Commodity Summaries*, January, 2012.

in alkali-rich, low-silica plutonic igneous rocks and pegmatites but most Nb is mined from carbonatites.

## Carbonatites.

*Carbonatites* are rare igneous rocks containing greater than 50% carbonate minerals. They are not formed by the melting of limestone rock. Most investigators believe carbonatites are produced by low degrees of partial melting of the mantle. The melt then experiences liquid immiscibility between an alkali plus carbonate-rich and silica-rich composition on cooling and/or decompression as the melt rises in the crust. The separated alkali plus carbonate-rich melt then precipitates Na, K, and Ca-rich carbonate minerals with some silicates on congealing giving rise to a carbonatite. Nb is preferentially enriched over Ta in the carbonate-rich melt, which leads to the presence of pyrochlore. Carbonatites are generally present in stable continental crust where continental extension and rifting has occurred. They are associated with Si-and alkali-rich igneous rocks. There are about 330 localities reported worldwide. These include locations in the East African rift; Kola peninsula in far northwest Russia; Oka and St. Honore in Quebec, Canada; the Fen complex in Norway; Jacupiranga in São Paulo, Brazil; and Mountain Pass, California. Carbonatite intrusions are typically 3 to 4 km diameter small, pipe-like plugs that extend to depths of about 3 to 13 km (**Figure 9.14**).

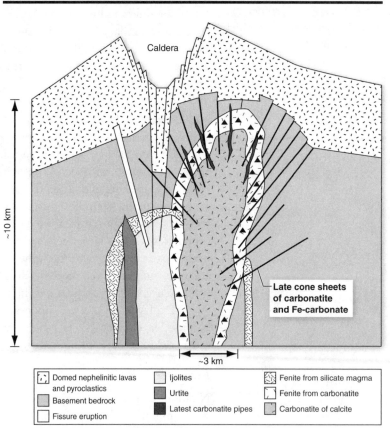

*Figure 9.14* Idealized carbonatite intrusion made up of calcite within an alkaline silicate volcanic complex. Early ijolite is cut by more evolved urtite. Ijolite is an igneous rock consisting essentially of nepheline and augite. Urtite is an igneous rock consisting essentially of nepheline with aegirine and feldspar. Fenite alteration aureoles formed when alkali-rich fluids from the crystallizing carbonatite or silicate magmas react with country rock. (Adapted from LaBas, J. M., 1977, Carbonatite-Nephelinite Volcanism: An African Case History, *Wiley Interscience*, London p. 264.)

## Production and Reserves

The two largest Nb mines extract pyrochlore from carbonatite intrusions near Araxá and Catalão, Brazil. Together these two mines supply 75% of the world market (**Table 9.7**). They have reserves that will last for more than 500 years at current world consumption rates. The mines are open pits containing about 3 wt% of $Nb_2O_5$. These mines produce only a small amount of tantalum. Brazil dominates niobium reserves with 93.3% and Canada has 6.6% of currently minable niobium. The U.S. does not currently mine its small low-grade deposits.

Unlike Nb, Ta resources are widespread. While some is present in carbonatites the major reserves are present in pegmatites. The single largest operating tantalum mine is the Mibra mine in Nazareno, Brazil. Other important pegmatite mines for Ta are the Kenticha mine in Ethiopia, Lovozero mine in Russia, Yichun mine in China, Pitinga mine in Brazil and Marropino mine in Mozambique. Because of their high-density Ta-containing minerals are also recovered as a weathering product of Ta bearing pegmatite veins from nearby stream gravels.

Ta production in 2011 was 790 metric tons with total world reserves of 120,000 metric tons. Reserves, therefore, appear adequate for the foreseeable future although no economic reserves are available in the U.S.

## Arsenic (As)

Arsenic is added as arsenates to produce fungicides and pesticides. Its main use is in the production of chromated copper arsenate (CCA) wood preservatives for producing pressure-treated green lumber including railroad ties and telephone poles. Up to 10 times the natural background level of arsenic has been found in the soil along the right-of-way of old railroad tracks. As a result railroad ties and telephone poles are now more often treated with creosote as some forms of As are potent poisons to humans. As is also used as an alloying agent to increase the strength of lead post and grids in lead-acid storage batteries.

Elevated As concentrations in drinking water are a health concern in many parts of the world (see the chapter on water quality). Seafood can contain naturally high levels. Most investigators conclude As in seafood is, however, a nontoxic form called "fish arsenic."

As occurs most commonly in arsenopyrite (FeAsS), but orpiment ($As_2S_3$) and realgar ($\alpha$-$As_4S_4$) are also found. These minerals are present in copper, lead, and nickel deposits where As is obtained as a byproduct. The major producer of

---

*Table 9.7* TANTALUM AND NIOBIUM PRODUCTION FOR 2011 AND RESERVES BY COUNTRY.*

| COUNTRY | Ta PRODUCTION | Nb PRODUCTION | Ta RESERVES | Nb RESERVES |
|---|---|---|---|---|
| Brazil | 180 | 58,000 | 65,000 | 2,900,000 |
| Mozambique | 120 | NA | 3,200 | NA |
| Rwanda | 110 | NA | NA | NA |
| Australia | 80 | NA | 51,000 | NA |
| Canada | 25 | 4,400 | NA | 200,000 |
| Other countries | 270 | 600 | NA | NA |
| World totals | 790 | 63,000 | 120,000 | 3,000,000 |

\* In metric tons.
NA = Not available.

*Data from*: U.S. Geological Survey, *Mineral Commodity Summaries*, January, 2012.

As is China, producing nearly half of the world's demand (**Table 9.8**). There has been no domestic production of arsenic in the U.S. since 1985. The United States imports about 20,000 metric tons of As-trioxide annually. World reserves are about 1 million metric tons with annual consumption of about 55,000 metric tons, a 180-year supply.

*Table 9.8* ARSENIC TRIOXIDE ($As_2O_3$) PRODUCTION FOR 2011 BY COUNTRY.*

| COUNTRY | PRODUCTION |
|---|---|
| China | 25 |
| Chile | 11.5 |
| Morocco | 8.0 |
| Peru | 4.5 |
| Russia | 1.5 |
| Belgium | 1.0 |
| Mexico | 1.0 |
| Other countries | 0.3 |
| World total | 52 |

* In thousands of metric tons.

*Data from*: U.S. Geological Survey, *Mineral Commodity Summaries*, January, 2012.

## Antimony (Sb)

Antimony's main use is to increase flame retardation in textiles, plastics, and rubber. It is also used as an alloying agent in lead to toughen it. These alloys are used in Pb plates in car batteries, as well as in Pb shot, bullets, and some Pb print type for presses. Some lead-free solders contain Sb. It is also added to some glass where it aids in the removal of microscopic bubbles. Like arsenic, many Sb compounds are toxic. A concern has developed because Sb had been found to leach from polyethylene terephthalate (PET) widely used in water and soft drink bottles.

Sb occurs in stibnite ($Sb_2S_3$). China produces 2/3 of the world demand of Sb from the huge Xikuangshan Sb deposits in Hunan Province. Stibnite ore with 2 to 3 wt% Sb was deposited with calcite in layers in black shale about 370 million years ago on a passive continental margin. The ore fluid is thought by some investigators to be related to deep-seated meteoric water that leached Sb and sulfur from Precambrian basement rocks that had elevated concentrations. Much of the remaining supply of Sb is produced in Bolivia, South Africa, Russia, and Tajikistan as given in **Table 9.9**. These other antimony deposits appear to be formed from low-temperature hydrothermal solutions at shallow depth similar to the

*Table 9.9* ANTIMONY PRODUCTION FOR 2011, RESERVES, AND RESERVE BASE BY COUNTRY.*

| COUNTRY | PRODUCTION | RESERVES | RESERVE BASE |
|---|---|---|---|
| China | 150 | 950 | 2,400 |
| Bolivia | 5.0 | 310 | 320 |
| Russia | 3.0 | 350 | 370 |
| South Africa | 3.0 | 21 | 200 |
| Tajikistan | 2.0 | 50 | 150 |
| Other countries | 6.0 | 150 | 330 |
| World totals | 169 | 1,800 | 3,800 |

* In thousands of metric tons.

*Data from*: U.S. Geological Survey, *Mineral Commodity Summaries*, January, 2012.

Xikuangshan deposits but fill fissures, joints, and irregular replacement bodies.

Given in **Figure 9.15** is the antimony production as a function of time from 1900. Production has increased about four-fold in an erratic pattern. The erratic pattern is generally attributed to production disruptions in China and normal economic cycles. While antimony was produced in the past in the U.S, its supply is now obtained from recycling and imports. As indicated in Table 9.9, antimony reserves will be a concern in 20 years or so given the increased usage as a function of time and with much production controlled by China.

# Bismuth (Bi)

## Use

Bismuth and most of its alloys expand when they solidify. This property allows bismuth-containing alloys to fill all the corners of a mold before hardening. A perfectly shaped replica of the item being replicated is produced. Because of the low melting point of Bi, some of it alloys solidify below 100°C. Low-melting bismuth alloys are in widespread use in emergency sprinkler systems in buildings. The alloy melts in the heat of a fire, the sprinkler becomes unplugged, and water is released which sprays on the fire.

The compound bismuth oxychloride (BiClO) is incorporated into cosmetics. It is slippery with a silky feel, and has good adhesion so it applies well to skin. In the medical industry the compound bismuth subsalicylate ($C_7H_5BiO_4$; Pepto-Bismol™ and Kaopectate™) is prescribed to treat nausea and heartburn. Because bismuth oxide ($Bi_2O_3$) has a yellow color it is sometimes added to paints and cosmetics.

## Resource Location

Bismuth is about twice as abundant as gold in the earth's crust. It occurs in the mineral bismuthinite (Orthorhombic $Bi_2S_3$), typically found with other sulfides. Bi is also found as the bismuthinite alteration product bismite (Monoclinic $Bi_2O_3$). China is the top supplier with 3/4 of world Bi production. Bi in China is obtained mainly as a byproduct of mining in which tungsten is the primary ore. However, it is also produced as a byproduct in some Sn and Zn ores. In other countries Bi is produced as a byproduct of Cu and Pb mining. World Bi reserves appear adequate far into the future (**Table 9.10**).

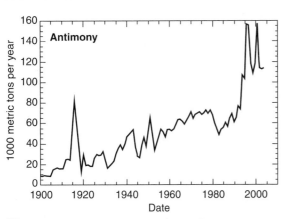

*Figure 9.15* World antimony production by year. (Reproduced from USGS.)

*Table 9.10* BISMUTH PRODUCTION FOR 2011, RESERVES, AND RESERVE BASE BY COUNTRY.*

| COUNTRY | PRODUCTION | RESERVES |
|---|---|---|
| China | 6.0 | 240 |
| Peru | 1.1 | 11 |
| Mexico | 1.0 | 10 |
| Bolivia | 0.1 | 10 |
| Canada | 0.1 | 5.0 |
| Other countries | 0.2 | 39 |
| World totals | 8.5 | 320 |

* In thousands of metric tons.

*Data from*: U.S. Geological Survey, *Mineral Commodity Summaries*, January, 2012.

# Germanium (Ge), Gallium (Ga), and Indium (In)

## Use

The principal use of all three of the metals germanium (Ge), gallium (Ga), and indium (In) is in the electronics industry. The development of the germanium transistor in 1948 increased the need for Ge. This was later replaced by high purity silicon in the 1970s because it has better electrical properties. $GeO_2$ has a high index of refraction and low optical dispersion of light so it is widely used in camera and microscope lenses and for the core of fiber-optic lines. The index of refraction measures the degree of change of the direction of light when it enters the substance from a vacuum. Presently about half the world's supply of Ge is used in fiber optics. Both Ge and $GeO_2$ are transparent to infrared light so they are also employed in both windows and lenses for night-vision goggles.

Gallium (Ga) is incorporated into compounds such as gallium arsenide (GaAs) and gallium nitride (GaN), which are electroluminescent. This means passing an electrical current through them causes light to be emitted. These compounds are manufactured into light-emitting diodes (LEDs). Because Ga expands on crystallizing into a solid it has been incorporated into metal alloys for various applications.

Indium (In) has a particularly low melting point, 156.6°C, so it is incorporated into low melting point alloys. Its primary application is to make transparent In-tin oxide that has good electrical conductivity and is used in liquid crystal displays, LCDs, and computer touchscreens.

## Resource Location

Up to 0.05 wt% Ge, up to 0.1 wt% Ga, and up to 1 wt% In occur as solid solutions in sphalerite (ZnS). They substitute for Zn with smaller amounts substituting for Cu in Cu-sulfides. Ga is also found in bauxites at an average 0.005 wt% level. Annually worldwide Ge production was 118 metric tons in 2011 (**Table 9.11**), with In production of 640 metric tons (**Table 9.12**). The primary sources for these metals are as byproducts obtained by treating flue dusts and slag from Zn and at times Cu smelters. For Ga, while some is obtained from sphalerite, the majority is obtained as a byproduct of Al production. In 2011, world primary production of Ga was estimated to be 216 metric tons, up by 1/2 from 2009 because of increasing demand. China, Germany, Kazakhstan, and Ukraine were the leading suppliers.

*Table 9.11* GERMANIUM PRODUCTION FOR 2011 BY COUNTRY.*

| COUNTRY | PRODUCTION |
|---|---|
| China | 80 |
| Russia | 5.0 |
| United States | 3.0 |
| Other countries | 30 |
| World total | 118 |

* In metric tons.

*Data from*: U.S. Geological Survey, *Mineral Commodity Summaries*, January, 2012.

*Table 9.12* INDIUM PRODUCTION FOR 2011 BY COUNTRY.*

| COUNTRY | PRODUCTION |
|---|---|
| China | 340 |
| Republic of Korea | 100 |
| Japan | 70 |
| Canada | 65 |
| Belgium | 30 |
| Brazil | 5.0 |
| Russia | 4.0 |
| Other countries | 30 |
| World total | 640 |

* In metric tons.

*Data from*: U.S. Geological Survey, *Mineral Commodity Summaries*, January, 2012.

It is difficult to determine reserves of these three metals as they are recovered as byproducts at such low levels during primary metal production. Not only are reserves dependent on the amount of primary metal processed and their concentration in the primary metal but also on the efficiency of extraction. For instance, there are over 1 billion kilograms of gallium in mineable bauxite, but it is not clear how to consider this in a reserve calculation. There is also a large amount that is recycled from scrap for these metals. It seems reasonable to conclude that with increased demand the market will drive improvements in extraction efficiency, usage, and recycling going forward to increase supply.

## Beryllium (Be)

Upper continental crust rocks contain 1.3 to 3 ppm beryllium (Be) on average. With the low atomic number for a metal of 4, beryllium is almost transparent to X-rays.

### Use

Be is primarily used as a hardening agent in alloys, in particular in beryllium copper (BeCu). BeCu is a weldable, machinable, nonoxidizing, acid resistant, and ductile copper alloy. It is, therefore, employed in high-tech applications even though it is relatively expensive. Other items that incorporate Be are golf clubs, wheel chairs, and dental appliances.

Given its unique properties, pure Be's most important application is probably for radiation windows, particularly for X-ray tubes. In 2011, technical pure Be cost about US $1,000 per kg. The high cost of Be is due in part to its high affinity for oxygen even at elevated temperatures and high melting point. It is therefore difficult to produce pure Be metal from the oxide mineral.

### Resource Location

Beryllium occurs in the minerals beryl [$Be_3Al_2(SiO_3)_6$] and bertrandite [$Be_4Si_2O_7(OH)_2$]. Pure beryl is colorless, but is frequently tinted by impurities to green, blue, yellow, red, or white. Beryl is found most commonly in granitic pegmatites, but also occurs in mica schists in the Ural Mountains and limestone in Colombia. Beryl is also often associated with tin and tungsten ore bodies, epithermal deposits, rhyolites, and Mo-W-Be skarns.

The Topaz and Hogs Back mines in the Spor Mountain area of Utah are the largest beryllium mines in the world accounting for nearly 90% of the world's output of 240 metric tons in 2011 (**Table 9.13**). *Rhyolite tuff* hosted bertrandite ore with 0.71 wt% BeO is mined. Tuff is solidified volcanic ash and cinder ejected during a volcanic eruption. The ore likely formed from Be- and F-rich solutions derived from the last stages of crystallization of rhyolitic magma reacting with high silica locations along fluid conduits in previously deposited tuff. Reserves in Utah are about 16,000 metric tons and worldwide resources are estimated to be more than 80,000 metric tons. Therefore, ample Be is available at current consumption rates far into the future.

### Health Concerns

Beryllium is quite toxic. It is responsible for chronic beryllium disease (CBD), an often-fatal lung scarring condition. Inhalation is the primary route of human exposure to beryllium. Occupational exposure in the Be mining and processing industries is of greatest concern.

*Table 9.13* BERYLLIUM PRODUCTION FOR 2011 BY COUNTRY.*

| COUNTRY | PRODUCTION |
|---|---|
| United States | 210 |
| China | 22 |
| Mozambique | 2.0 |
| Other countries | 1.0 |
| World total | 240 |

* In metric tons.

*Data from*: U.S. Geological Survey, *Mineral Commodity Summaries*, January, 2012.

## Rare Earth Elements

The *rare earth elements (REEs)* include the 15 lanthanides with atomic numbers from 57 to 71 as well as scandium (Sc) and yttrium (Y) (**Figure 9.16**). While not lanthanides, scandium and yttrium are included as rare earth elements because they are found in high concentrations with lanthanides in ores and have similar chemical properties. All the rare earth elements have an oxidation state of +3 as might be expected from their position in group 3 of the periodic table of elements. Inner-shell electron orbitals are filled as the atomic number increases in the REEs. These added inner-shell electrons together with their added protons cause increased electrostatic attraction between the electron cloud and nucleus so that the ionic radius of the elements decrease in size as the atomic number of the REE increases. Because the elements continue to have three outer-shell electrons that they can easily donate to other atoms, they all can become cations with an oxidation state of +3.

Exceptions to this observation are that Ce can also be oxidized to a +4 state in an oxidized environment and Eu can be reduced to a +2 oxidation state in a reducing environment. Based on their atomic number, the REEs are divided into light REEs from lanthanum to samarium plus scandium and heavy REEs from europium to lutetium plus yttrium.

### Use

REEs have found a multitude of uses in modern technological devices. Ce is one of the least expensive REEs and is enlisted to absorb ultraviolet radiation in glass and in the production of lighter flints as it catches fire easily. Nd and Y are incorporated into the Nd-YAG (yttrium aluminium garnet) laser used in medicine and manufacturing for cutting, welding, and engraving applications. Nd and Ce are utilized to decolor glass. La absorbs infrared radiation and improves alkali resistance of glass. It also produces the bright white light in carbon arc lamps. Eu produces a

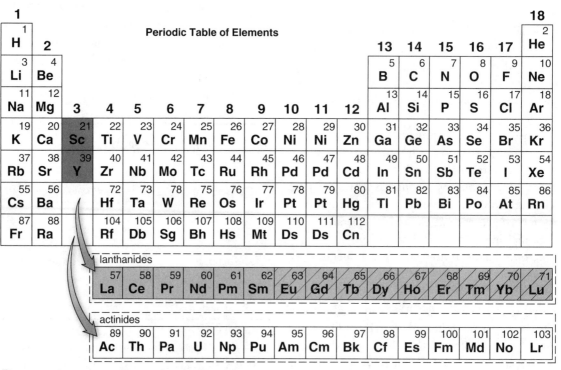

*Figure 9.16* Periodic table of the elements showing location of the rare earth elements in grey. The heavy rare earth elements are shown with hachured lines.

red phosphorescence in cathode-ray tubes (e.g., classic televisions). Light rare earths are incorporated into superconductors. Nd and Sm alloys can be made into strong permanent magnets. Dy is required to fabricate many of the battery systems and advanced electric motors in hybrid electric engines.

Catalysts for modifying emission exhaust from cars and in cracking hydrocarbons for petroleum production represents about 20% of the total market sales of rare earths worldwide. The Toyota Prius hybrid car uses about 1 kg of neodymium in its electric motor and 10 to 15 kg of lanthanum in its batteries as well as REEs in its exhaust's catalytic converter and braking system. Modern wind turbine designs use neodymium-iron-boron magnets in their electrical generators. It takes about 1 metric ton of neodymium per megawatt of generating capacity in these wind turbines.

### Resource Location

Due to their large cation radii (0.94 to 1.15 Å) and +3 oxidation state, REEs in minerals have large coordination numbers with oxygen. They, therefore, do not substitute for $Al^{+3}$ (radius = 0.41 to 0.61 Å) in minerals' structures, but form their own set of minerals instead. REE-rich minerals include carbonates such as bastnäsite $[(REE)CO_3F]$ and the phosphates of monazite $[(REE)PO_4]$ and xenotime $[YPO_4]$. REE-bearing minerals are found in pegmatites, carbonatites, and granites but tend to be most concentrated in carbonatites and associated alkaline rocks. REE-containing minerals are resistant to weathering and have high densities. Weathered carbonatites have lead to the concentration of REE-containing minerals in placer deposits and produced REE-absorbed clay ores in some areas.

As shown in **Figure 9.17** the U.S. was once the leading supplier of REE in the world after a mine at Mountain Pass, California in a carbonatite was put into production. U.S. production competed with production from REE-bearing monazite-rich veins in silica-undersaturated rocks in Steenkampskraal, South Africa as well as placer sand deposits discovered in India and Brazil. Today, the Indian and South African deposits are still mined, but they are small compared

to Chinese deposits, which currently account for 95% of the world production (**Table 9.14**). While the mine at Mountain Pass closed in 2002, it underwent modernization and went back in production in 2011. The U.S. Geological Survey in 2011 reported at least 1 million metric tons of light REEs are present in the Khanneshin carbonatite in Helmand Province, Afghanistan. If and when this world-class deposit will be exploited is not clear given the presence of the Taliban insurgents.

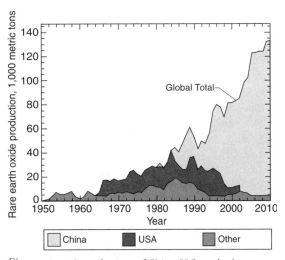

*Figure 9.17* Contributions of China, U.S., and other countries to the global yearly rare earth oxide production as a function of time.

*Table 9.14* RARE EARTH ELEMENT PRODUCTION FOR 2011 BY COUNTRY.*

| COUNTRY | PRODUCTION | RESERVES |
|---|---|---|
| China | 130,000 | 55,000,000 |
| India | 3,000 | 3,100,000 |
| Brazil | 550 | 48,000 |
| Malaysia | 30 | 30,000 |
| United States | 0.0 | 13,000,000 |
| World totals | 130,000 | 110,000,000 |

* In metric tons of the oxide.

*Data from*: U.S. Geological Survey, *Mineral Commodity Summaries*, January, 2012.

Bayan Obo in Inner Mongolia is the largest REE deposit yet discovered. Reserves are estimated at 55 million metric tons of REEs with grades of 3.0 to 5.4 wt% REE. This is half of the world's known REE reserves. The deposit occurs in an east-west trending mid-Proterozoic (1.6 billion to 900 million year old) rift zone along the northern margin of the Sino-Korean craton. A *craton* is the tectonically quiet interior of continental lithosphere undisturbed since the Precambrian era. In the case of the Sino-Korean craton it has been tectonically inactive since before the rift zone formed.

The stratiform dolomite marble hosted Bayan Obo deposit also contains abundant fluorite making it the world's largest fluorite deposit as well. The deposit appears to have formed by hydrothermal replacement of sedimentary carbonate rocks. The hydrothermal fluid was likely derived from nearby rift-related carbonatite intrusives.

## SUMMARY

Precious metals are scarce and have high economic values and include Au, Ag, Pt, Pd, Rh, Ru, Os, and Ir. The demand for precious metals is driven by their industrial use, the fabrication of jewelry, and their role in wealth accumulation. Specialty metals are important in high-technology manufacturing and include Cs, Re, Te, Nb, Ta, As, Sb, Bi, Ge, Ga, In, Be, and the rare earth elements.

Gold is most commonly found as an electrum of native gold and silver in quartz- or carbonate-rich veins. The California Mother Lode is a belt of gold-quartz veins in the foothills of the Sierra Nevada Mountains. The gold fields of South Africa are placer deposits found in ancient alluvial fans from weathered gold-rich veins that washed into the Witwatersrand Basin. The Homestake gold mine in South Dakota contains disseminated gold in metamorphosed volcanic rocks. Low-grade gold, as occurs at Carlin, Nevada, is recovered by a bulk mining cyanide leaching technique. Gold is mined in over 50 countries with the biggest producers being China, Australia, U.S., and then Russia.

Silver is found in its native form, as an electrum with gold and in the mineral argentite ($Ag_2S$). Its occurs in epithermal veins deposited from aqueous fluid during boiling such as occurred in the Comstock Lode in Nevada. Presently, most silver is obtained as a byproduct of mining of other metals with Peru, Mexico, and China the largest producers.

The platinum group metals Pt, Pd, Rh, Ir, Ru and Os have similar chemical and physical properties and occur together in the same ore deposits. They are obtained as a byproduct of nickel and copper mining. Large reserves occur in the Bushveld Complex, South Africa; the copper-nickel deposits of Norilsk, Russia; and the Sudbury Basin of Canada. The platinum group metals have outstanding catalytic properties, and are used for electrical contacts and thick- and thin-film circuits.

Cesium is obtained from a zeolite. Rhenium occurs as a solid solution of Mo in molybdenite. Tellurium is most often found as a telluride of gold and obtained as a byproduct of copper ore production. Minerals that contain niobium often also contain tantalum, such as columbite. This mineral is found in carbonatites, carbonate-rich igneous rocks. Arsenic commonly occurs in arsenopyrite (FeAsS) and is obtained as a byproduct of copper, lead, and nickel smelting. Antimony occurs in stibnite ($Sb_2S_3$) with 2/3 of the world's production from China. Bismuth is found in bismuthinite ($Bi_2S_3$), which is typically found with other sulfides, particularly tungsten sulfides. China also dominates the world production of Bi.

Germanium, gallium, and indium are used in the electronics industry. They are found as solid solutions in sphalerite (ZnS) substituting for Zn. Gallium is also found in bauxites. Beryllium occurs in beryl [$Be_3Al_2(SiO_3)_6$] and bertrandite in granitic pegmatites. The world's largest production is from a mine in rhyolitic tuff in Utah.

The rare earth elements (REEs) include the 15 lanthanides with atomic numbers from 51 to 71 plus scandium and yttrium. They generally have a +3 charge and occur in carbonate and phosphate minerals present in carbonatites and their associated alkaline rocks. Bayan Obo in China is the largest REE deposit.

## KEY TERMS

| | |
|---|---|
| alluvial fan | lode |
| carbonatite | pegmatites |
| craton | rare earth elements (REE) |
| electrum | retrograde solubility |
| epithermal | rhyolite tuff |
| lanthanides | |

## PROBLEMS

1. As shown in Figure 9.7 the production of gold in the world increased dramatically starting in 1980 to a level about twice as high by 2000. Why?

2. From **Table 9.15** below plot the starting DJIA (Dow Jones Industrial Average) for the given years. Compare with the price of gold shown in Figure 9.2. Much of the time they have opposite slopes as a function of time. Why?

*Table 9.15* STARTING DOW JONES INDUSTRIAL AVERAGE FOR THE GIVEN YEAR.

| YEAR | DJIA |
|---|---|
| 1975 | 616 |
| 1976 | 852 |
| 1977 | 1004 |
| 1978 | 831 |
| 1979 | 805 |
| 1980 | 838 |
| 1981 | 963 |
| 1982 | 875 |
| 1983 | 1046 |
| 1984 | 1258 |

3. Is antimony more or less abundant than rhodium in the earth's crust? What are the world's reserves in years of these metals at 2011 production levels? Why are antimony reserves so low relative to demand?

## REFERENCES

Bateman, A. M. 1950. *Economic mineral deposits*, 2nd ed. New York: John Wiley & Sons, Inc.

Brechin, G. and Kirshenbaum, N. W. 2004. Underground power generation at the Comstock lode and the Sutro tunnel, Nevada. *Bulletin of the Peak District Mines Historical Society* 15(4/5):104–110.

Brown, A. C. 1978. Stratiform copper deposits—Evidence for their post-sedimentary origin. *Mineral Science and Engineering* 10(3):172–181.

Buchanan, L. J. 1981. Precious metal deposits associated with volcanic environments in the Southwest in Dickinson, W. R. and Payne, W. D. (eds.) Relations of tectonics to ore deposits in the southern Cordillera. *Arizona Geological Society Digest* 14:237–262.

Clark, W. B., 1998. *Gold districts of California*. Bulletin 193, Sesquicentennial edition. California Gold Discovery to Statehood, California Division of Mines and Geology 1976.

Cohen, D. 2007. Earth audit. *New Scientist* 194(2605):34–41.

Lide, D. R. (ed.) 2009. *CRC Handbook of Chemistry and Physics*, 90th ed. Boca Raton, FL: CRC Press.

Evans, A. M. 1997. *An Introduction to Economic Geology and Its Environmental Impact*. Oxford, UK: Blackwell Science Publishing.

Gilluly J., Waters, A. C., and Woodford, A. O. 1959. *Principles of Geology*. San Francisco: W. H. Freeman and Company.

Haxel, G. B., Hedrick, J. B., and Orris, G. J. 2002. Rare earth elements—Critical resources for high technology. Fact Sheet 087-02. Reston, VA: U.S. Geological Survey.

Hyrsl, J. and Petrov, A. 2006. Famous mineral localities: Llallagua, Bolivia. *The Mineralogical Record* 37(2):117–162.

LaBas, J. M. 1977. *Carbonatite-nephelinite volcanism: An African case history*. London: Wiley Interscience, John Wiley & Sons, Inc.

Mackin, J. H., 1947, Some structural features of the intrusions in the Iron Springs district [Utah]: Utah Geological Society, Guidebook to the geology of Utah, no. 2, 62 p.

Misra, K. C. 2000. *Understanding mineral deposits*. Dordrecht, The Netherlands: Kluwer Academic Publishers.

Nelson, C. E. and Giles, D. L. 1985. Hydrothermal eruption mechanisms and hot spring gold deposits. *Economic Geology* 80:1633–1639.

Sangster, D. F. 1990. Mississippi Valley-type and SEDEX lead-zinc deposits: A comparative examination. *Transactions of the Institution of Mining and Metallurgy, Section B: Applied Earth Science* 99:21–42.

Taylor, S. R., 1964, Abundance of chemical elements in the continental crust: a new table, *Geochim. Cosmochim. Acta*, 28:1273–1285.

Theodore, T. G. 1977. Selected copper-bearing skarns and epizonal granitic intrusions in the southwestern United States. *Geol Soc Malaysia Bull* 9:31–50.

Turneaure, F. S. 1960. A comparative study of major ore deposits of central Bolivia, *Economic Geology* 55(Part 1):217–254; (Part II):574–606.

Tucker, R. F. and Viljoen, R. P. 1986. The geology of the West Rand Goldfield. In: Annhaeusser, C. R. and Maske, S. (eds.) *Mineral Deposits of South Africa* (pp. 649–598). Johannesburg: Geological Society of South Africa.

U.S. Geological Survey. 2011. *Mineral commodity summaries 2011*, Washington, DC: U.S. Government Printing Office.

# Part Three

# Life-Supporting Resources: Building and Industrial Material, Chemical Minerals, Fertilizer, and Gases

The most basic concerns of humans are for shelter, food, and the air we breathe. In this part these resources are considered. They include resources needed to construct buildings, minerals used for their unique properties to produce commercial goods, fertilizers required to grow food, and resources we obtain from the atmosphere.

Building materials have changed little over the years. We still use wood and stone to house ourselves, build our roads, and construct our walls. They have little monetary value per unit volume because of their abundance and widespread distribution, so transportation costs can be significant.

While metals are extracted from some minerals, for other minerals their unique properties are used directly rather than the metals they contain. These minerals are incorporated into consumer goods we purchase. These run the gamut from the greasy feeling mineral talc added to cosmetic powders, to clays added to candy bars to give them additional weight, to diamonds that sparkle when cut and are used to announce a marriage engagement.

Many chemical elements are obtained from evaporated water. These salts are easily redissolved to obtain their constituent compounds. As a result little energy is required to process them. They can be found in both terminal lakes and seawater evaporite deposits. Rivers bring compounds in diluted water to terminal lakes. They are concentrated by evaporation of $H_2O$ from the lake and precipitate as salts. The other environment where evaporites form is with the evaporation of seawater. This occurs in enclosed bays and along the seacoast. These environments also concentrate dissolved constituents by the evaporation of $H_2O$. Sea salts can then precipitate when their solubility limit is exceeded.

The fertilizer elements P and K are obtained from evaporites but nitrogen in fertilizer is currently obtained by processing the $N_2$ in air, and sulfur is extracted from petroleum during processing. These resources, needed to grow food, may be large in abundance in the earth and its atmosphere but they are difficult to get to all the farmers in the world who need them. This is because a large amount of energy is needed to turn $N_2$ in air to usable fixed nitrogen for fertilizer use and the salts of P and K and petroleum processing for S are not equally distributed across the earth's surface making them reasonably expensive.

Besides $N_2$ used to make nitrogen-based fertilizers the $N_2$ in air is also used industrially as a gas to produce an oxygen-free atmosphere for many industrial processes. Other gasses in air are used for their unique properties. The largest consumer of pure oxygen gas is the steel industry where excess carbon in the steel is reacted with $O_2$ to turn it into a gas that can escape the molten metal.

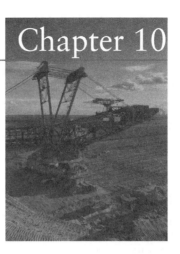

## Chapter 10

# Building Material and Industrial Minerals

Building material is any substance, natural or synthetic, used for a construction purpose. They can be organic (e.g., wood) or inorganic (e.g., mineral and rock material). Industrial minerals are distinguished from ore minerals because they are used for their unique properties rather than the elements or compounds they contain.

## Wood

There are more than $8 \times 10^9$ acres ($3.2 \times 10^7$ km$^2$) of forests and woodlands on the earth, about 1/3 of the earth's land surface. In the world as a whole wood is primarily consumed for fuel but with a significant usage in construction. The use of wood in industrial countries is stable with almost all of it produced in these countries on a sustained basis from plantations, that is, managed forests. The production technology has been improving and with a constant area of managed forestland the amount of wood produced is increasing. In the undeveloped tropical world the situation is different. In these countries 0.8% of forestland is being converted to agriculture and pasture land each year, up from 0.6% in 1980.

Trees grow by addition of new material to an outer circular layer just below the bark called the *cambium* (**Figure 10.1**). The sapwood is the living part of the tree just below the cambium that is transformed to heartwood, which is old sapwood rings that have died over time. A series of concentric rings in cross-section is then preserved in the heartwood. Each ring is the product of one year of tree growth. The wood in each ring is made up of connected long, thin tubular cellulose fibers. In softwoods these fibers

are about 3 mm long and 0.03 mm in diameter. In hardwoods the fibers are smaller in size.

Generally, softwoods, as the name implies, are easier to cut than hardwoods, are less dense, and have weaker wood fibers. Conifer trees produce softwoods, whereas hardwoods are the product of deciduous trees. Deciduous trees, those that shed their leaves each year, include basswood, birch, cherry,

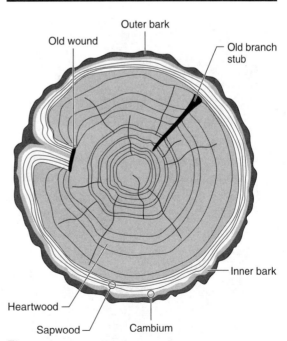

*Figure 10.1* Tree trunk showing the cambium where new cells are growing, the light-colored sapwood where water and dissolved nutrients are transported, and the inner dark-colored non-water-carrying heartwood that gives strength to the trunk.

elm, hickory, mahogany, maple, oak, pecan, poplar, and walnut. Note that some deciduous hardwoods such as basswood and poplar are softer than average softwood and the softwoods Longleaf Southern Pine and Douglas Fir are harder than some hardwoods.

## Manufacture of Lumber

Construction wood and wood processed into wood pulp for paper production is termed lumber or timber. Lumber is extensively used in construction because for a given weight, wood is stronger than steel.

Trees are cut and felled with chainsaws, harvesters, and feller bunchers (**Figure 10.2**). A harvester is a vehicle employed in cut-to-length logging operations for felling, delimbing, and bucking trees. Bucking is the process of cutting the tree into useable lengths. A feller buncher cuts and gathers several trees before felling them. The tree is turned

into logs by removing the limbs and cutting it into sections of optimal length. The felled tree or logs are moved from the tree stump, a small remaining portion of the trunk with attached roots, to a landing by a skidder or forwarder. A landing is a nearby location where logs are collected. Skidders drag logs while a forwarder carries logs clear of the ground.

Logs are transported to a sawmill or plywood plant from the landing by a truck, rail car, or barge or floated in water. At the sawmill logs are passed through a barker to remove the bark and rough protrusions. The logs are then cut into dimensional lumber. Dimensional lumber is lumber that is planed and cut into standardized widths, depths, and lengths. Bark, chips, and limbs are used to make fuel logs, soil conditioners, fiberboard, and a variety of paper products.

### Plywood

Plywood is a wood product made from thin sheets of wood with the grain of each sheet set at right angles and glued together. Plywood is often used instead of solid sheets of wood because of its increased strength and better resistance to warping, cracking, and shrinkage. Plywood is cheaper than regular sheets of wood because it can be fabricated in sheets wider than the trees from which it is made. It has replaced much dimensional lumber in construction.

### Paper

To make paper cellulose fibers in wood chips must be separated from the lignin. This is generally accomplished with the Kraft process, which uses a solution of sodium hydroxide (NaOH) and sodium sulfide ($Na_2S$) to break down the lignin so it can be washed out with the aid of steam from the cellulose fibers in the pulp. Bleaching the pulp also helps delignify it. The reactions that occur to break the lignin apart produce a large amount of heat. In most pulping operations this heat is used to heat water to generate electricity. Often chalk (soft porous calcium carbonate plates produced by single-celled algae called coccolithophores), kaolinite (see below), or titanium oxide (see chapter on abundant metals) is added to the pulp before it is dried to improve its whiteness and its ability for ink and paint absorption.

The countries producing the largest amount of paper in the world are given in **Table 10.1**.

*Figure* 10.2 A feller buncher that harvests trees by shearing them off near the ground.

*Table 10.1* PAPER AND PAPERBOARD PRODUCTION FOR 2010.

| COUNTRY | PRODUCTION |
|---------|------------|
| China | 92,600 |
| United States | 75,850 |
| Japan | 27,300 |
| Germany | 23,100 |
| Canada | 12,800 |
| Finland | 11,800 |
| Sweden | 11,400 |
| South Korea | 11,100 |
| Indonesia | 9,950 |
| Brazil | 9,800 |

\* In thousand metric tons.

*Data from*: Resource Information Systems, Inc (RISI)

This includes writing, wrapping, tissue, cigarette, and other industrial papers. There have been some environmental concerns as 35% of harvested trees are used for paper manufacture and some of this is from old growth forests. Also, bleaching of the pulp with chlorine produces dioxins, highly toxic organic compounds (see the chapter on water quality).

# Stone

Perhaps the most important construction material is rock, or what is termed stone when used in the construction industry. Some stone is used directly from the quarry where it is obtained while others are treated before usage. Often a classification based on its origin as igneous, sedimentary, or metamorphic stone is used. Large amounts of crushed stone are used as construction aggregate. Construction aggregate is a group of coarse particulate material used in construction including sand, gravel, and larger crushed stone.

## Building Stone

A building stone is any massive, dense rock suitable for use in construction. This includes dimension stone as well as construction aggregates. Stone that is selected from a quarry and fabricated to specific sizes or shapes is termed dimension stone. Quarries that produce building stone can typically change their quarrying techniques to produce either stone blocks or aggregates. Specific names dependent on usage of the blocks are given in **Table 10.2**. Dimension stone is produced by controlled light blasting together with diamond wire and diamond studded belt sawing. Aggregates are quarried after heavy blasting.

### Dimension Stone

Dimension stone is generally selected for its color and texture. Depending on the application, durability to decay and the change of appearance through time are also considered. It is often used as non-load-bearing facing of a building

*Table 10.2* STONE BLOCK TYPES, CHARACTERISTICS, AND USES.

| TYPE | DESCRIPTION | USE |
|------|-------------|-----|
| Rubble | Broken stone, of irregular size, shape, and texture | Sea walls, bridges, and foundations |
| Fieldstone | Irregular rough blocks | Freestanding walls and walkways |
| Rip-rap | Irregular large blocks | Used along rivers and seashores to protect embankments against erosion |
| Ashlar | Dimensional masonry stone of rectangular blocks of nonuniform size | Wall construction (Figure 10.3) |
| Monument stone | Stone blocks | Gravestones and monuments |

or for walls (**Figure 10.3**). The principal types of dimension stone are given in **Table 10.3**. Sandstone was commonly the dimension stone of choice in the nineteenth century but today in North America granite and marble are more common.

*Figure 10.3* Ashlar wall of randomly set rough-hewn blocks.

The major producers of dimension stone for export are the countries of Brazil, China, India, Italy, and Spain. The U.S. is the world's largest market for dimension stone, but the majority is obtained internally. Buildings constructed in an environmentally conscience manner that are *Leadership in Energy and Environmental Design (LEED)* certified require dimension stone be quarried within an 800 km radius of the building site. This increases the amount of domestic dimension stone that is used. Dimension stone is often reused after a structure is removed from a site.

## Construction Aggregates

A large amount of aggregate is used in construction. Aggregate can be large crushed rock, sand, gravel, or lightweight aggregate with some being heat-treated. Aggregate often serves to add strength to the overall material. Aggregate is combined with cement to form concrete and with asphalt or bitumen to produce roadway blacktop. Aggregate is also used in septic tank and surface water drains,

*Table 10.3* SOME COMMERCIAL BUILDING STONE TYPES AND CHARACTERISTICS.

| TYPE | DESCRIPTION |
| --- | --- |
| Granite | Coarsely crystalline igneous rock. Gabbro is considered black granite. |
| Argillite | A lithified clay and silt-rich sedimentary rock. It does not possess the slaty cleavage of a slate. |
| Sandstone | Sand-size particles of quartz or feldspar cemented with silica, iron oxide, or calcium carbonate. Brownstone is brown or reddish-brown sandstone due to the presence of iron oxide cement. |
| Quartzite | A metamorphosed quartz sandstone. The quartz grains are compacted and interwoven to form a tough dense rock. |
| Schist | A significantly metamorphosed, compacted, originally clay-rich rock; typically contains a preponderance of mica and elongated quartz and feldspar grains. |
| Limestone | Rock made up of calcite ($CaCO_3$) and/or dolomite [$CaMg(CO_3)_2$]. |
| Travertine | A special kind of limestone deposited in a bubbling hot spring. Gas bubbles become trapped creating a pitted surface on the stone. |
| Greenstone | Metamorphosed rock with a distinctive greenish color owing to the presence of the minerals chlorite, epidote, and/or actinolite. |
| Marble | Recrystallized limestone or serpentinite, a rock that often contains hydrous magnesium iron silicate minerals. Marble is an easily cut material that takes a polish. |
| Soapstone | Stone predominantly composed of metamorphosed talc that is stain proof with a greasy feel. |
| Slate | Fine-grained rock with aligned micas that can be split into thin slabs. Used in roof and blackboard construction but is being replaced by other non-rock materials. |

as subbase material under building foundations, for road construction, and as base material for railroad tracks. This is because aggregates have a high hydraulic conductivity as compared to most soils. Hydraulic conductivity is a measure of the ease that water can move through a rock or soil. High hydraulic conductivity allows water drainage while maintaining structural support.

## Sand and Gravel

The main source for sand and gravel is from ancient river channel, floodplain deposits, alluvial fans, shore line deposits, sand dunes, and glacial esker deposits as given in **Figure 10.4**. A *floodplain* is the flat area along the side of a river where flooding occurs when a river's water level exceeds its banks. An esker is a long ridge of sand and gravel in a channel produced by a stream under a glacier. In Europe, deposits of glacial origin in the ocean on submerged continental shelves are exploited by dredging.

In 2011 the U.S. produced sand and gravel for construction valued at almost US $6 billion from 4,000 companies located in all 50 states. This was 790 million metric tons at on average US $7.50 per ton. Another 30 million metric tons valued at over US $1 billion was mined for industrial use in such applications as hydraulic fracturing of oil and gas wells and to make glass. Sand and gravel quarries comprised the largest percentage of all mining operations, about half. Gravel is used mainly in construction, but is also used in landscaping and as a filtration media in a variety of industries. The uses of sand in the U.S. are outlined in **Figure 10.5**.

## Crushed Stone

After sand and gravel the largest volume of mineral commodities extracted worldwide from the earth is the aggregate termed crushed stone. Most have low value per ton so a local supply is important to keep transport costs to a minimum. In 2011 the U.S. produced over 1.1 billion metric tons worth US $11 billion or about US $10 per metric ton. Crushed stone in the U.S. is produced

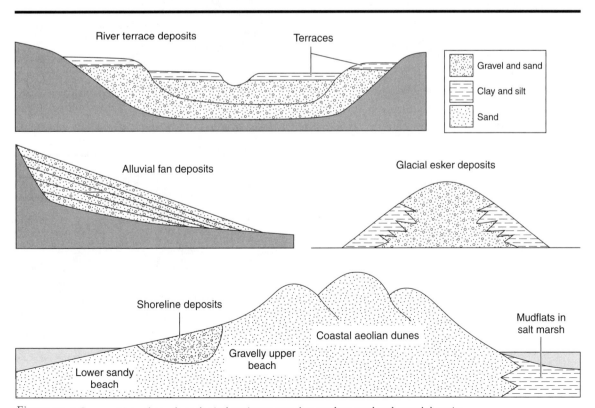

*Figure 10.4* Cross-sections through geological environments that produce sand and gravel deposits.

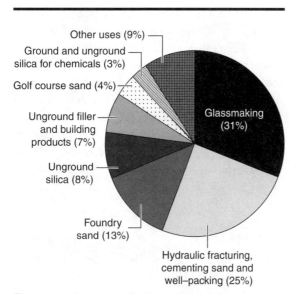

*Figure 10.5* Sand use in the United States in 2010.

in over 3,000 quarries and 83 underground mines in all 50 states. Of the total produced, 70% was limestone, 14% granite, and 7% trap rock. Trap rock is solidified lava that is crushed after quarrying to produce construction aggregate.

The dominant use of crushed stone is in road construction for tarmac, also called blacktop. John McAdam, a Scottish road-builder, in 1816 developed tarmac, a road paving material made of 10% tar and 90% crushed stone. It is used to build roads with a smooth hard surface that is curved slightly upward in the middle, to allow water drainage off the surface.

## Lightweight Aggregates

The lightweight aggregates pumice, vermiculite, and perlite are used chiefly in wallboards, plaster, and concrete. They are low density, easy to handle, and contain trapped air making them good thermal and sound insulators.

**Pumice.** *Pumice* is volcanic rock that has formed when gases in the magma escape as pressure is lowered on eruption, producing a solidified foam on cooling. The solidified foam contains vesicles in the form of isolated bubbles, tubes, and irregular cavities enclosed in glass (**Figure 10.6**). The isolated porosity is great enough that pumice floats on water.

Most deposits are found in layers of air fall material deposited near andesitic or rhyolitic volcanic vents. Deposits of pumice are crushed and sized after quarrying. About 20 million metric tons are used annually in the world. The largest producers in descending order are Turkey, Italy, and Greece.

The primary use of pumice is as an additive to concrete to decrease its weight and increase its thermal insulation properties. This includes making cinder or breeze blocks (**Figure 10.7**) for wall construction. Scoria is also used to make thermally-insulating concrete. It is also solidified magma foam but is denser than pumice with larger thicker-walled vesicles and it does not

*Figure 10.6* Pumice grain showing vesicles (gas bubbles) formed by escaping gas when the magma solidified. Enclosed bubbles make the grain light enough to float on water.

*Figure 10.7* Cinder blocks made from a slurry of portland cement and pumice.

float on water. Pumice is also used as an abrasive in toothpaste, polishes, skin exfoliants (pumice stones), hand cleaners, and in some pencil erasers.

**Vermiculite.** *Vermiculite* is a hydrous Mg-Fe aluminosilicate clay mineral with the chemical formula $(Mg,Fe,Al)_3(Al,Si)_4O_{10}(OH)_2 \bullet 4H_2O$. When biotite and phlogopite are altered vermiculite is sometimes formed. It is a 2:1 clay, meaning its mineral structure is made up of two *tetrahedral sheets* for every *octahedral sheet* (**Figure 10.8**). In the tetrahedral sheet four oxygen atoms are bonded together with an internal Al or Si into a tetrahedron and these tetrahedron share oxygens to form a sheet of tetrahedron as shown in **Figure 10.9**. In the octahedral sheet six oxygen atoms are bonded together with internal Al, Mg, and/or Fe atoms to form octahedrons as shown in **Figure 10.10**. These octahedrons also share oxygen atoms to form a sheet of octahedrons.

The four water molecules in the vermiculite formula reside in the interlayer between the tetrahedral sheets as given in the vermiculite mineral structure outlined in Figure 10.10. When heated to 1090°C this water turns into steam forcing an 8- to 12-times expansion of the interlayer between the sheets before the steam is lost. This forms dead air space in the vermiculite with the expansion process called exfoliation as shown for exfoliated vermiculite in **Figure 10.11**. It is routinely accomplished in commercial exfoliation furnaces near the final market for the product as transport of higher density raw vermiculite is much more economic.

Vermiculite is used as a plant-growing medium in combination with peat moss. The major use of exfoliated vermiculite is, however, as loose fill between joists in lofts and in walls of

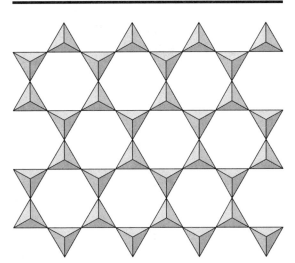

*Figure 10.9* Sheet of tetrahedron made up of silicon plus alumina atoms bonded to four oxygen atoms.

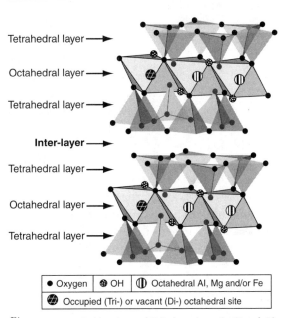

*Figure 10.10* A three-layered 2:1 clay where the Si and Al cations in the oxygen tetrahedron are not shown for clarity. The interlayer is where the $H_2O$ resides in the vermiculite clay structure.

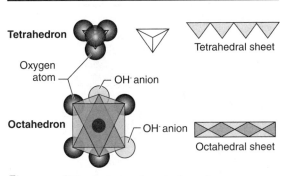

*Figure 10.8* Tetrahedron and octahedron of oxygen atoms showing how they combine into a sheet structure.

*Figure 10.11* Exfoliated vermiculite showing expanded layers.

buildings for heating, cooling, and sound insulation. Exfoliated vermiculite is also used as packaging material. Not only is it lightweight and clean, but it is easily poured around irregular shaped objects providing good protection against the shocks of handling. Exfoliated vermiculite readily absorbs liquids. It is therefore used in animal feed to carry nutrients, vitamins, and molasses and in plant food to carry fertilizers, herbicides, and insecticides. The weight of concrete is lowered and thermal insulation increased by the addition of exfoliated vermiculite.

Vermiculite deposits are found in many parts of the world, but currently operating major mines in descending order of amount of production are located in South Africa in the Phalaborwa of the northeastern Transvaal; near Korla in Xinjiang, China; in the eastern Appalachians of Virginia and South Carolina; at the Mud Tank mine in Queensland, Australia; in the Carajás region of the state of Pará, Brazil; and in the Buhera District, Zimbabwe. The world total production in 2011 was 580,000 metric tons.

Most commercial vermiculite mines operating today are in altered ultramafic rocks formed during the Precambrian. However, deposits at Libby, Montana are Triassic in age. Deposits typically occur at shallow depths. Lab experiments indicate phlogopite and biotite when reacted with 0.001 M $MgCl_2$ at room temperature partially alters to vermiculite. This suggests a supergene origin of the deposits with meteoric water reacting

with Mg-rich rocks such as igneous ultramafics. Reserves of vermiculite are large. In the U.S. alone there are 25 million metric tons that could supply the world usage for nearly 50 years. China and South Africa also have large deposits.

**Perlite.** *Perlite* is formed from amorphous volcanic glass with high water content, typically hydrated obsidian. Obsidian is high-silica glass found in solidified magma flows around rhyolitic volcanic vents. After the glass is quarried, it is crushed and then rapidly heated. At 850° to 900°C, it softens because it is a glass. Water trapped in the glass vaporizes and escapes. This causes the glass to expand to 4 to 20 times its original volume producing what is termed perlite, as shown in **Figure 10.12**. The perlite is a brilliant white, due to the reflectivity of its trapped air bubbles.

Perlite has many of the same uses as exfoliated vermiculite. Additional applications include its use as an abrasive in soaps, cleaners, and polishes. Its higher thermal stability leads to its use in refractory bricks, mortars, and pipe insulation. Perlite does not react significantly with water and is therefore used as a filtration medium for drinking and swimming pool water as well as industrial effluents.

About 1.9 million metric tons of perlite were produced worldwide in 2011 with the largest producing countries in descending order being Greece, U.S., Turkey, and Japan. World reserves are estimated to be at least 700 million metric tons.

*Figure 10.12* Perlite formed by heating obsidian.

**Diatomite.** *Diatomite*, also known as diatomaceous earth, is a lightweight aggregate composed of the fossilized remains of diatoms (**Figure 10.13**). Diatoms are a type of small algae with ornate siliceous tests, that is, tiny shells. These are easily crumbled into a fine white to off-white, very light, amorphous $SiO_2$ powder. Deposits occur when high amounts of silica are added to water and nutrients are available. These include coastal lagoons and lakes that have received significant amounts of volcanic ash from a nearby volcanic center.

Two-thirds of diatomite is used as a filtering medium, especially for swimming pools but also for beer and other alcoholic beverages. Because it is mildly abrasive it is also added to toothpaste and metal polishes. As a good absorbent for liquids, diatomite is used in cat litter and as a component to form sticks of dynamite.

About 1.8 million metric tons of diatomite were mined worldwide in 2011 with the largest producing countries in descending order being the U.S., China, Denmark, Mexico, and Japan. World reserves are estimated to be greater than 1,000 million metric tons.

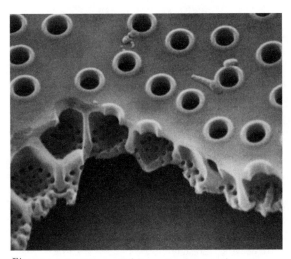

*Figure 10.13* Close-up of diatomite showing the pattern of pores in individual amorphous silica diatom cell wall remains.

# Cement, Plaster, and Bricks

Before they are ready for use in the construction industry some rock products are chemically processed and fired after mining. These include portland cement, plaster, and bricks.

## Portland Cement

*Portland cement* was developed in 1824 and takes its name from similarities with limestone in quarries near Portland, England. Traditional portland cement contains 60 to 66 wt% CaO and 0 to 5 wt% MgO from limestone and dolostone as well as 19 to 25 wt% $SiO_2$ from silica sand and clay together with 3 to 8 wt% $Al_2O_3$ from clay and 1 to 5 wt% $Fe_2O_3$ from iron oxides. These materials are crushed to a powder and heated first to 870°C to drive off $CO_2$ from the limestone and dolostone, then to 1480°C where incipient melting and fusion occurs producing a glass-like material. After cooling, 1% to 3% $SO_3^{2-}$ from gypsum ($CaSO_4 \bullet 2H_2O$) is added.

To use as cement, water is added to this material. A gel is formed by way of a complex series of chemical reactions, that are still only partly understood. These reactions can be represented by

$$2Ca_3SiO_5 + 7H_2O \rightarrow 3(CaO) \bullet 2(SiO_2) \bullet 4(H_2O)_{(gel)}$$
$$+ 3Ca(OH)_{2\,(solid)} + 173 \text{ kJ mol}^{-1} \quad [10.1]$$

where some of the Si is typically replaced by Al. As the gel sets, interlocking calcium silicate and calcium aluminate crystals are formed. These interlocking crystals give the portland cement its strength.

### Resource location

As a mixture of limestone, dolomite, silica sand, gypsum, and clay, the needed components of portland cement are abundant and typically locally available in areas where sedimentary rocks are present. Local sources tend to have a cost advantage as transportation can be a significant expense.

## Concrete

*Concrete* is an artificial rock made by binding together aggregate with a paste made of portland cement and water as shown in **Figure 10.14**. The aggregate is typically sand and gravel or crushed stone. Lightweight aggregates like pumice, perlite, or vermiculite can be used in place of some of the gravel and crushed stone to produce concrete with improved heat and sound insulating properties rather than strength. Concrete is often strengthened with rebar, iron, or steel rods to increase its tensile strength. Tensile strength is the ability to resist deformation by being pulled apart, that is, put under tension.

## Plaster

*Plaster* is produced by heating gypsum ($CaSO_4 \cdot 2H_2O$) to 150°C. A new compound is formed, plaster of Paris ($CaSO_4 \cdot 1/2\ H_2O$). The name comes from a gypsum deposit at Montmartre in Paris, France. Mixed with water plaster of Paris rapidly rehydrates and sets, producing an interlocking mass of gypsum crystals. Wet plaster is soft and is easily worked with hand tools. Dry plaster can be smoothed with sandpaper. This makes plaster a good finished surface that does not bear weight. Addition of organic additives slows the rapid setting times and expands its usage. In 1918 prefabricated wallboard was developed, termed plasterboard or drywall. Drywall is a panel of gypsum plaster pressed between two thick sheets of paper, and then dried in a kiln.

### Resource Location

Gypsum is an evaporite mineral precipitated out of seawater after a large degree of evaporation commonly found in marine tidal flats. Often gypsum is found in thick beds together with anhydrite ($CaSO_4$) and halite ($NaCl$) in large deposits from shallow tropical marine basins. Because many areas of the world have experienced tropical climates during geological times in the past due to continental drift, gypsum resources are significant and widely distributed. As given in **Table 10.4** China is by far the leading producer. World reserves are not accurately known but are quite large; likely at least a 100-year supply. In the U.S. gypsum resources appear adequate for the near future but is augmented by imports from Canada and China.

Between 2001 and 2007 China exported drywall that emitted $H_2S$ gas often creating a noxious odor and corroding copper in air conditioners, electrical wiring, plumbing, appliances, and electronics. The source of the $H_2S$ gas was unclear because the manufacturing process is not a part of the public record. It is possible that the

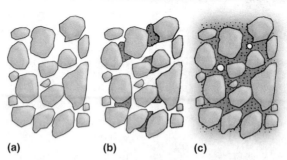

(a)          (b)          (c)

*Figure 10.14* Reactions to form concrete. (a) Right after water is added, grains of calcium silicate and water are present but no reaction has occurred. (b) Reaction of water and calcium silicate produces calcium silicate hydrate. (c) As the reaction proceeds it slows until small pockets of calcium hydroxide solution, the white areas, are left.

*Table 10.4* GYPSUM MINE PRODUCTION FOR 2011 BY COUNTRY.*

| COUNTRY | MINE PRODUCTION |
|---------|-----------------|
| China | 47 |
| Iran | 13 |
| Spain | 11.5 |
| United States | 9.4 |
| Thailand | 8.5 |
| World total | 148 |

\* In millions of metric tons.

*Data from*: U.S. Geological Survey, *Mineral Commodity Summaries*, January, 2012.

gypsum was mined from deposits formed in an anoxic environment where reduced sulfur gases were present. As it precipitated $H_2S$ could have been trapped in the gypsum crystals.

## Clay Products

Major uses in the U.S. of common clay, estimated by the U.S. Geological Survey, are 57% for bricks, 19% for lightweight aggregates, 14% for cement additives, and 10% for other uses. Many clays can take up water enabling them to be easily molded. If this water-rich mixture is baked by the sun or heated in a kiln the clays tend to fuse together into a continuous solid. Bricks are then rectangular blocks of fused clay. However, rectangular blocks of shale, concrete, and quarried stone are also termed bricks. Calcium-rich clays tend to produce a yellowish color when fired in a kiln while iron-rich clays produce a red color.

Most clays are mined from large open pit mines. They are then transported by truck or rail to a brick production plant. The material at the plant is crushed and sieved to produce a fine clay material. The fine clay material is mixed with water to create a thick mud. This is extruded into a square column that is cut to lengths. The 7% to 30% moisture is removed in drying kilns. After drying the bricks are subjected to burning. This involves increasing the temperature to 200°C to evaporate any free water. The temperature is then increased to dehydrate structural water in the minerals. At temperatures above ~600°C oxidation of all the Fe in the brick occurs. At temperatures in excess of ~870°C vitrification starts to occur. The glass produced seals the pores in the brick making it impervious to the absorption of water. The bricks are then slowly cooled to prevent cracking.

### Important Clays

Clays are hydrous minerals made up of layers of Si and Al atoms in tetrahedral (T) coordination with oxygen and layers of Al, Fe, and Mg in octahedral (O) coordination with oxygen. These layers can repeat in the sequence 1 tetrahedral + 1 octahedral + 1 tetrahedral in a T-O-T three-layered clay as shown in Figure 10.10 or in a sequence 1 tetrahedral + 1 octahedral units in a T-O two-layered clay as shown in **Figure 10.15**. Additionally, some clays can be a mixture of two and three-sheet layering. Clays are also

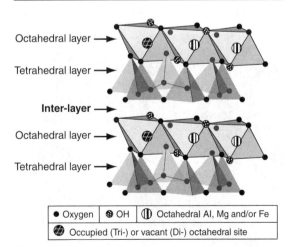

| ● Oxygen | ⊕ OH | ⫰ Octahedral Al, Mg and/or Fe |
|---|---|---|
| ⊛ Occupied (Tri-) or vacant (Di-) octahedral site | | |

*Figure 10.15* A two-layered clay where the Si and Al cations in the oxygen tetrahedrons are not shown for clarity.

characterized by the occupancy of cations within the three different atomic octahedral sites in the octahedral sheet. If two of the sites are occupied with atoms the clay is termed *dioctahedral* and if all three are occupied it is termed *trioctahedral*.

**Kaolinite.** Kaolinite [$Al_2Si_2O_5(OH)_4$] is a two-layered dioctahedral clay where a silica tetrahedral sheet is joined to an alumina octahedral sheet with no interlayer (**Figure 10.16**). It was first described from Kaoling, Jiangxi, China and is also called kaolin or china clay. Pure kaolinite is white in color. It forms from extensive chemical weathering of Al-rich rocks. These are typically feldspar-rich as feldspar contains significant Al and weathers easily. Elevated temperatures and the needed water for the leaching reactions are present in tropical rainforests.

The largest use of kaolinite is in the production of white glossy paper. It has good natural brightness and ink absorption characteristics. Less pure varieties are used to produce bricks for furnace liners. In its pure form it is the main ingredient for making porcelain. It is also used in paint as a supplement to the more expensive titanium oxide to increase whiteness. Although kaolinite is relatively inexpensive, it has a higher unit price than other clays because it is less abundant. Its price depends significantly on its purity.

Kaolinite is mined by two methods. The most common is to use high-pressure water jets on the clay layer in open pit mines. This loosens the soft clay and creates a liquefied clay mixture that is raised by a bucket elevator to a flume that transports it to the processing plant. The other mining method is to remove the overburden with a power shovel or dragline excavator (**Figure 10.17**). The exposed clay is then mined by a power shovel and transported to the processing plant by a conveyer belt, truck, or rail car if the distances are large.

Kaolinite resources are widespread, extremely large, and are varied in their origin, physical and chemical properties, and morphology. In decreasing order the U.S., Uzbekistan, Germany, Czech Republic, and Brazil are the major producers of kaolinite with a total world production of 33 million metric tons in 2011.

**Halloysite.** Halloysite [$Al_2Si_2O_5(OH)_4 \bullet 4H_2O$] is a two-layered dioctahedral clay generally formed by alteration reactions during soil formation on glassy volcanic rocks. It, like other clay minerals, is favored by alteration reactions in tropical and subtropical climates. It often occurs intermixed with kaolinite, montmorillonite, and

*Figure 10.16* Kaolinite showing its microscopic pseudo-hexagonal platelet habit that is aggregated into compact masses.

*Figure 10.17* Kaolinite mine in the state of Georgia.

other clay minerals. Halloysite has been employed as a petroleum-cracking catalyst but its main use is in the manufacture of porcelain and bone china ceramics. Resources are widespread and large throughout the world. In the U.S. mines in Utah and Georgia are significant.

**Illite.** Illite [$(K,H_3O)(Al,Mg,Fe)_2(Si,Al)_4O_{10}(OH)_2 \bullet H_2O$] is a three-layered dioctahedral clay produced as an alteration product of feldspar and mica during weathering and low-grade metamorphism. While present in most clay-rich sedimentary rocks, loose and fragmental material ejected from a volcano often weathers to produce high concentrations of illite. Illite is a nonexpanding clay, meaning it does not expand by absorbing layers of water in the interlayer between its tetrahedral layers. Bricks and tiles are produced with illite in mixtures with other clays.

**Smectites.** The smectite group of clays is made up of three-layered clays. Because they can have significant amounts of Mg and Fe substituting for Al in the octahedral layers, smectites can be either dioctahedral or trioctahedral. They can absorb water in their interlayer between their tetrahedral layers and expand several times their size or release water molecules and shrink. These are then expanding clays. The properties of a particular smectitic clay determines how it is used.

The most common smectite is montmorilloite [$(\frac{1}{2}Ca,Na)(Al,Mg,Fe)_4(Si,Al)_8O_{20}(OH)_4 \bullet nH_2O$]. Bentonite is the name given to a clay mixture made up largely of montmorillonite that is typically derived from weathering of volcanic ash. Because of its ability to absorb many times its volume in water, bentonite is sold as cat litter. Bentonite becomes sticky when wet and is used to bond sand to make molds. It is also used to bond iron ore powder in taconite as well as in animal and poultry feed products to transport and distribute these products more efficiently. Color molecules can be absorbed onto the charged bentonite surface leading to its use for oil decolorizing. Bentonite is added to an assortment of foods such as candy bars because it is cheap, adds bulk,

and has no taste. It is added to some yogurts to increase smoothness. Because it is an inexpensive way to add mass and has desirable expansive properties, it is often used in oil and gas well drilling fluids.

**Organoclays.** Organoclays are synthesized from normal clay but interlayer $Na^+$ and $K^+$ in the clay are replaced by large organic cations. These are used as liners in landfills to reduce transport of contaminants as oils in the water absorb on these clays. Organoclays can also be used for treating wastewater that contains oil, and for oil-spill control in general. Also, small amounts of amine-rich organoclay makes plastic products stronger, flame-retardant, and resistant to ultraviolet light and chemical damage. Amines are derivatives of ammonia that contain nitrogen.

# Glass

Glass is an amorphous, that is, noncrystalline solid material. Most is made up of oxides that contain a large concentration of $SiO_2$ and is referred to as silica glass. Most silica glass is used to allow transmission of visible wavelength light through the glass while blocking movement of air and heat from escaping an interior atmosphere (**Figure 10.18**).

Silica ($SiO_2$) is used to make glass because its electrons absorb energy in the ultraviolet region of light. If light is at a wavelength longer than ultraviolet, there is no absorption by the electrons. Since visible light has a longer wavelength than ultraviolet no absorption occurs and the light passes through the silica unaltered. Therefore, we can see objects outside because visible light is transmitted from the outside into the enclosed space. As the wavelength of light lengthens to infrared, the heat energy region, it begins to be absorbed again but in this case by the vibrational modes of Si-O bonds rather than by electrons. Thus, $SiO_2$ allows visible light to pass but absorbs ultraviolet and infrared wavelengths.

$SiO_2$ for glass is obtained from quartz crystals or sandstones. The high melting point of quartz,

*Figure 10.18* Glass skyscraper in Shanghai, China.

1713°C, can be lowered to 500°C with the addition of soda ($Na_2O$). The $Na_2O$ is derived from sodium carbonate ($Na_2CO_3$) or sodium nitrate ($NaNO_3$) (see the chapter on chemicals derived from evaporites). $Na_2O$ + $SiO_2$ glass is called water glass because it easily dissolves in water. Calcium oxide (CaO) or calcium plus magnesium oxide (CaO + MgO) is generally added as a stabilizer to the glass from crushed limestone ($CaCO_3$) or dolostone [$CaMg(CO_3)_2$] to produce common glass which is a soda-lime silicate. Sometimes a small percentage of alumina ($Al_2O_3$) is also added from feldspars to improve chemical resistance.

Borosilicate glass such as Pyrex™ contains 10% to 14% boron oxide ($B_2O_3$) from borax deposits (see the chapter on chemicals derived

from evaporites). It is resistant to corrosion and can withstand repeated heating and cooling. Lead crystal glass can contain up to 37% PbO. Lead increases the refractive index and light dispersion of glass. Therefore, light tends to separate into its spectra of colors similar to what happens in a prism. Leaded glass is used in chandeliers, for high-quality tableware, and for optical lenses.

## Selenium (Se)

Selenium is one of the most important coloring and decoloring agents in glass manufacture. Being a nonmetallic element that acts as an anion, selenides ($Se^{2-}$), selenites ($SeO_3^{2-}$), and selenates ($SeO_4^{2-}$) of metallic cation elements present in the glass are formed. Most of these are colorless leading to its ability to decolor glass. If free selenium atoms are present they produce a pink to red color in the glass.

Selenium is found replacing sulfur in sulfide minerals including pyrite to a small extent. Minerals that are selenides where $Se^{2-}$ occurs are rare. They can be found at low concentrations in some Cu, Ag, and Pb sulfide ore deposits. Most selenium is obtained as a byproduct of the processing of ores for their Cu, Ag, and Pb content. The price of selenium is often inversely related to the supply of the major product from which it is derived, Cu. As a byproduct of copper refining, selenium prices typically fall during periods of high copper production driven by higher Cu demand.

## Glass Production

In the U.S. in 2010, 15 million metric tons of $SiO_2$ were mined to make glass. Ordinary glass is produced by two different methods depending on its use. Plates of soda-lime glass are manufactured by floating the molten glass on a bed of molten tin. Glass containers are manufactured from a molten bulb of glass falling or being blown into a mold. As the molten glass cools it shrinks and solidifies. Uneven cooling causes stresses in the glass that leads to inherent weakness. An annealing oven can be used to heat the glass to relieve the stress and cool it slowly so little stress remains in the glass.

Laminated safety glass is fabricated by bonding a thin, tough transparent layer of polyvinyl butyral resin between two sheets of glass. This type of glass is employed extensively in the automotive and building industry. Heat-absorbing glass is made by adding controlled quantities of ferrous iron to the glass that absorbs most of the wavelengths of sunlight beyond the visible spectrum. A pale bluish green or gray coating can be applied to glass that decreases the transmission of light through the glass as well as glare and brightness. This absorbs 40% of the sun's infrared rays and approximately 25% of the visible rays reducing the amount of energy that enters a room. These coatings are common on windows for buildings where people are present. Insulating double- or triple-paned glass has a dead, dry air space of about 2/3 of a cm between glass panes. The air space decreases the transport of heat through the window.

Recently, superinsulated windows have been developed where the dead air space, in between the sheets of glass, is replaced by a vacuum and a reflective coating is added to the glass. This leads to a significant increase in the insulating value of the window making it comparable to a standard 2"×4" stud wall. The increased heat insulation is due to less heat transfer by conduction, convection by air molecules, and heat being radiated back by the coating. See the chapter on alternative energy for a discussion of heat transport mechanisms. The Empire State Building in New York City recently converted its old windows to superinsulated windows saving more than $400,000 per year in energy costs.

# Asbestos

*Asbestos* is a group of silicate minerals with long, thin fibrous crystals. Its name is based on its structure and not its chemical formula. Six different minerals can form asbestos. The two mineral

*Figure 10.19* Chrysotile asbestos showing fibrous habit.

types that produce asbestos are amphibole and serpentine (**Figure 10.19**). Amphibole asbestos has long thin fibers that form a chain-like crystal structure. The serpentine asbestos fibers are shorter, thicker, and curlier. The asbestos fibers are strong but flexible, have high heat resistance, and are stable in many corrosive environments. They also have good sound- and heat-insulating properties. Asbestos is used for air duct and wall insulation. Asbestos fibers can be separated then spun and woven to make a fire- and heat-resistant cloth used for such purposes as covering pipes. Asbestos is also incorporated into roofing shingles, floor tiles, brake pads, and patching compounds to increase their strength.

Serpentine asbestos dominates the world production. Approximately 90% to 95% of the asbestos contained in buildings in the U.S. is serpentine asbestos. The asbestos is found in Mg-rich altered ultramafic rocks. Commonly these rocks are exposed in the lower part of an old oceanic lithosphere section that has been thrust onto the continent in what is termed an ophiolite.

## Health Concerns and Demand

Microscopic asbestos fibers employed when constructing buildings become airborne if disturbed. If inhaled into the lungs they can cause lung and mesothelioma cancer as well as asbestosis, a scarring of the lungs. No safe level of minimum exposure to asbestos has ever been established. The International Labor Union estimates that 100,000 people die each year from work-related asbestos exposure.

Adverse publicity and lawsuits have lead to stricter worldwide asbestos regulations causing many companies to find substitutes for their asbestos-containing products. As a result usage has fallen off in most industrialized countries. Consumption, however, is increasing in Azerbaijan, China, India, Iran, Kazakhstan, Thailand, and Ukraine. Worldwide asbestos production peaked in 1976, but has been steady at about 2 million metric tons since the late 1990s as shown in **Figure 10.20**. There has been no production of asbestos in the U.S. since 2002. World production is dominated by Russia accounting for half of world production. With identified world resources of 200 million tons of asbestos there should be no problem meeting future world demand.

# Major Production Products

Major production products are valued because of a particular property it possesses including those of strength, density, flexibility, inertness, heat resistance, fluxing ability, and pigmentation.

## *Refractories*

Nonmetallic materials that can withstand high temperature and retain their strength are referred to as refractories. They are produced to line furnaces, kilns, and incinerators and to make crucibles. Refractories are typically manufactured into fire bricks from sandstone, diaspore ($Al_2O_3 \cdot H_2O$) and kaolinite clay [$Al_2Si_2O_5(OH)_4$]. Some are made from the aluminosilicate minerals sillimanite, andalusite, and kyanite ($Al_2SiO_5$), together with magnesia (MgO) obtained from magnesite ($MgCO_3$) and dolomite [$CaMg(CO_3)_2$].

The composition of a refractory used depends on the acidity it will be exposed to. Acid refractories

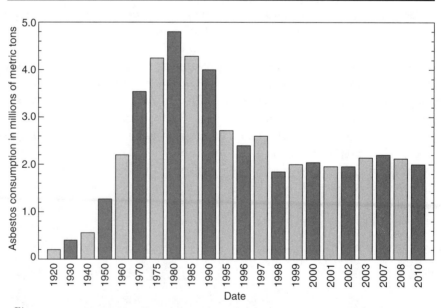

*Figure 10.20* Worldwide asbestos consumption for the indicated year. (Adapted from Virta, R.L., 2006, *Worldwide asbestos supply and consumption trends from 1900 through 2003*: USA Geological Survey Circular 1298, 80 p.)

typically have high silica and added zirconia ($ZrO_2$) obtained from zircon ($ZrSiO_4$). Neutral pH refractories typically have high alumina together with added chromium oxide ($Cr_2O_3$) obtained from chromite [$(Fe,Mg)Cr_2O_4$] or they contain a carbon binder. Alkaline refractories typically include high concentrations of magnesia (MgO), with some also containing chromium.

## Foundry Sand

Foundry or molding sand is sand used in a foundry. A foundry is a manufacturing facility that pours molten metal into a preformed mold to produce a metal casting. Typically this is iron and steel but aluminum, copper, brass, and bronze are also cast in foundries. There are approximately 3,000 foundries in the U.S.

There are two basic types of foundry sand: greensand, also called molding sand, that uses clay/glauconite as the binder material, and chemically-bonded sand. Most foundry sand is greensand. Greensand is quartz sand or sandstone that contains a small amount of organic material and 2 to 12 wt% glauconite mica plus clay that forms in calm anoxic marine environments. Glauconite [$(K,Na)(Fe^{3+},Al,Mg)_2(Si, Al)_4O_{10}(OH)_2$] is a very friable mica with a light to dark green color. The glauconitic clay acts as a binder that allows the sand to be molded into a particular shape. The organic material prevents the fusing of sand onto the casting surface. Chemically-bonded sand consists of 93 to 99 wt% silica sand and 1 to 3 wt% chemical organic binder.

## Fluxes

In smelting, flux is added to the melt to remove waste rock gangue by producing a separate Si-rich liquid slag phase that floats to the melt surface. The most common fluxing agent in smelting is limestone, used for its CaO. CaO lowers the density and eutectic temperature of the unwanted silica, producing a $CaO+SiO_2$ slag phase. Fluorite ($CaF_2$) (see the chapter on base metals) is sometimes used as a flux because it lowers the viscosity of the slag and promotes higher fluidity.

In metallurgy, a flux is a substance used to promote fusion. Fluxes are used with soldering, brazing, and welding compounds to decrease the melting point of substances, lower surface tension, and remove oxidized waste material from metals. Soldering and brazing is the joining of metal surfaces with a filler metal of different composition and lower melting point than the metal to be joined. Soldering is a lower temperature operation than brazing ($< 425°C$). Fluxes can be rosin, an organic compound made up of terpene compounds produced by pine trees. Lead-tin solder used in electronics contains about 1% rosin which helps the molten metal flow and reacts with any oxidation layer on the metal reducing it back to unoxidized metal. Brazing fluxes often contain borax used for the same purposes but more effective at higher temperatures.

Fluxes are also used in welding. Welding is the joining of metals by melting them together rather than using a filler metal as with soldering or brazing. Flux compositions vary greatly but most contain calcium carbonate as a major ingredient. Breakdown of the calcium carbonate creates $CO_2$ gas that precludes the oxidizing effect of atmospheric $O_2$ from the weld. Chloride and fluoride in the flux combine with and dissolve the oxides already on the surface of a metal, forming a runny liquid. This liquid coating protects the surface of the metal from any oxidation during the welding process.

## Mineral Fillers

In the most general terms, a mineral filler is a finely divided mineral that is used to modify the mechanical, electrical, and/or optical properties of a material in which it is dispersed. In many instances it is merely used to increase size or bulk. Mineral fillers are commonly chemically inert, inexpensive material that can be added to increase the amount or be used as an extender when applying the material. Some fillers are used as excipients. Excipients are inactive substances that enclose a drug or medication allowing it to be transported to the stomach unreacted.

Many mineral fillers are clay minerals. Kaolinite clay is added to paints to increase its covering power, desirable flow and suspension properties, and weather resistance. Coarse-grade kaolinite can provide a matte finish to a painted surface. Carbonate fillers differ by particle size, whiteness, and mineralogical and chemical purity. They are typically used in food and pharmaceutical additives because they dissolve in aqueous solutions.

The mineral filler talc is a silicate that is distinguished from almost all other minerals by its extreme softness. It is often referred to as soapstone because of its soapy, greasy feel. Talc is added to many products such as paper, plastic, paint, and rubber to promote a soft feel. Talc also finds uses as a cosmetic and for lubrication as outlined in **Figure 10.21**.

Wollastonite (CaSiO₃), because of its bright white color, is used primarily in ceramics including floor and wall tiles. It is also added to paint and paper as a white filler. Wollastonite forms when silica-rich hydrothermal solutions react with calcite and dolomite or during contact metamorphism of limestone by a felsic igneous rock. Mined ore contains at least 30 wt% wollastonite. Mine production in 2011 was 510,000 metric tons with China producing 300,000 metric tons.

Silica sand fillers are used in flooring, paint, and polymer systems. These include specialty floor, wall, and marine coatings, where textured and nonslip finishes are required. In plastics the major improvements imparted by silica include increases in stiffness, strength, temperature resistance, dimensional stability, surface hardness, and scratch resistance.

## Pigments

Pigments from mineral sources are enlisted dominantly in the paint industry. However, mineral pigments are also employed to change colors of inks, fabric, plastic, food, and cosmetics. Given in **Table 10.5** are the colors and compositions of some common mineral pigments.

## Graphite

Graphite is a chemically inert and highly refractory nonmetal with a very high melting point of nearly 4,000°C. It has a low density, 2.09 to 2.23 g/cm³, and is reasonably electrically and thermally conductive.

### Use

Graphite is most often fabricated into molds, bricks, and to lesser extent crucibles for foundry operations. These are mainly in alumina-graphite mixtures with 20 to 50 wt% graphite. A large amount of graphite is also utilized as a source of carbon in steelmaking for use in a heat-treatment hardening process. Implantation of carbon atoms into the surface layer of steel hardens it. Graphite is also incorporated into paints, coatings, lead pencils, and brake linings; used as a lubricant; and made into rocket nozzle insets. Alkaline and lithium-ion batteries have electrodes fabricated from graphite.

### Resource Location

Graphite occurs naturally in many metamorphic rocks, such as gneiss, marble, and mica schist from the metamorphism of original organic material. Mineable concentrations occur when concentrations of coal, kerogen, or other hydrocarbons in the

*Figure 10.21* Talc sold as a facial powder and as powder for pool cue lubrication.

*Table 10.5* SOME MINERAL PIGMENTS.

| MINERAL | COMPOSITION | COLOR |
|---------|-------------|-------|
| Barite | $BaSO_4$ | White when pure |
| Celadonite | Mica with iron in 2 oxidation states: $K(Mg,Fe^{2+})(Fe^{3+},Al)$ $[Si_4O_{10}](OH)_2$ | Dull gray-green to bluish green |
| Glauconite | Iron potassium mica: $(K,Na)(Fe^{3+},Al,Mg)_2(Si,Al)_4O_{10}(OH)_2$ | Green |
| Goethite (Yellow ochre) | Iron hydroxide: $FeO(OH)$ or $Fe_2O_3 \bullet H_2O$ | Yellowish to reddish to dark brown |
| Gypsum | $CaSO_4 \bullet 2H_2O$ | White |
| Hematite | $Fe_2O_3$ | Rust red to silvery-grey |
| Jarosite | Hydrous sulfate: $K(Fe^{3+})_3(SO_4)_2(OH)_6$ | Dark yellow to yellowish-brown |
| Lazurite | Feldspathoid: $(Na,Ca)_8(AlSiO_4)_6(SO_4,S,Cl)_2$ | Deep blue to greenish blue |
| Limonite | Mixture of hydrated iron oxides, mostly goethite | Shades of brown and yellow |
| Riebeckite | Na-rich amphibole: $Na_2(Fe,Mg)_5Si_8O_{22}(OH)_2$ | Dark blue |
| Rutile | Ground $TiO_2$ | Bright white |
| Shungite | Natural noncrystalline carbon | Black |
| Sienna | Limonite clay with ferric oxides | Yellow-brown |
| Umber | Clay mixture with ferric oxide and manganese oxides | Shades of brown |
| Vivianite (Blue Ochre) | Hydrated iron phosphate $Fe_3(PO_4)_2 \bullet 8(H_2O)$ | Deep bluish green |
| White clay | Low $Fe_2O_3$ clay | White |
| Zincite | $ZnO$ | White |

subsurface are subjected to regional or contact metamorphism. Well-ordered graphite requires these carbon-containing rocks to be subjected to 350° to 500°C. This is greenschist facies metamorphism, which occurs from about 8 to 50 km in depth in the earth.

There are three types of natural graphite produced; flake, vein that is sometimes called lump, and amorphous graphite. Flake graphite is typically found as discrete flakes ranging in size from 50 to 800 μm in diameter and 1 to 150 μm thick. Vein graphite occurs in fissures as massive platy intergrowths of fibrous crystalline aggregates of smaller size then flake graphite. Amorphous graphite is the lowest quality and most abundant. Amorphous graphite is really very small-sized

crystal and is not truly amorphous. Mineable concentrations are often found in weathered rocks because the weathering process does not affect the graphite but makes it easier to separate graphite from the gangue.

### Production and Reserves

The minimum mining grade of flake graphite is about 3 wt%. For amorphous graphite the minimum grade is nearer 45 wt%. Graphite is mined worldwide in both open pits and underground.

World production of natural graphite in 2011 was 925,000 metric tons with China mining 600,000 metric tons. World reserves are 77 million metric tons with 55 million metric tons present in China. Graphite is not mined nor are there significant reserves in the U.S. In 2009, the U.S. produced 118,000 metric tons of synthetic graphite. In addition to this, 42 kilotons of natural graphite and 82 kilotons of synthetic graphite were imported.

Synthetic graphite is produced by pyrolysis where hydrocarbons, such as petroleum coke and coal tar, are thermally decomposed in the absence of oxygen. A mixture of fine-grained crystalline graphite and cross-linking intercrystalline carbon is produced. This powdered graphite is mixed with a binder such as pitch or synthetic resins. It can then be pressed and molded into a desired shape.

# Diamonds and Other Abrasives

Abrasives are used to cut, grind, buff, drill, sand, and polish both metals and ceramics. Given the large number of applications they are a common material employed in a wide variety of domestic as well as industrial applications.

## Diamonds

Diamonds are the densest form of carbon, 3.52 g cm$^{-3}$. About 20% of diamonds found in the earth are considered gemstones with 80% used as abrasives. The formation of diamonds requires carbon to be subjected to pressures in the earth that occur below a depth of 140 km. It has been argued that diamonds form by oxidation of methane in a fluid or magma with reduction of Fe in the chromite present there by the reaction:

$$Fe_2O_3 + CH_4 \rightarrow C + 2H_2O + 2FeO.$$   [10.2]
(*in* chromite)   diamond   (*in* chromite)

Because the quartz polymorph coesite, rather than the higher pressure polymorph stishovite, is found as inclusions in diamonds, the maximum depth to which diamonds occur in the mantle is thought to be about 300 km.

Diamonds are present as xenoliths in kimberlites and some lamproites. They are also found in placer deposits developed from weathering of these rocks. *Xenoliths* are rock fragments or crystals found in a rock that are incorporated from the wall rocks where the magma that produced the rock formed or are incorporated from wall rocks during magma ascent toward the surface.

*Kimberlite* is K-rich ultramafic peridotite volcanic rock that occurs in a vertical pipe-like structure (**Figure 10.22**). It can contain up to 50% xenoliths. The magma is brought rapidly and violently to the earth's surface from diamond stability depths by sudden expansion of large concentrations of $CO_2$ gas in the magma. Lamproites also occur in pipe-like structures but are ultra-potassic, magnesian-rich, ultramafic mantle-derived volcanic rocks that do not contain extensive carbonates but are driven by the expansion of $H_2O$. The olivine-rich ones have compositions that approach kimberlite and are the ones that contain diamonds.

The magma that produces kimberlites and lamproites acts as an elevator that carries deep-formed mantle rocks and minerals upward. The speed has been calculated to be ~70 km hr$^{-1}$. It is argued that substantially slower speeds would allow the diamonds time to react to the more stable graphite phase. The ages of inclusions in diamonds are from 1.0 to 3.3 billion years leading researchers to suggest the diamonds are stable phases in the upper mantle that grow slowly over time.

More than 12,000 kimberlite deposits have been found worldwide, yet fewer than 1% contain enough diamonds to make them economically viable. Roughly half of all diamonds mined are found in Precambrian aged rocks in central and southern Africa (**Figure 10.23**). Blood diamonds are those that are mined in war zones and sold to finance insurgencies as has occurred in Angola and Côte d'Ivoire, Africa.

The largest diamond deposit ever mined was the open pit "Big Hole" at Kimberley, South Africa. From 1866 to 1914 up to 50,000 miners dug the hole 495 meters deep with picks and shovels yielding 2,700 kg or 14 million carats of diamonds, as 1 carat = 0.2 g. **Table 10.6** lists the countries that produced the most gem-quality diamonds and their estimated production

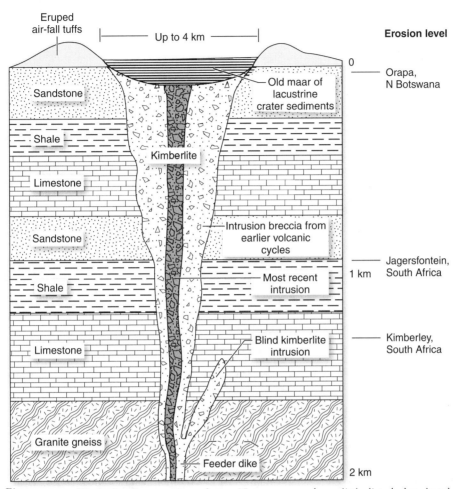

*Figure* 10.22 Kimberlite diatreme pipe and maar. Note country rocks are little disturbed or altered. A maar is a volcanic crater formed below ground level from groundwater producing explosive steam as it comes into contact with molten magma. After forming the maar typically fills with water forming a lake. (Adapted from Nixon. P.H., 1980. Kimberlites in the south-west Pacific: *Nature*, v. 287, p. 718–720.)

for 2010. Ranking in gem quality diamond production between countries is constantly changing as new discoveries are made and old mines are worked out. For instance, South Africa is no longer a large producer and commercial mining for diamonds did not begin in Canada, the third largest producer, until 1998. In the mid 1990s Australia was the largest producer, but now mines very few by comparison.

Industrial-grade diamonds are valued mostly for their hardness and heat conductivity rather than clarity and color as in the gemstone market. Diamonds can be produced synthetically in a high temperature and pressure apparatus that reaches the pressure and temperature conditions of diamond stability in the earth's mantle. Synthetic diamonds can have high clarity and color making them difficult to distinguish from natural gems. Most, however, is grit and powder manufactured in a bulk process that is impregnated on cutting blades and abrasive surfaces.

About 11,000 kg of diamonds are mined annually, with a total value of nearly US $9 billion. Almost 1 million kg of diamonds are synthesized annually, nearly 100 times the natural market. China is by far the leading producer

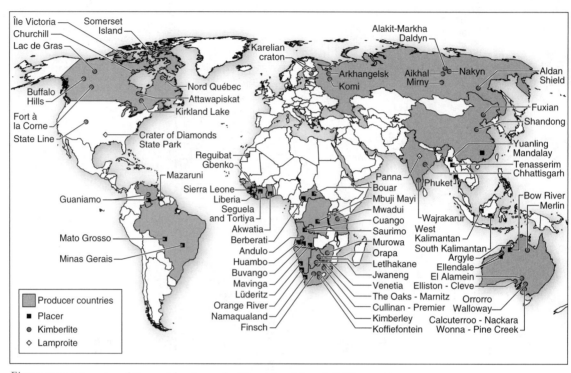

*Figure 10.23* Location of diamond-bearing rocks. (Reproduced from Info Diamond.)

accounting for 88% of the synthetic market. What limits the production of gem-quality synthetic gem diamonds is the cost to produce large clear single diamonds. The mining and distribution of gemstone diamonds are controlled by a limited number of interests that include the African producer countries; the mining companies of De Beers, Rio Tinto, BHP Billiton, and Zao Alrosa; and the individuals, Lev Leviev and Harry Winston.

## Other Natural Abrasives

After diamond, corundum ($Al_2O_3$) is the second hardest natural substance. It is, therefore, used as an abrasive. Emery, a rock made up mainly of corundum mixed with other hard nonsilica containing oxide minerals, is also used as an abrasive. Large emery deposits occur on the Greek island of Naxos.

Garnet is the abrasive used in sandpaper and sandblasting applications. Garnet resources are large, occurring in Al-rich metamorphic rocks.

They are also present in heavy metal sands and some gravel deposits. In 2011, 1.4 million metric tons was mined worldwide. Industrial garnets are obtained from a number of different countries including the U.S. with India and China producing 83% of the world total. Natural garnet abrasives are, however, being replaced by synthetic ones like zirconia alumina a mixture of zirconium oxide ($ZrO_2$) and aluminum oxide ($Al_2O_3$).

# Barite

While barite ($BaSO_4$) is the main source of elemental barium (Ba), 80% of the barite mined is incorporated in drilling mud utilized for drilling wells for oil and natural gas recovery. In the U.S. it is closer to 95%. Barite has the high density, ~4.5 g cm$^{-3}$, needed to increase the drilling mud fluid pressure at depth so a well blowout does not occur. Outside the petroleum industry barite is used as a filler, extender, and to increase the weight of paints, plastics, and rubber. It is also used as a white pigment in glassmaking to produce special

*Table 10.6* ESTIMATED PRODUCTION OF GEM-QUALITY DIAMONDS FOR 2011 BY COUNTRY.*

| COUNTRY | PRODUCTION |
|---|---|
| Botswana | 25,000 |
| Russia | 17,800 |
| Angola | 12,500 |
| Canada | 11,800 |
| Congo | 5,500 |
| South Africa | 3,500 |
| Namibia | 1,200 |
| Guinea | 550 |
| Ghana | 300 |
| Central African Republic | 250 |
| Sierra Leone | 240 |
| Brazil | 200 |
| Guyana | 144 |
| China | 100 |
| Australia | 100 |
| Tanzania | 77 |

* In thousands of carats.

*Data from*: U.S. Geological Survey, *Mineral Commodity Summaries*, January, 2012.

glass, in ceramics for glazes, and with porcelain manufacture to produce enamels.

Barite is found in some hydrothermal veins and deposits as well as occurring in karst terrains. However, most is derived from black shale and chert-hosted, stratiform marine deposits occurring on continental shelves. These deposits appear to have formed either by biogenetically modified sedimentary-exhalative (SEDEX), cold seep, or ocean circulation processes. If by a SEDEX process hydrothermal black-smoker brines are thought to leach Ba from underlying sediments and release it together with reduced sulfur into an anoxic seawater layer on the seafloor. Precipitation of barite occurs on oxidation of sulfur as the oxic-anoxic boundary moves to shallower water on the continental shelf. In a cold seep process, a tensional fracture along a transform fault in a continental shelf is thought to release cold Ba-rich water into normal sulfate-rich seawater causing barite to precipitate. The ocean circulation model argues that coastal upwelling of oxygen depleted and nutrient-rich seawater along the edge of the continental shelf causes a rain of large amounts of planktonic debris. This causes barite to precipitate from Ba plus sulfate in seawater in the photic zone. The barite produced settles and then dissolves in a reducing bottom water, but reprecipitates where anoxic bottom waters mix with normal oxygenated sulfate-rich seawater.

In China, barite deposits are present on the continental shelf of the Yangtze platform, both in the Qinling region in the north and the Jiangnan region in the south. Indian production is predominantly from a single sedimentary deposit located on basement rocks in Andhra Pradesh. In the U.S. deposits are found in marine rocks along the western margin of the North American platform in the Nevada barite belt. In Morocco barite is present in vein and karst deposits in the western Jebilet Mountains thought to have been deposited during Atlantic rifting of northwest Africa from North America.

Barite world reserves are estimated to be 240 million metric tons. China is now the world's largest barite-producing country mining 4 million metric tons of the total of 7.8 million metric tons produced worldwide in 2011. This is followed by India with 1.1 million metric tons, Morocco with 650,000 metric tons and the U.S with 640,000 metric tons.

# Zeolites

Zeolites are a group of hydrous Al-silicates joined in a framework that produces large structural channels of 2.2 to 9.0 angstroms in diameter as shown in **Figure 10.24**. Over 40 naturally occurring zeolites have been described and over 135 artificial zeolite frameworks synthesized. The most common zeolites and their chemical formulas are given in **Table 10.7**.

Zeolites are useful as molecular sieves. The sizes of the channel ways in zeolites are large enough to allow small molecules to pass through the mineral but not larger ones. Because of their open structure

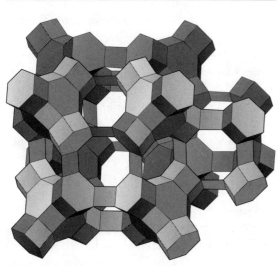

*Figure* 10.24 Zeolite structure showing open channels. The structural units are composed of linked tetrahedra of $AlO_4$ and $SiO_4$ in a three-dimensional array. (Courtesy of Michael Treacy and the IZA Database of Zeolite Structures.)

*Table* 10.7 COMMON ZEOLITES.

| NAME | FORMULA |
|---|---|
| Chabazite | $CaAl_2Si_4O_{12} \cdot 6H_2O$ |
| Clinoptilolite | $(Na,K,Ca)_{2-3}$ $Al_3(Al,Si)_2Si_{13}O_{36} \cdot 12H_2O$ |
| Erionite | $(Na_2,K_2,Ca)_2Al_4Si_{14}O_{36} \cdot 15H_2O$ |
| Ferrierite | $(Na,K)_2Mg(Si,Al)_{18}O_{36}(OH)$ $\cdot 9H_2O$ |
| Heulandite | $(Ca,Na)_{2-3}$ $Al_3(Al,Si)_2Si_{13}O_{36} \cdot 12H_2O$ |
| Laumontite | $Ca(AlSi_2O_6)_2 \cdot 4H_2O$ |
| Mordenite | $(Ca,Na_2,K_2)Al_2Si_{10}O_{24} \cdot 7H_2O$ |
| Phillipsite | $(Ca,Na_2,K_2)_3Al_6Si_{10}O_{32} \cdot 12H_2O$ |

zeolites have large surfaces with high ion exchange capacities. As a result they are also employed as ion exchange mediums such as in water softeners.

Most natural zeolites are employed to produce lightweight concrete and in asphalt to lower the mixing and laying temperature by decreasing asphalt viscosity. Other important uses are in animal feed, pet litter, and for water purification. Recent increased demand for zeolites has been driven by need for water purification in municipal drinking water and water parks, swimming pools, aquariums, and aquaculture environments.

Zeolites form from alteration of volcanic ash and tuff deposited in lake environments. Alkaline, that is, high pH conditions, develop in the water. At high pH, waters can dissolve and remove high concentrations of silica from the ash and tuff. During alteration reactions glass and feldspars are transformed to the low-Si silicates, zeolite and feldspathoid. Feldspathoids are a group of uncommon low-silica containing minerals that resemble feldspars but have a somewhat different structure. Natural zeolites rarely have a constant composition and are commonly intermixed with other minerals in deposits.

## Production

In 2011 about 2.8 million metric tons of zeolites were mined. Zeolites are obtained by open pit mining. The mined material is crushed, dried, purified, and milled. The zeolite material is typically shipped in bags or transported in bulk. When uniformity and purity are important synthetic zeolites are manufactured. The production of synthetic zeolites is somewhat greater than 12,000 metric tons per year.

Synthetic zeolites have found a wide use in catalytic cracking of oil in the production of petrochemicals. Synthetic zeolites are also used extensively in laundry detergents because of their cation-exchanging capacity. Most modern laundry detergent powders that do not contain phosphate contain zeolites. Zeolites soften water by removing the ions $Ca^{2+}$ and $Mg^{2+}$ from water that allows soap to work properly. This occurs by exchanging $Ca^{2+}$ and $Mg^{2+}$ in the water for $Na^+$ ions on the surface of the zeolite. The extent of its ability to do this is termed its cation-exchange capacity. Zeolites have a high cation-exchange capacity. Like other silicates, zeolites are insoluble in the water and are removed in the wash water as finely dispersed crystals of ~4 microns.

Zeolites can be manufactured from kaolinite by shock heating it and adding it to a NaOH solution at 70° to 100°C. If the Si content is increased by adding amorphous silica, a Si-rich zeolite is formed. Synthetic zeolites can also be produced from sodium aluminate and sodium silicate high pH solutions. A sodium aluminosilicate gel is formed from which zeolite can grow from seed crystals.

# Gemstones

Gemstones are valued for their color, luster, transparency, durability, and rarity. They are divided into precious and semiprecious. Precious gemstones include diamond, ruby, sapphire, emerald, and pearl. Semiprecious gemstones are all the others. Precious and some semiprecious gemstones are characterized in **Table 10.8**.

*Table 10.8* PRECIOUS AND COMMON SEMIPRECIOUS GEMSTONES.

| NAME | COMPOSITION AND COLOR |
|---|---|
| **PRECIOUS** | |
| Diamond | Pure C (clear), trace boron (blue), trace nitrogen (yellow) |
| Emerald | $Be_3Al_2(SiO_3)_6$ with trace $Fe^{3+}$ and $Fe^{2+}$, green beryl |
| Pearl | 95% $CaCO_3$, 5% complex proteins and organic polymers, produced by oysters covering foreign objects both naturally and by human addition (cultured) |
| Ruby | $Al_2O_3$ with trace Cr, producing pink to blood-red corundum |
| Sapphire | $Al_2O_3$ with trace Ti, producing blue corundum, with trace V purple corundum, and trace Fe a pale yellow to green corundum |
| **SEMIPRECIOUS** | |
| Agate | $SiO_2$, cryptocrystalline quartz with colored growth bands |
| Amethyst | $SiO_2$ with trace $Fe^{3+}$ and $Al^{3+}$, a violet to purple variety of quartz |
| Aquamarine | $Be_3Al_2(SiO_3)_6$ with trace $Fe^{2+}$, a turquoise beryl |
| Catseye | $BeAl_2O_4$, chrysoberyl with inclusions of fine, slender parallel mineral fibers producing a catseye effect |
| Jade | $NaAlSi_2O_6$, green jadeite, which is a pyroxene or is $Ca_2(Mg,Fe)_5Si_8O_{22}(OH)_2$, nephrite, a green amphibole |
| Jasper | $SiO_2$, microcrystalline quartz with a trace of colored minerals produced during co-deposition from supersaturated hydrothermal solutions |
| Opal | Small hydrous $SiO_2$ spheres displaying regular packing in a solid. Colors are produced by the size of spheres at the correct half wavelength in the diffraction grating produced |
| Onyx | $SiO_2$, cryptocrystalline quartz, striped agate with white and black, brown, or red alternating bands |
| Peridot | $(Mg,Fe)_2SiO_4$, olive green forsterite olivine with the coloring agent $Fe^{2+} < 15\%$ |
| Topaz | $Al_2SiO_4(F,OH)_2$, a blue, orange, yellow, green, or pink fluorine-rich silicate. Pink, red, and violet tones are due to Cr impurities; others depend on the ratio of F to OH and minor impurities |
| Turquoise | $CuA_{16}(PO_4)_4(OH)_8 \bullet 4H_2O$, blue-to-green cryptocrystalline secondary Cu-mineral |

Gemstone production was an over $80 billion market in 2011. Diamonds account for 95% of this. The U.S. dominates the demand, accounting for 35% of sales of unset gem-quality diamonds. The U.S. is expected to continue dominating global gemstone consumption. Gem diamond production in the U.S. is zero except for the lucky tourist that finds a gem-quality diamond at the Crater of Diamonds State Park in Arkansas.

Since about 4000 BC gemstones have been cut and polished. The El-Dorado Topaz, found in the topaz mines of Minas Gerais, Brazil in 1984, is the largest faceted gemstone in the world weighing 37 kg when discovered. It has an emerald cut and weighs 31,000 carats or 6.2 kg after being cut.

## Shapes and Cuts

The cut shape of a gemstone is the outline or form that any gemstone exhibits when viewed from the top. Common gemstone shapes include square, rectangle, round, triangle, octagonal (square or rectangular), pear, heart, marquise, and oval as shown in **Figure 10.25**. The cut is how the shape is faceted. Sometimes one refers to a pear cut or a cushion cut which clearly refer to shapes so the two terms can be confused.

For diamonds a round shape with a brilliant cut is commonly done. The diamond is cut into two cones, face to face, with the upper cone more severely truncated than the lower. This cut produces the most complete return of light of any gem cut. A brilliant cut has 58 facets if it includes a facet called the culet, cut into the bottom cone point (**Figure 10.26**).

## Pearls

Pearls ($CaCO_3$) are produced by a combination of aragonite and a protein secreted by oysters around a small foreign object. This produces a spherical mother of pearl object. While valuable pearls occur in the wild they are rare. The pearl industry is built on raising cultured pearls by placing a sand grain in a young oyster. More recently a sphere of calcium carbonate shell is introduced to speed the process of coating the object. Pearls are usually harvested every 2 to 4 years. Three types of oysters are used to produce cultured pearls; the white-lipped South Sea pearl oyster, *Pinctada maxima* (50% of total), the

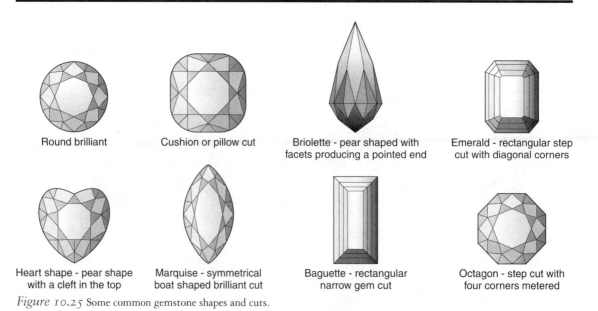

Round brilliant

Cushion or pillow cut

Briolette - pear shaped with facets producing a pointed end

Emerald - rectangular step cut with diagonal corners

Heart shape - pear shape with a cleft in the top

Marquise - symmetrical boat shaped brilliant cut

Baguette - rectangular narrow gem cut

Octagon - step cut with four corners metered

*Figure 10.25* Some common gemstone shapes and cuts.

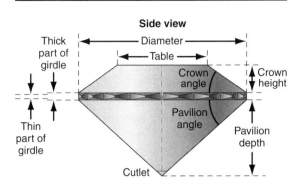

**Side view**

**Top view**

*Figure 10.26* Terms describing the brilliant cut for gemstones. (Modifed from Tolkowsky, M., 1919, *Diamond Design, A Study of the Reflection and Refraction of Light in a Diamond*, E. F. N. Spon, Ltd, London, 97 p.)

Akoya pearl oyster, *Pinctada fucata* (26% of total), and the black-lipped pearl oyster, *Pinctada margaritifera* (24% of total). For a given size and color the value of the pearls in jewelry is determined by its luster, lack of surface flaw, and symmetry.

## Turquoise

The semiprecious gemstone turquoise is a hydrated phosphate of Cu and Al with a composition of $CuAl_6(PO_4)_4(OH)_8 \bullet 4(H_2O)$. Turquoise is probably one of the oldest gemstones known. Robin's egg-sized blue turquoise was worn by pharaohs and Aztec kings.

Turquoise is formed by percolation of meteoric or groundwater through Al-rich rock in the presence of Cu. It is associated with Cu deposits as a secondary mineral in arid environments. For thousands of years the finest intense blue turquoise in the world was found in Persia. This changed during the late 1800s and early 1900s when miners discovered or rediscovered significant deposits of high-quality turquoise in the western and southwestern U.S. The majority of the world's finest-quality turquoise presently comes from the U.S., the largest producer of turquoise.

## Garnet

Some garnets are considered to be semiprecious gemstones. They have the general chemical formula $A_3B_2(SiO_4)_3$, where A is $Ca^{2+}$, $Mg^{2+}$, $Fe^{2+}$, or $Mn^{2+}$, and B is $Al^{3+}$, $Fe^{3+}$, or $Cr^{3+}$, or in rare instances, $Ti^{3+}$. The formulas and names of common garnet species are given in **Table 10.9**.

*Table 10.9* COMMON GARNET FORMULA.

| GARNET | FORMULA |
| --- | --- |
| Uvarovite | $Ca_3Cr_2Si_3O_{12}$ |
| Pyrope | $Mg_3Al_2Si_3O_{12}$ |
| Grossularite | $Ca_3Al_2Si_3O_{12}$ |
| Almandite | $Fe_3Al_2Si_3O_{12}$ |
| Andradite | $Ca_3Fe_2Si_3O_{12}$ |
| Spessartite | $Mn_3Al_2Si_3O_{12}$ |

Rare species of garnet that are considered gemstones have uncommon substitutions. Most of these are named for particular colors of the mineral. Hessonite is a fine orange, cinnamon brown, or pinkish variety of grossularite due to manganese and iron substitutions. Tsavorite is dark-green grossularite due to vanadium and/or chromium impurities. Melanite is a black titanium bearing variety of andradite. Demantoid is a bright-green variety produced by impurities of iron and chromium. Malaya™ is the name for a pyrope-spessartite that varies in color from red, through shades of orange and brownish orange to peach and pink due to iron and manganese impurities. Rhodolite is a purplish red pyrope-almandite solid solution garnet from North Carolina.

## Summary

One-quarter of the earth's surface is forest and these are either of conifer softwoods or deciduous hardwoods. The use of wood is stable in industrial countries with almost all of it produced in these countries on a sustained basis from managed forests. In developing countries 0.8% of forestland is being converted to agriculture land and pastures each year.

Rock, or what is termed stone in the construction industry, is used directly from a quarry or is treated before usage. Building stone may be dimension stone or aggregate. Brazil, China, India, and Spain are the major producers of dimension stone. The largest volume of mineral resources mined is sand and gravel followed by crushed stone; these are second only to fossil fuels in value. Crushed stone is used dominantly in road construction. Sand and gravel is mined from ancient river channel, river floodplain, alluvial fans, shorelines, sand dunes, and glacial deposits. Lightweight aggregates include pumice, vermiculite, perlite, and diatomite.

Concrete is an artificial rock made up of portland cement and aggregate. Portland cement is dominantly ground limestone and dolostone from which the $CO_2$ is driven off by heating together with sand and clay. Plaster is a product derived by heating gypsum.

Clays are most often used to produce bricks. They are hydrous sheet silicates made up of layers of Si and Al atoms in tetrahedral coordination with oxygen together with layers of Al, Mg, and Fe in octahedral coordination with oxygen. They can have a two-layered, octahedral + tetrahedral or three-layered, tetrahedral + octahedral + tetrahedral structure. Clays are termed dioctahedral if two of the three octahedral sites are occupied by cations and trioctahedral if all three are occupied. Kaolinite is a two-layered dioctahedral clay used primarily to produce white glossy paper.

Glass is made from $SiO_2$ derived from quartz or sandstones together with other oxides such as $Na_2O$, $CaO$, $B_2O_3$, and/or $PbO$. Asbestos is amphibole or serpentine in a fibrous crystal habit. Due to health concerns asbestos consumption has fallen in many countries but is increasing in China, India, and other industrializing nations. Refractories are materials that can withstand high temperatures and are often made from sandstone, diaspore, or kaolinite. Foundry sand is used to make molds for casting metals.

Fluxes such as limestone and fluorite are used to promote melting. Some minerals such as kaolinite and talc are used as fillers. These are chemically inert and inexpensive. Mineral pigments are used to change colors of inks, fabrics, and other consumer items.

Diamonds are valued both as gemstones and used for abrasives. They occur in kimberlites. These are igneous rocks brought from great depths in the mantle to the surface by very rapid expansion of $CO_2$ and $H_2O$. Most diamonds are very small and synthetically produced for use as abrasives. China dominates the synthetic diamond market.

Barite ($BaSO_4$) is added to oil and gas drilling mud to increase its density. China is the world largest producer from deposits of the continental shelf of the Yangtze platform. Zeolites are a group of hydrous Al-silicates having a framework with large structural channels. They are produced from altered volcanic ash and tuffs deposited in lake environments.

Precious gemstones include diamond, ruby, sapphire, emerald, and pearl with diamonds accounting for 95% of the market. A variety of cuts are used to refract the light so it returns to the observer's eye. Turquoise and some garnets are considered semiprecious gemstones and obtain their color by a variety of elemental impurities.

## KEY TERMS

| | |
|---|---|
| asbestos | octahedral sheet |
| cambium | perlite |
| concrete | plaster |
| diatomite | portland cement |
| dioctahedral | pumice |
| floodplain | tetrahedral sheet |
| kimberlite | trioctahedral |
| Leadership in Energy and | vermiculite |
| Environmental Design (LEED) | xenolith |

## PROBLEMS

1. If a tree is 1 meter in diameter and the average width of a ring in the tree is 2 millimeters, how old is the tree?

2. Calculate the percentage of world gypsum production that is from China and from the U.S. Are these larger or smaller than the percentage of world population in these countries using the numbers in the introductory chapter?

3. a. If a solid rock weighs 2.8 metric tons per cubic meter, what is its density in $g/cm^3$?

   b. If rock is crushed into uniform sizes, the presence of open space between the particles causes the crushed rock to be lighter, now 1.7 tons per cubic meter. How much porosity, that is, % of empty space is present?

4. If pumice floats on water and the density of the glass in pumice is 2.6 $g/cm^3$ what is the minimum pore space? How far above the water's surface would the pumice extend for your calculated pore space?

5. Determine the pressure range over which diamonds are stable in the earth. While the mantle's density increases with depth, assume it has an average value of 4.0 g per cubic cm for this calculation. Remember: Pressure = density × acceleration of gravity × height. (Make sure you use the correct units.)

## REFERENCES

Bush, A. L. 1976. *Vermiculite in the United States*. In the Proceedings of the 11th Industrial Minerals Forum, Montana Bureau of Mines Special Publication 74:146–155.

Clark, D, 2011, Forest Products Annual Market Review 2010–2011, Geneva Timber and Forest Study Paper 27, United Nations Pub, Geneva, 174 p.

Hindman, J.R. 2006 Vermiculite. In Kogel, J. E., Trivedi, N. C., Barker, J. M., and Krukowski, S. T. (eds.) *Industrial minerals and rocks*, 7th ed. (pp. 1015–1026). Littleton, CO: Society for Mining, Metallurgy, and Exploration, Inc.

Nixon, P. H. 1980. Kimberlites in the southwest Pacific. *Nature* 287:718–720.

Tolkowsky, M. 1919. *Diamond design: A study of the reflection and refraction of light in a diamond*, London: E. & F. N. Spon, Ltd.

U.S. Geological Survey, 2012, Mineral commodity summaries 2012, U.S. Government Printing Office, 198 p.

Virta, R. L. 2006. *Worldwide asbestos supply and consumption trends from 1900 through 2003*. U.S. Geological Survey Circular 1298. Available only online from pubs.usgs.gov/circ/2006/1298/.

## Chapter 11

# Chemicals from Evaporation of Water and Gaseous Elements from Air

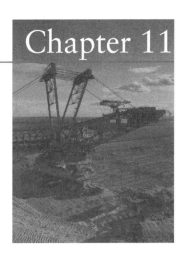

Elements used in agriculture and for the chemical industry primarily obtained from the evaporation of water are considered in this chapter. Most are salts of $Cl^-$, $SO_4^{2-}$, and $HCO_3^-$. These include the agriculturally important elements, nitrogen (N), phosphorous (P), potassium (K), and sulfur (S). Elements and molecules that are obtained as gases from air are outlined at the end of this chapter.

## Resources from Evaporites

Evaporites are rocks formed by the evaporation of water. Evaporite minerals are very soluble and outcrop only in arid terrains. There are two different types of evaporite environments: marine and nonmarine. In marine environments the evaporating fluid is seawater or modified seawater whereas in nonmarine environments the fluid is continental river water and/or groundwater. Given in **Figure 11.1** are the areas in the U.S. that are underlain by evaporites. Note that most of these are marine environments located where there were major incursions of seawater onto the U.S. landmass in the geological past.

### Nonmarine Evaporates

Consider first the nonmarine evaporite environments. These occur in closed sedimentary basins where river water and/or groundwater flow into a terminal lake or playa in a desert environment. Playas, which are also referred to as saltpans if surface salts are present, are dry lakebeds that can temporarily be covered with water.

While typical river water and groundwater have a low concentration of dissolved constituents, over time the water in a terminal lake can become a brine as the $H_2O$ is evaporated from its surface but the dissolved constituents continue to be added. Minerals can then become saturated in the brine and precipitate out of solution. This is currently happening in the Danakil Depression, Ethiopia; Death Valley and Searles Lake, California; and the Simpson Desert in western Australia. Also drainage basins feeding into extremely arid environments, including the deserts of Chile and Nambia and in parts of the Sahara, have recently formed saltpans that over time will accumulate large volumes of continental salts.

Continental surface waters are typically bicarbonate that is $HCO_3^-$ rich, derived from the absorption of atmospheric $CO_2$ with significant $Ca^{2+}$ and contain lesser amounts of $Na^+$ and $Mg^{2+}$. The cations are present from reaction of atmospheric $CO_2$ and soil organic acid-charged waters with minerals in surface sediments. Two typical reactions for calcite and sodium-rich feldspar to produce typical $HCO_3^- + Ca^{2+} + Na^+$ surface water are

$$CO_2 + H_2O + CaCO_3 \rightarrow Ca^{2+} + 2HCO_3^- \qquad [11.1]$$
$$\text{calcite}$$

and

$$2CO_2 + 3H_2O + 2NaAlSi_3O_8 \rightarrow Al_2Si_2O_5(OH)_4$$
$$\text{albite} \qquad \text{kaolinite}$$
$$+ 4SiO_2 + 2Na^+ + 2HCO_3^-. \qquad [11.2]$$
$$\text{quartz}$$

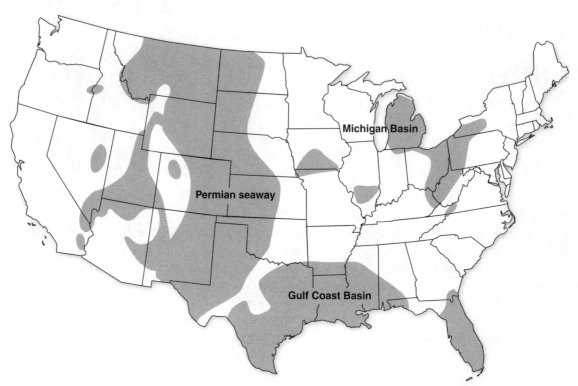

*Figure 11.1* Areas underlain at some depth by evaporites. (Adapted from Blatt, H. and Tracy, R.J., *Petrology: Igneous, Sedimentary, and Metamorphic*, 2nd edition, 1996, Freeman, 529 p.)

Except where the watershed transverses pre-existing marine evaporates, on a molar basis $SO_4^{2-}$ and $Cl^-$ are significantly lower than $HCO_3^-$ in most river water.

## Trona

The continental surface water can be sequestered in a terminal lake. On evaporation, the water, which typically contains some Mg, first saturates and precipitates Mg-calcite [$(Ca,Mg)CO_3$]. With the loss of $Ca^{2+}$ and $Mg^{2+}$ further evaporation in the lake saturates the remaining solution with sodium carbonate, *trona*. Trona precipitates by the reaction:

$$3Na^+ + 2HCO_3^- + 2H_2O \rightarrow Na_3CO_3(HCO_3)\cdot2H_2O + H^+$$
$$\text{trona} \qquad [11.3]$$

and forms a layer on the lakebed as shown in **Figure 11.2**.

The mining of trona and production of soda ash ($Na_2CO_3$) from evaporites typically has a cost advantage over production of $Na_2CO_3$ from NaOH, which has historically demonstrated more dramatic price fluctuations. Thirty percent of domestic trona is used to produce $Na_2CO_3$ for Na-glass manufacture. Trona is also used in the production of sodium bicarbonate ($NaHCO_3$), which is also known as baking soda. This is an essential ingredient in the beverage, coating, detergents, food, dialysis, and personal care industries.

The largest known trona deposits in the world are in 25 major beds at depths ranging from about 270 m to 1200 m below the surface in the Green River Formation in southwestern Wyoming. These were deposited in a broad, shallow ancient lake called Lake Gosiute.

## Thenardite

Thenardite ($Na_2SO_4$), or natural sodium sulfate, is another common nonmarine evaporite mineral. The needed sulfate ($SO_4^{2-}$) in the water to

*Figure 11.2* A lakebed near the Yukon River in Yukon Flats, Alaska showing trona precipitation. A pressure transducer has been placed on the lakebed to measure the height of water when the lake fills.

precipitate thenardite is derived from oxidation of sulfides, principally pyrite in the soils through which the water moves by a reaction like

$$2FeS_2 + 2H_2O + 3O_2 \rightarrow 4SO_4^{2-} + 2Fe^{2+} + 4H^+. \quad [11.4]$$
pyrite

Thenardite is found in Searles Lake, California; Espartinas in Madrid Province, Spain; and the Bilma Oasis of Niger, as well as other arid environments. It is used as a source of sodium to produce soda ash and various kinds of Na-glass and paper.

## Boron (B)

Boron is considered a metalloid having properties intermediate between a metal and nonmetal. It is relatively rare making up only 0.001% of the earth's crust.

**Use.** Fiberglass and borosilicate glass production has driven the global demand for boron. Another major use of boron is to make sodium perborate ($NaBO_3$), or bleach. Boron is also incorporated into boric acid ($H_3BO_3$). This is an antiseptic that is also used as an insecticide to control cockroaches. *Borax* [$Na_2B_4O_5(OH)_4 \cdot 8H_2O$] softens water in laundry detergents and is used as a flux for metal welding. Amorphous boron produces a green color in pyrotechnics.

**Resource location.** Concentrations of boron are found in nonmarine evaporite deposits formed in playa lakes in arid regions. These develop on rhyolite tuff deposits where volcanic hydrothermal solutions have brought boron to the playa. Boron combines with alkali and alkaline earth elements in the brine to form the borate minerals borax and ulexite [$NaCaB_5O_6(OH)_6 \cdot 5H_2O$] during brine evaporation. These can be later altered to colemanite [$CaB_3O_4(OH)_3 \cdot H_2O$], where Ca-rich waters are present. Boron is also present in sassolite ($H_3BO_3$), also called natural boric acid that precipitates from hot spring waters.

**Production and reserves.** The Bigadiç borate evaporite deposits in Turkey and the Death Valley evaporite deposits in California are the world's largest producers of boron. The Death Valley ore was originally hauled out of the desert by 20-mule teams, giving rise to the 20-Mule Team® borax cleaner (**Figure 11.3**). World boron production in 2011 was 4.3 million metric tons of $B_2O_3$. The worldwide commercial borate deposits are estimated to contain reserves of 210 million metric tons of $B_2O_3$. World boron resources are, therefore, adequate for the foreseeable future.

## Lithium (Li)

Lithium (Li) is an alkali metal that is highly reactive and combustible. Estimates of the average upper continental crustal content of Li range from 20 to 40 ppm.

**Use.** Lithium is a supplement taken to control the symptoms of mania associated with bipolar disorder. Lithium oxide is a widely used flux for processing molten silica and is a common component of ovenware. The red color in pyrotechnics is produced by lithium compounds. As a strong base lithium carbonate is combined with fat to produce lithium stearate for soaps and high-temperature greases. Recently lithium has received considerable attention because lithium ion batteries are rapidly becoming the battery of choice for electric vehicles (EVs) (**Figure 11.4**).

*Figure 11.3* A 20-mule team borax wagon on the move in Death Valley, California.

*Figure 11.4* A 40-volt, 11-cell lithium ion battery used in electric and hybrid vehicles. Typically 200 to 400 cells are combined into one powerful circuit that produces about 300 volts.

Production of 60 million plug-in EVs per year with lithium ion batteries would require 420,000 metric tons of lithium each year, six times the current production.

The lithium ion battery uses lithium cobalt oxide ($LiCoO_2$) as the positive cathode and a highly crystallized carbon as the negative anode in an organic solvent, $C_6$, containing lithium ion complexes. During discharge the following reaction occurs at the cathode:

$$LiCoO_2 + C_6 \rightarrow Li_{1-x}CoO_2 + C_6Li_x^{x+} + xe^- \qquad [11.5]$$

where $Co^{3+}$ in $LiCoO_2$ is partial oxidized to $xCo^{+4}$ with $(1-x)Co$ remaining as $Co^{+3}$ to produce $Li_{1-x}CoO_2$. The $x$ electrons, $xe^-$, are transmitted through a wire to produce electrical power. During recharge, $C_6Li_x^{x+}$ cations move to the cathode. At the cathode they combine with supplied electrons to reconstitute $LiCoO_2$ in a reverse of reaction [11.5].

**Resource location.** Spodumene [$LiAl(SiO_3)_2$], a pyroxene, and lepidolite [$K(Li,Al)_3(Si,Al)_4O_{10}$ $(F,OH)_2$], a mica found in pegmatites, contain significant concentrations of lithium. Lithium-containing pegmatites are relatively rare. Most of the world's lithium is either mined from pegmatites or extracted from brines.

Nonmarine brines and evaporites can contain significant concentrations of Li. The lithium is derived mainly from the leaching of volcanic rocks. The Great Salt Lake, Utah has from 30 to 60 ppm Li. Brine pools associated with salt flats, or solars in the Spanish language, in volcanic terrains of Bolivia, Argentina, Chile, Tibet, and China also have high Li contents. The government of Bolivia is moving ahead with plans to tap potentially huge lithium reserves in a brine pool below Salar de Uyuni in southwest Bolivia. It is the world's largest salt flat with a Li content of 3,000 ppm.

Lithium is also present in geothermal brines associated with volcanic terrains in Wairakei, New Zealand (13 ppm Li); Reykjanes, Iceland (8 ppm Li); and at El Tatio in Chile (47 ppm Li). Also oilfield brines in North Dakota, Wyoming,

Oklahoma, East Texas, and Arkansas can have up to 700 ppm Li. Li can also be obtained from the clay mineral hectorite [$Na_{0.3}(Mg,Li)_3Si_4O_{10}(F,OH)_2$], with a Li content up to 12,000 ppm. Hectorite is found as an alteration product of high-Si volcanic ash and tuff in hydrothermal deposits. The largest known deposit in the U.S. is in the McDermitt caldera complex that straddles the Nevada/Oregon border where hectorite is mined from a series of elongate lenses.

**Production and reserves.** There is dispute in the analysis of lithium's geological resource base. The concern is whether there is sufficient lithium available to sustain EV manufacture in the amount required. Most investigators believe there are with sources in pegmatites, continental brines, geothermal brines, oilfield brines, and hectorites. Given in **Table 11.1** are estimates of Li production and reserves from the U.S. Geological Survey. Note that Chile is the leading producer in the world and even if production increased by a factor of six ample reserves exist for over 60 years.

---

*Table 11.1* LITHIUM PRODUCTION FOR 2011 AND RESERVES BY COUNTRY.*

| COUNTRY | PRODUCTION | RESERVES |
|---------|------------|----------|
| Chile | 12.6 | 7,500 |
| Australia | 11.3 | 970 |
| China | 5.2 | 3,500 |
| Argentina | 3.2 | 850 |
| United States | W | 38 |
| Portugal | 0.82 | 10 |
| Zimbabwe | 0.47 | 23 |
| Brazil | 0.16 | 64 |
| World totals | 34 | 13,000 |

\* In thousands of metric tons of Li.

W = Withheld to avoid disclosing company proprietary data.

*Data from*: U.S. Geological Survey, *Mineral Commodity Summaries*, January, 2012.

## Marine Evaporites

In marine environments, a basin that is partially isolated from open seawater is required to produce a concentrated seawater brine. If evaporation of $H_2O$ is greater than the influx of new seawater into the basin, ions will concentrate in the water and salts precipitate. To get a significant thickness of evaporite salts there needs to be a way to continue to introduce seawater to evaporate into the basin. Examples of such environments include the Black Sea, Red Sea, and Mediterranean Sea. There are also totally isolated basins that were once open to seawater, such as the Caspian Sea and Dead Sea, that contain marine evaporites in the subsurface.

Seawater brines and precipitated salts also form in the sabkha regions of Saudi Arabia and Iran. A *sabkha* is a flat area along an ocean coast subjected to periodic flooding and evaporation of seawater. In these environments seawater inundates the sabkha during high tide and/or storm surges. The seawater percolates through the sediments and $H_2O$ evaporates in the heat, producing salt crusts. In some areas, shallow salt lakes can be produced that later evaporate. Brines are also trapped underground in aquifers in the pores of rock when the sediments surrounding them solidify into rock. These brines can change their character by reacting with the rocks they are enclosed in or from dissolving earlier precipitated evaporite minerals.

Minerals precipitate out of evaporating seawater in the reverse order of their solubilities, so that the order of precipitation is:

1. Limestone ($CaCO_3$) and dolomite [$CaMg(CO_3)_2$]
2. Gypsum ($CaSO_4 \bullet 2H_2O$) and anhydrite ($CaSO_4$)
3. Halite (i.e., common table salt, NaCl)
4. Potassium and magnesium salts.

The abundance of minerals formed from seawater precipitation is in the same order as this precipitation sequence. Thus, in the geological record limestone and dolomite are more common than gypsum plus anhydrite, which is more common than halite deposits, and halite is more common than the potassium and magnesium salts.

Limestone, dolomite, gypsum, and anhydrite are used mainly as building material and are discussed in another chapter. Halite and the Mg-K salts are discussed below.

Given in **Figure 11.5** is the sequence and quantity of minerals per 1,000 g of $H_2O$ that precipitate out of seawater as a function of $H_2O$ remaining after, on average, 0.1 g of calcite has precipitated. The determination shown in Figure 11.5 assumes the precipitated minerals can react with the remaining brine so for instance early formed gypsum ($CaSO_4 \cdot 2H_2O$) dehydrates to produce anhydrite ($CaSO_4$) as more $H_2O$ evaporates. Gypsum begins to precipitate when seawater volume has been reduced to 30% of its original volume. At 10% of the original volume halite begins to precipitate. Note that the final assemblage contains

Mg-K salts that are less than 10% of the amount of halite that has precipitated, but greater than the moles of anhydrite.

Early-formed minerals often do not react with the brine to produce the final assemblage given in Figure 11.5. Also, if fresh seawater is introduced into the brine after precipitation has started, the relative composition of the evaporite produced can change. Some basins become totally closed to seawater and the introduction of freshwater changes the mineral crystallization in the basin. As a result there are many different sequences of salts that can precipitate in seawater evaporites.

Seawater evaporite deposits occur throughout Phanerozoic time but are rare in sedimentary sequences of Precambrian age (see time scale on inside back cover for ages). Significant deposits

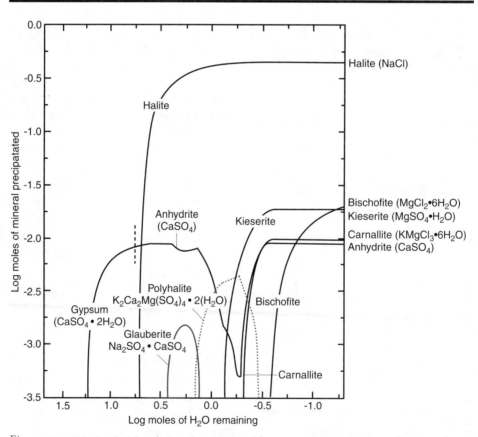

*Figure 11.5* Calculated moles of minerals precipitated from evaporating seawater as a function of moles of $H_2O$ remaining after initial calcite precipitation. Left side of diagram starts with 55.51 moles (1000 g) of $H_2O$ in seawater. (Modified from Harvie, C.E., Weare, J.H., Hardie, L.A. and Eugster, H.P., 1980, Evaporation of seawater; calculated mineral sequences, *Science*, v. 208, p. 498–500.)

are preserved from the Cambrian, Permian, and Triassic. They tend to be closely associated with shallow marine shelf carbonates and mudrocks deposited during times when the continents were flooded with seawater.

## Halite

Halite (NaCl), or table salt, is made up of the ions $Na^+$ and $Cl^-$ and readily dissolves in water. It is a common inexpensive industrial material.

**Use.**    Table salt is part of human and animal diets accounting for about 20% of its use. Food can also be preserved with NaCl as it dehydrates and kills microbial organisms. Animal hides are curried with NaCl by dehydrating the hide without disturbing the skin structure. Large amounts of salt are used to deice highways and in home water softeners to remove the hardness caused by $Ca^{2+}$ and $Mg^{2+}$ in water.

About 60% of halite is used in the chemical industry to produce such compounds as sodium hydroxide (NaOH). NaOH is an important ingredient for soap manufacture. Na in NaCl is employed to produce soda carbonate ($Na_2CO_3$), whose uses include the production of common soda-lime glass for windows and glass containers. Cl in NaCl is processed to make hydrochloric acid (HCl) to produce vinyl chloride ($C_2H_3Cl$) for PVC plastic production as well as to make chlorine bleach (NaClO). Industrially, elemental chlorine is usually produced by the electrolysis of NaCl brine. $Cl^-$ ions are oxidized at an anode to produce chlorine gas by the reaction:

$$2Na^+ + 2Cl^- + 2H_2O \rightarrow Cl_2 + H_2 + 2NaOH. \qquad [11.6]$$

Along with the chlorine, this process also creates $H_2$ gas and sodium hydroxide. Hydrochloric acid is typically made by the combustion of the product hydrogen and chlorine gases to form hydrogen chloride (HCl) gas that absorbs water to form a 32% by weight solution of hydrochloric acid.

## Resource location, production, and reserves.

Halite is obtained by mining underground evaporite deposits (**Figure 11.6**), evaporation of seawater, water from salt ponds, salt lakes and

*Figure 11.6* Salt production in Thailand. During the dry season, water is injected into near surface salt beds and brine is pumped out into about 25 cm deep ponds. It takes about 10 days for evaporation to precipitate most of the salt in the brine.

*Figure 11.7* Terraced Inca salt evaporation ponds outside Maras, Peru feed by a saltwater spring which taps subsurface evaporite deposits.

underground brines (**Figure 11.7**). Nearly every country on the earth has either salt deposits or solar evaporation operations of various sizes to obtain halite. World halite production was 290 million metric tons in 2011 up from 280 million metric tons in 2010. Reserves are extremely large with ample reserves in most halite exporting countries. The ocean contains virtually an unlimited supply of halite.

## Strontium (Sr)

**Use.** The major use for strontium (Sr) at present is to produce glass for cathode ray tubes (CRTs) for color television faceplates that contain ~8 wt% SrO to block X-ray emission. As digital television sets become more prevalent this use will decrease. Sr is also employed to produce special low-cost magnets with good resistance to demagnetization and in refining zinc. Flares and fireworks use strontium carbonate ($SrCO_3$) and strontium nitrate [$Sr(NO_3)_2$], which burn a bright red color.

**Resource location, production, and reserves.** The amount of Sr in seawater varies from 5.8 to 10 ppm. With evaporation *celestite* ($SrSO_4$) can precipitate out of a seawater brine given the right conditions. However, in most cases Sr is incorporated at the thousands of ppm level in precipitating gypsum and anhydrite. Celestite appears to form during reactions in the sediments with loss of the Sr from gypsum and anhydrite producing a Sr-rich meteoric groundwater. This groundwater mixes with $SO_4^{2-}$ from brines and precipitates the celestite.

The U.S. hasn't mined celestite since 1959. China is the world's leading producer of strontium. In 2011 China shipped 210,000 metric tons of Sr in Sr-carbonate, over half of the world's production. Even though world reserves of Sr are estimated to be only 6.8 million metric tons, an 18-year supply at current production rates, world resources of Sr have been estimated to exceed 1 billion tons.

## Bromine (Br)

**Use.** Forty percent of bromine is used in fire retardants as brominated ($BrO_3^-$) polymers that reduce the flammability of products such as circuit boards and clothes. Drilling muds account for 24% of Br use in the compounds calcium bromide ($CaBr_2$, 3.4 g/cm³) and zinc bromide ($ZnBr_2$, 4.2 g/cm³) used to increase drilling fluid density (see the chapter on petroleum). Bromine use in pesticides account for 12% of the total but is being phased out because of its potential harmful effects.

Bromate ($BrO_3^-$) is a suspected human carcinogen if ingested. Soft drink producers in the U.S. put brominated vegetable oil in citrus-flavored sodas but it is not an approved additive in over 100 countries.

**Resource location.** Bromine, whose average concentration in seawater is 65 ppm, can be recovered as a byproduct during seawater evaporation to produce table salt. However, most bromine is recovered from brines where it has very high concentrations. In the U.S. it is obtained from Br-rich Smackover Formation brines in Arkansas that have a Br content between 4.0 and 4.6 g per liter. In the Michigan basin, bromine is extracted from brines as a byproduct of the production of magnesium. In Israel, Br is extracted from Dead Sea brine that contains up to 4.2 g per liter. China produces bromine from underground brine in Shandong Province with a concentration of 0.18 to 0.20 g per liter.

**Production and reserves.** Israel with 200,000, China with 155,000, and Jordan with 75,000 metric tons dominate the world production of 460,000 metric tons produced worldwide in 2011. The world bromine resources are virtually unlimited. At 65 ppm seawater contains almost 100 billion metric tons of bromine.

## Iodine (I)

**Use.** Iodine is a trace element needed by the human body to make thyroid hormones. Therefore, it is important that the human body ingest adequate iodine. Global iodine consumption continues to increase with the recognition of the importance of iodine as a dietary ingredient in a well-balanced diet.

In the U.S. utilization of iodine is greatest in the pharmaceutical and sanitation industries. In pharmaceuticals iodine compounds can act as reducers in oxidation-reduction reactions and have catalytic effects which help their absorption. Iodine is also introduced into the human body where it acts as an X-ray contrast media for medical diagnostics.

Iodine solutions are a disinfectant similar to chlorine solutions, but unlike chlorine iodine solutions do not need to be rinsed from surfaces to avoid ingestion. With environmental concerns about bromine-based herbicides there has been an increased use of iodine in herbicides.

**Resource location, production, and reserves.** There is 0.41 ppm I in seawater. This could be removed by electrolysis. However, as with bromine there is a cheaper source in iodine-rich brines. At present, I is extracted from waste brines brought up during natural gas production and as a byproduct of nitrate production.

Caliche, when considering resources, is a rock-containing gravel and sand cemented by sodium nitrate ($NaNO_3$) but can contain some niter ($KNO_3$). It is found in arid areas of South America from the evaporation of water rich in organic matter. Caliches in Chile are mined for nitrate, which also contain small amounts of sodium iodate ($NaIO_3$) and sodium iodide ($NaI$). During the production of pure sodium nitrate by a leaching technique, the sodium iodate and iodide are extracted.

In 2011, 29,000 metric tons of I was produced worldwide. The leading producer was Chile with 18,000 metric tons from nitrate. This was followed by Japan with 9.8 thousand metric tons from the Minami Kanto gas field east of Tokyo where waste brine contains 2,400 ppm I. Third was U.S. production from the Anadarko Basin gas field in northwest Oklahoma with 300 ppm I in waste brine. Another iodine resource is in seaweeds of the *Laminaria* family. These plants extract and accumulate I from seawater and contain 0.45 wt% in the dry seaweed. Reserves of I appear ample into the foreseeable future, as 15 million metric tons have been identified worldwide.

## Potassium (K)

Potassium (K) is quite common in the upper continental crust being the seventh most abundant element. The common minerals K-feldspar ($KAlSi_3O_8$) and mica such as muscovite [$KAl_3Si_3O_{10}(OH)_2$], as well as their weathering products including illite [$KAl_2(AlSi)_4O_{10}(OH)_2$], contain abundant K. However, these are not good sources of elemental K because of the large amount of energy required to break the Si-O bonds in silicates to liberate the K. Potassium is, therefore, obtained from seawater brines and evaporites.

Note in **Table 11.2** that Borax and Mono Lakes in California are saline but their salts are produced by evaporation of fresh water. On the other hand,

*Table 11.2* COMPOSITION OF SOME SALT LAKES.*

| ION | BORAX LAKE, CA | MONO LAKE, CA | GREAT SALT LAKE, UT | CASPIAN SEA | DEAD SEA |
|---|---|---|---|---|---|
| $Cl^-$ | 5,945 | 15,100 | 112,900 | 131,180 | 208,020 |
| $SO_4^{2-}$ | 22 | 7,530 | 13,590 | 31,300 | 540 |
| $HCO_3^-$ | 6,668 | 26,430 | 180 | 200 | 240 |
| $Na^+$ | 6,199 | 21,400 | 67,500 | 131,180 | 34,940 |
| $K^+$ | 322 | 1,120 | 3,380 | 2,260 | 7,500 |
| $Ca^{2+}$ | nil | 11 | 330 | 8,510 | 15,800 |
| $Mg^{2+}$ | 31 | 32 | 5,620 | 29,160 | 41,960 |
| Total salinity | >19,400 | 71,900 | 203,500 | 350,070 | 315,040 |

*By weight in ppm.

the Great Salt Lake in Utah and the Caspian and Dead Seas have the high $Mg^{2+}$ and $K^+$ that is a hallmark of earlier evaporation of seawater. Further evaporation of these seawater brines produces potassium and magnesium evaporite minerals as outlined for seawater above. These minerals are an important source of potash particularly for fertilizer. They are, however, a less important source of Mg metal than from magnesite and dolomite (see the chapter on abundant metals). However, carnallite, a magnesium-rich evaporite mineral, is still Russia's most significant source of Mg metal.

From the 400 ppm K in seawater the K-Mg salt carnallite ($KMgCl_3 \bullet 6H_2O$) precipitates. If fresh seawater is added during the evaporation process after some of the salts have precipitated the salts sylvite (KCl), langbeinite [$K_2Mg_2(SO_4)_3$], and kainite ($KMgSO_4Cl \bullet 3H_2O$) form. These potassium-bearing salts make up what is called *potash*, a term that includes all potassium-bearing salts.

Fertilizers account for about 94% of world potash consumption. Manufactured potassium compounds are largely potassium sulfate ($K_2SO_4$) and potassium nitrate ($KNO_3$) for use as fertilizers. China is the world's largest potash consumer, accounting for 20% of world demand. Potassium is essential for plant growth but little K ends up in the edible portion of the plant. K facilitates sugar movement through plants, and boosts resistance to stresses such as drought and disease. Most of the potash that is not used in fertilizers is used in livestock feed supplements where it increases growth and milk production. The rest of the potash is used in manufacturing. This includes $K_2CO_3$ for production of glass, KOH for soaps, $KNO_3$ for explosives, and KCl in dietary supplements, drugs, and chemical dyes.

**Resource location.** Because K-salts are quite soluble, deposits are only preserved in surface deposits in arid regions or more commonly in strata below the surface that have not interacted with groundwater. Canada is the largest potash producer in the world as indicated in **Table 11.3**. Most production is from the very large Williston Basin evaporite deposits in southern Saskatchewan.

*Table 11.3* POTASH PRODUCTION FOR 2011 AND RESERVES BY COUNTRY.*

| COUNTRY | PRODUCTION | RESERVES |
|---|---|---|
| Canada | 11.2 | 4,400 |
| Russia | 7.4 | 3,300 |
| Belarus | 5.5 | 750 |
| China | 3.2 | 210 |
| Germany | 3.3 | 150 |
| Israel | 2.0 | 40 |
| Jordan | 1.4 | 40 |
| United States | 1.1 | 130 |
| World totals | 37 | 9,500 |

*In millions of metric tons of $K_2O$.

*Data from*: U.S. Geological Survey, *Mineral Commodity Summaries*, January, 2012.

Deposits in this basin extend underground to eastern Montana and western North and South Dakota. Russia has the second largest deposits. These are the Verkhekamskoye deposits of potash-magnesium salt evaporites in the Upper Kama Basin in the Perm region.

The U.S. has subsurface potash deposits in southeastern New Mexico and western Texas in the Salado Formation of the Permian basin. A shallow inland sea deposited thick layers of evaporites over the area. K-bearing salt layers in the evaporites near Carlsbad, New Mexico can reach 4 m in thickness. K-rich salts also occur in the evaporite deposits in the Paradox Basin of southeastern Utah and southwestern Colorado at depths between 0.8 and 1.8 km. Solution-mining techniques are used to extract the salts (**Figure 11.8**).

**Production and reserves.** Most potash mines today are underground mines as deep as 1.4 km. Some deposits closer to the surface are mined as strip mines and by solution mining operations. Potassium is also produced through evaporation of saline lake and subsurface brines.

*Figure 11.8* Intrepid Potash, Inc. potash evaporation ponds near Moab, Utah. Water from the Colorado River is injected into the underground mine. The water dissolves evaporite layers over 900 m below the surface. The potash-rich brine is pumped up from depth to 1.6 km² of surface ponds to evaporate the water and precipitate the salts that include potash minerals.

Estimated world resources total about 250 billion tons of $K_2O$ in potash. Three countries, Canada, Russia, and Belarus, hold nearly 90% of the known world reserves. The U.S. produced 1.1 million metric tons of potash in 2011, mostly in New Mexico. About 8 million tons per year is imported to the U.S., primarily from Canada. Humankind should have no problems obtaining potash far into the future.

# Elements for Agriculture

There is some concern as to whether the world's food production can keep pace with increased population growth. Local food shortages continue to occur. These are, however, due mainly to drought or other unusual weather conditions or political conflicts. Increased food production with increased population growth has occurred historically because of greater use of irrigation. This increased both available cropland and its productivity. More recently genetic engineering has made plants more productive and disease resistant. Also, the increased use of fertilizers has promoted more rapid and heartier plant growth. There is, however, data to indicate that in the world as a whole,

as is clearly the case in the U.S., that the increase in fertilizer usage will not increase the production of food by the same degree as in the past. More and more acreage has reached maximum growth rates independent of increases in water or fertilizer application.

A plant requires, in descending order, the elements, H, O, C, N, P, K, and S. The first two are supplied by water from soil solutions and the third by $CO_2$ from air through photosynthesis. The remaining elements are obtained from soil mineral matter. The reason the abundant nitrogen in air is not used by most plants is that plants require negative valence or charged nitrogen. Generally plants do not have the ability to break the strong N–N bond in uncharged $N_2$ gas in air needed to change nitrogen's valence.

If the elements needed by plants are inadequate for plant growth they can be added to the soil by fertilizer. The four primary nutrients necessary for plant growth are, then, nitrogen, phosphorous, potassium, and to a lesser extent, sulfur. Given on a bag of fertilizer are numbers, such as 10-30-20, which are the concentrations of N, P, and K, respectively. The weight percent of nitrogen is reported directly. However, P is reported as the mass fraction of phosphorus pentoxide ($P_2O_5$) and K is reported as the mass fraction of potassium oxide ($K_2O$). A 10-10-10 bag of fertilizer, therefore, contains 10 wt% elemental nitrogen, 4.4 wt% elemental phosphorus, and 8.3 wt% elemental potassium.

## Major Agricultural Crops

Given in **Figure 11.9** are the major crops grown in the world as a whole and in the U.S. Note that sugar cane is the largest crop grown worldwide but is not grown in the U.S. On the other hand, hay and soybeans are two of the largest crops grown in the U.S. but little is grown on average worldwide. The different crops require different application rates of fertilizer to grow optimally, as outlined in **Table 11.4**. There is a small nitrogen application rate needed for soybeans relative to the other crops. This is because soybeans and other legumes

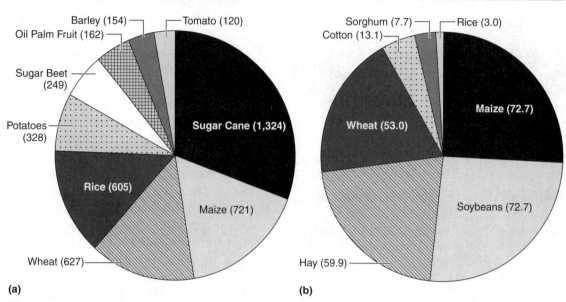

*Figure* 11.9 (a) Top agricultural products in the world in million of metric tons for 2004. (b) Harvested area in million acres of the major crops grown in the United States in 2000. (Data from: the United Nations Food and Agriculture Organization (FAO) and the U.S. Environmental Protection Agency (EPA)

*Table* 11.4 ANNUAL WORLD FERTILIZER USAGE* BY CROP TYPE AND PERCENTAGE OF TOTAL FERTILIZER USAGE AND AVERAGE APPLICATION RATE.[†]

| CROP TYPE | N | $P_2O_5$ | $K_2O$ | % TOTAL USAGE | APPLICATION RATE |
|---|---|---|---|---|---|
| Cereals | 38.2 | 16.2 | 7.0 | 64.8 | 102 |
| Oilseeds | 3.5 | 2.8 | 2.4 | 9.2 | 85 |
| Vegetables | 2.5 | 1.2 | 0.9 | 4.9 | 242 |
| Sugar beet/cane | 2.1 | 1.1 | 1.3 | 4.7 | 216 |
| Roots/tubers | 2.1 | 1.1 | 1.1 | 4.5 | 212 |
| Fibers | 2.8 | 1.0 | 0.4 | 4.4 | 144 |
| Fruits | 1.8 | 0.8 | 0.8 | 3.6 | 163 |
| Tobacco, beverages | 0.9 | 0.4 | 0.6 | 2.0 | 153 |
| Legumes | 0.7 | 0.8 | 0.3 | 1.9 | 39 |
| All | 54.6 | 25.4 | 14.8 | 100 | 109 |

*In millions of metric tons.
[†]In kilogram per hectare. Hectare = 10,000 square meters.

*Data from*: Food and Agriculture Organization of the United Nations, 2006.

are able to obtain their nitrogen requirements from $N_2$ in air.

Of the *cereal crops*—maize, rice, wheat, barley, sorghum, millet, oats, and rye—the three grains maize, wheat, and rice together accounted for 87% of food production worldwide and use 65% of fertilizer. They are staple foods that can be stored for use throughout the year. These grasses require high amounts of fertilizer to do well but are a rich source of vitamins, minerals, carbohydrates, fats, oils, and protein and are a majority of the daily sustenance of people in most developing countries.

Of fruits, tomatoes are grown more than any other type, 130 million metric tons worldwide in 2008 with 34 million metric tons grown in China alone. Second in amount are grapes, with 60 million metric tons worldwide, then citrus fruits, apples, and bananas. About 80% of grapes are used in making wine, 13% are consumed as table grapes, with the rest dried for raisins or made into juice and jelly.

Given in **Table 11.5** are the crops with the highest average fertilizer application rates. Note that bananas are heavy feeders requiring almost twice the application rate of sugar beets, the second highest. This is because bananas are such fast growers (**Figure 11.10**). Banana plants are often mistaken for trees because the thick, sturdy pseudostem looks a lot like a trunk. A pseudostem is a false stem composed of concentric rolled leaf blades that surround the growing point. The banana plant can get quite tall, almost 8 m high, with leaves nearly 3 m long.

## Fertilizer Resources

Our understanding of needed elements for plant nutrition developed in the nineteenth century. Guano, that is bird manure, was recognized as fertilizer and harvested from islands off the coast of Peru starting in 1808 and shipped to Europe until the 1880s. Peruvian guano was replaced by thousands of years of guano and saltpeter accumulations preserved in the Atacama Desert in the Tarapacá Region of Peru and the Antofagasta Region of Bolivia. This is now

northernmost Chile due to a change in the countries' boundaries after the "Saltpeter War," which ended in 1884.

*Table 11.5* CROPS WITH HIGHEST AVERAGE APPLICATION RATES ($N + P_2O_5 + K_2O$).*

| RANK | CROP | APPLICATION RATE |
|------|------|------------------|
| 1 | Banana | 479 |
| 2 | Sugar beet | 254 |
| 3 | Citrus | 252 |
| 4 | Vegetables | 242 |
| 5 | Potato | 243 |
| 6 | Oil-palm | 242 |
| 7 | Sweet potato | 225 |
| 8 | Tobacco | 225 |
| 9 | Tea | 225 |
| 10 | Sugar cane | 216 |

*In kg/ha per crop harvested (ha = hectare = $10^4$ m²).

*Data from*: Soh, Kim Gai, 1997, *Fertilizer demand and crops*: International Fertilizer Industry Association, published as supplement of CONTACT No. 14, June, 75 p.

*Figure 11.10* A grove of banana plants.

*Saltpeter*, a nitrogen fertilizer, is also sometimes called niter. It is mainly K-nitrate ($KNO_3$), with some Na-nitrate ($NaNO_3$) and Ca-nitrate [$Ca(NO_3)_2$]. Saltpeter is formed from ammonia ($NH_3$) produced by decaying organic matter. This nitrogen is oxidized by bacteria to produce $NO_3$. Together with K, Na, or Ca supplied by weathered rocks, saltpeter is precipitated out of solution. Saltpeter from the Atacama Desert as a source of nitrogen was replaced by nitrogen from air synthesized into ammonium ($NH_4$). This occurred in the late 1920s when the Haber-Bosch process for nitrogen fixation, developed in 1909, was commercialized (see below).

Given in **Table 11.6** are the compositions of various fertilizers. Note that depending on the fertilizer, it can supply significant quantities of needed N, P, and/or K. For instance, bat guano is a good source of all three elements while animal urea supplies only nitrogen. Typically, the K in fertilizer is supplied by mined potash from seawater evaporite deposits as discussed, but seaweed and the mineral glauconite [$(K,Na)(Fe^{3+},Al,Mg)_2(Si,Al)_4O_{10}(OH)_2$] are also used.

## Nitrogen (N)

**Use.** There are two different major usages of nitrogen as a resource. Nitrogen gas is used because it is a relatively inert gas. This usage is examined at the end of the chapter when the extraction of gases from air is discussed. The other use of nitrogen is by plants. Nitrogen is vital for plants as both amino acids and chlorophyll contain nitrogen.

While some plants such as soybeans and other legumes obtain their needed nitrogen symbiotically from atmospheric or soil air ($N_2$) with the aid of fungi, nearly all other plants need fixed nitrogen (**Table 11.7**). *Fixed nitrogen* is positively or negatively charged nitrogen. This nitrogen has changed its charge, that is, has been fixed from the zero charged nitrogen present in air in the

*Table 11.6* COMPOSITION OF FERTILIZERS IN TERMS OF ELEMENTAL NITROGEN (N), PHOSPHORUS OXIDE ($P_2O_5$), AND POTASH ($K_2O$) IN wt%.*

| MATERIAL | N | $P_2O_5$ | $K_2O$ |
|---|---|---|---|
| Seaweed (Kelp) | 0.6 | trace | 1.3 |
| Bat guano | 10 | 4.5 | 2.0 |
| Dried cow manure | 1.3 | 0.9 | 0.8 |
| Urea (Urine of animals) | 46 | 0.0 | 0.0 |
| Glauconite (Greensand) | 0.0 | 1.4 | 4.0–9.5 |
| Apatite (Rock phosphate) | 0.0 | 38–40 | 4.5 |
| Bone meal (Ground bones) | 2.0 | 22 | 0.0 |
| Fish meal (Ground fish waste) | 10 | 6.0 | 0.0 |
| *Ammonium nitrate* ($NH_4NO_3$) | 34 | 0.0 | 0.0 |
| *Monoammonium phosphate* [$(NH_4)_3PO_4$] | 11 | 52 | 0.0 |
| *Nitrophosphate* [$3(NH_4)_2HPO_4$] | 21.2 | 53.3 | 0.0 |
| *10-10-10 fertilizer* | 10 | 10 | 10 |

*Artificial fertilizers are given in italics.

*Table 11.7* TOTAL FIXED NITROGEN CONSUMED IN AGRICULTURE AND IN FEED AND PASTURES BY COUNTRY.*

| COUNTRY | TOTAL CONSUMED | FEED AND PASTURE |
|---|---|---|
| China | 18.7 | 3.0 |
| United States | 9.1 | 4.7 |
| France | 2.5 | 1.3 |
| Germany | 2.0 | 1.2 |
| Brazil | 1.7 | 0.7 |
| Canada | 1.6 | 0.9 |
| Turkey | 1.5 | 0.3 |
| United Kingdom | 1.3 | 0.9 |
| Mexico | 1.3 | 0.3 |
| Spain | 1.2 | 0.5 |
| Argentina | 0.4 | 0.1 |

*In million tons per year.

*Data from*: United Nations Food & Agriculture Organization, 2009.

diatomic gas $N_2$. While plants need negatively charged nitrogen, positively charged nitrogen is easily converted to a negative charge by bacteria in the soil as in the reaction to produce ammonia ($NH_3$):

$$H^+ + H_2O + KNO_3 \rightarrow NH_3 + K^+ + 2O_2. \qquad [11.7]$$

Note that because H has a valence of +1 the nitrogen in ammonia has a valence of −3.

**Production and reserves.** Fixation of N requires that the strong bond between N atoms in $N_2$ be broken. Nitrogen is commercially fixed from $N_2$ in air by combining it with hydrogen gas obtained from natural gas in the *Haber-Bosch process* given by the reaction:

$$N_{2(g)} + 3H_{2(g)} \rightarrow 2NH_{3(g)}. \qquad [11.8]$$

The problem with using this reaction is that it is slow at room temperature and strongly exothermic.

Exothermic reactions produce heat. Because it is strongly exothermic if the temperature of reaction is increased, to accelerate the forward reaction rate, the reverse reaction becomes rapidly more favored. In this case, the forward reaction dramatically slows and little ammonia can be produced as outlined in *Le Chatelier's Principle*. This states that when a chemical equilibrium experiences a change in concentration, temperature, or pressure the equilibrium shifts to counteract the imposed change and a new equilibrium is established. To overcome the production problem Haber and Bosch used a reasonably high pressure (150 to 250 atm) to favor the forward reaction because 4 moles of gas is being converted to 2. With temperatures between 400° and 550°C over an iron catalyst composed of treated magnetite ($Fe_3O_4$), they got about a 15% conversion of $N_2$. The process can then be repeated to get nearly 100% $NH_{3(g)}$ production.

The ammonia gas produced can be converted into the fertilizer compound of urea [$(NH_2)_2CO$] by reacting it with $CO_2$:

$$CO_{2(g)} + 2NH_{3(g)} \rightarrow (NH_2)_2CO + H_2O. \qquad [11.9]$$

The needed quantities of carbon dioxide are typically obtained during the manufacture of ammonia from the burning of coal and also from $CO_2$ derived from the production of petroleum from crude oil. Another fertilizer, ammonium sulfate [$(NH_4)_2SO_4$] is produced by reacting ammonia with sulfuric acid ($H_2SO_4$) in the reaction:

$$H_2SO_4 + 2NH_{3(g)} \rightarrow (NH_4)_2SO_4. \qquad [11.10]$$

This then adds sulfur as well as nitrogen to the fertilizer.

The fertilizer compound ammonium nitrate ($NH_4NO_3$) is manufactured by reacting ammonia with nitric acid ($HNO_3$):

$$HNO_3 + NH_{3(g)} \rightarrow NH_4NO_3. \qquad [11.11]$$

Ammonium nitrate can decompose explosively to the gases $N_2$, $O_2$, and $H_2O$ by reaction at temperatures above 210°C:

$$NH_4NO_3 \rightarrow N_2 + O_2 + 2H_2O. \qquad [11.12]$$
$$rapid$$

The decomposition reaction for ammonium nitrate greatly increases product volume. This expansion removes the product gases from the reactants speeding the reaction. Reaction [11.12] is also highly exothermic producing heat that increases the decomposition rate and expands the gases to produce a blast concussion. Ammonium nitrate mixed with fuel oil that is ignited to increase the temperature and start the reaction was used as the explosive in the 1995 Oklahoma City and the 2003 Bali bombings. Ammonia can also be used to produce TriNitroToluene [$C_6H_2(NO_2)_3CH_3$], the explosive know as TNT.

China, India, Russia, and the U.S. are the primary producers of reduced N compounds as given in **Table 11.8**. Half of the total U.S. ammonia production capacity is in Louisiana, Oklahoma, and Texas. These are regions with large reserves of natural gas ($CH_4$) used to obtain the needed $H_2$ for ammonia preparation. Therefore, the cost of fertilizers depends to a large extent on the cost of natural gas. With virtually unlimited atmospheric nitrogen

*Table 11.8* ANNUAL AMMONIA PRODUCTION FOR 2011 BY COUNTRY.*

| COUNTRY | PRODUCTION |
|---|---|
| China | 41 |
| India | 12 |
| Russia | 11 |
| United States | 8.1 |
| Trinidad and Tobago | 5.6 |
| Indonesia | 4.8 |
| Canada | 4.1 |
| World total | 136 |

*In millions of metric tons.

*Data from*: U.S. Geological Survey, *Mineral Commodity Summaries*, January, 2012.

and significant natural gas supplies in the producing countries there should be an adequate supply of reduced nitrogen into the foreseeable future.

## Phosphorous (P)

Phosphorous on average makes up 0.23 wt% of the earth's crust. Currently 70% of the demand for P is for fertilizer with most of the rest used in chemical applications. These chemical applications include phosphates in toothpaste, detergents, glass, pesticides, steel production, pyrotechnics, and incendiary bombs. P is a needed element to sustain life. In most cases, P is the limiting nutrient for plant growth. The growth of crops, therefore, depends on an adequate supply of P that requires its introduction in fertilizer, as farming tends to deplete it in most soils.

In 1927, Erling Johnson developed a commercial process for producing nitrophosphate [$3(NH_4)_2HPO_4$] fertilizer. He reacted sedimentary phosphorite rock, found on Nauru and Banaba Islands in the South Pacific with nitric acid ($HNO_3$). Phosphorite contains the mineral *apatite* [$Ca_5(PO_4)_3(OH)$], where the $OH^-$ group can be replaced by $F^-$ or $Cl^-$. The reaction produced phosphoric acid ($H_3PO_4$) and calcium nitrate crystals [$Ca(NO_3)_2$] on cooling by the reaction:

$$2Ca_5(PO_4)_3(OH) + 20HNO_3 \rightarrow 6H_3PO_4$$
$$+ 10Ca(NO_3)_2 + 2H_2O. \qquad [11.13]$$

The phosphoric acid was neutralized with ammonia ($NH_3$) that together with nitrate crystals produced nitrophosphate by the reaction:

$$Ca(NO_3)_2 + 4H_3PO_4 + 8NH_3 \rightarrow CaHPO_4 + 2NH_4NO_3$$
$$+ 3(NH_4)_2HPO_4. \qquad [11.14]$$

The phosphorite deposits on Nauru and Banaba Islands were depleted in the 1980s but other deposits are available worldwide.

**Resource location.**    Sources of phosphorous on earth include:

1. Guano from seabird droppings (limited),
2. Marine sedimentary deposits of phosphorite (large), and
3. Igneous apatite [$Ca_5(PO_4)_3(F,Cl,OH)$] (large, but low grade, and therefore, less profitably obtained).

Therefore, most phosphorous is produced from marine phosphorite deposits, which contain fluorapatite [$Ca_5(PO_4)_3(OH,F)$] and carbonate fluorapatite [$Ca_{10}(PO_8)_6(OH,F)_{2-2x}(CO_3)_x$] in rocks with at least 8 wt% $P_2O_5$. These deposits appear to form within 40° of the equator. Phosphorites typically occur in layers a few centimeters to tens of meters thick interbedded with layers of organic-rich shale and chert. Most investigators believe that the vast majority of phosphorites form on the outer continental shelf often at the break between the continental shelf and the continental slope.

As shown in **Figure 11.11** the model of deposition involves upwelling, cold, nutrient-rich waters with elevated P concentration that occur on the western side of continents in the zone of dominantly eastern winds. The nutrient-rich waters produce high organic productivity with high phytoplankton growth in the photic zone at the surface of the ocean. Phytoplankton such as diatoms contain about 0.6 wt% P. Upon death, the organic material fluxes to the seafloor. Some of the organic matter produced reacts with any oxygen in the water creating anoxic conditions in the water column. Besides the dead phytoplankton flux with its P, mass mortalities of fish can occur in the water column due to the lack of $O_2$ over extended areas. Fish bones have a high concentration of P (see Table 11.6).

Bacteria in the sediments oxidize the deposited organic carbon. This transforms it to $HCO_3^-$ and produces $CH_4$ (methane). Other bacteria denitrify the organic matter turning organic nitrates into $N_2$. The produced gases escape into the water column. The remaining P is then concentrated and transformed to apatite.

Economic deposits in the geological record appear to form during times of sea level rise with reworking of the shelf deposits of apatite to collect major concentrations in bays and around structural highs.

**Production and reserves.** The fluorapatite [$Ca_5(PO_4)_3F$] in phosphate rock is not very soluble and provides little dissolved phosphorus to plants, except in some moist acidic soils. This is why fluorapatite is converted to superphosphate [$Ca(H_2PO_4)_2$] + $CaSO_4$ fertilizer produced by reacting fluorapatite with sulfuric acid:

$$2Ca_5(PO_4)_3F + 7H_2SO_4 \rightarrow 7CaSO_4 + 3Ca(H_2PO_4)_2 + 2HF \qquad [11.15]$$

or triple-superphosphate, a pure $Ca(H_2PO_4)_2$ fertilizer produced by reacting flourapatite with phosphoric acid:

$$Ca_5(PO_4)_3F + 7H_3PO_4 \rightarrow 5Ca(H_2PO_4)_2 + HF. \qquad [11.16]$$

The $Ca(H_2PO_4)_2$ dissolves much more readily than fluorapatite.

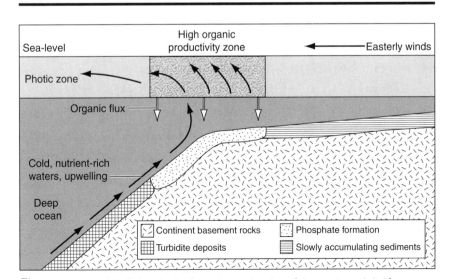

*Figure 11.11* Model of phosphorite formation at the edge of the continental shelf.

Deposits of apatite are widely distributed throughout the world and range in age from recent to Precambrian. The largest mining operations occur in China, U.S., Morocco, the western Sahara, and Russia as given in **Table 11.9**. Greater than 17% of the total world production of P occurs in the U.S. Florida and North Carolina deposits account for 85% of this production mined using surface mining methods. These unconsolidated phosphorite sediments occur in thin beds that extend over large areas including lower grade offshore deposits on the continental shelf (**Figure 11.12**). Also in the U.S. is the Phosphoria Formation that extends over more than 350,000 km$^2$ in parts of Colorado, Nevada, Idaho, Utah, and Wyoming (**Figure 11.13**). It is estimated to contain 2 billion metric tons of phosphorite with average composition ~80% apatite. However, much of it is not near the surface and would be expensive to mine. The near surface layers that contain about 25% $P_2O_5$ are mined near the Wyoming-southeastern Idaho border.

While there is no substitute for phosphorus in agriculture, extensive phosphate resources have been identified on the continental shelves and on seamounts in the Atlantic and Pacific Oceans. Resources and reserves appear adequate into the foreseeable future although the easily obtained deposits will likely be mined out in the next 50 years or so. With 2011 world production of 191 million metric tons per year and world reserves of 71 billion metric tones, availability of phosphate rock should not be a problem. However, the third largest producer is

*Table 11.9* PHOSPHATE ROCK[†] PRODUCTION FOR 2011 AND RESERVES BY COUNTRY.[*]

| COUNTRY | PRODUCTION | RESERVES |
|---------|-----------|----------|
| China | 72 | 3,700 |
| United States | 28.4 | 1,400 |
| Morocco and Western Sahara | 27 | 50,000 |
| Russia | 11 | 1,300 |
| Brazil | 6.2 | 310 |
| Jordan | 6.2 | 1,500 |
| Egypt | 6.0 | 100 |
| Tunisia | 5.0 | 100 |
| Israel | 3.2 | 180 |
| Syria | 3.1 | 1,800 |
| Australia | 2.7 | 250 |
| South Africa | 2.5 | 1,500 |
| World totals | 191 | 71,000 |

[*]In millions of metric tons.
[†]24% to 34% $P_2O_5$ concentration.

*Data from*: U.S. Geological Survey, *Mineral Commodity Summaries*, January, 2012.

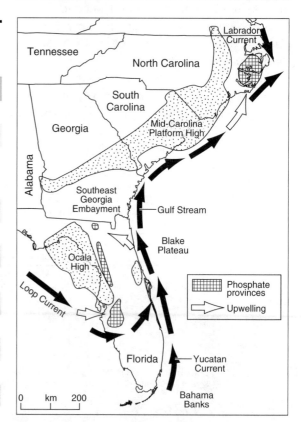

*Figure 11.12* Paleogeographic reconstruction of the southeastern U.S. continental margin during the Mid to Late Miocene transgression showing development of phosphate-rich sediments at sites of oceanic upwelling from currents along the coast shown by arrows

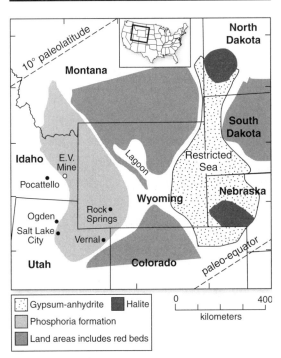

*Figure 11.13* Paleogeographic map showing the extent of the Permian Phosphoria Formation. It likely represents a shallow subtidal to intertidal environment near the paleoequator. An inland sea is thought to have existed to the west of the formation and land masses to the east that contained a restricted sea at the time the Phosphoria was deposited. (Adapted from Piper, D. Z., 2001, Marine Chemistry of the Permian Phosphoria Formation and Basin, Southeast Idaho, *Econ. Geol.*, v. 96, p. 599–620.)

Morocco, that together with the disputed western Sahara, is home to over 70% of identified world reserves of marine sedimentary phosphate.

## Sulfur

**Use.**  There are two ways sulfur is significant to agriculture. It is a plant nutrient and sulfuric acid is used to process the mineral fluorapatite into phosphate fertilizers as outlined in reaction [11.15]. The production of phosphate fertilizers consumes 60% of the total agricultural sulfur used with the remainder used directly as a plant nutrient.

Sulfur has been increasingly recognized as essential for plant nutrition. It is a component of amino acids, proteins, chlorophyll, fats, and other compounds found in plants. Historically the increase in the use of fertilizers containing little or no sulfur together with new environmental regulations limiting the amount of atmospheric sulfur from industrial emissions that falls on land surfaces has led to a decrease in soil sulfur content resulting in soil sulfur deficiencies in many areas.

Crop response to sulfur applications will vary depending on the type of soil and the sulfur requirement of the crop. This is high for alfalfa and low for soybeans and corn. Sulfur can be applied directly as elemental sulfur, sulfur-clay mixes, ammonium sulfate, potassium sulfate, and superphosphates.

**Production and reserves.**  Sulfur in ocean water occurs as $SO_4^{2-}$. In rocks it is present in its native form, in sulfate minerals such as gypsum ($CaSO_4 \cdot 2H_2O$) and as metal sulfides such as pyrite ($FeS_2$). Native sulfur is mined at a few locations worldwide. It is found associated with hot springs in volcanic areas in many parts of the world. Native sulfur is also present along with gypsum and anhydrate in salt domes along the U.S. Gulf of Mexico coast, both on and offshore (**Figure 11.14**) and in evaporites from eastern Europe and western Asia. Microorganisms obtain energy by converting gypsum in these deposits to native sulfur by the reaction:

$$2CO_2 + 2CaSO_4 \cdot 2H_2O \rightarrow S_2 + 2CaCO_3 + 4H_2O + 3O_2.$$
$$\text{gypsum} \qquad\qquad \text{calcite} \qquad\qquad [11.17]$$

Until recently most of the world's sulfur supply was obtained by the *Frasch process*. This is a method of mining deep-lying native sulfur, invented by Herman Frasch. It involves heating water to ~170°C under pressure and forcing it downward into the deposit in order to melt the sulfur (melting point = 115°C). The sulfur is lifted to the surface by means of compressed air. The mixture of sulfur and water is then discharged into bins and the sulfur allowed to cool and solidify.

Currently sulfur recovered as a byproduct of petroleum and natural gas refining has surpassed the Frasch process as the world's largest source of sulfur. Production from refining should increase as the demand for low-sulfur fuel continues.

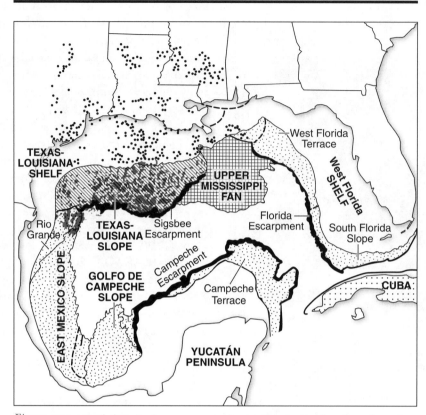

*Figure 11.14* Gulf of Mexico bottom topography showing the location of continental and oceanic salt domes with black dots. They become extremely numerous on the Texas-Louisiana slope as the Sigsbee escarpment is approached. (Adapted from Martin, R., G., and Bouma, A., H., 1978, Physiography of the Gulf, in, Bouma, A., Moore, G., and Coleman, J., eds. Framework, *Facies, and Oil-Trapping Characteristics of the Upper Continental Margin*, Amer. Assoc. Petrol. Geol., Tulsa, Oklahoma, p. 3–19.)

In petroleum refining, sulfur in the oil is reacted with hydrogen to form $H_2S$ while sulfur in natural gas occurs naturally as $H_2S$. This gas is cooled and catalyzed in a converter to produce sulfur vapor. Most of this is used to make sulfuric acid ($H_2SO_4$).

In 2011, 69 million metric tons of sulfur was extracted worldwide from the earth. With 9.6 million metric tons, China was the world's largest producer. The U.S. was second producing with 8.8 million metric tons and Canada third with 7.1 million metric tons. Canada is the largest sulfur exporter in the world. It has large stockpiles of sulfur in the sulfur-rich Athabasca oil sands in Alberta, Canada. Reserves of sulfur in crude oil, natural gas, coal, and sulfide ores are extremely large and the future supply of sulfur should not be a problem.

## Environmental Concerns of Fertilizers

Overfertilization of crops and the subsequent run-off of excess fertilizer in water contribute to nitrogen and phosphorous accumulations in watersheds termed *eutrophication*. Adding phosphorus and nitrogen to water promotes heavy growth of aquatic plants and algal mats at the surface of fresh and saltwater, respectively. Upon their death these plants and algae sink toward the bottom of the water mass where the reaction:

$$CH_2O + O_2 \rightarrow CO_2 + H_2O \qquad [11.18]$$

occurs. This leads to the deoxygenation of the water, the hallmark of eutrophication.

Eutrophication affects the water near the bottom of the Gulf of Mexico along the Louisiana-Texas coast where dissolved $O_2$ < 2 ppm in an area the size of New Jersey. This is mainly a consequence of fertilizer runoff into the Mississippi River being transported to the Gulf. Because fish require elevated oxygen levels in water to survive, this low oxygen zone is termed the Gulf of Mexico Dead Zone.

Nitrate levels above 10 ppm in drinking water can cause methemoglobinemia, a decrease in oxygen carrying capacity of hemoglobin. This is a particular problem for children under 6 months of age where it is called blue baby syndrome and can lead to death.

Industrial and sewage discharges of phosphate to water have been reduced greatly in the past 25 years. Most phosphorus enters bodies of water through soil erosion from agricultural lands. Improved planting methods, fertilizer management, and soil conservation are being used to reduce this phosphorus runoff.

# Gaseous Elements and Molecules from Air

Besides supplying $O_2$ for animals like mankind and $CO_2$ for plants, the earth's atmosphere is a resource for some important pure gases used in industry.

## *Use*

Pure nitrogen gas ($N_2$) obtained from air is used to produce a wide variety of chemical compounds, including ammonia ($NH_3$). Nitrogen gas is also used to protect substances during processing from the oxidation effects of the oxygen in air. Liquid nitrogen is produced for freeze-drying foods and refrigeration.

The steel industry is the largest consumer of pure oxygen gas. This gas is also obtained from air. Oxygen is reacted with excess carbon in the steel producing CO gas. The CO gas is then reacted with iron oxides to reduce iron to pure iron metal. Sewage is oxidized with oxygen gas

to destroy it and with carbon containing solid waste to incinerate it. Tanks of oxygen gas are the oxidant in oxyhydrogen and oxyacetylene torches for metal welding. Liquid oxygen is a propellant in some rockets. Oxygen is needed to crack, that is break apart, hydrocarbons by partially combusting them to produce $CO_2$ and $H_2O$.

About 700,000 metric tons of argon (Ar) gas are produced worldwide every year from air. Argon gas is often used when an inert gas is required. In particular, argon is the cheapest alternative when $N_2$ is not sufficiently inert. Argon is the filler gas used in most light bulbs. Ar atmospheres are maintained to exclude $O_2$ during the manufacture of high-quality stainless steels and the production of impurity-free silicon crystals in the semiconductor industry. As a dry, heavier than air gas, Ar is used as a filler gas for the space between glass panels in high-efficiency multipane windows.

Neon's major use is in making neon advertising signs. Neon, unlike other inert gases, discharges electricity at normal currents and voltages. It is used in fluorescent lamps because when an electric current is passed through neon gas-filled tubes a bright orange-red colored light is produced. If trace quantities of mercury are added a brilliant blue color is produced. Neon lights are popular on vehicles because the light has the ability to penetrate fog. Liquid neon is used as a cryogenic refrigerant. It has high cooling capacity with a heat capacity over 40 times greater per unit volume and is less expensive than liquid helium.

Because it can produce a very bright light for a very short period of time, krypton is used in flashbulbs for high-speed photography. Krypton is also used to improve brightness with Ar in fluorescent lights. If added to the nitrogen atmosphere in incandescent lights, krypton can extend their lifetime.

Xenon is used in strobe lights to produce intense, short bursts of light. Although expensive to extract from air, xenon is also used for sedation of the critically ill and in applications where a high molecular weight noble gas is needed.

## Production

Dry air contains:
- 78.08 vol% nitrogen ($N_2$)
- 20.95 vol% oxygen ($O_2$)
- 0.93 vol% argon (Ar)
- 0.039 vol% carbon dioxide ($CO_2$)
- 0.0018 vol% neon (Ne)
- 0.00052 vol% helium (He)
- 0.00018 vol% methane ($CH_4$)
- 0.00011 vol% krypton (Kr)
- 0.000055 vol% hydrogen ($H_2$)
- 0.00003 vol% nitrous oxide ($N_2O$)
- 0.00001 vol% carbon monoxide (CO)
- 0.0000087 vol% xenon (Xe).

Generally, the gaseous elements of interest in air are extracted in a large commercial plant as shown in **Figure 11.15**. These are all the listed gases except carbon dioxide ($CO_2$), helium (He), methane ($CH_4$), hydrogen ($H_2$), nitrous oxide ($N_2O$), and carbon monoxide (CO). The multielement gases are more easily produced from the single element gases. For helium and hydrogen gas a cheaper source is found in natural gas (see below).

The gases are separated by first passing air through filters to remove dust and other particulate matter. The air is then exposed to an alkali metal, Na or K. This absorbs water and carbon dioxide in the air. This dry air is then compressed to ~200 atm. Because compression raises the temperature of the air, it is next cooled by refrigeration. The cooled, compressed air is passed through winding coils in an empty chamber. A portion of the air is allowed to expand into the chamber. This sudden expansion absorbs heat from the coils, cooling the compressed air. The compression, refrigeration, and expansion process is repeated until the air has been cooled to a temperature of −247°C. The temperature needs to be only −196°C if neon is not separated.

At −247°C the gases in the air are transformed into liquids except for He and $H_2$. The mixture is vented to release the Ne and $H_2$ as gases. The vented liquid is then fractionally distilled to separate the various components found in liquid air. This process relies on the fact that the different components will be transformed from liquid to gas at different temperatures. See Figure 2.26 in the chapter on petroleum for a distillation tower involving crude oil processing. The liquid air is slowly warmed. As the temperature increases the substances with the lowest boiling point become a gas first and can be removed from the remaining liquid: neon at −246.05°C, nitrogen at −195.8°C, argon at −185.7°C, oxygen at −182.95°C, krypton at −152.30°C, and then xenon at −108.13°C.

# Gases from Petroleum

## Helium (He)

Helium (He) is generated underground by the radioactive decay of uranium and thorium. While much of this makes its way to the atmosphere some is trapped underground in petroleum and transfers to any associated natural gas phase. *Crude helium* (> 50% helium together with $N_2$ and small amounts of other gases) can vary from 0% to 4% by volume of natural gas. It is extracted by using low-temperature liquefaction. This involves cooling the natural gas and liquefying it. This allows the escape of the crude helium gas from the liquefied natural gas. $N_2$ and other contaminants are then removed from the escaped crude helium gas to produce gaseous or liquid helium using either pressure swing adsorption (PSA) or a cryogenic distillation method as described above. PSA uses a selectively absorbent bed for contaminants from

*Figure 11.15* Modern plant for air liquefaction and gas separation in Mossel Bay, South Africa showing four small and two large fractional distillation towers.

condensed crude helium to produce helium that's at least 99.997% pure.

Crude helium is concentrated in large quantities in natural gas fields under the U.S. Great Plains, in particular in southwest Kansas and the panhandles of Texas and Oklahoma. The majority of the helium-bearing natural gas sources are located in the U.S., Qatar, Russia, Algeria, Canada, China, and Poland.

Objects that float in the air, such as balloons and blimps, are filled with helium. Liquid helium (–269°C) is used in cryogenics, in particular to cool superconducting magnets employed in magnetic resonance imagers (MRI). Helium is also a purge gas that can remove all other gas from a container without causing reaction. Tanks of helium gas are employed to produce a protective oxygen free atmosphere during welding. Analytical instruments such as gas chromatographic separators typically use helium as the carrier gas.

## Hydrogen (H₂)

Hydrogen gas (H₂) is produced from methane gas by reacting it with steam:

$$CH_4 + H_2O \rightarrow CO + 3H_2. \qquad [11.19]$$

The CO produced is further reacted with steam to generate more H₂ by the reaction:

$$CO + H_2O \rightarrow CO_2 + H_2. \qquad [11.20]$$

Large amounts of H₂ are consumed in both the petroleum and chemical industries. The single largest application is for hydrogenating of cracked long chained hydrocarbons in petroleum to produce gasoline-length chains. Also significant concentrations of H₂ are used to make ammonia ($NH_3$) and hydrochloric acid (HCl). Electrical power stations use hydrogen with its high specific heat to cool their generators as they produce frictional heat while operating.

## SUMMARY

A plant requires from soil these elements in descending order: N, P, K, and S. Sugar cane is the largest crop grown worldwide and tomatoes the most abundant fruit. Maize (corn), wheat, and rice account for 87% of world cereal crop production.

Plants required fixed nitrogen as in $NH_4$. Fixed N is produced from $N_2$ in air by the Haber-Bosch process. $NH_{3(g)}$ is often reacted with nitric acid to produce ammonium nitrate ($NH_4NO_3$). Decomposition of ammonium nitrate is highly exothermic, which expands the gas produced and leads to a blast concussion.

The majority of phosphorous is used to make fertilizer. Sources include guano, phosphorite, and igneous apatite. Most is derived from phosphorite deposits formed from digenesis of organic matter produced by upwelling nutrient-rich water on the edge of continental shelves. Production is dominated by China with the U.S. and Morocco contributing a substantial amount. Resources appear adequate for the foreseeable future.

Potassium is found in a large number of natural minerals. It is, however, utilized from K-salts found in evaporite mineral deposits precipitated out of seawater, termed potash. Canada is the world's largest producer of potash. There is no problem with K reserves far into the future.

Sulfur is a plant nutrient and also used to produce sulfuric acid to process phosphorite rock. Native sulfur is mined from salt domes but most is now produced as a byproduct of crude oil and natural gas processing. The U.S. is the world's largest producer. The resource base of sulfur is extremely large.

Evaporites are deposits formed by evaporation of water and occur in outcrop only in arid terrains. These deposits can form from either seawater or riverwater. Nonmarine evaporites from fresh water produce trona [$Na_3CO_3(HCO_3) \cdot 2H_2O$], thenardite ($Na_2SO_4$), and borax [$Na_2B_4O_5(OH)_4 \cdot 8H_2O$]. Li is obtained from nonmarine brine pools and the mineral hectorite.

Evaporation of seawater produces limestone and dolomite, followed by gypsum and anhydrite, then halite, and finally potassium and magnesium salts. Halite, or table salt, is a common industrial mineral where the Cl is used to make HCl acid. Sr is obtained from the evaporite mineral celestite and substitutes for Ca in gypsum and anhydrite. Br is obtained from seawater and seawater brines. I is derived from I-rich seawater brines.

Gaseous elements and molecules are extracted from air. Dry air contains 78% $N_2$, 21% $O_2$, 0.9% Ar, and lesser amounts of other gases, which include Ne, He, and Kr. $N_2$ is used in the production of $NH_3$ and liquid $N_2$ for freeze-drying foods. The steel industry uses $O_2$ to produce CO gas for the reduction of iron oxides to pure Fe metal. Ar is used as a filler gas in light bulbs while Ne is used to produce fluorescent lamps. The other gases found in air are separated and used for their unique properties. Helium is obtained from natural gas as this is a less expensive source. It has found a great use in lighter than air balloons and as liquid He in cryogenics.

## KEY TERMS

| | |
|---|---|
| apatite | Frasch process |
| borax | Haber-Bosch process |
| celestite | Le Chatelier's principle |
| cereal crops | potash |
| crude helium | sabkha |
| eutrophication | saltpeter |
| fixed nitrogen | trona |

## PROBLEMS

1. Determine the wt% of N, P, and K as well as the mole ratios of N and K to P in a bag of 30-20-10 fertilizer.
2. Using the values in **Table 11.2** calculate the molar ratio of $Cl^-$ to $HCO_3^-$ in Borax Lake, CA.
3. Using the vol % of the four main components in dry air calculate their concentration in ppm by weight. Assume air is an ideal gas where pressure × volume = number of moles × gas constant × absolute temperature.
4. Verify the statement in the book that "At 65 ppm seawater contains almost 100 billion metric tons of bromine." There are 1.31 billion $km^3$ of seawater in the ocean and its average density is about 1.05 kg $m^{-3}$.

# REFERENCES

Blatt, H., Tracy, R. J., and Owens, B. 1996. *Petrology: igneous, sedimentary, and metamorphic*, 2nd edition. San Francisco: W. H. Freeman and Company.

Cathcart, J. B., 1989, The phosphate deposits of Florida, with a note on the deposits in Georgia and South Carolina, USA, in *Phosphate rock resources*, A. J. G. Notholt, R. P. Sheldon and D. F. Davidson, *eds*, Cambridge Univ. Press, 2:62–69.

Food and Agriculture Organization (FAO) of the United Nations. 2006. Fertilizer use by crop. *FAO Fertilizer and Plant Nutrition Bulletin* 17.

Harvie, C. E., Weare, J. H., Hardie, L. A., and Eugster, H. P. 1980. Evaporation of seawater; Calculated mineral sequences. *Science*, 208:498–500.

Martin, R. G. and Bouma, A. H. 1978. Physiography of the Gulf of Mexico. In Bouma, A. H., Moore, G. T., and Coleman, J. M. (eds.) *Framework, facies, and oil-trapping characteristics of the upper continental margin*. Amer Assoc Petrol Geol Stud Geol 7: 3–19.

Piper, D. Z. 2001. Marine chemistry of the Permian Phosphoria Formation and Basin. *Southeast Idaho Econ Geol* 96:599–620.

Soh, K. G. 1997. *Fertilizer demand and crops*: International Fertilizer Industry Association, presented at the IFA Agro-economics Meeting, Beijing, China, 1997.

Steinfeld, H., Gerber, P., Wassenaar, T., Castel, V., Rosales, M., and de Haan, C. 2006. Livestock's long shadow: Environmental issues and options (p. 87, Table 3.3). Rome: Food and Agriculture Organization of the United Nations. Available online at www.fao.org/docrep/010/a0701e/a0701e00.htm.

U.S. Geological Survey, 2012, *Mineral commodity summaries 2012*, U.S. Government Printing Office, 198 p.

# Part Four

## Water and Soil Resources

*"We abuse water and soil because we regard them as commodities belonging to no one. When we understand they are part of a global community to which we belong, we may begin to treat them with more respect."*

— *modified after Aldo Leopold, 1949*

Water, along with soils, can be considered our most important natural resources (**Figure PT**4.1). All living things on the earth require water for survival. Without soil people can't grow the food needed to sustain them. They are both under threat and need to be protected and carefully managed. What are the best management practices to conserve them? To answer this question one must understand the interactions of soil with water.

When water resources are considered it is generally freshwater that is of concern. Water is unevenly distributed on the earth and is not a fixed resource like the other resources considered in this text. Its supply

*Figure PT*4.*1* Water resources like this stream need to be protected to ensure its quality and availability is not compromised.

is variable depending on both location and time. The time frame can be seasonal or over much longer time frames if global climate change is considered. Water resource problems deal with both supply and pollution concerns from a flux of water as outlined in a general sense in the water cycle (see introductory chapter). It is, therefore, necessary to understand what influences not only the composition of freshwater but also its supply.

The world's supply of unpolluted freshwater is steadily decreasing due to global warming and aquifer depletion as humankind's demand for it increases. In some areas the problem is becoming acute. This occurs because of increased population densities and the stress that needed increased agricultural production puts on this water as well as its soils. Also as humankind increases the amount of meat it consumes this increases the amount of agricultural production that is required per person. This is a particular problem because the major use of freshwater is to grow food. If its availability is limited in an area, the amount of food that can be grown is also limited.

Soil formation is directly dependent on the presence of water. In arid desert environments little soil is present. Environments where large amounts of water have fluxed through the solid surface layer of the earth produce thick soils. The best soils for the production of food are produced by intermediate fluxes of water. Primary igneous and metamorphic minerals must be broken down with the aid of water to produce a mature soil, but the amount of water must not be so much that valuable nutrients are washed away.

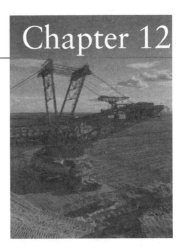

# Chapter 12

# The Distribution and Movement of Water

## Global Distribution of Water

As planets go, the earth is unusual because it has a large ocean covering 71% of its surface to an average depth of 3,800 m. Of the water on the earth 96.5% is in the ocean with nearly an additional 1% in saline groundwater. This means that only 2.5% of the earth's water is fresh. From a resource perspective, this is the water of importance. **Table 12.1** indicates that about 69% of freshwater is tied up in ice caps, glaciers, and permanent snow with another 30% in groundwater. This

*Table 12.1* LOCATION AND ESTIMATED SIZE OF RESERVOIRS OF WATER ON EARTH.

| WATER SOURCE | VOLUME (km³) | FRESH WATER % | TOTAL WATER % |
|---|---|---|---|
| Oceans, seas, and bays | 1,338,000,000 | – | 96.5 |
| Ice caps, glaciers, and permanent snow | 24,064,000 | 68.7 | 1.74 |
| Groundwater | 23,400,000 | – | 1.7 |
|    Fresh | 10,530,000 | 30.1 | 0.76 |
|    Saline | 12,870,000 | – | 0.94 |
| Soil moisture | 16,500 | 0.05 | 0.001 |
| Ground ice and permafrost | 300,000 | 0.86 | 0.022 |
| Lakes | 176,400 | – | 0.013 |
|    Fresh | 91,000 | 0.26 | 0.007 |
|    Saline | 85,400 | – | 0.006 |
| Atmosphere | 12,900 | 0.04 | 0.001 |
| Swamp water | 11,470 | 0.03 | 0.0008 |
| Rivers | 2,120 | 0.006 | 0.0002 |
| Biological water | 1,120 | 0.003 | 0.0001 |
| Totals | 1,386,000,000 | 100 | 100 |

*Data from:* Gleick, P. H., 1996: Water resources. In *Encyclopedia of Climate and Weather*, ed. by S. H. Schneider, Oxford University Press, New York, vol. 2, pp. 817–823.

leaves about 1% of freshwater in permafrost, soil moisture, lakes, rivers, and the atmosphere combined. While humankind does not directly use water in ice caps and glaciers, it is still an important resource in two ways.

The first is that many rivers are supplied primarily by melt water from glaciers (**Figure 12.1**). In particular, in Asia the Himalayan glaciers feed rivers like the Brahmaputra, Ganges, Indus, Mekong, and Yangtze. These rivers are the freshwater supply of nearly half the world's population. In the U.S., melt water from the snow pack in the Sierra Nevada Mountains feeds streams that are the freshwater source for much of California. Melt water from Rocky Mountain snows feeds aquifers and streams on which many communities east of the Rockies depend. With global warming the melt water produced during the summer has been increasing from increased melting of ice in mountain glaciers. However, at present with a smaller area of ice and snow cover the amount of summer melt water is starting to decrease with time. Also the peak discharge that used to occur in middle/late summer is now occurring in late spring/early summer.

The second way this ice is important is that the current melting of ice is causing sea level throughout the world to slowly rise. This is leading to increased flooding of many coastal landmasses. By 2050 it is forecasted that increased sea level will start to cause significant seawater invasion of coast freshwater aquifers (see below).

## Hydrologic Cycle

Water on the earth is in a constant state of motion moving from one location to another. This flux of water for the whole earth was described in Figure 1.6 in the introductory chapter where a system of reservoirs and connected fluxes was considered and is termed the water cycle. An important concept of the water cycle is the *residence time* of water in a particular reservoir. The residence time is the average length of time water spends in the reservoir. At steady state the size of the reservoir does not change with time. The residence time at steady state is calculated from the size of the reservoir divided by the flux into or out of the reservoir:

$$\text{Residence time (yr)} = \frac{\text{Reservoir size (g)}}{\text{Flux (g yr}^{-1})}. \quad [12.1]$$

Average residence times for the ocean and reservoirs of freshwater are given in **Table 12.2**.

*Figure 12.1* Hubbard Glacier, Alaska photographed calving, that is breaking away a mass of ice to form an iceberg.

*Table 12.2* RESIDENCE TIMES OF WATER IN THE INDICATED RESERVOIR.

| RESERVOIR | AVERAGE RESIDENCE TIME |
|---|---|
| Oceans | 3,200 years |
| Glaciers | 20 to 100 years |
| Seasonal snow cover | 2 to 6 months |
| Soil moisture | 1 to 2 months |
| Groundwater: shallow | 100 to 200 years |
| Groundwater: deep | 10,000 years |
| Lakes | 50 to 100 years |
| Rivers | 2 to 6 months |
| Atmosphere | 9 days |

Steady state should not be confused with equilibrium. For instance, water is transported between the ocean and the atmosphere by evaporation and precipitation. If these two reservoirs were in equilibrium the atmosphere would need to be saturated with water, and therefore at 100% humidity, which typically it is not. Also at equilibrium, the net transfer of water between the reservoirs would be zero, which it is not.

The movement of water transports a large amount of thermal energy. The water is heated at the equator, evaporates, and water vapor moves toward the poles where heat is released when it rains or snows. Water vapor, unlike dry air, has a high heat capacity. As a result water vapor, and not dry air, can transport large amounts of thermal energy. Heated water transports energy in both the atmosphere (2/3 of the total) and the ocean (1/3 of the total heat transfer). The ocean heat transfer is why the United Kingdom, warmed by the North Atlantic Current, has a milder climate in winter than interior Canada at the same latitude.

## Rainfall Patterns

Water in the atmosphere is carried by air masses. These deflect to the right in the Northern Hemisphere and to the left in the Southern Hemisphere due to the *Coriolis effect* on a rotating earth. To understand this, imagine you are 50 km above the earth at the equator observing an air mass on the surface of the rotating earth. At the equator the air mass and the ground below are moving from west to east or to the right as you face north at the speed of the earth's daily rotation. At the equator this is:

$$\frac{\text{Circumference of the earth}}{1 \text{ day}} = \frac{40,000 \text{ km}}{24 \text{ hr}} = 1,667 \text{ km/hr.}$$

[12.2]

As **Figure 12.2** shows if you make this same observation at a more northerly latitude this rotation speed decreases because the more northerly the latitude the smaller the circular distance around the earth becomes, decreasing to zero at the North Pole.

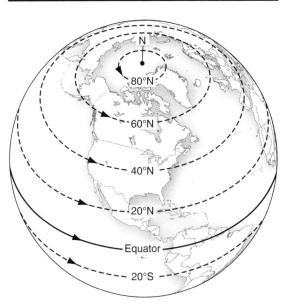

*Figure 12.2* Circles of latitude on a rotating earth showing the greater distances traveled, in the same amount of time, during a rotation of the earth the nearer the equator one is.

Consider that the air mass at the equator, which is moving to the east at the speed of the rotating earth at the equator, now moves directly north. It is still moving to the east at 1,667 km/hr but the land below it is moving more slowly to the east. An observer on the earth's surface will, therefore, see the air mass veering to the east as it goes north even though relative to the reference frame of the air mass it is moving straight north.

Now consider an air mass at the North Pole. As it travels south the land below is traveling to the east at greater rates. To an observer on the earth's surface the air mass appears to move to the west because the air mass is moving to the east at a slower rate. This is shown with the arrows in **Figure 12.3**. Because of the earth's rotation, when view from a location on the earth's surface air masses deflect to the right as one is facing in the direction they are moving in the Northern Hemisphere and to the left in the Southern Hemisphere. This is the Coriolis effect.

Shown in **Figure 12.4** are the general idealized surface wind patterns that move air masses on the earth. Note that the patterns in the Southern Hemisphere are similar to the Northern

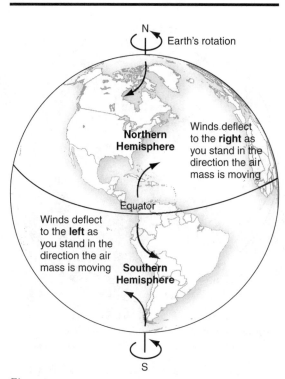

*Figure 12.3* The Coriolis effect causes an air mass to deflect to the right in the Northern Hemisphere and to the left in the Southern Hemisphere.

Hemisphere but in an opposite sense because of the Coriolis effect. Water-laden air convects heat from the equator at 0° latitude to the North Pole at 90°N latitude and brings cold air from the pole to the equator. Three interconnected convection cells, the *Hadley*, *Ferrel*, and *Polar cells*, as shown in the figure rather than one large convection cell, transport the air. This occurs because the heights of the cells are only ~12 km, limited to the top of the troposphere, while the distance from the equator to the pole is about 10,000 km. With this large length-to-height ratio a single cell becomes unstable producing the three-cell configuration. Consequently, the thickness of the convection cells is greatly exaggerated in Figure 12.4. These cells control wind patterns and therefore water precipitation patterns on the earth.

At the equator, in the Hadley cell, heat from the sun expands the air and evaporates a large amount of water from the ocean. This low-density, high humidity air rises. Because air cools

on ascent, it can hold less water. Large amounts of precipitation, therefore, occur at the equator producing climatic conditions that maintain the earth's rainforests. Near 30°N latitude air in the atmosphere sinks as high pressure develops from more dense cooling air. Sinking air warms as it compresses. Its humidity decreases because it can hold more water and precipitation is unlikely. Large deserts are located near the 30°N latitude. Similar events occur in the Southern Hemisphere.

Cold dense, and therefore high pressure, air sinks at the North Pole and is transported to the south. As the air moves southward it warms, expands, and becomes a low-pressure air mass that rises near 60°N latitude. Parts of this low-pressure air mass break away from the front established at 60°N and are transported into the Ferrel cell, producing stormy low-pressure weather systems. These weather systems are generally transported from west to east because of the west to east component of the prevailing surface winds in the Ferrel cell. This is the zone of the Westerlies, as shown in Figure 12.4.

Mean annual rainfall from 1950 to 2000 at locations on the earth's surface is outlined in **Figure 12.5**. Rainfall is averaged for 50 years because it can be variable from year to year at a particular location. Consistent with the idealized model of low and high pressures given in Figure 12.4, areas of high precipitation are centered on the equator and the areas of low precipitation occur at 30°N and S latitude and at the poles. The correlation is not perfect because of other factors that are involved. For instance, the southeastern U.S. and southeastern China have significant rainfall that would not be anticipated by the pattern of air circulation given in Figure 12.4. This, in part, is due to the hurricanes and typhoons that break the normal weather pattern in these regions by bringing oceanic water westward and then on shore. This occurs because these regions lie to the west of an ocean and are in the zone of the Easterlies.

The average annual precipitation on the earth is about 1.0 m per year. The wettest place on the earth is Lloro, Colombia near the equator at 5.5°

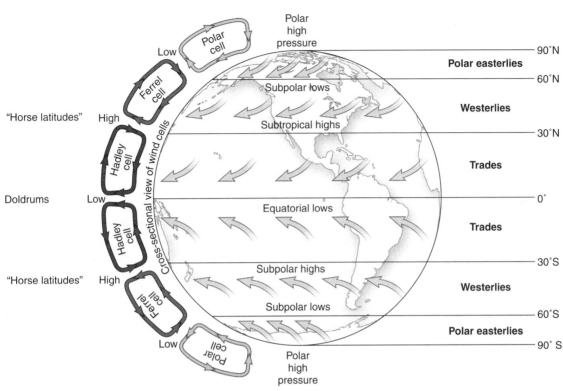

*Figure 12.4* Idealized general global surface wind pattern on the earth which includes the Polar easterlies, Westerlies and Trades. These winds are part of the three convection cells present in each hemisphere: the Hadley, Ferrel, and Polar cells.

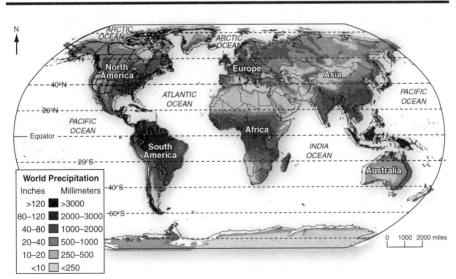

*Figure 12.5* Mean annual precipitation amounts on the earth in mm per year, from 1950 to 2000. (Data from: GWSP Digital Water Atlas (2008). Map 36: Annual River Discharge (V1.0). Available online at http://atlas.gwsp.org.)

North latitude with 14.3 m of rain per year on average. In a given year this can be surpassed by Cherrapunji and Mawsynram in northeastern India because of seasonal monsoon rains. A *monsoon* is a seasonal sea breeze developed by heating the continental landmass and increasing its temperature in the summer relative to the ocean. This occurs because of the greater ability of water to retain heat. The low atmospheric pressure produced over the landmass from the expanded high temperature air draws in moist ocean air. Large amounts of rain fall as the moist ocean air cools on ascent over the continent. The driest place on earth is in the Dry Valleys of Antarctica where there has been no precipitation for nearly 2 million years. Their cold climate means little water is contained in air masses and the sinking winds from the adjacent mountains move the air masses away from the valleys.

**Figure** 12.6 gives the average annual precipitation in the U.S. from 1961 to 1990. The average rainfall is about 3/4 of a meter or 25% less than the worldwide average. Note the significant amount of rain in the southeastern U.S. and drier conditions in the west except for the Pacific Northwest. This occurs in part because moisture-rich air from the Gulf of Mexico moves north into the zone of the Westerlies, precipitating water as it moves to the east. In the Pacific Northwest a large amount of rain falls because low-pressure systems from the Pacific Ocean encounter the Cascade Mountains. The cooling of the air masses as they rise over the Cascades causes them to supersaturate with moisture and

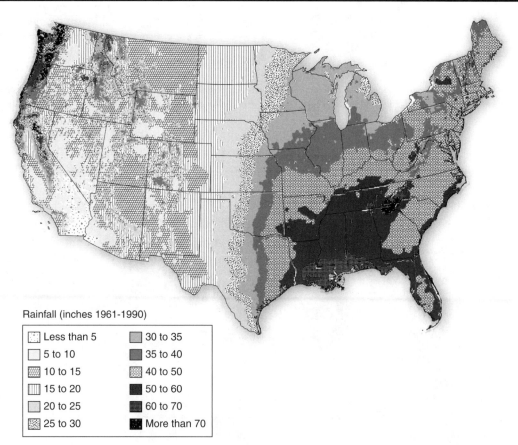

Rainfall (inches 1961-1990)

| | |
|---|---|
| Less than 5 | 30 to 35 |
| 5 to 10 | 35 to 40 |
| 10 to 15 | 40 to 50 |
| 15 to 20 | 50 to 60 |
| 20 to 25 | 60 to 70 |
| 25 to 30 | More than 70 |

*Figure* 12.6 Average annual precipitation in the U.S. from 1961 to 1990. (Data from: NOAA and NRCS.)

produce rain and snow. The low amount of moisture that falls east of the Sierra Nevada and Rocky Mountains is attributed to loss of water from air masses on cooling when rising over these mountains. As the air masses descend on the eastern side of the mountains they warm, lowering their relative humidity, and giving rise to rain shadow deserts.

These precipitation patterns are punctuated by episodic weather events that can cause dramatic changes in yearly water precipitation amounts. The tropical eastern Pacific Ocean warms during an *El Niño event* and cools during a *La Niña event* that occur roughly once every five years. El Niño's pool of warm ocean water causes thunderstorms along South America's west coast. During a La Niña, drought conditions develop along South America's west coast. In the northern U.S., winters during an El Niño are warmer and drier than average whereas the southern U.S. experiences wetter winters. During a La Niña the reverse occurs between the northern and southern U.S. The weather changes are not confined to the Western Hemisphere as El Niño and La Niña are both worldwide weather events.

Particularly in the U.S. Midwest, persistent, multiyear droughts and flooding have important implications for managing water resources. A drought can be defined as the deficiency of normal precipitation over a prolonged period of time, usually for at least one season. Droughts are temporary conditions that change with time. Drought leads to reduced surface and groundwater levels often with the loss of crops, rangeland vegetation, and increased fire risk. The Midwestern Dust Bowl in the 1930s occurred during a multiyear drought. In 2012 as well as in the early 2000s dry conditions in the U.S. Midwest combined with depleted well water caused considerable problems for agricultural producers. These unpredictable precipitation patterns argue for multiyear water storage systems together with effective water distribution systems.

Increased global temperatures are having a large effect on world climate. Rainfall patterns are changing with more heavy downpours, increased evaporation from the ocean, and increased water loss from plants. More hurricanes and typhoons are occurring, permafrost is thawing, and changes in growing seasons are taking place with some areas having more seasonal flooding and others more severe droughts.

## Evapotranspiration

The rain that falls on a land surface either evapotranspirates, runs off the earth's surface into a stream, lake, or the ocean, or infiltrates into the ground below the earth's surface. *Evapotranspiration* is the sum of water released to the atmosphere by evaporation from the land surface and water transpired by plants. Transpiration occurs when plants release water vapor into the atmosphere through pores on their leaves and stems that was originally absorbed by their root system. This allows a continuous transport of nutrients in the water from the roots to the leaves.

The amount of evapotranspiration at various locations on the earth are given in **Figure 12.7**. On most land surfaces the amount of water transpired by plants is larger than the amount evaporated from wet surfaces. A large mature tree may transpire a number of cubic meters of water through its leaves on a hot, dry day. The rate of transpiration depends on the number and size of plants, availability of water to the root system, temperature, the degree to which leaf pores are open, and the humidity of the atmosphere surrounding the leaf. In the Amazon transpired water is effectively recycled in rain leading to the rain forest climate present.

The rate of evaporation of water also depends on a number of factors. These include the temperature of the air and water as higher temperatures increase evaporation. The humidity of the air is also important with the rate of evaporation higher in drier air. The greater the air flow across the wet surface the greater the ability to transport the water away and bring in new air that can absorb water.

To evaporate water requires a significant amount of heat, about 2,500 joules per gram. This explains

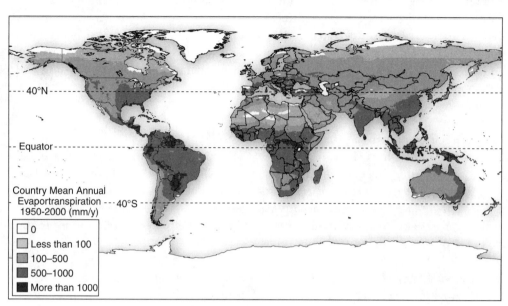

*Figure 12.7* Mean annual evapotranspiration between 1950 and 2000. (Data from: GWSP Digital Water Atlas (2008). Map 37: Annual River Discharge (V1.0). Available online at http://atlas.gwsp.org.)

why people perspire to stay cool and use fans to speed the process. Spraying a fine mist of water in the air will make it evaporate faster by increasing the air-water surface area.

# Groundwater

Groundwater is water present in the subsurface. This includes moisture in the soil as well as water below the water table if one exists. The *water table* is the upper surface in the earth below which all pore spaces in the sediments and rocks are filled with water. The bottom surface of this water-filled section consists of nonporous rocks. Water can also exist in *aquifers*. These are rock and sediment units below the surface that can hold and transport water. If the aquifer is bounded by a water table as the upper surface this would be referred to as a water-table aquifer.

## Vadose Zone

Above any water table there exists a layer that extends to the earth's surface where pores and fractures contain air ± water. This layer is termed the unsaturated zone because it is unsaturated

with water, zone of aeration because it contains air, or *vadose zone* (**Figure 12.8**). The vadose zone typically consists of three layers. Nearest the earth's surface is a belt of soil moisture or soil water. This is water that has been captured from a recent rainstorm or is present from humankind's addition. As a function of time, this layer loses its water as it is transpired by plants and evaporates into the atmosphere or as it slowly percolates downward through pores and fractures to a water table below.

## Capillary Fringe

Just above the water table there is a layer consisting of a fringe of interconnected water-filled pores in a layer where pores are not filled with water. In the water-filled pores, water from below the water table has risen and wetted the mineral surfaces it contacts. Water molecules are more attracted to the charged mineral surface than to other water molecules. This produces a surface tension that pulls water upward until the weight of the lifted water present is too great to be overcome by the surface tension force.

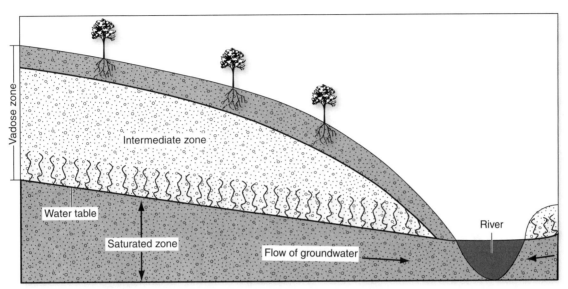

*Figure 12.8* Cross-section through the earth showing the vadose zone and saturated zone below the water table. Water in the vadose zone includes soil moisture, the intermediate zone, and a capillary fringe just above the water table.

This is similar to what occurs in a narrow diameter tube. The height water rises in a narrow tube when one end is immersed in water increases as the diameter of the tube decreases (**Figure 12.9**). With smaller pore diameters there is more mineral surface for the water to be attracted to relative to the water that rises in the interconnected pores. Therefore, what is termed a *capillary fringe* in silty material can be 1 meter high but is only 1 cm in coarse sand with its larger pore size.

The area between the capillary fringe and the belt of soil moisture is an intermediate vadose zone. This zone contains little water except at times where water is being actively transported through the zone to the water table and capillary fringe.

## Aquifers

When considering groundwater, the primary unit from a resource perspective is the aquifer. Aquifers are underground layers of sediments or rock that can store and transmit water in significant quantities (**Figure 12.10**). They have a significant *porosity*, that is void spaces, and this void space is interconnected and generally filled with

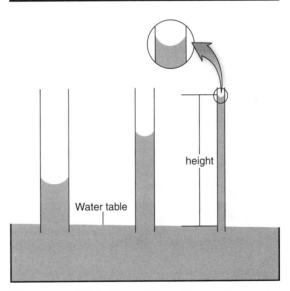

*Figure 12.9* Tubes of three different diameters immersed in water. Surface wetting produces a meniscus and "pulls" water up from the water table into the soil above. The interconnected pores act as capillary tubes. As the diameter of the tubes, that is pores, decreases with decreasing grain size the height of water infiltration due to capillary action increases.

water. This interconnectivity allows water to be transmitted through an aquifer. The ability of an aquifer to transmit water is termed its *permeability*. Layers that can't store or transmit significant quantities of water have low permeability and are known as confining layers or *aquicludes* (**Table 12.3**).

As shown in **Figure 12.11** two important kinds of aquifers are water-table and confined aquifers. In a water-table aquifer the upper boundary of the aquifer is a water table. In a confined aquifer the upper boundary is a confining layer. On the right side of the figure, water from the surface is infiltrating downward through the pores in sediments and rocks until the water table is reached filling the pores at the water table with water. The added water increases the height of the water table unless it is transported through the aquifer more rapidly

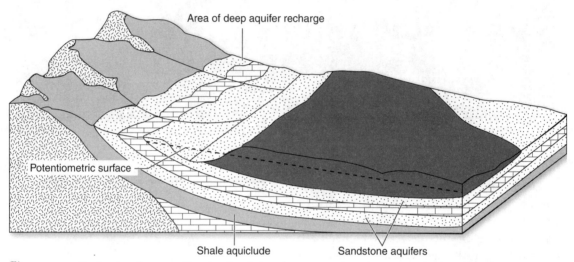

*Figure 12.10* Regional sandstone aquifers showing their recharge areas where they outcrop in elevated areas on the earth's surface.

*Table 12.3* TERMS USED IN THE STUDY OF AQUIFERS.

| TERM | DEFINITION |
|---|---|
| Aquifer | Rock or sediment permeable to water below the water table or a confining layer that contains crevices and pores full of water. (Water can take on the order of >200 years to be transported from the location of surface recharge). |
| Water-table aquifer | An aquifer whose top surface is the water table. |
| Confined aquifer | An aquifer that is overlain by a bed that does not allow significant passage of water, a confining layer. |
| Aquiclude | A low permeability rock unit that does not allow the passage of water. |
| Recharge area | Area where rainfall and surface water can infiltrate into an aquifer. |
| Discharge area | Area where water can discharge from an aquifer. |
| Artesian well | Well from which water flows freely without pumping. |
| Potentiometric pressure surface | The surface that defines how high water from a confined aquifer will rise in a standing pipe from the aquifer due to the aquifer's fluid pressure. |

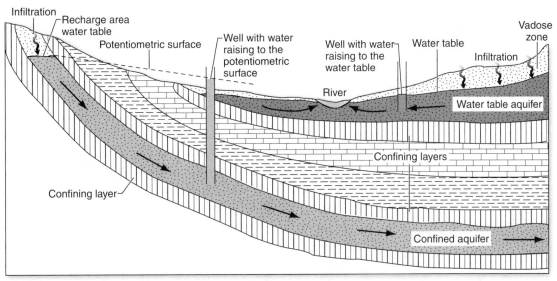

*Figure 12.11* Confined and water-table aquifers in somewhat folded sedimentary rocks. Note the water levels in wells penetrating the confined and water-table aquifer. The arrows give the flow direction of water in the aquifers with time.

than it receives water. This water is typically discharged into a river.

The confined aquifers shown in Figures 12.10 and 12.11 receive their water from an elevated recharge area and the water flows downward and to the right. Because the potentiometric surface of the confined aquifer is above the land surface, water in the well drilled into the aquifer flows freely to the surface without pumping, resulting in an artesian well. The *potentiometric surface* is the surface that water in an aquifer would rise to if a vertical pipe is connected upward from the aquifer at each location in the aquifer. This rise of water occurs because all the water in the aquifer is interconnected. Its pressure at any point in the aquifer is given by the elevation of the water table in the recharge area minus a small pressure loss due to transport of water from higher to lower pressure in the aquifer as it moves.

## Ogallala Aquifer

The Ogallala Aquifer is a large water-table aquifer located in the western Great Plains of the U.S. (**Figure 12.12**). It covers parts of eight states from South Dakota to Texas. Erosion of the Rocky Mountains 2 to 6 million years ago provided river and wind-blown sediments to a basin to the east producing the Ogallala Formation. The Ogallala Formation varies from a couple of meters to 300 meters in thickness consisting of coarse-grained sedimentary rocks at the bottom, grading upward into finer-grained sedimentary rocks at the top. With the melting of Pleistocene continental ice sheets the influx of melt water into the formation created the Ogallala water-table aquifer. The *Pleistocene* is the geological epoch starting 2.6 million years ago and lasting until 12,000 years ago consisting of the most recent time period of repeated glaciation. Present day recharge to the Ogallala Aquifer is negligible because of the dry climate in the region.

The Platte and other rivers in the watershed are below the water table of the Ogallala Aquifer in places and therefore the rivers extract water from the aquifer. A *watershed* is a term used for a region bounded by divides where water drains into a particular body of water. Humankind is presently extracting more water from the Ogallala Aquifer than the entire flow of the Colorado River, nearly all for agriculture. Depending on location the water table is dropping 15 cm to 1 meter a year, a situation that cannot be sustained. If present trends continue the aquifer will not be able to supply sufficient water in many areas in 25 to 40 more years.

### Nubian Sandstone Aquifer System

The Nubian Sandstone Aquifer System extends over 2.2 million km² beneath the northeastern Sahara desert in Egypt, Libya, Sudan, and Chad (**Figure 12.13**). It is the largest aquifer system in the world holding 150,000 km³ of water in the Dakhla Basin of Egypt and the Kufra Basin of Libya, Chad, and Sudan. Withdrawal by Egypt = $1,030 \times 10^6$ m³/yr, by Libya = $2,370 \times 10^6$ m³/yr, by Sudan = $400 \times 10^6$ m³/yr, and Chad withdraws very little. The high Libya water withdrawal is due to the Great Man-Made River project started in 1984. It consists of more than 1,300 wells that supply a network of underground pipes and aqueducts carrying $6.5 \times 10^6$ m³/day of freshwater to Tripoli and other Mediterranean coastal cities in Libya.

The Nubian Sandstone Aquifer System consists of both Nubian Sandstone and post-Nubian sediments. It has only a small recharge because of the arid climate in the region. Water is stagnant with little or no flow and is thus termed fossil water. The average age of the aquifer water is greater than 20,000 years with some as old as 1 million years. In the south, water is contained in medium- to coarse-grained Nubian Sandstone overlying Precambrian basement. The sandstone is flat lying to gently dipping and confined by overlying

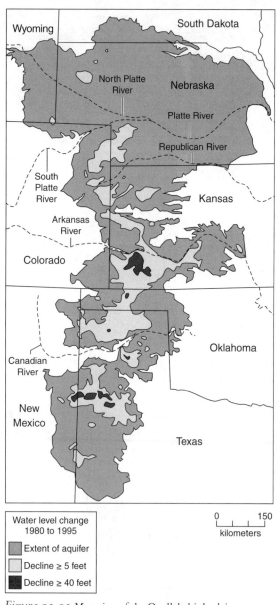

*Figure 12.12* Map view of the Ogallala high plains aquifer. (Data from: the U.S. Geological Survey.)

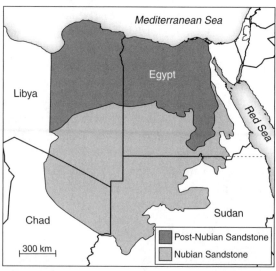

*Figure 12.13* The Nubian Sandstone Aquifer System (NSAS) of northeastern Africa consisting of Nubian and post-Nubian sandstone.

alluvial deposits. North of the 26th latitude is the unconfined post-Nubian aquifer consisting of younger Tertiary continental sands, gravels, and silt deposits together with some carbonate rocks. The Tertiary is the geological period from 65 to 2.6 million years ago just before the Pleistocene.

Although the Nubian Sandstone Aquifer System contains a large amount of water only about 10% to 15% is easily extracted. Concerns have arisen because once it is used the water in the aquifer system will be gone, much like the petroleum in the region. However, at present extraction rates this is not likely to happen very soon.

## Quaternary Aquifer System of Northern China

Chinese topography is elevated on the Qinghai-Tibet Plateau in the west of the country with average elevation of 4,500 m. Traversing east the elevation decreases where, from north to south, are located the Inner Mongolia, Loess, and Yunnan-Guizhou Plateaus with average elevations ranging from 1,000 to 2,000 m. East of these plateaus are the coastal plains where elevations decrease to 50 m above sea level on average.

The coastal plain Hai River basin south of Beijing consists of river overbank and coastal deposited sediments of Quaternary age (2.6 million years to present). The coastal sediments were deposited in deltas, lagoons, shoreline bays, and beaches. These sediments have been shed from the Taihang Mountains on the eastern edge of the Loess Plateau forming an aquifer system termed the Quaternary Aquifer of North China (**Figure 12.14**). The aquifer system is recharged by runoff from the Taihang Mountains.

High demand is currently being placed on the Quaternary Aquifer of North China due primarily to increased grain production, but also increased municipal water consumption and water supplied for industrial use. As a result the aquifer is being

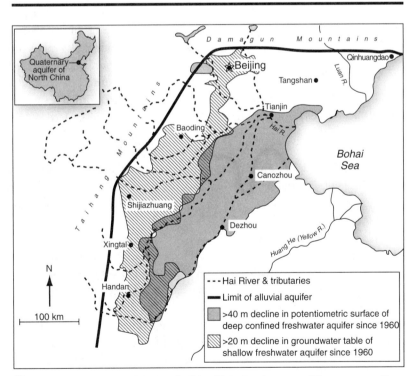

*Figure 12.14* Map view of the North China Plain showing areas of Quaternary aquifer depletion. (Adapted from Foster, S., Garduni, H., Evans, R., Olson, D., Tian, Y., Zhang. W., and Han, Z., 2004, Quaternary Aquifer of the North China Plain-assessing and achieving groundwater resource sustainability, *Hydrogeology Journal*, v. 12, pp. 81–93.)

rapidly depleted. This is causing the water table of the shallow water-table aquifer to be lower about 0.5 m each year and the potentiometric surface of a somewhat deeper confined aquifer to decrease about 1 m/year.

In 1988 extraction of groundwater increased and exceeded recharge by 8,800 million m³/year. Since then the water quality has been deteriorating. China's grain production peaked in 1998. The decrease in grain production since 1998 has been attributed to less water availability from aquifers like the Quaternary Aquifer of North China. Continued nearly uncontrolled extraction of water from beneath the North China plain will lead to severe future water and food shortages in the region.

## Groundwater Supply Problems

A crisis concerning water can occur if the supply of water in an area is less than its demand. In some areas in the world demand has increased leading to difficulties of supply and in others the supply has decreased either due to depletion of the source of water or droughts brought on by climate change.

### Overdraft and Ground Subsidence

Although population in the U.S. increased almost 90% from 1950 to 2000, total freshwater withdrawals increased 127% during the same period. Total water withdrawals, water withdrawn for thermoelectric power, and water withdrawals for irrigation peaked about 1980 in the U.S. and then stayed relatively constant to the present. However, water withdrawals for the public water supply and self-supplied domestic withdrawals have increased steadily since 1950. Self-supplied industrial water use is the only category to decline consistently during the past two decades.

If the supply of water to a water-table aquifer is less than its discharge the water table will be lowered, decreasing the size of the aquifer. Discharge from the aquifer can be natural (i.e., to a river) as shown in Figure 12.12, or it can be from wells as shown in **Figure 12.15**. If the amount of

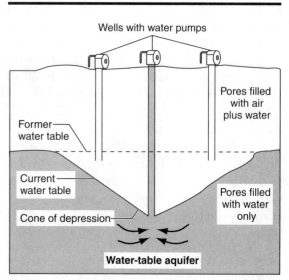

*Figure 12.15* Cross-section through the earth showing the possible effects of extracting water from a water-table aquifer.

water extracted from wells is substantial relative to the aquifer's recharge, the water table around the wells can be lowered. This produces a *cone of depression* of the water-table surface causing nearby shallower producing wells to become dry.

Removal of groundwater can lead to the subsidence of the land surface as the water pressure that helped hold open pore spaces is decreased. This lowering of the land surface is permanent as has occurred in many parts of the San Joaquin Valley in California. Therefore, recharging the aquifer until the groundwater returns to its original levels does not result in recovery of the land-surface elevation. In some cases 15% to as much as 25% of an aquifer's thickness is lost.

Depletion of aquifers is a significant problem in the developing world as major metropolises grow. A number of cities including those in **Table 12.4** have been affected. They extract more water from beneath the city than the local aquifer's naturally recharge. Note that under Long Beach and Los Angles, California it was the extraction of petroleum rather than water that has caused the subsidence. Most of the affected cities are located on the coast where large amounts of sediments have been

*Table 12.4* MAXIMUM SUBSIDENCE AND EXTENT OF SUBSIDENCE AREA IN SOME MAJOR CITIES DUE TO WATER EXTRACTION.*

| CITY | MAXIMUM SUBSIDENCE (cm) | AREA AFFECTED (km²) |
|------|------|------|
| Long Beach/Los Angeles, CA | 900 | 50 |
| San Joaquin Valley, CA | 880 | 13,500 |
| Mexico City, Mexico | 850 | 225 |
| Tokyo, Japan | 450 | 3,000 |
| Osaka, Japan | 300 | 500 |
| Houston, TX | 270 | 12,100 |
| Shanghai, China | 263 | 121 |
| Niigata, Japan | 250 | 8,300 |
| New Orleans, LA | 200 | 175 |
| Taipei, China | 190 | 130 |
| Las Vegas, NV | 180 | 200 |
| Davis, CA | 120 | 100 |
| Bangkok, Thailand | 100 | 800 |
| Venice, Italy | 22 | 150 |
| London, England | 30 | 295 |

* Also includes Long Beach/Los Angeles, CA where petroleum was extraction.

deposited. Many of these aquifer-forming sediments have remained unconsolidated and are easily shifted. In cities like New Orleans, Bangkok, and London the aquifers below the city are in river sediments and in Las Vegas, NV and Davis, CA, basin fill sediments contain the aquifer. In the case of Mexico City it is lake sediments. Differential changes in the elevation and slopes of stream beds, drainage ditches, and sewer pipes occur during land subsidence. The amount and direction of the flow of water can change. Land subsidence can also shift buildings, roads, bridges, oil and water wells, and levees causing them to fail.

**Sinkholes.** A *sinkhole* is a depression or hole in the surface topography caused by surface rocks catastrophically filling a cavity below. These depressions can be several hundred meters in width and depth (**Figure 12.16**). Sinkholes are common across the earth's surface. They have been observed to form both suddenly and gradually. Formation of sinkholes occurs by continual removal of slightly soluble bedrock, limestone, or salt beds by water below the land surface. For limestone the dissolution reaction can be written as:

$$H_2O + CaCO_3 \rightarrow Ca^{2+} + HCO_3^- + OH^-. \qquad [12.3]$$
$$\text{limestone}$$

When surface waters undersaturated with calcite are transported through limestone, calcite in the limestone dissolves. Over time the porosity increases and large open spaces filled with water are produced in the rock. These can coalesce via continued dissolution producing a water-filled cavity in the formation. Generally, collapse of the cavity occurs with a lowering of the water table so the water no longer helps support the overlying rock. In some instances the cavities are not too extensive and the material between cavities is strong enough to support the cavity roofs. In this instance if the water table drops below the dissolving limestone a set of interconnected limestone caverns is produced.

A *karst* topography forms from the production of sinkholes (**Figure 12.17**). The landscape typically contains hundreds of sinkholes in a confined area with vertical shafts and disappearing streams. From above it looks pockmarked with few surface streams visible because most water moves in the subsurface.

### Saltwater Incursions

Encroachment of salt water into fresh groundwater aquifers can be a problem for land areas along the ocean. Shown in **Figure 12.18** is a cross-section through a typical coastal region with a barrier island. Barrier islands are relatively narrow ridges of old submerged coastal sand dunes that parallel the shoreline. They formed with the rise in sea level with melting of the ice sheets after the last ice age in the Pleistocene.

Note the boundary between fresh and saline groundwater. If significant freshwater is withdrawn from wells on the land surface, saline groundwater will move inland. This occurs due to saltwater's greater density so that the pressure exerted by a column of saltwater is greater than that of a column of freshwater of the same height. Potentially, saltwater can be drawn into the freshwater well. Also note a freshwater lens is present below the surface of the barrier island producing

*Figure 12.16* Sinkhole formed in Guatemala City, Guatemala.

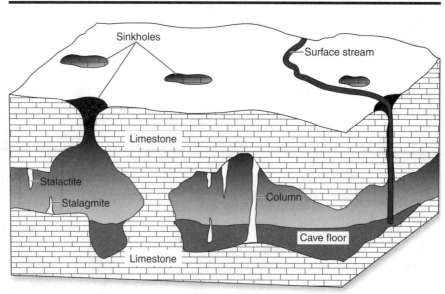

*Figure 12.17* Diagram illustrating cave formation and features in a karst terrain. (Adapted from Christopherson, R. W., 2003, *Geosystems: An Introduction to Physical Geography, 7th edition*, Pearson Education, Inc., New Jersey, 660 p.)

a local water-table aquifer supplied by rainfall. This is water "floating" on seawater due to its lower density. Similar freshwater-lens aquifers are present on ocean islands like the Hawaiian Islands.

# Rivers

Rivers have two sources of water as shown in **Figure** 12.19: *runoff* that has moved down the land surface after a rainstorm into the river and

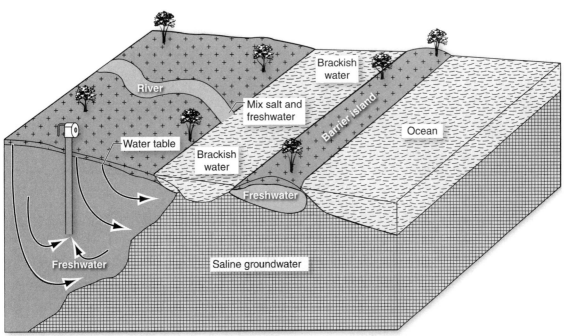

*Figure 12.18* Subsurface saltwater/freshwater boundary along the sea coast. Arrows on the cross-section show the general transport direction of freshwater that arrives at the water table from atmospheric precipitation.

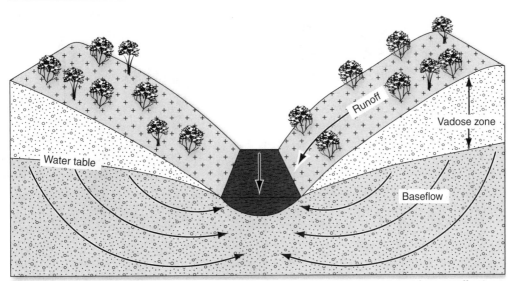

*Figure 12.19* Cross-section through the earth showing routes of water addition to a river from runoff and that which has infiltrated the vadose zone to the water table and become baseflow.

*baseflow,* which is water below the water table that is moving slowly through pores in the rocks to the river. Because of baseflow, water continues to be added to a river long after surface runoff from a rainstorm has stopped.

Rivers are natural courses of water connected into watersheds, also called drainage basins, that flow into an ocean, a sea, lake, or another river. Therefore, the drainage basin is the land surface from which all the surface runoff drains downhill defining the watershed for the river. With a length of 6,700 km from its headwaters in Burundi to its mouth at the Mediterranean Sea, the Nile River in Africa is the longest river on earth. The Amazon River in South America releases 6,300 km³ of water a year into the Atlantic Ocean at its mouth. This is the largest discharge of water of any river. Discharge is the volume of water that moves past a particular location as a function of time. Shown in **Figure 12.20** is the 103,500 km² watershed of the Wabash River, located in the midwestern U.S. The discharge of the Wabash is 32 km³ of water a year, about 0.5% of the discharge of the Amazon River.

The water in a river is typically confined in a channel. The river channel generally contains a single stream of water that meanders as it moves downstream. However, when sediment eroded into a river is large compared to the water transported, a braided river with several interconnecting channels of water is generally present. A braided river forms more often near the river's beginning in what is termed its headwaters as headwaters typically arise in mountainous areas where abundant loose material is present.

Most rivers in their lower reaches have levees along their banks (**Figure 12.21**). A *levee* is a natural or artificial embankment or wall on the edge of the river channel. Natural levees are produced when rivers overflow their banks during a flood. As soon as the floodwater overflows the bank and is no longer confined, its velocity decreases dramatically. The floodwater drops its load of coarse sediments producing a levee. Floodplains often exist next to levees on riverbanks. These are nearly flat areas adjacent to the river that experience periodic flooding. Much fine-grained sediment carried in floodwaters that is not deposited on the levee is deposited on the floodplain during flooding.

Given in **Figure 12.22** is a map of the world's largest rivers divided into groups of 1.5–15, 15–150, and more than 150 km³ per year of discharge. The largest 10 are listed in **Table 12.5**. As might be

*Figure* 12.20 Watershed of the Wabash River with its tributaries in the midwestern U.S. shown in the shaded area.

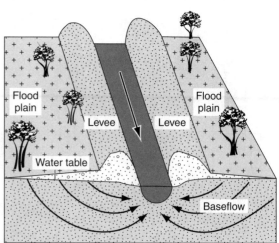

*Figure* 12.21 Cross-section through river sediments showing those deposited to produce a levee and those deposited in the flood plain.

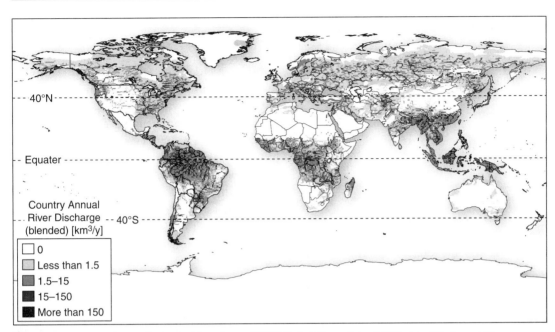

*Figure 12.22* Annual river discharge. (Data from: GWSP Digital Water Atlas (2008). Map 39: Annual River Discharge (V1.0). Available online at http://atlas.gwsp.org.)

*Table 12.5* DISCHARGE AMOUNTS OF THE 12 LARGEST RIVERS ON EARTH.

| RANK | RIVER | LOCATION | AVERAGE WATER DISCHARGE* (1000 m³/s) |
|------|-------|----------|--------------------------------------|
| 1 | Amazon | S. America | 209 |
| 2 | Zaire (Congo) | Africa | 41.0 |
| 3 | Orinoco | S. America | 33.0 |
| 4 | Yangtze | Asia | 30.2 |
| 5 | Rio Negro | S. America | 28.0 |
| 6 | Madeira | S. America | 24.4 |
| 7 | Yenisei | Asia | 19.6 |
| 8 | Brahmaputra | Asia | 19.3 |
| 9 | Paraná | S. America | 17.3 |
| 10 | Lena | Asia | 16.9 |
| 11 | Mississippi | N. America | 16.8 |
| 12 | Madre de Dios | S. America | 15.5 |

* There is some discrepancy among investigators about the average discharge of these rivers.

expected many of the rivers with the largest discharge are located in the tropics where rainfall is high. Note that the Amazon River, the world's largest river, is five times larger than the second largest river, the Zaire River. However, the Zaire is the deepest river in the world with some depths in excess of 230 m.

## River Sediments

Rivers shape the landscape through which they flow by eroding rocks and sediments and transporting and depositing them further downstream. A river can erode in a number of different ways. The material suspended in the water, the suspended load, can grind against the river bottom and its flanks. Turbulent water can open and enlarge joints and cracks through which it flows. This produces dislodged fragments that can be swept away in the water.

Rivers carry material both in their dissolved and solid load. The solid load is all the material carried that is not dissolved in solution. Some rocks such as limestone and evaporate minerals like halite can be dissolved in the water and carried downstream. Most material from silicate rocks is, however, carried downstream by the solid load (**Figure 12.23**). This includes material transported by traction. Traction is the process of moving rocks and gravel-sized material by dragging or rolling it along the riverbed by the moving water. Another way solid load is carried is by *saltation*. In saltation rock fragments are bounced along the river bottom by an incoming fragment dislodging a stationary fragment on the riverbed. The dislodged fragment is carried up into the moving water and carried downstream where it dislodges another fragment and the process repeats itself. Some of the solid load of the river is small enough, typically clay-sized material, and settles so slowly to the bottom of the river, that in most rivers it is carried in suspension. The suspended load can increase to silt-sized material in a fast moving river. During times of flooding the higher velocities of the water carry even larger-sized material in suspension.

A river will start to drop its solid load when the water velocity decreases. The velocity of a river is controlled by its gradient and the amount of water in the river. The gradient is the steepness of the river bottom. The steeper the gradient the greater is the water velocity. Also the more water a river carries the greater is its velocity. Friction, which slows the water velocity, per unit volume of water carried is less the more water that is present. This friction is due to water interacting with the sides and bottom of the river channel.

One thing that determines how much water a river carries is the season of the year. In dry seasons less water enters the river from runoff so its volume decreases. Evaporation can be important, particularly in dry climates, as water is lost as a function of time it spends in the river. The permeability of the rocks in the riverbed is also a factor if the water table is below the river bottom so river water is being lost by transport downward from the river channel. Decreases in the river gradient often occur as a river enters a lake or ocean or if the river flows on rock that is not easily eroded. All these factors lower water velocity and cause the river to deposit material.

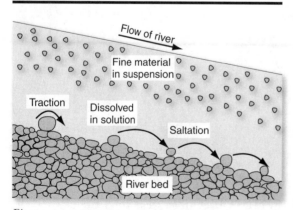

*Figure 12.23* The various ways a river can transport material.

## Hydrograph

Except for the typically small amount of water that enters a river by direct precipitation, the water in a river is supplied by runoff and baseflow. The amount of these two contributions can be determined with a *hydrograph*. A hydrograph is a record of the height or discharge of water in a river as a function of time at a particular location. The river height can be converted to discharge from knowledge of the cross-section dimensions of the river basin at the location of the hydrograph station and the velocity of the water. Charts are typically available to give the conversion for a particular hydrograph location. Measurement of other variables such as the water pH as a function of time

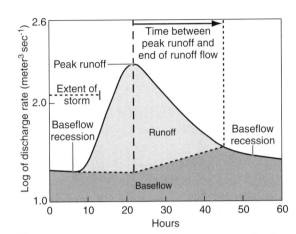

*Figure 12.24* Storm hydrograph showing the contribution of water in a river due to runoff into the river and that contributed by baseflow from groundwater.

would be called the hydrograph pH. Given in **Figure 12.24** is a hydrograph record for discharge changes in a river during a local storm.

As runoff from a storm enters the river the discharge rate down river increases when this additional water arrives at the hydrograph station. In most rainstorms in a basin it takes some time after the rainfall stops before the peak runoff reaches the recording station. With no runoff from rain, rivers are fed by baseflow as groundwater is discharged into the river. As groundwater is lost from the subsurface, the baseflow decreases leading to a slow decrease in its discharge into the river as a function of time, termed baseflow recession.

## Floods and Flood Control

A flood occurs when increases in the amount of water present submerges dry land. Flooding on a river or a lake occurs if runoff is added to a river or lake faster than it can be removed. In the case of a river, water overflows a levee or the levee breaks while on a lake its banks overflow.

Besides runoff another way to add water rapidly to a river or lake is by failure of a dam or other structure that holds water. Along the seashore flooding can occur because of unusual high tides caused by winds blowing water ashore because of the presence of an atmospheric low-pressure system such as a hurricane. Flooding also occurs

when a tsunami wave, caused by an earthquake, transports water on shore as occurred in Japan after the very large Tōhoku earthquake in March 2011.

On most rivers, flooding every year or two is a natural event. Flooding can make soil in the floodplain more fertile by providing nutrients in which it is deficient. However, property and vegetation are typically destroyed by floods and lives of humans and animals are endangered. Because the velocity of water increases as the banks of a river fill to flood stage, soil erosion is more prevalent. The removed soil leads to soil deposition further downstream. This can cause problems in drainage and interfere with farming after the floodwater has retreated. Rivers that are prone to flood are typically carefully managed. Flood control techniques for rivers include construction of artificial levees, dams, spillways to release water to uninhabited areas along the river, and retention ponds to hold extra water during times of flooding.

As societies become increasingly urbanized the change in land use means there is less area for precipitation to infiltrate into the ground and more water runs off city streets into rivers. This increases the volume of water in rivers during storms. Urban stream heights rise more quickly during storms and have a higher peak discharge rate than do rural streams of the same size. Also the total volume of water discharged during a flood event tends to be larger for urban rivers than for rural streams as more water runs off rather than infiltrating the land surface becoming groundwater. Humankind's straightening of river channels, which increases a river's gradient, increases the velocity of the water and peak discharge. This can cause more flooding both up and down river as water in a river is interconnected along its length.

Consider stream flow in Mercer Creek in Bellevue, Washington as opposed to nearby Newaukum Creek. The flow in the creeks from their watersheds is shown in the stream hydrographs in **Figure 12.25**. These are similar to the stream hydrograph in Figure 12.24. Mercer Creek flows through urban areas while Newaukum Creek flows through undeveloped land. Note that peak discharge is greater and comes earlier on Mercer Creek because of significant urban development

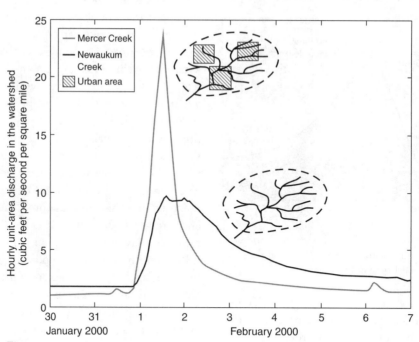

*Figure* 12.25 Hourly normalized discharge as a function of time for Newaukum and Mercer Creeks for a one-day rainstorm starting late in the day on January 31, 2000. (Data from: U.S. Geological Survey Fact Sheet 076-03.)

in the watershed so less water infiltrates into the ground than in the undeveloped Newaukum Creek watershed. Discharge for the week following the rainstorm, however, was greater in Newaukum Creek as baseflow was greater.

## Dams

Dams are barriers for impounding water. Generally, dams higher than 15 to 20 m are considered large and those over 150 m in height are termed major dams. The 300-m high earth-filled Nurek Dam, located on the Vakhsh River in the central Asian republic of Tajikistan, is the tallest dam in the world. The Aswan Dam in Egypt on the Nile River holds the most water in its reservoir. The Itaipu Dam on the Paraná River in Brazil/Paraguay produces the most hydroelectric power per year while the Three Gorges Dam in Sandouping, China has the world's largest hydroelectric generating capacity. It produces less power per year than the Itaipu dam because for six months each year little water flows from the Yangtze River into the reservoir

behind the dam. The total hydroelectric capacity is not used when the dam is actively storing water.

Dams are built not only to create hydroelectric power and control flooding but also to impound water for irrigation of crops and water supplies for towns and cities. Other considerations that determine the nature of a dam are its ability to create recreation areas, habitat for wildlife, and improved ship and barge navigation.

The surrounding aquatic ecosystems are changed both upstream and downstream of dams. Sediment becomes trapped behind the dam and the lower amount of suspended sediment downstream and the control of flood events does not allow for normal changes in riverbanks and floodplains. This causes a loss of wildlife habitat down river from the dam. For instance, the Aswan Dam on the Nile has decreased soil fertility downstream by limiting flooding onto the floodbanks along the river. Dams along the Columbia and Snake Rivers in the U.S. Pacific Northwest impede salmon migration.

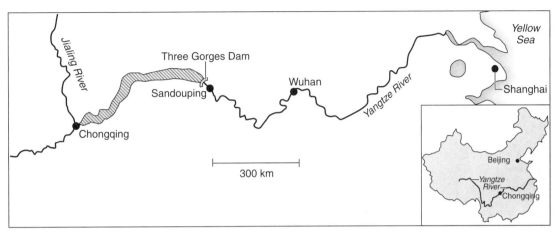

*Figure 12.26* Map of the Yangtze River and Three Gorges Dam with the inset showing its location in China. The hatched area shows the surface extent of the water behind the dam.

The Glen Canyon Dam on the Colorado River is filling with silt and some have argued the useful lifespan of the dam is less than 100 years. If the silting causes dam failure it is likely the flood wave produced will overflow the Hoover Dam down river. Small local earthquakes have occurred from the increased stress in the earth caused by the presence of the additional mass of water and sediments in a dam reservoir. Such earthquake epicenters have been reported to occur behind the Lake Kariba Dam in Zambia, Koyna Dam on Lake Shivajisagar in India, and the Nurek Dam in Tajikistan.

The Three Gorges Dam in China flooded many archaeological, historic, and cultural sites and caused significant ecological changes to the surrounding area with the submergence of hundreds of factories, mines, and waste dumps. Some 13 cities and 140 towns were flooded and 1.3 million people needed to be relocated (**Figure 12.26**). The dam became fully operational in 2011. It produces about 80 TWh (terawatt hours) of electricity per year and increases the Yangtze's shipping capacity. However, the prime purpose of the Three Gorges Dam is to reduce flooding and control discharge downstream, to avoid major problems of the seasonal flow of the Yangtze River with a November to May dry season. During the dry season water is now supplied from the rainy season water runoff stored behind the dam (**Figure 12.27**).

*Figure 12.27* Three Gorges Dam across the Yangtze River by the town of Sandouping in China.

## Water Transportation and Diversion Systems

Canals are excavated from the ground surface and filled with water to connect water bodies for transportation, sanitation, and water supply. The 162-km long Suez Canal is a sea level artificial transportation waterway in Egypt connecting the Red Sea and the Mediterranean Sea. The Panama Canal is an 80-km long canal connecting the Pacific and Atlantic Oceans through a number of artificial lakes and three sets of locks. The longest canal on the earth at almost 1,800 km in length is the Grand or Imperial Canal in China

*Figure 12.28* The Beijing-Hangzhou Grand Canal contains 24 locks. The bed of the canal only varies from 1 m below sea level at Hangzhou to 42 m above at its summit in the mountains of Shandong.

*Figure 12.29* Tennessee-Tombigbee waterway linking the Mississippi River system to Mobile, Alabama and the Gulf of Mexico.

(**Figure 12.28**). It is a nearly north-south water-way between Beijing and Hangzhou. The canal serves as the main transportation artery for goods between northern and southern China as most rivers in China traverse from the highlands in the west to the lowlands in the east.

The United States has an extensive interconnected inland water transportation system in the Mississippi River and its tributaries. A 377-km long shipping canal with 10 locks, the Tennessee-Tombigbee (Tenn-Tom) waterway was completed in 1985 to connect the Tennessee and Tombigbee rivers (**Figure 12.29**). It provides a low-cost and energy-efficient trade link between the Sunbelt states and the rest of the 7,200 km of navigable waterways in the Mississippi River system. It also serves as an alternate route to that of the Mississippi River, from mid-America to the Gulf of Mexico.

## Drinking Water Supplies

Furnishing clean plentiful freshwater is a problem for cities. The growth of a city can be limited by the available freshwater. A city's water supply is dependent upon the city's location. Cities like Bombay, India impound rainwater from the annual monsoon rainstorm season to provide freshwater that is used throughout the year. New York City has nearby watersheds to supply water. Los Angeles, given its dry climate, would not be a large metropolis without the water supply system implemented by William Mulholland as outlined below.

**New York City water supply system.** The New York City municipal water supply system is one of the most extensive in the world. It has a storage capacity of over 2 trillion liters and delivers 4.5 billion liters per day of drinking water. It first obtained water from the nearby Croton River Watershed (**Figure 12.30**). This was later expanded to watersheds in the Catskills and finally the Delaware River 200 km away. It meets the needs of over 9 million people with 95% of the total water moved by gravity feed.

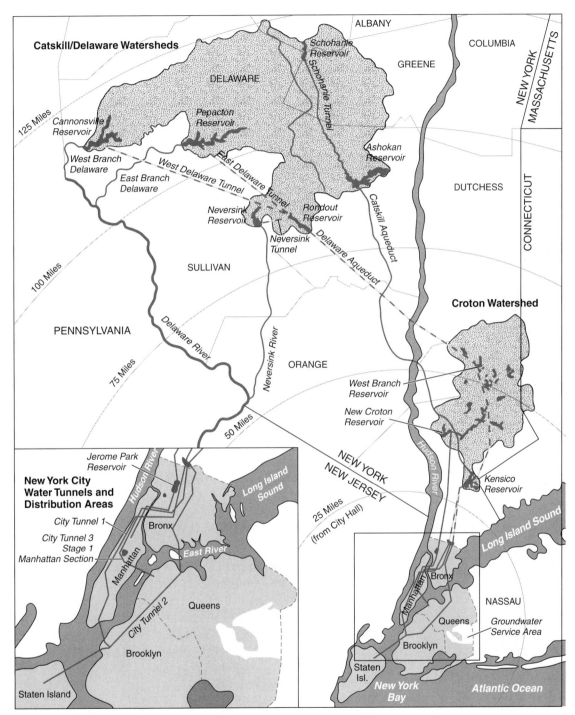

*Figure* 12.30 New York City water supply system consisting of aqueducts from 18 reservoirs in the Catskill Mountain/ Delaware River and Croton River Watersheds. (Reproduced from New York City Department of Environmental Protection Map: New York City's Water Supply.)

Unfortunately there is significant leakage of water from the supply system that is currently in the process of being repaired.

### Southern California water supply system.

At the start of the twentieth century it was clear that the limiting factor for the growth of Los Angeles in Southern California was availability of freshwater. Owens Valley in eastern central California had a large amount of runoff that originally fell on the Sierra Nevada Mountains and drained to the east. The rights to this water were owned by local farmers. William Mulholland, the superintendent of the Los Angeles Department of Water and Power (LADWP), recognized that a gravity-fed aqueduct could be used to transport the Owens Valley water to Los Angeles. Through purchases, intimidation, and bribery Mulholland obtained enough water rights in Owens Valley to enable the Los Angeles Aqueduct to be built between 1908 and 1913 (**Figure 12.31**). It consists of a 360-km long 3.7 m diameter steel pipe. The water removed by the Los Angeles Aqueduct that allowed the Los Angeles

area to prosper devastated the ecosystem of Owens Lake and the ability to farm the arid Owens Valley.

The LADWP serves over 4 million customers and provided more than 760 billion liters of freshwater in 2010. The Los Angeles Aqueduct supplies 48% of the water. Water from the Colorado River through the Colorado River Aqueduct and from the San Joaquin-Sacramento River east of San Francisco through the California Aqueduct accounts for another 40% of the water. Local groundwater supplies 10% and 2% is contributed from recycled water.

## Colorado River Watershed

The Colorado River Watershed includes portions of seven U.S. states and is 632,000 km$^2$ in area as shown in **Figure 12.32**. Major tributaries include the Green River in Wyoming, Utah, and Colorado; the Gunnison and Yampa Rivers in Colorado; the San Juan River in Colorado, New Mexico, and Utah; and the Gila River in Arizona and New Mexico. The Colorado River flow has been changed dramatically from its natural state owing to demands for water in the arid region. For instance, large dams and water diversions for irrigation have decreased the Colorado River's discharge at its mouth in the Gulf of Baja to minimal amounts.

In 1922 the seven states of Colorado, New Mexico, Utah, and Wyoming in the upper basin and Nevada, Arizona, and California in the lower basin signed the Colorado River Compact outlining water rights that was modified in 1928 leading to the "The Law of the River." It divides the river into an upper and lower basin each stipulated to receive $9.25 \times 10^{12}$ liter/year of water (**Table 12.6**). A treaty with Mexico in 1944 gave this country $1.85 \times 10^{12}$ liter/year of Colorado River water. The amount of water allocated was based on an expectation that the river's average flow was $20.4 \times 10^{12}$ liter/year. Specific allotments for the states have been disputed. Concerns by the state of Arizona lead to the funding by the U.S. Congress in 1968 of the Central Arizona Water Project (see below).

Most present analyses conclude that the compact was negotiated in a period of abnormally high precipitation, and that the recent drought

*Figure 12.31* The water supply system for Southern California showing the supply through the Los Angeles, California and Colorado River Aqueducts.

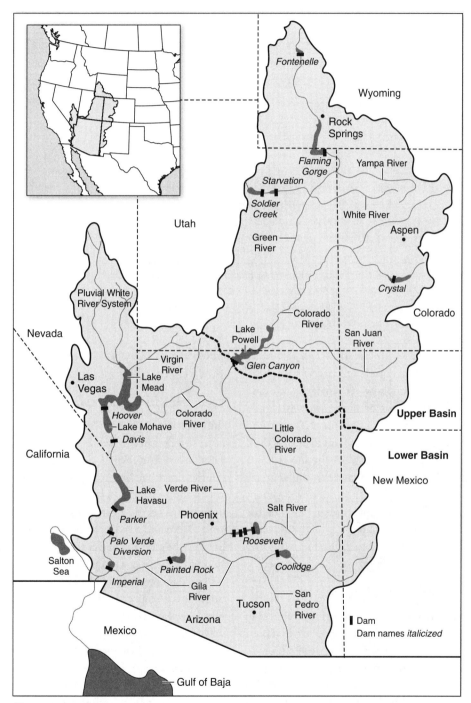

*Figure 12.32* Colorado River Watershed showing the location of the major tributaries and dams. (Modified from LaRoe, E.T., G.S. Farris, C.E. Puckett, P.D. Doran, & M.J. Mac (Eds.) *Our living resources: a report to the nation on the distribution, abundance, and health of U.S. plants, animals, and ecosystems*, (pp. 149–153). Washington, D.C.: U.S. Department of the Interior, National Biological Service.)

*Table 12.6* COLORADO RIVER WATER ALLOTMENTS

| UPPER BASIN | % OF ALLOTMENT | $10^{12}$ LITERS/YR |
|---|---|---|
| Colorado | 51.75 | 4.79 |
| Utah | 23.00 | 2.13 |
| Wyoming | 14.00 | 1.30 |
| New Mexico | 11.25 | 1.04 |
| Arizona | 0.70 | 0.06 |
| LOWER BASIN | % OF ALLOTMENT | $10^{12}$ LITERS/YR |
| California | 58.70 | 5.43 |
| Arizona | 37.30 | 3.45 |
| Nevada | 4.00 | 0.37 |

The current specific annual allotments in the lower basin were established in 1928 as part of the Boulder Canyon Project. The current specific annual allotments in the upper basin were established by the Upper Colorado River Basin Compact of 1948.

conditions in the region are, in fact, a return to historically typical precipitation patterns. In 2007 a set of guidelines on how to allocate Colorado River water in the event of shortages was signed by the U.S. Secretary of the Interior. The guidelines state that in a severe shortage, $8.6 \times 10^{12}$ liter/year needs to be delivered to the lower basin for allocation. At the present time the scarcity of water is decreasing agricultural development in the watershed. In the upper basin 90% of the water and in the lower basin 85% of the water is used for irrigating crops. Much of this is heavily water-consumptive alfalfa grown for cattle feed.

### Central Arizona Water Project

The Central Arizona Water Project is mainly an open 540-km long aqueduct that carries about $1.85 \times 10^{12}$ liter/year of water from the Colorado River at Lake Havasu to major population centers in central and southern Arizona. The system also includes tunnels, pipelines, and pumping plants as shown in **Figure 12.33** as well as the construction of dams to hold water at a total cost of US $4 billion. It was substantially completed in 1994 as the last section extended it to south of Tucson, Arizona.

Shown in **Figure 12.34** is the granite reef aqueduct of the project at Phoenix, Arizona.

### Northern River Reversal, Russia

The Caspian Sea is shrinking and becoming increasingly salty because of diversion of freshwater from the Volga and Kur Rivers, which empties into the sea for industrial, agricultural, and residential uses. The Aral Sea is also shrinking and becoming increasingly salty because the Amu Darya and Syr Darya Rivers that feed it were diverted by the former Soviet Union starting in the 1960s for irrigation projects to grow cotton in the surrounding desert. These terminal lakes are called seas because their waters have always been salty like seawater.

Russia has major rivers that flow northward and drain into the Arctic Ocean. Given the tremendous need for freshwater it has been suggested to reverse their flow southwards toward populated agricultural areas that lack freshwater in what was called the Northern River Reversal Project. This major water-diversion project would bring Pechora, Northern Dvina, and Onega River water and possibly Ob River water to the Caspian and Aral seas

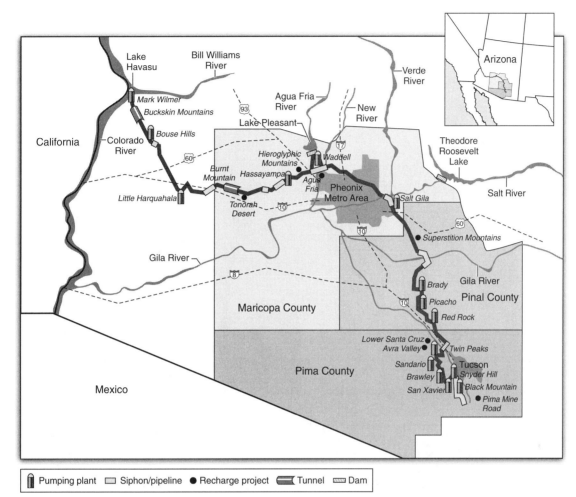

*Figure* 12.33 Central Arizona Water Project. (Adapted from Arizona Department of Water Resources, Central Arizona Project.)

(**Figure 12.35**). Research and planning work on the project started in the 1930s but the highly controversial project was abandoned in 1986.

Its pros and cons have been hotly debated. The environmental effects of diverting so much water to arid regions and away from the Arctic Ocean are not well known. Because of global warming there has been increased rainfall input into the Russian northern flowing rivers. This could make the Arctic Ocean less salty and interfere with circulation of the ocean. Some have argued that this is another reason for diversion of northern flowing freshwater to the south.

*Figure* 12.34 The granite reef aqueduct at Phoenix, AZ.

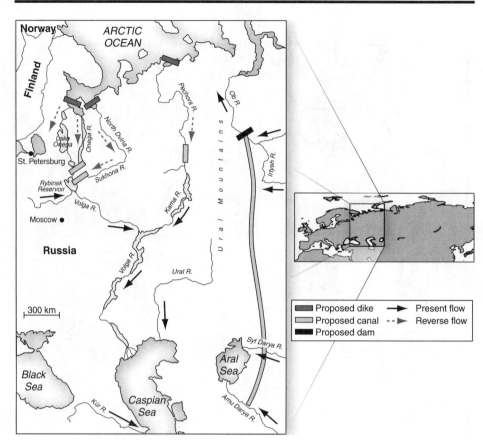

*Figure 12.35* Map of eastern Europe showing the rivers that flow northward into the Arctic Ocean and the location of dikes, dams, and canals that are proposed to be built to reverse their flow to bring this freshwater to the Caspian and Aral Seas. (Adapted from Craig, J. R., Vaughan, D. J., and Skinner, B. J., 2001, *Earth Resources and the Environment,* 3rd edition, Prentice Hall, Boston, 508 p.)

## SUMMARY

Only 2.5% of the earth's water is fresh. It is this freshwater that is of concern as a resource. Two thirds is tied up in ice caps and 30% is in groundwater. The freshwater is part of the hydrologic cycle where the residence time at steady state in a particular reservoir is given by:

$$\text{Residence time (yr)} = \frac{\text{Reservoir size (g)}}{\text{Flux (g yr}^{-1})}.$$

Rainfall on the earth depends on the surface wind patterns. These winds are influenced by the Coriolis effect. This effect causes winds on the rotating earth to veer to the right in the Northern Hemisphere and to the left in the Southern Hemisphere. Air masses rise at the equator because of the greater intensity of sunlight and therefore less dense air and sink at the poles because the cold air masses present are dense. Rather than be transported between the equator and the poles in a single convection cell, a three-cells configuration is stable.

Near the equator the rising air cools and produces rain. Near 30°N and S latitude air masses sink and are warmed; the decreasing humidity and lack of rain produces deserts at these latitudes. El Niño and La Niña events break the normal weather patterns on average about every five years.

Evapotranspiration is the sum of water evaporated from the earth's surface as well as that transpired from plants. In most regions transpiration is greater than evaporation. Groundwater is water present in the subsurface. Typically a water table exists below the surface where the pores in the rock are filled with water. Above the water table is the vadose zone where pores contain air as well as water. Extending upward from the water table is a capillary fringe where water is pulled upward by the wetting of the walls of the pores. Aquifers are layers of sediment or rock that can store and transmit water. They have significant porosity and permeability. A water-table aquifer has a water table as an upper surface. A confined aquifer is present between two layers that do not transmit water.

Rivers occur in watersheds and carry solid material as well as water. The solid material is transported by traction, saltation, and in suspension. The solid load is deposited when the water velocity decreases. Hydrographs are used to determine the extent of runoff as opposed to baseflow that enters a river. A flood occurs when water overflows the levees of a river. On most rivers, flooding every year or two is a natural event. Humankind has increased the potential for flooding by creating urban spaces that decrease water infiltration so runoff increases. Dams are built to control flooding, create hydroelectric power, and impound water for irrigation and municipal water supplies.

A cone of depressed water levels can occur in a water-table aquifer due to pumping from a well. In a confined aquifer the cone reduces water pressure around the extraction point decreasing the potentiometric surface. The Ogallala water-table aquifer of the western U.S. Great Plains, the Nubian Sandstone Aquifer System of northeastern Africa, and the Quaternary Aquifer System of Northern China are decreasing in size. This is occurring because recharge is less than humankind's extraction of water, primarily due to need of water for irrigation.

A boundary in the earth exists between fresh groundwater and seawater at the coastline. A freshwater well can become saline as seawater intrudes into the coastal aquifer as freshwater is removed. The removal of groundwater can lead to the subsidence of the land surface. Sinkholes can develop from continual removal of slightly soluble limestone or salt beds.

Canals are used to connect bodies of water for transportation, water supply, and sanitation. These include the Suez, Panama, Grand, and Tennessee-Tombigbee Canals. The nature of the water supply systems for a city depends on its location.

The Colorado River shares its water with seven states and Mexico. Problems occur because discharge is less than the water allocated to each state and to Mexico. The Central Arizona Water Project takes water from the Colorado River and transports it to central Arizona. The Northern River Reversal Project in Russia has been abandoned, at least for now, because of the cost and likely environmental damage.

## KEY TERMS

| | |
|---|---|
| aquiclude | monsoon |
| aquifer | permeability |
| baseflow | Pleistocene |
| capillary fringe | polar cell |
| cone of depression | porosity |
| Coriolis effect | potentiometric surface |
| El Niño event | residence time |
| evapotranspiration | runoff |
| Ferrel cell | saltation |
| Hadley cell | sinkhole |
| hydrograph | vadose zone |
| karst | watershed |
| La Niña event | water table |
| levee | |

## PROBLEMS

1. Using the values in **Tables 12.1** and **12.2** calculate the amount of water that leaves the ocean and that enters the atmosphere each year. Which is larger? Why?

2. Using the values in the text how long would the water last in the Nubian Sandstone Aquifer if only 10% could be extracted and withdrawal continued at the present day rates?

3. Assume you have cut a cube, 10 cm on each side, out of sandstone and it weighs 2.0 kg. If it is made up of quartz grains with a density of 2.65 g cm$^{-3}$ what is its porosity?

4. Pressure = force per unit area. If the water pressure at a point in a confined aquifer is 100,000 pascals (= Newton per m$^2$) how high above the point will the potentiometric surface be for the aquifer? Assume the water has a density of 1.0 g cm$^{-3}$. Remember that force = mass × acceleration and the acceleration of gravity is $g$ = 9.81 m s$^{-2}$.

## REFERENCES

Christopherson, R. W. 2003. *Geosystems: An introduction to physical geography*, 7th ed. Upper Saddle River, NJ: Pearson Education, Inc.

Craig, J. R., Vaughan, D. J., and Skinner, B. J. 2001. *Earth resources and the environment*, 3rd ed. Boston: Prentice Hall, Inc.

Foster, S., Garduni, H., Evans, R., Olson, D., Tian, Y., Zhang. W., and Han, Z. 2004. Quaternary Aquifer of the North China Plain—Assessing and achieving groundwater resource sustainability. *Hydrogeology Journal* 12:81–93.

Gleick, P. H. 1996. Water resources. In S. H. Schneider (ed.) *Encyclopedia of climate and weather*, Vol. 2 (pp. 817–823). New York: Oxford University Press.

Martin, R., 2013, *Earth's Evolving Systems: The History of Planet Earth*, Sudbury, MA: Jones & Bartlett Learning.

Solley, W. B., Pierce, R. R., and Perlman, H. A. 1998. *Estimated use of water in the United States in 1995* (U.S. Geological Survey Circular 1200). Available online from http://water.usgs.gov/watuse/pdf1995/pdf/circular1200.pdf.

Starnes, W. C. 1995. Colorado River Basin fishes. In LaRoe, E. T., G. S. Farris, C. E. Puckett, P. D. Doran, and M. J. Mac (eds.) *Our living resources: A report to the nation on the distribution, abundance, and health of U.S. plants, animals, and ecosystems* (pp. 149–153). Washington, DC: U.S. Department of the Interior, National Biological Service.

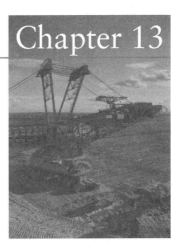

## Chapter 13

# Water Quality, Usage, and Law

Access to clean drinking water is a basic concern of humankind. In the world today about 880 million people drink unhealthy water. This is almost one in eight people. It is estimated that over 3.5 million people die each year from water-related diseases. Freshwater shortages occur in 88 developing countries that are home to half of the world's population. In these countries, 80% to 90% of all diseases and 30% of all deaths directly result from poor water quality.

## Water Composition and Quality

The composition of both surface and ground-water changes naturally as a function of time and location. Given in **Table 13.1** are reported compositions of some natural waters at the indicated locations. While rain and snow have about 1 to 2 ppm and pure spring water has about 30 ppm of dissolved solids, some surface waters can contain substantial amounts of dissolved constituents. The Dead Sea between Jordan and Israel and the Great Salt Lake in Utah have about 300,000 ppm and seawater has 35,000 ppm of dissolved solids. Water is considered drinkable if it has less than about 500 ppm of dissolved material.

### Rainwater

Analysis (a) and (b) of Table 13.1 give the composition of dissolved matter in rainwater. Rainwater

dissolves solid matter in the atmosphere as droplets of water nucleate on atmospheric particulates. Over the ocean and coastal areas these are primarily salt particles of $NaCl$ or $MgSO_4$ from seawater contributing $Na^+$, $Cl^-$, $Mg^{2+}$, and $SO_4^{2-}$ to the rainwater. Over land surfaces silicate dust particles contribute $Ca^{2+}$, $Mg^{2+}$, $K^+$, $Na^+$, and $SiO_2$. Rain also changes its composition due to the incorporation of the dissolved gases $CO_2$, $NH_3$, $SO_2$, and $NO_2$ obtained during transit through the atmosphere. As a result, rainwater composition varies through time and with location.

**Figure 13.1** shows the average pH of rainfall at sites in the continental U.S. Note that rainfall generally has a pH < 6 or is somewhat acidic. This occurs due to the absorption of $CO_2$ and other acids in the air mass through which the rain falls. $CO_2$ produces carbonic acid in rainwater by the reaction:

$$CO_2 + H_2O \rightarrow H_2CO_3. \qquad [13.1]$$

The carbonic acid ($H_2CO_3$) then dissociates to some extent via the reaction:

$$H_2CO_3 \rightarrow H^+ + HCO_3^-. \qquad [13.2]$$

producing an acidic solution. Pure $H_2O$ in contact with uncontaminated air has a pH of about 5.6 and therefore is slightly acidic. When the pH is below this value, reactions with other acidic components in air have occurred.

Many coals contain significant sulfur. The abundance of sulfur in coals is controlled principally by

*Table 13.1* COMPOSITION OF WATER FROM THE GIVEN SOURCE AND LOCATION AND THEIR pH*.

| | Rainwater from Menlo Park, CA (a) | Rainwater from North Carolina and Virginia (b) | Mississippi River at Cape Girardeau, MO (c) | Rhine River as it leaves the Alps (d) | Stream draining igneous rocks in Washington Cascades (e) | Jump-Off Joe Creek, Oregon, wet season, Nov., 1990 (f) | Jump-Off Joe Creek, Oregon, dry season, Sept., 1991 (g) | Groundwater from volcanic rocks, New Mexico (h) | Groundwater, Sierra Nevada with short residence time (i) | Groundwater, Supai limestone, Grand Canyon (j) | Great Salt Lake, Utah (k) | Seawater (l) |
|---|---|---|---|---|---|---|---|---|---|---|---|---|
| **Calcium** | 0.8 | 0.65 | 47 | 40.7 | 1.68 | 14 | 22 | 6.5 | 3.11 | 144 | 241 | 412 |
| **Magnesium** | 1.2 | 0.14 | 14 | 7.2 | 0.24 | 13 | 17 | 1.1 | 0.7 | 55 | 7,200 | 1,350 |
| **Sodium** | 9.4 | 0.56 | 11 | 1.4 | 0.16 | 8 | 14 | ~37 | 3.03 | ~27 | 83,600 | 10,770 |
| **Potassium** | 0.0 | 0.11 | 4.0 | 1.2 | 0.31 | - | 0.5 | ~3 | 1.09 | ~2 | 4,070 | 399 |
| **Bicarbonate** | 4 | 5 | 138 | 114 | 5.4 | 104 | 129 | 77 | 20 | 622 | 251 | 140 |
| **Sulfate** | 7.6 | 2.2 | 64 | 36 | 1.3 | 4.7 | 1.3 | 15 | 1.0 | 60 | 16,400 | 2,712 |
| **Chloride** | 17 | 0.57 | 12 | 1.1 | 0.06 | 8.5 | 33 | 17 | 0.5 | 53 | 140,000 | 19,354 |
| **Silica** | 0.3 | - | 6.8 | 3.7 | 0.7 | 24 | 30 | 103 | 16.4 | 22 | 48 | 6 |
| **TDS** | 38 | 4.7 | 254 | 207 | 10 | 120 | 180 | 222 | 36 | 670 | 254,000 | 35,000 |
| **pH** | 5.5 | 4.5 | 7.5 | - | 6.9 | 7.7 | 7.0 | 6.7 | 6.2 | - | 7.4 | 8.1 |

* In milligrams per liter. Note one liter is equal to 1,000 cubic centimeters.

A dash (–) indicates that the component was not detected or the water was not analyzed for this constituent. A tilde (~) means "approximately."

Adapted from the Advameg, Inc.

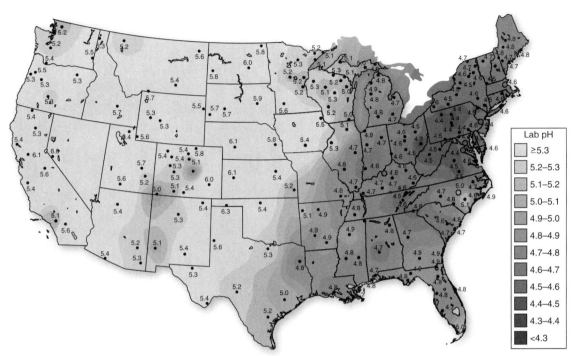

*Figure 13.1* Average pH of rainwater in the U.S. at the indicated locations in 2007. (Adapted from the National Atmospheric Department Program/National Trends Network.)

how much seawater sulfur the peat that formed the coal has reacted with. The $SO_2$ released during coal combustion combines with atmospheric $O_2$ to produce sulfuric acid ($H_2SO_4$) by the reaction:

$$SO_2 + 0.5O_2 + H_2O \rightarrow H_2SO_4 \rightarrow 2H^+ + SO_4^{2-}. \quad [13.3]$$

Also the burning of petroleum both in factories and vehicles emits $NO_2$ into the atmosphere. This produces nitric acid ($HNO_3$) by the reaction:

$$2NO_2 + 0.5O_2 + H_2O \rightarrow 2HNO_3 \rightarrow 2H^+ + 2NO_3^-. \quad [13.4]$$

As there is generally a higher concentration of sulfur in the coals that are burned than in nitrogen compounds in refined petroleum, atmospheric sulfuric acid is the main contributor to *acid rain*.

If the pH of rain is above 5.6 the rain has likely condensed on carbonate dust particles in the atmosphere and these have reacted with the acid in the rain decreasing $H^+$ by the reaction:

$$CaCO_3 + H^+ \rightarrow Ca^{2+} + HCO_3^-. \quad [13.5]$$

Note in Figure 13.1 the low pH of rain centered in western Pennsylvania, eastern Ohio, and northern West Virginia. Almost half of all the $SO_2$ and $NO_2$ emitted to air in the U.S. is produced in the Ohio River valley where large amounts of coal are burned. $SO_2$ and $NO_2$ are carried by winds northeasterly toward the Mid-Atlantic states. Here acids are produced in the atmosphere decreasing the pH of rain to 4.5 or less.

## Rivers

Rivers carry rainwater, groundwater, and the products of surface and soil weathering including organic matter to lakes and the ocean. The weathered material is carried as both dissolved and suspended material. The consequences of the suspended material were discussed previously. In this chapter the dissolved constituents will be considered. The major source of dissolved material in rivers is the sediments and rocks the water encounters along its path from where it falls as precipitation to where it empties into a lake or the ocean. This can be by the weathering of organic matter, carbonates such as limestone or dolomite, silicate rocks, and/or evaporites.

The composition of the dissolved material carried in a river varies along the length of the river as well as seasonally. Given in analyses (c) to (g) in Table 13.1 are the dissolved compositions of four rivers. Most river water composition tends to have $Ca^{2+}$ as the major cation and $HCO_3^-$ as the major anion. During dry periods of the year groundwater contributes more to river water composition whereas in the rainy season, runoff from the land surface controls the water composition. Because most rivers have urban and other developed areas along their banks, pollution from humankind is also a major factor in controlling water compositions.

## Lakes

Lakes receive their water from rivers, groundwater, and to a lesser extent from the atmosphere and runoff from the land surface. Many lakes have compositions similar to the rivers that flow into them. Because they can have large surface areas with the atmosphere, evaporation of water can be significant. As the water evaporates, dissolved constituents can remain in the lake and the waters then supersaturate with minerals that precipitate out of solution. Often this is calcite ($CaCO_3$) or gypsum ($CaSO_4 \cdot 2H_2O$). This is particularly true of lakes in hot, dry climates where evaporation is extensive as well as in terminal lakes. Recall that a terminal lake is one in a closed hydrologic basin where water flows into the lake but leaves only by evaporation. The materials precipitated in terminal lakes are important natural resources (see chapter on evaporites). The Great Salt Lake in Utah given in analysis (k) in Table 13.1 is a terminal lake. Many lakes have concentrations of dissolved constituents that change with depth. To monitor the changes in water composition with depth samples can be taken with a Kemmerer sample bottle as shown in **Figure 13.2**.

*Figure 13.2* A Kemmerer bottle that has sampled lake water to obtain samples at a specified depth below the winter ice for chemical and biological analysis. It uses a messenger weight which is dropped down via a rod to release a catch that opens then closes end covers on the bottle.

## Groundwater

Groundwater contacts a large amount of rock surfaces and moves more slowly than rivers. As a result, groundwater often contains more dissolved material than river water as given in analyses (h) to (j) in Table 13.1. Typically groundwater has a total dissolved solids (TDS) content of less than 250 milligrams per liter. However, some groundwater that is made up of large portions of trapped seawater or groundwater that contacted salt deposits underground can be quite salty. These produce saltwater aquifers.

## Seawater

Given in analysis (l) in Table 13.1 is the composition of average seawater. Note that the dominant constituents of seawater are Na and Cl. Because Na has a molecular weight of 22.99 g and Cl a molecular weight of 35.45 g, the number of moles in 1,000 g = 1,000 cm³ of water for

$$Na = \frac{10,770 \times 10^{-3}\text{g per 1,000 cm}^3}{22.99 \text{ g per mole}} = 0.469 \text{ moles}$$

and for

$$Cl = \frac{19,354 \times 10^{-3}\text{g per 1,000 cm}^3}{35.45 \text{ g per mole}} = 0.546 \text{ moles}.$$

Therefore seawater is, to a first approximation, a half-molal NaCl solution and contains an order of magnitude less $MgSO_4$. Note the greater abundance of Na than Ca in seawater, but the opposite in river water supplied to the ocean. This occurs because the precipitation of minerals like calcite ($CaCO_3$) in the ocean decreases the amount of $Ca^{2+}$ in seawater relative to the amount of $Na^+$ that enters it. The major ion composition of seawater is also influenced to a large extent by reactions with hot rocks at mid-oceanic ridges where new ocean crust is forming.

## Desalination of Water

As there is plentiful salt water on the earth, a way to meet the need for additional freshwater is to produce it from saltwater in a process called desalination. There are two different approaches to the desalination process to remove salt from water, *distillation* and membrane filtration. They each presently account for about half of the worldwide capacity of 27 billion liters per day of freshwater produced from saltwater.

Distillation is the oldest method of desalination. In distillation the saltwater is boiled and the steam vapor is condensed into freshwater. Rather than doing this in a single chamber, multistage flash distillation (MSF) is used because it is up to six times more energy efficient than single-chamber distillation. Commonly up to 15 different stages are used (**Figure 13.3**).

Cold seawater is piped through a series of coils in chambers and then heated. The heated seawater enters the first chamber where freshwater vapor is released, cooling and somewhat concentrating salt in the remaining seawater brine. Freshwater from the vapor condenses on cold seawater containing coils. The freshwater from the coils is then collected as shown in **Figure 13.4**. The heated somewhat concentrated brine is passed to the next chamber and the process is repeated. A variation of this process is multiple-effect evaporation (MEE). Brine water is boiled in a sequence of vessels, each held at a lower pressure than the one before. Because the boiling temperature of water decreases as pressure decreases, the vapor boiled off in one vessel can be used to heat the next, and only the first vessel at the highest pressure requires an external source of heat. While more energy efficient than MSF, MEE develops scaling, that is salt precipitation problems, on the heat transfer tubes

*Figure 13.3* Ten multistage saltwater distillation units with brine recirculation producing 26,700 m³/day of freshwater located at Al Khobar, Saudi Arabia.

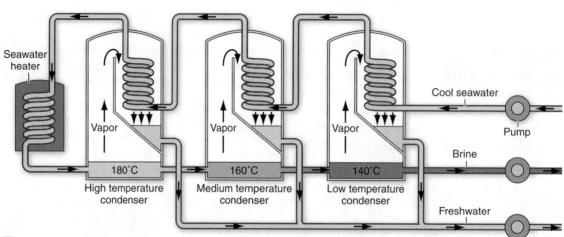

*Figure 13.4* Multistage flash evaporation (MFE) saltwater/brine distillation facility.
(Adapted from the U.S. Department of the Interior, Office of Saline Water, Research and Development Report 580,
*Evaluation of the Vertical Tube Evaporator and the Multistage Flash Desalination Processes*, 1970.)

and is not nearly as commonly employed as MSF distillation.

Membrane filtration methods for obtaining freshwater are of two general types. In one a voltage is produced between two membranes so that one membrane has a positive and the other a negative charge. Ions in saltwater of opposite sign are retained as they pass through the membranes. Water passing through a series of membranes will finally become drinkable freshwater. The other type of filtration employs a *reverse osmosis* membrane. Pressure pushes saltwater through a water-permeable membrane that allows smaller water molecules but not the larger ions with their hydration spheres to pass through it (**Figure 13.5**). Normally in osmosis water will be transferred through a water-permeable membrane from where $H_2O$ is at high concentration (low salt side) to where $H_2O$ is in low concentration (high salt side). When pressure is applied to the water with the higher ion concentration, the water will move in the opposite direction through the membrane, leaving the ions behind. This is known as reverse osmosis. Reverse osmosis techniques for transforming seawater and salty groundwater into freshwater are starting to overtake multistage distillation as the dominate

desalination process. Not having to heat the water saves significant energy costs.

There are other techniques to obtain freshwater in smaller amounts. One is a solar still where, similar to a greenhouse, solar energy passes into a container through a clear glass panel to heat saltwater. Freshwater evaporates from the heated saltwater and condenses on the cooler glass panel. The condensed freshwater runs down the panel and is collected. In most environments about 1 m² of land surface is required to produce about 4 liters of freshwater per day.

## Drinking Water Contaminants

All sources of drinking water contain some dissolved constituents. These can be beneficial, of no consequence, or contaminants. A beneficial constituent like Mg can be detrimental if its concentration is too high. Constituents that are contaminants are not considered a health risk unless their levels in the water are too high.

The World Health Organization (WHO) sets up guidelines for drinking-water quality, outlining international standards for drinking-water safety. The European Union incorporated drinking-water standards in Council Directive 98/83/EC. Water quality and protection issues

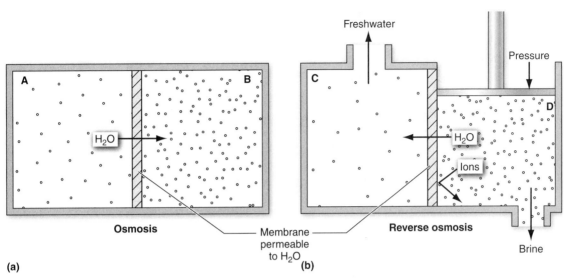

*Figure 13.5* Water-filled chambers with dissolved ions represented as dots. (a) In osmosis, water is passed through a semipermeable membrane from where H₂O is at high concentration (side A with lower ion concentration) to where it is at lower concentration (side B with higher ion concentration). (b) In reverse osmosis, pressure forces water through the membrane in the opposite direction so that side C has a higher H₂O concentration (lower ion concentration) and therefore becomes fresher.

in the U.S. are handled by the Environmental Protection Agency (EPA).

Contamination in potential drinking water can be from point or nonpoint sources. An example of a possible point source would be from a pipe emptying arsenic-containing wastewater from an electronics fabrication plant into a river. A nonpoint source is nitrates or phosphates from fertilizers used on farmland carried by surface runoff into a stream. Nitrates and phosphates are contaminates because their presence in water stimulates the growth of contaminating algae.

Together with states agencies, the EPA is involved with permitting the discharge of pollutants into navigable waters in the U.S. The EPA works with the Army Corps of Engineers to issue permits for the filling and dredging of wetlands and the Coast Guard to coordinate clean up of chemical and oil spills into bodies of water. The EPA is also responsible for setting guidelines and standards to control surface discharge of specific water pollutants as well as underground injection of wastes to protect the purity of groundwater.

The EPA reviews the scientific literature and determines the maximum amount of a constituent that is safe in drinking water. This is termed the maximum contaminant level (MCL). MCLs of 90 possible contaminants in drinking water are outlined in Appendix D. By law drinking water with constituents above the MCL requires some sort of remediation to lower the concentrations to within acceptable limits.

## Microorganisms

Pollutant microorganisms enter waterways from untreated or poorly treated sewage from sewers and storm drains, septic tank leaks, runoff from farm fields, and boats dumping sewage. Two important microorganisms of concern are *Giardia lamblia* and *Cryptosporidium* (**Figure** 13.6). Both are pathogens, that is, infectious agents that colonize and reproduce in the human small intestine and cause diarrheal illness. They are not transmitted by insects, but rather are excreted in feces that are transmitted in water containing untreated sewage. They are transferred to a new host by drinking contaminated water. *Giardia lamblia* and

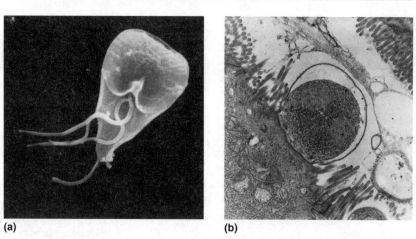

(a)                                                    (b)

Figure 13.6 (a) *Giardia lamblia*, a flagellated protozoa parasite. (b) A number of cryptosporidium. These are spherical protozoan parasites.

*Cryptosporidium* become a problem where there is inadequate sanitation or treatment of drinking water. Fecal-oral transmission is also common, and often happens among infants at day-care centers.

*Giardia lamblia* causes gastrointestinal disturbances. The parasite attaches to the lining of the small intestine, where it interferes with the body's absorption of fats and carbohydrates. Normally the illness lasts for one to two weeks, but chronic infections lasting months to years have been reported. It is common in the former Soviet Union, Mexico, Southeast Asia, and western South America. It can survive for weeks in cold water and is present in wilderness regions in North America requiring the boiling of water for at least one minute before drinking.

*Cryptosporidium* causes a diarrheal illness that can last for two months. It is one of the most common worldwide waterborne diseases. It has been estimated that *Cryptosporidium* accounts for 20% of all cases of childhood diarrhea in developing countries. There have been six major outbreaks of cryptosporidiosis identified in the U.S. One outbreak in Milwaukee, Wisconsin in 1993 infected more than 400,000 people. In all six cases local utilities met all state and federal drinking water standards. The problem is that *Cryptosporidium* is small and can pass through many drinking water filtration systems as well as being resistant to chemical treatments such as chlorination. *Cryptosporidium* has been found to be 240,000 times more resistant to chlorination treatments of water than *Giardia lamblia*. It is typically treated by reacting the water with ozone ($O_3$), an effective disinfectant for *Cryptosporidium*.

## Human-Made Chemicals

Human-made chemicals have been found in drinking water for many years. An investigation in 2008 by the U.S. Geological Survey (USGS) found 130 human-made chemicals in natural waters. These included disinfectants, pesticides, gasoline species, and household-use solvents. The investigation found most of the concentrations were below levels that would be harmful to humans. Most of the human-made chemicals assessed in the USGS investigation are unregulated in drinking water and are not required to be monitored or removed from municipal water. About 2/3 of the 130 chemicals remained in the water even after water treatment. Concern has been expressed about possible harmful synergistic effects between the various chemicals.

## Inorganic Chemicals

Inorganic chemicals typically are those that do not contain carbon and include metals like Pb and Hg as well as metalloid elements like arsenic. Some metals in water like Ca and Mg are needed in the human body. However, even with Ca and Mg, an excess in water produces "hard" water which limits the formation of lather with soap.

Surface and underground water can become acidic by interacting with human-made material buried or released into the water, by natural organic acids in soils, and with $CO_2$ in the atmosphere. The increased acidity in water reacts with minerals and buried waste increasing the concentration of inorganic chemicals in water. In this way inorganic chemicals in water may increase their concentrations to unsafe levels.

Arsenic (As) and nitrate ($NO_3^-$) in water are two inorganic chemicals of particular concern. Other inorganic chemical contaminants of importance are listed in Appendix D. Arsenic is a poison even at very low levels. The enforceable maximum contaminant level (MCL) for As allowed by the EPA in drinking water was lowered from 50 ppb (parts per billion) to 10 ppb in 2001. 10 ppb is also the standard of the European Union.

### Arsenic (As).

Arsenic occurs at high levels in groundwater of many countries. This is a particular problem in Bangladesh and parts of Southeast Asia where 50 million people are estimated to drink water with As levels above 10 ppb. Some locations in the U.S., such as Fallon, Nevada, have groundwater with As concentrations in excess of 80 ppb (**Figure** 13.7). The estimated lethal accumulation dose of As is between 70 and 200 mg (70,000 and 200,000 ppm). Lower doses can cause stomach aches, nausea, vomiting, and diarrhea.

High concentrations of As are found in sulfide minerals. It can enter drinking water as a result of being leached from sulfides in mine wastes. There are also industrial sources of arsenic in the environment both from sulfide metal smelting operations and from the use of chromated copper arsenate (CCA). Concerns about CCA in railroad

ties is pointed out in the chapter on specialty metals where As production is discussed.

As is present at higher concentrations in sedimentary rocks than in other rock types. High As is particularly a problem in aquifers present in alluvial sediments. The alluvial sediments allow water to react with a large amount of minerals that contain As weathered from sulfides present in exposed mountainous areas. Most of the time As is absorbed on iron hydroxides (FeOOH) in these sediments and doesn't enter the aquifer's water. However, decay of organic matter in the sediments has been shown to reduce As absorption on iron hydroxides. This leads to desorption of arsenic from iron hydroxide surfaces and transfer into groundwater. This is what is happening in Bangladesh where human-generated organic wastes are entering aquifers.

### Nitrate ($NO_3^-$).

Nitrate enters the environment mainly from the use of nitrate-containing fertilizers. These dissolve into runoff water and can make their way to local water supplies. Nitrate that enters the human body by drinking water is transformed to nitrite ($NO_2^-$). Nitrites oxidize Fe atoms in blood from $Fe^{2+}$ to $Fe^{3+}$ rendering it unable to carry oxygen. Infants are particularly susceptible to nitrate toxicosis. Consuming water with greater than 10 ppm nitrate can cause them to develop hypoxia. This is also referred to as "blue-baby syndrome" as the blood is less able to color the skin pink. Hypoxia can lead to unconsciousness, coma, and ultimately death.

## Organic Chemicals

Organic chemicals are typically carbon-containing compounds. They can be volatile or nonvolatile. *Volatile organic chemicals* (*VOCs*) are carbon-containing compounds that evaporate when exposed to air at normal temperatures. These contain short chains of carbon atoms or are made up of a single benzene ring and its attached functional groups. A benzene ring is a common organic structure made up of a ring of six carbons with alternating single and double bonds.

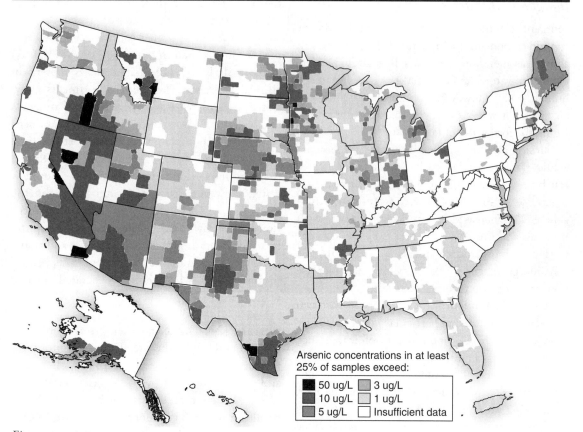

*Figure 13.7* U.S. arsenic occurrence map by county from data compiled by the U.S. Geological Survey from 31,000 water samples. Counties with less than five wells are considered to have insufficient data. Arsenic has been found at levels of health concern in the tap water of tens of millions of Americans in 25 states. A potentially fatal added cancer risk above 1 in 10,000 if a liter of water with 1.0 ppb = 1.0 µg/liter is consumed each day. (Data from: U. S. Geological Survey.)

VOCs are present in liquid fuels, degreasers, solvents, metal and furniture polishes, cosmetics, drugs, and dry cleaning chemicals. Common VOCs include trichloroethylene, benzene, toluene, and xylene (**Figure 13.8**). The EPA estimates that VOCs are present in 1/5 of the nation's drinking water supplies. Important nonvolatile organic chemicals include *polyaromatic hydrocarbons* (*PAHs*) and *polychlorinated biphenyls* (*PCBs*) as shown in **Figure 13.9** as well as the unintended by-products of water and paper chlorination processes, dioxins.

**Dioxins.** *Dioxin* is the common name given to polychlorinated dibenzene compounds (**Figure 13.10**). They are supertoxic. The MCL is set at only 0.00003 ppb although some argue that no

safe level exists in water. The most common is <u>t</u>etra-<u>c</u>hlorinated <u>d</u>i-benzene <u>d</u>ioxin, which is designated as 2,3,7,8-TCDD. The numbers give the locations of the four chlorine atoms in the compound. Dioxins are not intentionally made but occur as a by-product of many chemical production, manufacturing, and combustion processes that use or produce Cl in the presence of organic molecules. This includes incineration as well as bleaching during paper manufacture. When plastics that contain chlorine, particularly polyvinyl chlorides (PVCs), are burned in the presence of wood or paper, some dioxins are produced.

Exposure to dioxins can cause cancer and severe reproductive and developmental problems

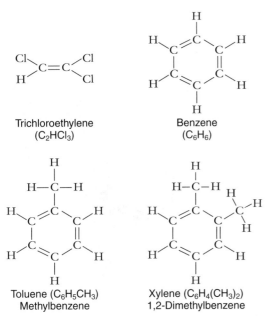

Figure 13.8 Structural formulas of trichloroethylene, benzene, toluene, and xylene.

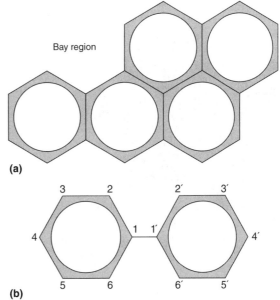

Figure 13.9 (a) Structure of benzol[a]pyrene, a polyaromatic hydrocarbon (PAH) showing the chemically active bay region. In this case the five fused benzene rings (referred to as phenyl groups when present in a structure) are shown as a circle enclosed in a hexagon rather than the hexagon of carbons with attached hydrogen atoms. (b) Structure of polychlorinated biphenyls (PCB) indicating the positions from 2 to 6 on each phenyl group of the possible 10 locations of Cl substitution for H in the phenyl groups.

because of their ability to damage and interfere with the body's immune and hormonal systems. Of particular concern are increased birth defects, loss of pregnancies, and decreased fertility. Dioxins are fat-soluble, persistent, and bioaccumulating. A substance bioaccumulates in an organism by increasing its concentration from continued low doses from water or eating other organisms containing the substance over time. As a result dioxins increase up the food chain as organisms eat other organisms that have accumulated them. Dioxins are a particular concern when eating freshwater fish (**Figure 13.11**).

# Global Water Usage

Some uses of freshwater by humankind are consumptive such as drinking and irrigation. Others are nonconsumptive such as the transport of goods, recreation, and sanitation. Hydroelectric plants use large amounts of freshwater and return most of it to its source. Although nonconsumptive use does not diminish the supply of water available for subsequent usage, the use can change

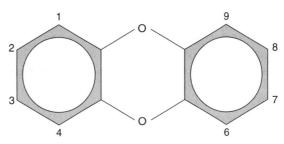

Figure 13.10 Structural formula of dioxins (polychlorinated dibenzene compounds). The numbers indicate where Cl atoms can replace the H atoms in the structure.

the composition or temperature of the water. Also nonconsumptive use, if large compared to water availability, can change the natural flow dynamics of the water.

*Figure 13.11* A sign posted along the Tittabawassee River near Midland, Michigan warning fishermen to limit fish consumption because of dioxin contamination.

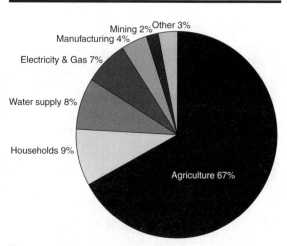

*Figure 13.12* Global freshwater use by sector in 2005. (Adapted from the Climate Institute.)

Like peak oil and peak metal production, it has been argued that humankind is reaching "peak freshwater," at which time populations will encounter natural limits to availability of freshwater. The volume of freshwater resources worldwide is around 35 million $km^3$. The world's 7 billion people are using a little over half of all the accessible freshwater contained in rivers, lakes, and underground aquifers. The daily drinking water requirement per person is only 2 to 4 liters, but it takes 2,000 to 5,000 liters of water to produce one person's daily food needs.

Freshwater use has been growing at more than twice the rate of population increase in the last century. Freshwater withdrawals have, therefore, increased three-fold over the last 50 years with the demand for freshwater increasing by 64 $km^3$ a year. It is estimated that in 2010 a little over 4,300 $km^3$ was withdrawn from water sources with 2,500 $km^3$ of this freshwater consumed. The nonconsumed water is used principally as cooling water for such industries as electric power generation. After use, other than that which has evaporated, it is then returned to the water reservoir from which it was taken.

Estimates of global freshwater use by sector are given in **Figure 13.12**. Note that agriculture consumes 2/3 of all the water used. The relative amounts of water withdrawal for consumptive use by region on earth are outlined in **Figure 13.13**. Water consumption tracks the size of the population in that 60% of the world's population lives on the continent of Asia and a little less than 70% of the world's total freshwater is consumed there. However, there are regions where water consumption does not track population well as 14.5% of the world's population lives on the continent of Africa but it consumes only 8% of the freshwater used in the world and North America is home to 5% of the world's population but accounts for 10% of the world's water consumption. Much of this is due to agricultural production as North America is a net exporter of food and Africa is a net importer. From the perspective of water use, food can be considered concentrated water that is traded on the world market.

## United States Water Usage

Given in **Figure 13.14** is the consumptive and nonconsumptive uses of freshwater in the U.S. Note that when nonconsumptive uses such as for thermoelectric power generation are considered, its usage is of similar size to that for irrigation in agriculture. The public water supply accounts for only 12% of the freshwater withdrawals.

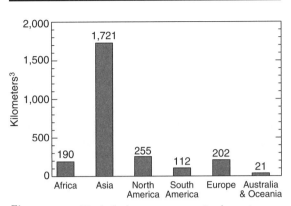

*Figure 13.13* Yearly freshwater consumption by region projected to 2010. (Data reported by Shiklomanov, I.A. 1998. *"Assessment of water resources and water availability in the world."* Report for the Comprehensive Assessment of the Freshwater Resources of the World, United Nations. Data archive on CD-ROM from the State Hydrological Institute, St. Petersburg, Russia.)

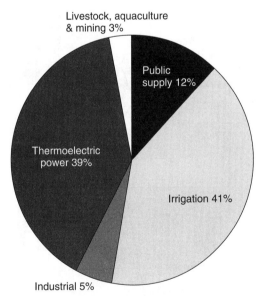

*Figure 13.14* Water usage by type in the U.S.

## Crop Irrigation

Water extracted for crop irrigation is about 40% of total freshwater withdrawals in the U.S. and about 60% worldwide. One-third is returned to the earth's surface with 2/3 lost by evaporation from wet surfaces and transpiration from plants. The average application rate of water on farmland has been 75 cm per year in the U.S. Surface water

accounted for 58% and groundwater 42% of these withdrawals. Irrigated farmland is located where there is insufficient precipitation to support crops without supplemental water (**Figures 13.15 and 13.16**). This is typically where annual precipitation is less than 50 cm per year. Therefore, the majority of water withdrawals were made in the western U.S. One-half of the total water withdrawn for irrigation is used in California, Idaho, Colorado, and Nebraska. Withdrawals for irrigation have stabilized between 500 and 520 billion liters per day between 1985 and 2010. This stabilization is a balance between climatic change

*Figure 13.15* Gravity furrow system, where small, shallow channels are used to guide water downslope across the field. Plants are typically grown on the ridge between the furrows.

*Figure 13.16* Lateral-move crop irrigation system being used near Fort Calhoun, Nebraska.

that has decreased available precipitation and advances in irrigation efficiency.

### Thermoelectric Power

Thermoelectric power is the production of electricity by heating water to produce steam to rotate turbines that run electric generators. It accounts for 39% of freshwater withdrawals. The majority of water is used to condense the produced steam back into liquid water and to cool this hot water. The water needs to be cooled before it is released back into the environment because of the damage to wildlife that heated water can cause. Typically the hot water is sprayed within a cooling tower.

Some of the water spray evaporates inside the tower. The evaporation absorbs heat and cools the remaining water.

Shown in **Figure 13.17** are the source, use, and disposition of freshwater used in the U.S. Note that over 3/4 of the water used is from surface water while the remainder is groundwater for a total of 340 billion gallons or 1,290 liters per day. With about 310 million people in the U.S. the per capita consumption is 1,100 gallons or 4,160 liters a day. Of the water used 71% is returned to the earth's surface and groundwater while 29% is consumed by evaporation and transpiration mainly from the water used for crop irrigation.

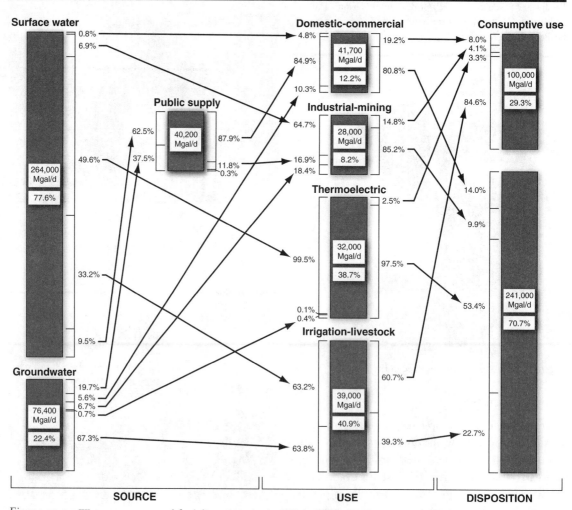

*Figure 13.17* Water source, use and final disposition in the U.S. in 1995 given in percent and millions of gallons a day (Mgal/d). (Modified from U.S. Geological Survey Circular 1200.)

# Water Law

Water law outlines the right to use, control, and own water. It incorporates common law concerns, which are based on previous practices and court cases. Similar to property law, it considers claim issues, but is influenced by environmental law in that water pollution control and water resource conservation and management are considered. Water is generally mobile with its supply varying by location, year, and season. A particular volume of water can satisfy many needs, simultaneously, for widely different purposes. For instance, a river can be used by municipalities for water supply and wastewater disposal while others use it for transport of goods, fishing, or recreational activities. Because water can both define and flow across national boundaries its regulation needs to consider international law including treaties and conventions. Water law differs greatly from country to country depending on how abundant freshwater is.

As the supply of water changes with time, and is generally in motion it is difficult to regulate using laws based on ownership. In regions where there is plentiful water relative to the users, water law is generally not contentious. In arid areas where farming is by irrigation and water is in short supply conflicts can often develop.

## Water Law in the U.S.

In the U.S. water law is generally passed by state legislatures and differs between the water-rich eastern U.S. and the relatively water-poor western U.S. Water law gives *riparian rights* for usage in the eastern U.S. where the abundance of water is not typically a problem. Riparian rights allow reasonable water usage to every landowner along a lake or river unless it interferes with the rights of other landowners. It gives ownership of land to the center of the body of water unless known to be owned by someone else or is navigable. Navigable water is that used for commercial transportation and is controlled by the state if wholly intrastate and by the federal government if not. Riparian rights allow water to flow onto your property in its natural quantity and quality. It requires any changes from its natural course to not affect the rights of other landowners. These rights have been modified locally to accommodate the water needs for cities and public utilities. These include the right of municipal governments to use eminent domain to obtain their water. Eminent domain for water is the power of the government to seize a citizen's right to water with due monetary compensation, but without the owner's consent.

### Surface Water

During their settlement, western U.S. states did not pass laws to control water use from rivers and other water sources. Water was treated like air as though it belonged to no one. In the absence of laws, people simply took water for their use; that is, they appropriated it. This practice was legalized and became known as the *Doctrine of Prior Appropriation*, having grown out of mining claim law. In the West, water law varies by state but is generally by prior appropriation. Prior appropriation gives absolute rights to use water independent of land ownership even if its use causes ill effects to others. The first person to use (i.e., divert) a quantity of water from a water source for a "beneficial" use has the right to continue to divert that quantity of water for that purpose. Using water to maintain a lake and the wildlife that depends on it, while originally not considered a "beneficial" use, has been accepted in some jurisdictions as "beneficial" more recently. Any subsequent persons can use the remaining water for their own "beneficial" purposes provided that they do not impinge on the rights of any previous users. By being a user one establishes right of usage.

Each water right has a yearly quantity and an appropriation date. A water right can be sold to any party, and retains its original appropriation date and type of usage. As outlined earlier in the text, Los Angeles bought the water rights of water running off the Sierra Nevada Mountains into the Owens Valley. This appropriated water is transported to Los Angeles by an aqueduct. A number of lawsuits concerning the ownership of this water have been decided with Los Angeles continuing to maintain rights to the Owens Valley water.

Most states where prior appropriation is the law have over time developed a list of preferential usage of water for times of shortage. Domestic and municipal usage generally has the highest preference, followed by water for economic activities (e.g., agriculture and mining) and finally recreational activities. California, Kansas, Nebraska, North and South Dakota, Oklahoma, Oregon, Texas, and Washington have water laws that have recognized both riparian and prior appropriative water rights. This came about because riparian rights were historically recognized, but the state has changed to an appropriative system. The riparian rights were converted to appropriative rights when the system changed. Generally, riparian landowners were able to claim a water right by a certain time and incorporate it into the state's prior appropriation system. These initial riparian rights tend to be superior in claim to any prior appropriative rights even if the water was not put to beneficial use until much later.

In 1908 the U.S. Supreme Court in Winters *vs.* United States held that when the United States sets aside public land for Indian reservations, military bases, and national parks, forests, or monuments, it also implicitly reserved sufficient water to satisfy the purposes for which it was dedicated. These rights must meet a "primary purpose" and "minimal needs" requirement. Before 1908 water rights were considered a state matter.

Statutes passed by both the states and the federal government are constantly modifying water law. Not only is new water law made with regard to federal legislation on water quality such as the Federal Water Pollution Control Act of 1948, Water Quality Act of 1965, Safe Drinking Water Act of 1974, and Clean Water Act of 1977, but also with an awareness of wildlife habitat preservation such as the Endangered Species Act of 1973, North American Wetlands Conservation Act of 1989, and the Water Resources Development Act of 2000. The result of the legislation is that present law considers freshwater resources in terms of environmental impact as well as supply, demand, and contamination.

### Groundwater

Regulating groundwater is even more difficult than surface water. The withdrawal of groundwater depends on both how much is present and how fast it can be replenished. Its quality can change with time. Groundwater can either be privately or publicly owned. Groundwater rights are a particular concern in the western U.S. where more groundwater is used. The laws regulating groundwater vary depending on the state. Some states divide groundwater into that which is hydrologically connected to surface water as with water-table aquifers and water that is not as with regional confined aquifers. The former falls under the rules governing surface water and the latter is considered separately. Unconnected underground water has no impact on the withdrawal of surface water; this water is considered independently of surface water rights.

As an understanding of the dynamics of underground water is improving water law has been changing. For instance "conjunctive" conditions between surface and groundwater and "safe yield" are often recognized. Safe yield or sustainable yield is the amount of water that can be withdrawn without having an impact on other parts of the water cycle. This can't be equated with average natural recharge and is difficult to determine because groundwater extraction causes time-dependent recharge and discharge effects and needs to be considered on the basis of hydrologic principles of mass balance.

Western states are putting the control of groundwater as well as surface water in the hands of water commissions. A system involving private ownership of water rights that is currently evolving is the liability rule known as the "reasonable use doctrine." Reasonable use states water may be used for a suitable and beneficial purpose as long as it does not lead to unreasonable interference with another's use of the water.

## International Water Law

There are 261 river basins covering 45% of the earth's land surface (excluding Antarctica) that are shared by more than one country. Mutually acceptable arrangements for use of transboundary freshwater, even between friendly and cooperative neighboring countries, is typically difficult. International law is based on the historic practices of nations. It is codified in international agreements, resolutions at international assemblies, decisions made by international courts and arbitrators, and often by unilateral acts of nations. This makes the law decentralized and in many cases conflicting.

When it suits their interests, countries often claim absolute control of the use of water in a river that passes through their boundaries. However, countries further down the river typically argue for the absolute integrity of the river in that upriver countries can do nothing that affects the quantity or quality of water in the river. A compromise known as the rule of equitable utilization has been implemented in some situations. This recognizes a watershed as a legal and managerial unit. Restricted sovereignty of the water is given to each country in the basin, but each country recognizes the right of all the other countries in the watershed to use water from the river. Uses by a country need to be managed so as not to interfere unreasonably with similar uses by other watershed countries.

The United Nations (UN) Convention of 1997 on the Non-Navigational Uses of International Watercourses codified the rule of equitable utilization. The problem is, what is equitable? Factors that must be taken into account are found in Article 6 of the UN Convention and include:

1. The geographic, hydrographic, hydrologic, climatic, ecological, and other factors of a natural character;
2. The social and economic needs of the watercourse nations concerned;
3. The effects of the use or uses of the watercourse in one watercourse nation on other watercourse nations;
4. The existing and potential uses of the watercourse;
5. The conservation, protection, development, and economy of use of the water resources of the watercourse and the costs of measures taken to that effect; and
6. The availability of alternatives, or corresponding value, to a particular planned or existing use.

Ratification by enough countries to make the UN Convention of 1997 enforceable never occurred.

Every three years starting in 1997 the World Water Forum has been held. It is the largest international event in the field of freshwater concerns. The successive World Water Forums have served as stepping-stones toward global collaboration on water problems (**Figure 13.18**). In 2000 at the 2nd World Water Forum 159 national delegations met and produced the Ministerial Declaration, "Water Security in the 21st Century." The importance of freshwater to the life and health of both people and ecosystems throughout the world was outlined. The Declaration states ecosystems are under threats from "pollution, unsustainable use, land-use changes, climate change, and many other forces." It then argues that these threats will hit the poor first and hardest. As a result governments and international institutions need to provide water security. By water security what was meant is that "every person has access to enough safe water at an affordable cost to lead a healthy and productive

*Figure 13.18* 6th World Water Forum in Marseille, France in 2012. The Forum brought elected officials from 173 countries together to create understanding for the urgency of water-related issues.

life and that the vulnerable are protected from the risks of water-related hazards."

The International Law Association founded in Brussels in 1873 studies, clarifies, and develops both public and private international law. In 2004 it adopted the Berlin Rules on Water Resources. These rules, however, have no mechanism of enforcement. They require countries to take appropriate steps to both sustain and manage their freshwater resources, and minimize environmental harm. The Berlin Rules contain regulations for nations to follow with respect to water within their boundaries as well as water they share with other countries. In wartime, countries are not permitted to take action that may result in a shortage of life-sustaining water for civilians or cause undue ecological damage. An exception is a nation being invaded that is compelled by military emergency to disable its own water supply.

Poisoning of the water necessary for survival is in all cases forbidden.

For over the last 30 years in Europe, European Union law has increasingly replaced national legislation concerning water use and protection. This includes the Bathing Water Directive on bacteriological contamination, the Surface Water Directive on drinking water, the Dangerous Substances Directive on pollution caused by dangerous substances discharged into water, the Groundwater Directive for the protection of groundwater against pollution, and the Urban Waste Water Directive to control municipal wastewater problems including the algal blooms in the North Sea. In 2000 the Water Framework Directive overlaid these previous directives. This Directive requires member countries to achieve good qualitative and quantitative use status of all water bodies and the production of river basin management plans across Europe.

## SUMMARY

The composition of both surface and groundwater changes as a function of time and space. Rainwater changes it composition from dissolution of particles in the atmosphere and by incorporating gases such as $CO_2$, $NH_3$, $SO_2$, and $NO_2$. The $CO_2$ naturally found in uncontaminated air gives rainwater a pH of 5.6. $SO_2$ and $NO_2$ released from burning coal and combusting petroleum, respectively, produces acid rain.

Rivers carry rainwater, groundwater, and the products of surface and soil weathering to lakes and the ocean. Most rivers have $Ca^{2+}$ as the major cation and $HCO_3^-$ as the major anion. Lakes receive water from both rivers and groundwater with a smaller amount from the atmosphere and runoff. They tend to have compositions similar to the rivers that flow into them. Groundwater contacts a large amount of rock surfaces and moves more slowly than rivers. As a result, groundwater often contains more dissolved material. Seawater is, to a first approximation, a half-molal NaCl solution that contains an order of magnitude less $MgSO_4$.

All drinking water contains some dissolved constituents. Contamination in drinking water in the U.S. is regulated by the Environmental Protection Agency. Two harmful microorganisms found in water are *Giardia lamblia* and *Cryptosporidium*. Human-made chemicals have been found in drinking water for many years but most are at a level too low to be harmful to humans. Inorganic chemicals of significance are arsenic from both natural and pollution sources and nitrate from fertilizers. Organic chemicals of importance include volatile organic chemicals (VOCs), polyaromatic hydrocarbons (PAHs), and polychlorinated biphenyls (PCBs). Of particular concern are dioxins as their maximum acceptable concentration level (MCL) is only 0.00003 ppb. These compounds are fat-soluble, persistent, and bioaccumulating.

Saltwater is converted to freshwater by distillation or membrane filtration. Typically multistage flash distillation, charged membrane, and reverse osmosis methods are employed. Global freshwater usage has grown at twice the rate of population increase with 2/3 for agriculture. In the U.S. water usage for thermoelectric power generation is of similar size to that for agriculture.

Water law in the U.S. is considered on a state-by-state basis. It gives riparian rights in the East whereas the Doctrine of Prior Appropriation is the basis for water law in the West. Groundwater is more difficult to regulate than surface water. However, with an increased understanding of the dynamics of the flow of groundwater western states use water commissions to set policies.

International water law is based on historic practices of nations. The United Nations has set guidelines to be considered. The World Water Forum attempts to move toward global collaboration on water issues. In Europe, European Union water law has replaced national laws for water usage and protection.

## Key Terms

| | |
|---|---|
| acid rain | polyaromatic hydrocarbon (PAH) |
| *Cryptosporidium* | polychlorinated biphenyl (PCB) |
| dioxin | reverse osmosis |
| distillation | riparian rights |
| Doctrine of Prior Appropriation | volatile organic chemical (VOC) |
| *Giardia lamblia* | |

## Problems

1.  If the maximum permissible concentration level of dioxin in drinking water = 0.00003 ppb, what is the maximum number of dioxin molecules allowed in a 100 g glass of drinking water? 2,3,7,8-TCDD has a chemical formula of $C_{12}H_4Cl_4O_2$. Remember that the number of items per mole, Avogadro's number = $6.022 \times 10^{23}$.

2.  Considering Figure 13.17, if all the domestic plus commercial freshwater needed in the U.S. was produced from seawater what would the daily energy cost be? Energy cost for seawater desalination is about 50 cents per cubic meter. 1 U.S. gallon = 0.003785 cubic meters.

3.  Using the values in Appendix D, put the following in order from lowest to highest MCL in terms of moles per liter of water: antimony, arsenic, and beryllium.

## References

Advameg, Inc. 2010. *Water Encyclopedia Science and Issues*. Available online at URL: www. waterencyclopedia.com/En-Ge/Fresh-Water-Natural-Composition-of.html. Accessed 01 Aug 2012.

Brandon, Leigh. 2009. *Are Hidden Internal Stressors Making You Sick? - Part 3*. Available online at URL: http://leighbrandon.typepad.com/brandos_blog/2009/06/are-hidden-internal-stressors-making-you-sick-part-3.html. Accessed 01 Aug 2012.

Climate Institute. 2008. *Water and Climate Change*. Available online at URL: www.climate. org/topics/water.html. Accessed 01 Aug 2012.

Craig, J. R., Vaughan, D. J., and Skinner, B. J. 2001. *Resources of the earth origin, use, and environmental impact*, 3rd ed. Upper Saddle River, NJ: Prentice Hall, Inc.

National Institute for Agricultural Research (INRA) of France. 2007. Breeding: stimulating the immune system in newborn animals. *Live from the labs* No. 16. Available online at URL: www.international.inra.fr/partnerships/with_the_private_sector/live_from_the_labs/breeding_stimulating_the_immune_system_in_newborn_animals. Accessed 01 Aug 2012.

Shiklomanov, I. A. 1998. *Assessment of water resources and water availability in the world*. Report for the Comprehensive Assessment of the Freshwater Resources of the World, United Nations. Data archive on CD-ROM from the State Hydrological Institute, St. Petersburg, Russia.

Starnes, Wayne C. 1995. Colorado River Basin Fishes. In LaRoe, E. T., Farris, G. S., Puckett, C. E., Doran, P. D., Mac, M. J. (eds.) *Our living resources: A report to the nation on the distribution, abundance, and health of U.S. plants, animals, and ecosystems* (pp. 149–153). Washington DC: U. S. Department of the Interior, National Biological Services.

USA Department of Agriculture, Economic Research Service. 2012. *Irrigation & Water Use*. Available online at URL: www.ers.usda.gov/briefing/wateruse/glossary.htm.

U.S. Geological Survey (USGS), 1996, Estimated use of water in the United States in 1995, Circular 1200, (URL: water.usgs.gov/watuse/pdf1995/pdf/circular1200.pdf).

U.S. Geological Survey (USGS), 2008, Man-Made Chemicals Found in Drinking Water at Low Levels, (URL: www.usgs.gov/newsroom/article.asp?ID=2086#.UBqfnUQmb_s).

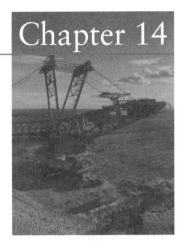

## Chapter 14

# Soil as a Resource

*"Essentially all life depends on soil…. there can be no life without soil and no soil without life; they have evolved together."*
— *Charles E. Kellogg—USDA*

*"… the life expectancy for a civilization depends on the ratio of initial soil thickness to the net rate at which it loses soil."*
— *David R. Montgomery in Dirt: the Erosion of Civilizations*

Civilizations have repeatedly degraded the soil on which their agricultural production is based. Some have even collapsed due to deterioration and loss of their soil. One of the primary reasons the Roman Empire sought new lands was because of the loss and deterioration of their soils. The migrations and decay of the Mayan civilization is often attributed to crop loss from soil deterioration. Depletion of nutrients in the soils of the southern U.S. was a factor that brought on the U.S. Civil War (Bagley, 1942). In order for modern societies to avoid similar situations requires an understanding of the capacity of soil to grow food.

The other important aspect of soil is its mechanical properties. Soils are porous but retain water. They can be very hard or quite plastic. This becomes important when considering such activities as constructing building foundations or earthen dams.

## Composition and Description

Soil is the dynamic layer on the earth's land surface that can contain living matter and is capable of supporting organic growth. As a life support system soils can be considered along with water as one of humankind's most basic and important natural resources. Soils form as a result of physical, biological, and chemical interactions of rocks with the hydrosphere, atmosphere, and biosphere. In contrast, regolith is the term used for broken unconsolidated material devoid of living matter as in the Moon's regolith. About 68% of the earth's ice-free land surface is covered with soil.

### Soil Characteristics

How does one describe a soil? Both the physical and chemical properties of the soil are important. The physical properties include the texture or size of soil particles. The texture of the soil is important because particle size influences a soil's drainage and nutrient holding capacity. Soils are formed from the disaggregation of rocks into its constituent grains and rocks typically do not have mineral grains larger than 2 mm. Therefore, common soils do not have particles larger than 2 mm. The larger particles in a soil tend to be equigranular minerals like feldspar and quartz while the smallest particles are platy clay minerals. In-between sized particles in soil tend to be equigranular.

This has lead to a three division classification of a soil's texture: sand, silt and clay as given in **Table 14.1**. Note that the size boundaries differ somewhat depending on who is describing the soil. Also the term clay is used here as a size term but as discussed both in the section on building materials and below clay is also used as a mineral composition term. This is rarely a problem as the finest fraction in a soil nearly always has a clay mineral composition. In the field the difference between sand and silt is that when a sample is rubbed between the fingers sand feels gritty while silt feels like ground flour. Shown in **Figure 14.1** is a triangular diagram, often referred to as a ternary diagram, used for naming soils depending on its texture. In this type of diagram the closer a soil sample is to an apex, the more of that component the soil contains.

Soil structure is the way individual particles of sand, silt, and clay are assembled, that is, aggregated together. There are five major classes of soil particle aggregation: platy, prismatic, columnar, granular, and blocky as shown in **Figure 14.2**. Natural soil aggregates are called *peds* whereas aggregates reworked and partially destroyed by plowing are called clods. The peds of a soil determine the ability of water penetration, moisture retention capabilities, susceptibility to erosion, and ease of plowing.

Another feature of soil is color, typically given by the *Munsell soil color index* (Munsell, 1929) for a moist soil. This index describes color based on the hue of the color, value of its brightness, and its colorfulness or color purity. Color often reflects the composition of the soil. For example, the very dark colors that approach black often indicate organic material (as in humus, see below) or reduced compounds of manganese in the soil, whereas red colors indicate oxidized Fe is present.

Soil consistency is a term used to describe the extent to which soil particles are held together and its resistance to deformation and rupture. It depends on the water content of the soil. Terms like moist or dry, loose or noncoherent, soft, hard, friable, plastic, sticky, and firm are used. For wet soils its stickiness is described by the terms given in **Table 14.2**. Plasticity of a soil is the extent

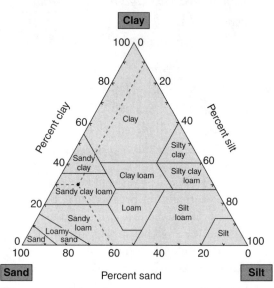

*Figure 14.1* Nomenclature for soil textural classes based on its wt% of sand, silt, and clay. The solid circle is a soil with 60% sand, 10% silt, and 30% clay that is classified as a sandy clay loam. Loam is a desirable soil for growing most crops and grasses as it retains nutrients, water, and soil air.

*Table 14.1* SOIL TEXTURE TERMS FOR SOIL SCIENTISTS AND GEOLOGISTS.

| PARTICLE | DIAMETER SOIL SCIENTIST* | DIAMETER GEOLOGIST† |
|----------|--------------------------|---------------------|
| Sand | 2.0–0.05 mm | 2.0–0.0625 mm |
| Silt | 0.05–0.002 mm | 0.0625–0.0039 mm |
| Clay | < 0.002 mm | < 0.0039 mm |

* U.S. Department of Agriculture scale.
† Udden-Wentworth scale.
mm = millimeter.

it will permanently deform without breaking. It can be determined by forming a string/rope of soil material. If no string/rope can be formed it is nonplastic. Degrees of plasticity are characterized on how the string/rope can be broken and reformed or more formally by its liquid limit of cohesion as given in **Table 14.3**. The greater the clay content of a soil the greater is its plasticity.

Chemical properties used to describe soils are its mineralogy including its content of gypsum and sulfides, as well as concentrations of nitrogen, phosphorous, potassium, and soluble salts. Soil pH is also considered (**Table 14.4**). pH is typically measured in a mixture of 5 g of soil to which 5 ml of distilled water is added. The mixture is stirred vigorously for 5 seconds and let stand for

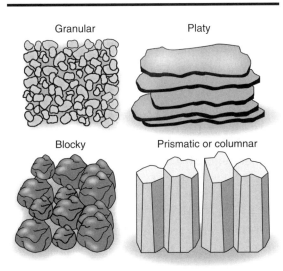

*Figure 14.2* Basic clay-rich soil structures. Note that prismatic and columnar are usually distinguished by their tops. In columnar soils the tops on the columns are very distinct and normally rounded while prismatic soils have tops that are indistinct and nearly flat.

*Table 14.2* STICKINESS OF SOIL.*

| SOIL ADHERENCE | DESCRIPTION |
|---|---|
| Non-sticky | Little or no adherence to fingers after finger separation |
| Slightly sticky | Adheres to both fingers after finger separation but with little stretching of soil |
| Moderately sticky | Adheres to both fingers after finger separation with some stretching of soil |
| Very sticky | Adheres firmly to both fingers after finger separation with significant stretching of soil |

\* Determined when a small blob of soil is first pressed between the fingers.

*Table 14.3* PLASTICITY OF SOIL.

| PLASTICITY | wL* |
|---|---|
| Low | ≤ 35% |
| Intermediate | 35%–50% |
| High | 50%–70% |
| Very high | 70%–90% |
| Extremely high | > 90% |

\* Where wL is the liquid limit of cohesion of soil from a plastic material to a liquid in percent water content.

*Table 14.4* USDA SOIL pH TERMINOLOGY.

| TERM | pH RANGE |
|---|---|
| Ultra acid | < 3.5 |
| Extremely acid | 3.5–4.4 |
| Very strongly acid | 4.5–5.0 |
| Strongly acid | 5.1–5.5 |
| Moderately acid | 5.6–6.0 |
| Slightly acid | 6.1–6.5 |
| Neutral | 6.6–7.3 |
| Slightly alkaline | 7.4–7.8 |
| Moderately alkaline | 7.9–8.4 |
| Strongly alkaline | 8.5–9.0 |
| Very strongly alkaline | > 9.0 |

10 minutes. The slurry is swirled before the pH electrode is inserted and pH determined. By applying an alternating current between electrodes in a soil water the electrical conductivity (EC) can be measured. This is done by measuring the change in this alternating potential by two additional nearby electrodes to indicate the amount of salt more soluble than gypsum in the soil water. This is, however, often difficult because of the typically small amount of water present, much of which is absorbed on the mineral surfaces in the soil.

## Soil Classification

A grouping system, the Comprehensive Soil Classification System, is in general use for naming soils similar to the taxonomic system that is used for naming organisms. From most general to most specific, the six categories of the classification system with the number of different types given in parentheses are:

Order (12)
Suborder (47)
Great group (185)
Subgroup (>1,000)
Family (>5,000)
Series (>15,000).

With closer examination of soils, new series and some new families are continuing to be added.

The "Houston Black" series soil is the Texas state soil and occurs on over 6,000 km² in what is called the Blackland Prairie of Texas that extends from north of Dallas south to San Antonio. Other official state soils can be found at the U.S. Department of Agriculture (USDA) website: http://soils.usda.gov/gallery/state_soils/. Nearly all Houston Black is cultivated for growing cotton, sorghum, or corn. It formed on weakly consolidated Ca-rich clays and calcite-rich mudstones. When dry it produces cracks ranging from 1 to 10 cm wide to a depth of 30 cm or more.

In the Comprehensive Soil Classification System it is a fine, smectitic, thermic, Udic Haplusterts. "Fine" refers to the subsoil containing 30% to 60% clay-size particles. "Smectitic" refers to the dominant clay mineral in the soil, which in this case is the clay mineral, smectite (see the chapter on building materials for a discussion of smectites). "Thermic" implies that the average soil temperature is between 15° and 22°C. Udic is the subgroup name and Haplusterts is the great group name. Udic Haplusterts soils are distinguished because if they are not irrigated they have cracks in normal years that are 5 mm or more in width, through a soil thickness of 25 cm or more within 50 cm of the mineral soil surface, for less than 150 cumulative days per year. The suborder is Usterts of the soil order Vertisol. Uderts are Vertisols with one of more faint horizons, that is layers, and are dry for an appreciable period of time during the year.

Given in **Table 14.5** are descriptions of the 12 different soil orders in the general classification system. A soil is placed in a soil order based on the presence or absence of diagnostic features found in the soil.

# Soil Distribution

Given in **Figure 14.3** is the distribution of soils orders on earth. Soils differ widely from one climatic zone to another. This implies climate plays a very important role in soil formation. For instance, the highly weathered oxisols develop near the equator while low temperature gelisols exist in polar regions. Climate change modelers can, therefore, use the past distribution of soils in the geological record to understand climate change over long time frames (**Box 14.1**).

# Soil Development

The major factors that control the development of a soil are the nature of the parental material, time of development, climate, presence of organisms, and the slope of the ground surface. The ground surface slope controls how much water will infiltrate into the soil as opposed to running off as well as the amount of sun the soil receives. It also influences the stability from erosion of the top layers of soil. The organisms, including vegetation, living in and on the soil respond to the

*Table 14.5* DESCRIPTIONS AND EXTENT OF COVERAGE OF THE 12 WORLD SOIL ORDERS.

| ORDER | DESCRIPTION |
| --- | --- |
| Ultisols | Clay-rich soil typical of subtropical, humid regions. Temperatures are generally > 8°C. They are common in the southeastern U.S. and Southeast Asia. These soils are acidic, pH ~5.6, and infertile, with a thick clay horizon. Ultisols occupy 8.1% of the earth's ice-free land area. |
| Alfisols | Soils of temperate humid and subhumid areas that receive more rainfall than is typical for Mollisols. They are highly productive forest soils, about equal in productivity to Mollisols. Alfisols have a thin A layer and a substantial B layer containing a grey, brownish, or reddish horizon of clay not darkened by humus. They cover 10.1% of the global ice-free land area and about 13.9 % of U.S. land area. |
| Spodosols | Soils of temperate moist climates. They are sandy conifer forest soils found in New England and Scandinavia with little silicate clay material. They form on acidic, pH ~4.9, sandy parent material with a subsurface accumulation of humus material containing Fe and Al in the B zone. They occupy ~4% of the earth's ice-free land area. |
| Mollisols | Dark-colored soils typical of prairie grasslands. They are highly productive and have a deep organic-rich topsoil, ~25 cm. They are the most common soil order in the U.S., 21.5% of the land area, typical of the Great Plains and common in mountain valleys. They occupy 7% of the global ice-free land area. |
| Andisols | Soils formed on volcanic ash or other volcanic ejecta mainly in tropic regions. They have high water-holding capacity and the ability to "fix" (and make unavailable to plants) large quantities of phosphorus. Globally, they are the least extensive soil order accounting for only 1% of the earth's ice-free land area. |
| Vertisols | Montmorillonite-rich soils that swell and shrink with changes in water content producing deep wide cracks when dry. This prevents formation of distinct, well-developed horizons in these soils. Vertisols are not common on a world scale occupying 2.4% of the earth's ice-free land area but are common in India and Texas. |
| Aridisols | Light-colored soils rich in carbonates. They are dry-land soils that typically have no soil moisture available for plants. These soils are not put into agricultural production unless irrigation is used. They are common in the western U.S. Aridisols are the third most abundant soil type worldwide covering 12% of the global ice-free land area. |
| Inceptisols | New soils that have weakly developed horizons and still contain much unweathered material. They are common in cool to very warm, humid and subhumid regions. They cover 17% of the global ice-free land area, the largest percentage of any soil order. |
| Entisols | Soils developed on recent deposits that lack horizons, or they are soils that are not very well developed because the climate is not conducive to soil formation. Another possibility is that the horizons were removed by uplift in regions subjected to tectonic stresses. They are nearly as abundant as Inceptisols occupying 16% of the global ice-free land area, making them the second most common. |
| Gelisols | Very cold climate soils found at high-latitude in polar regions or at extremely high elevations. They have no B horizon with the A horizon containing permafrost within 2 meters of the soil surface. Decomposition of organic materials occurs very slowly so these soils store large amounts of organic carbon. They cover 9.1% of the global ice-free land area. |
| Histosols | Organic-rich acidic soils that form peat. Defined as having > 40 cm of organic matter in the upper 80 cm of the soil. They are common in wet, swampy or marshy areas that are permanently waterlogged especially in cold climates such as bogs, moors, and peat fields. They have poor drainage and are often low in fertility. Histosols occupy 1.2% of the earth's ice-free land area. |
| Oxisols | The oldest or most highly weathered soils. They occur in warm, moist tropical regions, and are most common near the equator in South America and Africa. These soils have a red or yellowish color, due to the high concentration of iron. They are dominated by $Fe^{3+}$ and $Al^{3+}$ oxides/hydroxides such as bauxite and clays with no primary minerals remaining. Oxisols are very infertile and require regular additions of lime and fertilizer to grow crops. Oxisols occupy 7.5% of the earth's ice-free land area. |

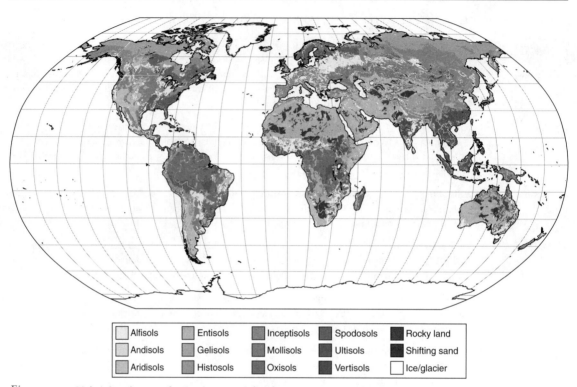

| | Alfisols | | Entisols | | Inceptisols | | Spodosols | | Rocky land |
| Andisols | | Gelisols | | Mollisols | | Ultisols | | Shifting sand |
| Aridisols | | Histosols | | Oxisols | | Vertisols | | Ice/glacier |

*Figure 14.3* Global distribution of soil orders. (Modified from USDA-NRCS, Soil Science Division.)

## Box 14.1 CLIMATES

Climate is the behavior of the atmosphere at the earth's surface averaged over numerous years, typically at least 30, in a particular area or region. This is distinguished from weather that is characterized by short-term events, on the order of days or less. Climate is characterized by average temperature, humidity, precipitation, and other meteorological measurements as a function of the time of year. Climate is affected by latitude, elevation, distance from the ocean, and terrain. The climate of a region will determine what plants will grow there, and what type of animals will be found. The Köppen climate classification system is the most widely used way of classifying the world's climates. There are five different climate groups in the system, A to E.

A—*Moist tropical climates* known for their high temperatures (> 18°C) year round and for large amount of year-round rain (≥ 60 mm in all 12 months). They usually occur within 5° to 10° latitude of the equator.

B—*Dry climates* characterized by very low precipitation and a very large daily range of temperatures. Two subgroups are:

BS—*Semiarid* or *steppe climates* where precipitation is somewhat below potential evapotranspiration. Potential evapotranspiration is the sum of evaporation from wet surfaces and water transpired by plants if ample plant water was continuously available.

BW—*Arid or desert climates* where precipitation does not sustain vegetation except in some cases where scrub bush is present.

C—*Temperate humid mid-latitude climates* found where land-water interactions are important. These climates have warm (average > 10°C), dry summers and cool, wet winters (coldest month average between –3°C and 18°C).

D—*Continental climates* occur in the interior regions of continents. Total precipitation is not very high and seasonal temperatures vary widely. They have average temperature > 10°C in summer, and below –3°C in winter.

E—*Cold polar climates* include areas where permanent ice and tundra are present. For only about four months or less of the year do they have above freezing temperatures. Sometimes alpine climates are separated from this group and given the climate group designation **H** for highlands.

In the Köppen climate classification system subgroups are designated by a second, lowercase letter that follows the climate group capital letter designations. These distinguish specific seasonal characteristics of temperature and precipitation as outlined in **Table B14.1**.

Given in **Figure B14.1** is one analysis of the climate regions of Australia. The climate of Australia varies widely. Most of Australia is closer to the equator than any part of the U.S. Therefore, northern Australia has a tropical climate and southern Australia a temperate climate. However, the dominant climates are desert and semiarid grasslands in central Australia. The temperature difference between winter and summer isn't large as it is a small continent and surrounded by water, which moderates the temperature of incoming weather patterns.

*Table B14.1* SUBGROUP DESIGNATIONS IN THE KÖPPEN CLIMATE CLASSIFICATION SYSTEM.

| SUBGROUP | CHARACTERISTICS |
| --- | --- |
| f | Moist with significant precipitation in all months and no dry season. This letter is usually used with A, C, and D climates. |
| m | Rainforest climate with short, dry monsoon season. Only used with A climates. |
| s | Dry season in the summer. |
| w | Dry season in the winter. |
| a | Hot summers where the warmest month is > 22°C. Used with C and D climates. |
| b | Warm summers where the warmest month is < 22°C. Used with C and D climates. |
| c | Cool, short summers with less than four months > 10°C. Used with C and D climates. |
| d | Very cold winters with the coldest month > –38°C. Used with D climate. |
| h | Dry-hot with a mean annual temperature > 18°C. Used with B climate. |
| k | Dry-cold with a mean annual temperature < 18°C. Used with B climate. |

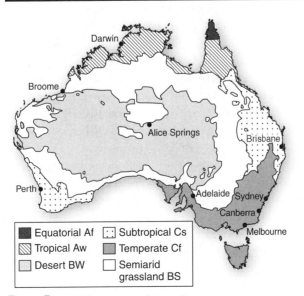

*Figure B14.1* Climate map of Australia.

climatic conditions in an area. Therefore, biotic factors and climatic soil development factors are interrelated.

Different types of vegetation that grow on a soil affect its development differently. For instance, the organic layer in soils developed under forested areas depends on whether the trees are deciduous or conifers. Deciduous forest recycle nutrient to the soil more rapidly. This retards leaching of basic cations and soil acidification. Soils developed under conifers are, therefore, more acidic.

Soils develop in layers or soil horizons from the breakdown and chemical weathering of primary minerals to secondary minerals together with any material transported by wind and water to or from the land surface where the soil is developing. These processes result in zones of accumulation and removal within the soil profile, and its generally layered appearance. Weathering involves the processes that decompose minerals and volcanic glass in rocks at the earth's surface to minerals that are closer to equilibrium in the earth's surface environment. This involves physical and biological breakdown as well as chemical alteration.

## Physical Properties

Rocks are ruptured by stresses to produce soils, which occur during pressure release, cyclic freezing plus thawing of any water present, temperature changes, abrasion, and root wedging. Pressure release takes place during the uplift of rocks that have formed at depth in the earth's crust. As the rocks above are eroded and removed, these deep-seated rocks are brought to the surface. The vertical stress on the rock decreases. Rocks then expand vertically and fracture horizontally. The rocks fragment into large flakes in a process termed *spalling* (**Figure 14.4**). Gravity plays a roll in the breakdown of these flakes to produce soil as mass movement such as landslides down a slope causes rock disaggregation.

Cyclic freezing then thawing of water in rocks can be an effective fracture mechanism. Liquid

*Figure 14.4* Spalling of large flakes of rock at Indian Rock, Yosemite National Park, CA.

water penetrates into a rock along a fracture. When it freezes the volume increases by about 7% when ice is formed. This causes great tensional stress in the rock leading to fracture. The new fracture allows deeper penetration of the water and the process is repeated, often on a daily basis. Temperature changes between day and night also cause rocks to expand and contract. Dark-colored grains in the rock expand more than light-colored grains because their greater absorption of sunlight results in greater heating expansion. This can lead to rupturing along grain boundaries and mineral cleavage planes over time.

Abrasion occurs by the rubbing of a rock surface by a frictional force. The abrasion can be due to rock or grain movement caused by gravity, wind, water, or glacial ice. The rubbing between surfaces dislodges and loosens material from the surface of the rock and splits minerals along their cleavage.

Root wedging occurs where roots intrude into crevices in rocks (**Figure 14.5**). As they grow and expand, roots exert stress on the sides of the crevice, which can break the rock apart. Trees are particularly effective at this as they have large, extensive root systems.

*Figure 14.5* Root wedging in a large boulder in northern Georgia.

## Soil Organisms

Animals living in the soil such as earthworms influence soil development by their mixing activities termed *bioturbation*. An acre of living topsoil, on average, contains approximately 400 kg of earthworms, 1,000 kg of fungi, 700 kg of bacteria, 60 kg of protozoa, and 400 kg of arthropods and algae, and even small mammals in some cases (Pimentel et al., 1995).

Most native soils are covered with a layer of growing plants or plant litter throughout the year. Below the litter is a mixture of soil organisms, decomposed plant residue, and dead roots. The organisms present facilitate release of nutrients from plant residue and mineral surfaces. They recycle these nutrients over and over again with the seasonal death and decay of each generation of plant. Where earthworms are found in soil their burrows enhance water infiltration. Soil aeration is also increased which stimulates aerobic bacteria. This leads to increased nutrient cycling.

Bacteria are the most numerous type of soil organism as every gram of productive topsoil typically contains over 1 million. These are microscopic simple single-celled organisms that lack a nucleus. There are many different types of bacteria in soil, each performing a different function. Many bacteria make nutrients available to plants.

Some of these release nitrogen, others release sulfur, phosphorus, or trace elements from organic matter in the soil to the soil solution. A different set of bacteria break down minerals in the soil, releasing potassium, magnesium, calcium, or iron. Some types of bacteria produce plant growth hormones, which stimulate root growth. Nitrogen-fixing bacteria transform nitrogen gas ($N_2$) in soil air into nitrate making it available for plant use.

Many different species, sizes, and shapes of fungi can be present in soil. These include single-celled yeasts, molds, and mushrooms. Fungi can colonize on large pieces of organic matter, for instance, wood that they decompose. Some fungi produce plant hormones. They live either on or in plant roots and act to extend the reach of root hairs into the soil. These fungi are particularly important in promoting plant growth in less fertile soils.

Fungi are more related to animals than plants as their cell walls are made of chitin that is also found in the exoskeletons of insects. Chitin is a tough substance consisting of a number of sugar molecules bonded together. Also fungi store energy in the molecule glycogen, which occurs in animal muscles and liver cells rather than in the starch found in plants.

## Chemical Properties

Weathering also involves chemical alteration processes. Primary igneous and metamorphic silicate minerals become unstable and dissolve in the water present in rock cracks and pores. New secondary minerals, which are more stable at the earth's surface, commonly clays, then grow from the pore water solution. Because of the need for pore solution, the rate of chemical alteration has a lot to do with access of the rocks to water.

The water that soaks into the ground produces a series of alteration layers or horizons in the soil determined to a large extent by solubility differences between minerals and chromatographic reactions of separation as shown in **Figure 14.6**. *Chromatographic separation* of layers occurs in soils because each component in the fluid has a different solubility and is absorbed by mineral surfaces differently.

Mineral surfaces in contact with pore solutions carry a pH-dependent surface charge. The repeating arrangement of atoms in the mineral has unsatisfied bonds at its surface that attach charged aqueous species of opposite sign from the pore solution. For silicate minerals, surfaces are negative in mildly acid to higher pH waters as large concentrations of negatively charged surface oxygen are exposed to the solution and the pH is high enough to keep many from becoming neutralized by attaching $H^+$ from solution. Cations adsorb to this negative surface. This is the major retention mechanism for heavy metals such as $Cd^+$, $Pb^{2+}$, and $Zn^{2+}$ in soils and is the basis of the general chromatographic behavior observed.

Poorly absorbed components of high solubility present in the fluid move downward more rapidly through the soil compared to strongly absorbed, low solubility components. This causes poorly absorbed soluble components to be concentrated near the bottom of the soil or to be leached out of the soil entirely and strongly absorbed components to remain concentrated near the top of the soil. In a typical soil this is complicated by slow rates of dissolution and precipitation reactions as well as the fact that organic matter is concentrated in the top layer.

A simple example of chromatographic separation occurs when a drop of black ink is put on a piece of blotter paper. As the organic and inorganic material in the paper absorb the black ink, the ink separates into rings of colors because of the different absorption properties of the different colored pigments in the ink to the material in the paper.

## Chemical Reactions in Soils

A number of different kinds of chemical reactions occur as acid-rich pore water reacts with rocks during the formation of a soil. Slight acidic (pH ~5.7) water is produced in the atmosphere due to the reaction of the rain with atmospheric $CO_2$ through which it falls. Additional carbonic acid ($H_2CO_3$) is then added to the water by reactions of $O_2$ in the

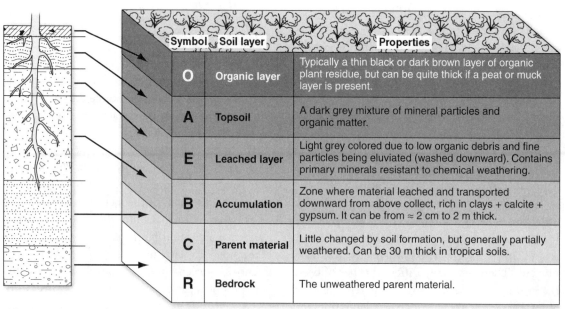

| Symbol | Soil layer | Properties |
|--------|-----------|------------|
| O | Organic layer | Typically a thin black or dark brown layer of organic plant residue, but can be quite thick if a peat or muck layer is present. |
| A | Topsoil | A dark grey mixture of mineral particles and organic matter. |
| E | Leached layer | Light grey colored due to low organic debris and fine particles being eluviated (washed downward). Contains primary minerals resistant to chemical weathering. |
| B | Accumulation | Zone where material leached and transported downward from above collect, rich in clays + calcite + gypsum. It can be from ≈ 2 cm to 2 m thick. |
| C | Parent material | Little changed by soil formation, but generally partially weathered. Can be 30 m thick in tropical soils. |
| R | Bedrock | The unweathered parent material. |

*Figure 14.6* General soil properties, layer development, and symbols for a typical soil. Note that most soils do not have every different kind of soil horizon shown here.

water with any organic matter ($CH_2O$) in the top layer of the soil by root respiration. Root respiration is the oxidation of organic matter by a plant to generate energy for root growth and ion uptake and can be represented by the reaction:

$$CH_2O + O_2 \rightarrow energy + CO_2 + H_2O \rightarrow H_2CO_3. \qquad [14.1]$$

The organic matter present in soil can contain reduced sulfur (e.g., $H_2S$) compounds, which are catalyzed by microorganisms to produce sulfuric acid ($H_2SO_4$). This occurs by combining it with oxidized species (e.g., $O_2$) in the water by the overall reaction:

$$H_2S + 2O_2 \rightarrow H_2SO_4 \rightarrow 2H^+ + SO_4^{2-}. \qquad [14.2]$$

Organic matter also contains organic acids such as *humic* and *fulvic acids*. These are complex organic acids formed when water passes slowly through humus made up of partially decomposed organic matter.

The acidic water produced in the soil undergoes hydrolysis reactions with any silicate minerals present, consuming the acid. Hydrolysis reactions are reactions that add water to the primary minerals of unaltered rock and produce secondary minerals, typically clays. For instance, the reaction of Na-feldspar with acidic water to produce the clay *montmorillonite* can be written as:

$5NaAlSi_3O_8 + 4H^+ + nH_2O \rightarrow NaAl_4(AlSi_7)O_{20}(OH)_4 \bullet nH_2O$
Na-feldspar           montmorillonite

$$+ 4Na^+ + 8SiO_2. \qquad [14.3]$$

Oxidation reactions also occur as $Fe^{2+}$ in primary silicates is oxidized to red-colored $Fe^{3+}$ in Fe-hydroxides by atmospheric oxygen dissolved in the soil solution. A typical reaction is iron olivine (fayalite) reacting to produce the oxidized iron oxide, goethite:

$$Fe_2SiO_4 + 1/2\,O_2 + H_2O \rightarrow 2FeOOH + SiO_2. \qquad [14.4]$$
fayalite               goethite

Surface ion exchange reactions also occur between a mineral surface and the soil solution such as the $Na^+$ for $K^+$ exchange in micas:

$$KAl_3Si_3O_{10}(OH)_2 + Na^+ \rightarrow NaAl_3Si_3O_{10}(OH)_2 + K^+. \qquad [14.5]$$
muscovite            paragonite

Surface ion exchange reactions are particularly important with clay minerals because their layered structure exposes a large surface with many exchange sites to the soil solution.

## Mineral Weathering Rates

The rates of mineral weathering in soil formation are controlled by four main factors. First, is the presence of soil water. If no soil water is present mineral breakdown reactions do not occur at an appreciable rate. If soil water occurs only at mineral triple junctions and does not often move it will reach its solubility limit and mineral breakdown reactions stop occurring. This is why soil-weathering rates are low in desert environments.

The second controlling factor is soil temperature. Mineral breakdown reactions occur about twice as rapidly with each 10°C rise in temperature. This is one of the reasons why rock weathering in the tropics with its elevated temperatures is so extensive.

The third controlling factor is soil water pH. As shown in **Figure 14.7** for albite and is true of most alumina + silica containing minerals found in soils, their dissolution rates are pH dependent. They dissolve more rapidly at both high and low pH relative to near neutral pH.

The final factor of importance in controlling mineral breakdown in soils is the strength of the bonds between silicon and oxygen atoms in the mineral structure. This bond strength can be characterized by the mean electrostatic oxygen site potential of silicon in the structure. Electrostatic oxygen site potential is determined by summing the charge over bond distances between oxygen and silicon atoms in the mineral. Bonds formed at high temperatures where silicon and oxygen atoms have lower coordination and are further apart are weaker. This is why the dissolution rate of silicates decreases as one descends Bowen's reaction series, which gives the sequence of minerals that crystallize out of silicate magmas with decreasing temperature as shown in **Figure 14.8**.

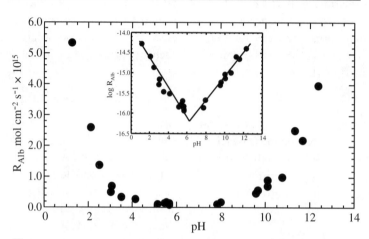

*Figure 14.7* Dissolution rate of albite in moles per square cm per second as a function of pH at 25°C. The inset is the same data plotted as a function of log dissolution rate. (Data from: Chou, L. and Wollast, R., 1985, Steady-state kinetics and dissolution mechanisms of albite. *Am. J. Sci.*, v. 285, pp. 963–993.)

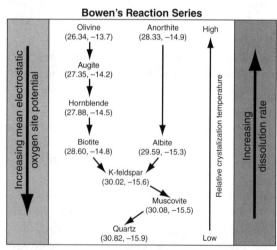

*Figure 14.8* Bowen's reaction series, the dissolution rates of silicate minerals, and the strength of bonding in the mineral in terms of electrostatic oxygen site potential. Bond strength and dissolution rate are designated as (site potential in volts, log dissolution rate in mole $cm^{-2}$ $s^{-1}$ at pH = 8) under the name of the mineral. (Adapted from Brady, P.V. and Walther, J.V., 1989, Controls on silicate dissolution rates in neutral and basic pH solutions at 25°C: *Geochim. Cosmochim. Acta*, v. 53, p. 2823–2830.)

## Clays

Silicate clays are common reaction products that develop in most soils. Both organic matter and silicate clay have small sizes and, therefore,

large surface areas per mass in soils. The surfaces of these clay particles carry either a negative or positive charge depending on the pH of the soil water as outlined above. Positive ions in solution are attached to negative surfaces and vice versa. Because clays in soils have very large surface areas they control the movement and retention of both contaminants and nutrients in a soil.

The cation absorption ability of a soil is characterized by its cation exchange capacity (**Table 14.6**). The pH-dependent *cation exchange capacity* (CEC) is defined as the extent to which a soil can adsorb and exchange cations. It is expressed in milliequivalents per 100 g of soil. Milliequivalents are 1/1000 of a mole (a milli) of cation charge. If 1 millimole of $Ca^{2+}$ could be adsorbed per 100 g of soil it would have a cation exchange capacity of 2.

### Metal Oxides and Hydroxides

Also found in the clay fraction of soils are metal oxides and hydroxides. These are formed when Si is leached from the soil by large amounts of infiltrating slightly acidic to near neutral pH water. At these pHs Si is about three orders of magnitude more soluble than $Al^{3+}$ and $Fe^{3+}$ that are retained in mineral structures. Very common in oxisols and ultisols is the mineral gibbsite ($AlOH_3$). The iron oxides goethite [FeO(OH)] and hematite ($Fe_2O_3$) are also found in highly weathered soils.

In the clay fraction of volcanic soils *allophanes* are also present. These are amorphous hydrous aluminosilicate solids of variable composition derived from the glassy volcanic material present.

As amorphous solids they are not minerals but do possess some short-range order between atoms. Also present in many volcanic soil's clay fraction is the mineral imogolite [$Al_2SiO_3(OH)_4$]. Imogolite is generally only several μm in length occurring in bundles of two to several hundred very small tubes with diameters of about 20 Å.

*Table 14.6* CATION EXCHANGE CAPACITY OF SOME CLAYS.*

| MINERAL | CEC |
|---------|-----|
| Allophane | ~70 |
| Chlorite | 10–40 |
| Glauconite | 11–20+ |
| Halloysite (2H$_2$O) | 5–10 |
| Halloysite (4H$_2$O) | 40–50 |
| Illite | 10–40 |
| Kaolinite | 3–15 |
| Montmorillonite | 70–100 |
| Palygorskite | 20–30 |
| Vermiculite | 100–150 |

* In milliequivalents per 100 g.

*Data from:* Carroll, D., 1959, Ion exchange in clays and other minerals, *Geological Society of America Bulletin*, v. 70, #6, p. 749–780. at pH=7

# Soil and Land Management

Land and therefore its soil can be put to many different usages. Shown in **Figure 14.9** is the use of land in the U.S. as well as the world as a whole. Special-use land accounts for recreation and wildlife areas, e.g., National Parks (10.7% of land). Also considered special-use is land for transportation (1.2% of land) that includes rural highways and national defense use, e.g., the Nevada National Security Site and military bases (0.8% of land). Miscellaneous areas of little surface use by humankind comprise land such as marshes, swamps, tundra, and bare rock in mountains. Note the largest uses of land are for forests, grasslands, and cropland. These proportions are affected somewhat by Alaska, which, relative to the contiguous 48 states, has smaller amounts of cropland and grasslands but larger areas of forest, special-use, and miscellaneous land. Although a concern on a local level, the argument that home building

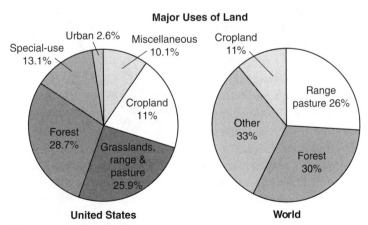

**Major Uses of Land**

**United States**

**World**

*Figure 14.9* Land use in the U.S. and worldwide. (Adapted from USDA Economic Research Service, 2002, and for the world as a whole adapted from Cunningham, W. P., Cunningham, M. A., and Saigo, B. W., 2003, *Environmental Science: A Global Concern*, 7th ed., McGraw–Hill, New York.)

is crowding out other important land uses at the national level does not appear to be correct judgment as **Figure 14.9** shows only 2.6% of land in the U.S. is put to urban use.

In the world as a whole, open range and pasture make up about 26% of the world's land surface similar to the percentage in the U.S. More than half of this is used at least occasionally for animal grazing. Cropland in the world as a whole is only 11% of the land surface, a little over half of what it is in the U.S. This is consistent with the observation that the U.S. is a net food exporter.

Forests make up 30% of the earth's current land surface. This is down from nearly twice the amount before humans starting altering the land surface turning much forest into crop and pasture land. About 40% of forests worldwide are old growth, having never been disturbed by humankind. Forests are disappearing at a rate of about 0.18% per year but this is slowing because increased loss in tropical rain forests is offset to a significant degree by planting of new trees in more temperate climates and natural forest expansion in many area. The earth's land surface that is formally protected in parks, refuges, and preserves is 4%. Besides the U.S. many nations, including Brazil, Costa Rica, and Zaire, have set aside large tracts of forest as national preserves.

## Land Reclamation and Restoration

Land reclamation or restoration is the process of changing the character of land by making it more useful to humans. This includes making new dry land along the ocean, lakes, and rivers, typically from original marshland. Land that has undergone desertification can also be restored to grow crops. Land reclamation also refers to the returning of land to its natural state after it has been disturbed by such activities as strip mining for coal.

Reclaiming land originally under water is done because of the need for new land for human activities. The classic example is in the Netherlands, which contains the delta region of the Rhine River. About 27% of the country is land that was originally below sea level (**Figure 14.10**). Land reclamation is a part of the development history

of most large metropolitan areas along a body of water. Some of the particularly noteworthy examples include Dublin Bay, Ireland; the Foreshore of Cape Town, South Africa; the Odaiba of Tokyo, Japan; Lake Texcoco in Mexico City, Mexico; the Marine Facade in Saint Petersburg, Russia; and Jätkäsaari in Helsinki, Finland. In the U.S. the cities of Boston, Chicago, Los Angeles, New Orleans, New York, and Seattle have all reclaimed a significant amount of wetlands for urban development. Recent recognition of the importance of wetlands means wetland reclamation projects will be under greater scrutiny in the future.

Land reclamation in the desert involves obtaining water, either by digging wells or transporting it from a water-rich area. The soil must first be stabilized from loss by winds. Planting rows of trees perpendicular to the predominant wind direction or constructing windbreaks can do this. Nitrogen-fixing plants such as clover, beans, and yellow mustard are then grown. These plants are

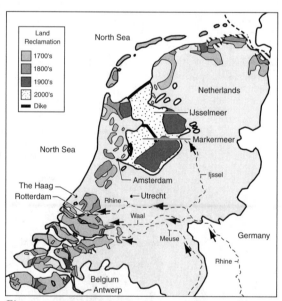

*Figure 14.10* Reclaimed land in the Rhine River delta in the Netherlands. The dikes across the Zuiderzee arm of the North Sea allowed the formation of the shallow (< 6 m) Ijsselmeer and Markermeer freshwater lakes from inflow of the Ijssel River after the seawater was pumped out. Instead of reclaiming the area for dry land, it is maintained as a municipal and agricultural freshwater supply. The lakes also serve as a recreation area.

plowed under to add nitrogen and structure to the soil before food crops are planted.

## Seawater Greenhouses

In desert areas near the ocean such as in Jordan, seawater greenhouses have been constructed. Seawater is pumped or gravity fed to a modified greenhouse. In the direction of the prevailing wind seawater is trickled down a honeycombed lattice. This humidifies and cools air passing through the lattice. This moist air enters a plant-growing greenhouse. Some of the moisture condenses on any plants present. The air then exits the greenhouse to keep the greenhouse cool. The exhausted high humidity air helps in the growing of more hardy plants in the desert downwind of the greenhouse.

Infrared heat energy is trapped in the roof of the greenhouse, but plant-growing visible light is allowed to pass through. This supplies the needed light for photosynthesis but keeps the heating effect inside the greenhouse to a minimum. The infrared energy from the roof is used to heat seawater that is sent to an evaporator where some of the hot seawater evaporates. The moisture-laden air produced is sent to a condenser. In the condenser incoming seawater cools the air, condensing its $H_2O$ to freshwater for the plants. Only a small amount of electricity is required to run the pumps of the seawater greenhouse and this can be supplied by solar panels.

## Farmland

In virtually every country food is of great concern. Citizens demand of their government a supply of food at a reasonable price. Problems exist because in many countries farmable acres have declined but crop yields are no longer increasing as they have in the past 100 years. The demand for meat in most countries is increasing with time. Meat production requires more farmland be used to raise animal feed. This increases the land per individual needed over a purely vegetarian diet.

For land devoted to farming you cannot control climate or land slope but you can control the amount of soil amendments and nature of tillage and practice crop rotation. Farmers who grow vegetables for their own consumption, or those that farm a large number of $km^2$ of corn or soybeans for international markets, try to achieve maximum yields from the soil on their land. Successful farming requires an understanding of the characteristics of soil. The nature of the soil determines whether rainfall runs off its surface or infiltrates with crop nutrients to plant roots. Soil gives physical support to plants as well as the roots underground. Soil's ability to hold and release both contaminants and fertilizers controls the quality of the water as well as the type of plants that grow in the soil. A soil is said to be fertile if it can grow the desired plants productively. Fertile soils have the properties given in **Table 14.7**.

## Crop Irrigation

Irrigation of soils is important in many regions. Only about 16% of the world's cropland is irrigated, yet irrigated land produces 30% of the world's food. The water used needs to be low in salinity; plants are sensitive to varying degrees

---

*Table 14.7* CHARACTERISTICS OF A FERTILE SOIL

1. Rich in major plant nutrients including nitrogen, phosphorous, and potassium.

2. Contains sufficient trace nutrients.

3. Contains organic matter.

4. Has a soil texture of a loam that can hold and release water.

5. Soil pH is between 5.5 and 6.2.

6. Contains microorganisms that support plant growth.

to the soil water salinity. Consider the schematic of the plant root hair in soil water depicted in **Figure 14.11.**

There are large organic molecules (the grey circles in Figure 14.11) contained in the root hair solution. The total pressure in the root hair solution is made up of the pressure of water molecules plus the pressure of organic molecules. Smaller water and nutrient molecules can enter the root through a semipermeable membrane but the larger organic molecules can't leave. Water and nutrient molecules continue to enter the root even though the total pressure inside the root is greater than total pressure outside. This occurs because water pressure is greater outside and the water pressure on both sides of the root membrane attempt to equalize. The flux of water into the root produces an *osmotic pressure* in the root that helps raise nutrients and water to the leaves in the plant. The osmotic pressure that drives water into the plant is large because soil waters are typically almost pure $H_2O$ while the fluid inside roots has low amounts of water being made up mostly of organic molecules.

The most common ions in soil water are the cations $Ca^{2+}$, $Na^+$, $K^+$, and $Mg^{2+}$, and the anions $HCO_3^-$, $Cl^-$, and $SO_4^{2-}$. If increased salts appear in the soil solution then the concentration and therefore $H_2O$ pressure of soil water decreases. This decreases the osmotic pressure in the plant and results in difficulty for water and nutrients to rise toward the leaves.

Soil water also needs to have a low level of $Na^+$ given by the sodium adsorption ratio (SAR):

$$SAR = \frac{[Na^+]}{\sqrt{\frac{1}{2}\left([Ca^{2+}] + [Mg^{2+}]\right)}} \qquad [14.6]$$

where [  ] denotes the concentration of the bracketed ion in millimoles per liter. A SAR value of 15 or greater indicates a problem. $Na^+$ in water exchanges for $Mg^{2+}$ and $Ca^{2+}$ on clay surfaces. The clay surfaces become more negatively charged as a +2 ion is replaced by a +1 ion. As the clay layers in the mineral are forced open by the repelling negative charges soil dispersion occurs. This cloud of fine clay material in the soil solution clogs small pores restricting root growth and water movement. A high SAR soil is extremely sticky when wet and becomes very hard and forms large lumps when dry. Gypsum ($CaSO_4 \bullet 2H_2O$) can be added to the soil to decrease the SAR by adding $Ca^{2+}$ to the soil solution.

Furrow irrigation, where water is distributed by small parallel channels in the direction of the steepest slope, is commonly used to apply water to small fields. However, sprinkler systems are now more commonly used on large fields. Many sprinkler systems are configured to pivot about a central point irrigating a circular area (**Figure 14.12**). Others are lateral traveling systems. Due to concerns about water usage there has been increasing use of Low Energy Precision Applicator (LEPA) designs that direct the water to the plant, rather than into the air where much of it evaporates.

Despite their high initial cost of installation, another trend in irrigation is the use of micro-irrigation. Developed in the arid lands of Israel,

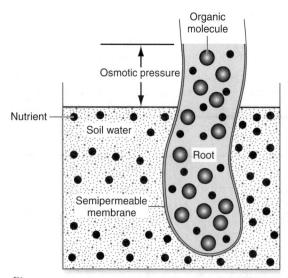

*Figure 14.11* Schematic diagram of a root hair in soil water showing the development of osmotic pressure.

*Figure 14.12* Different crops can be grown under a single valley circular pivot irrigation system by dividing the crop circle into segments as shown in this field in Kenya. In the upper left barley is growing while in the upper right it is jatropha. Jatropha is a succulent plant whose seeds can be used to make biodiesel (see biodiesel in the chapter on alternative energy).

each micro-irrigation system is constructed for the specific crop and the horticultural requirements for a given field. Most often drip or trickle emitters are used to supply the water directly to the plants roots. Besides saving water, micro-irrigation systems require less energy to run than sprinkler systems. They also control the development of weeds and can be used to supply fertilizers to the plants. Micro-irrigation systems are particularly beneficial in hilly terrains as no surface water runoff is produced.

# Degradation of Soils

Soil degradation continues to be an important problem because it impacts agricultural productivity and therefore food security. Degradation of soil also lowers environmental quality and therefore the quality of life. Some soil scientists argue that soil is a nonrenewable resource on a human time-scale. They contend some adverse effects, such as soil erosion and desertification, are irreversible.

## *Desertification*

Desertification is the degradation of arid and semi-arid land due to loss of water by climatic changes or human activities. Most desertification can be directly related to human activities. Much of this is caused by regional human overpopulation and the stress of needed increased food production. Overgrazing of grasslands by animal herds, lowering the water content of the ground due to excessive extraction of groundwater, and the diversion of water from rivers for farming as well as raising livestock causes the desertification by increased evapotranspiration (**Figure 14.13**).

The major impacts of desertification are the reduction of biodiversity, and the diminished ability to grow crops, raise livestock, and produce fuelwood, as these depend on water availability. Note in **Figure 14.14** the areas at risk of

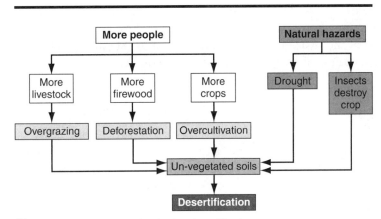

*Figure 14.13* The events that lead to desertification.

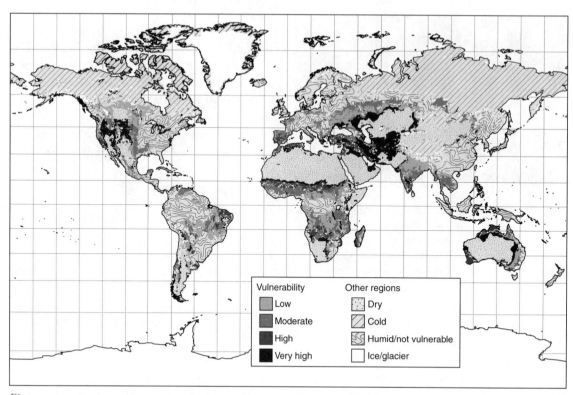

*Figure 14.14* Regions of desert vulnerability. (Modified from USDA-NRCS, Soil Science Division.)

desertification in the Sahel region along the southern edge of the Sahara desert, in the western U.S. around the Mojave and Sonoran deserts, and along the deserts regions in central Australia.

Most people living on arid and semiarid land are in developing countries. These people have lower per capita income and higher infant mortality rates than the rest of the world. The situation is particularly bad in the borderland surrounding the deserts of Africa and Asia. Desertification, therefore, has a devastating effect on the poorest populations of the world.

## Soil Erosion

Soil loss is one of the world's most important environmental problems (**Figure 14.15**). Serious degradation from erosion has occurred on 40% of the world's cropland. It is estimated that in the U.S. corn belt, 1 kg of topsoil is lost for each 2/3 of a kg of corn harvested. Soils can be eroded by wind or water. These forces affect both natural environments and agricultural land. Present U.S. soil erosion averages 14 metric tons per hectare ($10^4$ m$^2$) annually with 7.7 tons caused by water and 6.2 tons by wind. One ton of soil per hectare is equal to a layer 0.073 mm deep. This rate is often faster than the rate at which new soil is formed. Erosion by wind is negligible in the U.S. east of the Mississippi River but is extreme in arid western regions, as in Nevada.

### Erosion by Wind

Dry soil particles are picked up by wind, carried, and removed from a region in what is termed *soil deflation*. These particles are typically less than 0.2 mm in diameter. On return to the earth's surface the particles impact soil and rocks causing erosion. Deflation and soil movement typically occurs

*Figure 14.15* Severe soil erosion in a wheat field near Washington State University.

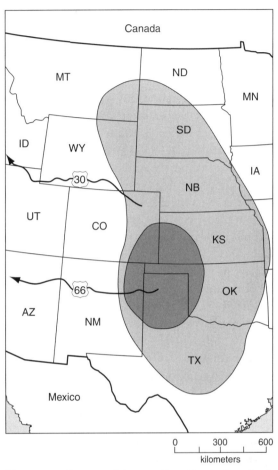

*Figure 14.16* Extent of the area affected by the Dust Bowl between 1930 and 1936. The darker grey area gives the extent of extensive damage and in lighter grey the area of some damage from the dust. The arrows give the routes of migration of the farmers who left the area after their farms were destroyed.

where there is little or no rooted vegetation to hold the soil and where soil is exposed to wind. Sand dunes and beach deposits are prime examples.

In the U.S. Midwest in the years before 1930 a period of above average rainfall occurred in the normally dry region. The rains lead farmers to plow the natural deep-rooted grasses to plant extensive wheat fields. From 1930 to 1936 a period of severe droughts occurred causing the loss of wheat and the exposure of barren plowed soil. Winds blew the soil into large clouds of dirt. The amount of dirt in a storm often caused visibility to be reduced to a couple of meters or less. Centered on the panhandles of Texas and Oklahoma the impacted area was termed the Dust Bowl (**Figure 14.16**). Because tens of millions of acres

of farmland became useless, hundreds of thousands of farmers left and headed west on Highway 66 and Highway 30.

## Erosion by Water

Water has the ability to detach and move soil grains both by the force of rain splash as well as the downslope transport of water on the earth's surface. The effectiveness of rain splash to move soil grains depends on the intensity of the rain, the slope of the land surface, and the kind of soil. Rain splash is most important in high intensity rainstorms produced by upward moving air masses in the world's equatorial regions. It is less important in the low intensity rains of the weather fronts that move across northern Europe and the U.S.

The downslope transport of soil in water running off the land surface depends on surface slope, the extent of vegetation, the type of soil, and the amount of rain that infiltrates the soil as opposed to running off the land surface. As the amount of water that infiltrates in a given time is limited, runoff depends on the intensity of the rain. The infiltration is decreased if the soil is already holding water in its pores before it rains.

The higher the velocity of the moving water, the more soil particles it can move. Soil movement first creates narrow, shallow incisions in the soil termed *rills*. These rills either grow in size or, alternatively, are filled with particles carried by the water. This self-organization produces a number of larger rills that can increase in size to become gullies before the water enters a stream channel (**Figure 14.17**).

The increase of soil loss during a rainstorm after forest clearing is typically large. In the Ivory Coast it was found that forested slopes lost 0.03 metric tons of soil per year per hectare ($10^4$ m$^2$), cultivated slopes lost 90 metric tons per year per hectare, and cleared slopes lost 138 metric tons per year per hectare. After clearing slopes for crops it was shown that Madagascar lost 400 metric tons per year per hectare due to soil erosion. The iron-rich clays in Madagascar's soils, which entered the country's rivers, turned them blood-red in color. At their mouths, the rivers in Madagascar have colored the Indian Ocean red to such an extent that astronauts in space have noted that it looks like "Madagascar is bleeding to death." (See Earth Observing Systems Project (NASA), Photo: Betsiboka Estuary, Madagascar, April 12, 2004.).

### Techniques to Control Soil Erosion

There are many techniques to control soil erosion by wind and water. First, contour plowing where plowing follows elevation contours can be done (**Figure 14.18**). This slows the flow of water downslope. Contour plowing is most effective on

*Figure 14.17* An erosion gully in a Kansas field.

slopes from 2% to 10%. Conservation tillage can be employed, which leaves a significant amount of crop residue on the soil surface. This also slows surface water movement and therefore reduces the amount of soil eroded by water. An extension of this technique is to put mulch on the soil surface or plant a soil retention crop such as hay between other crops on the slope. Where fruit trees are grown a cover crop of legumes with their extensive surface root system can be planted to reduce erosion. Fertilizers are often applied to enhance plant root growth that helps hold the soil in place. Terracing of growing fields is often used on slopes greater than 10% to control soil erosion. This includes farmland in the mountains of Nepal, Peru, and the Philippines (**Figure 14.19**). For wind erosion a vegetation wind buffer can be grown to protect the acreage.

*Figure 14.18* Contoured fields planted across the hill slope along a contour rather than up and down the slope.

*Figure 14.19* Banaue Rice Terraces in the Philippines carved out of the rock by the indigenous Ifugao mountain people about 2,000 years ago. They are irrigated by water from the rain forests above the terraces.

## Salinization

*Salinization* is a process where water-soluble salts accumulate in the soil, particularly at the soil surface. It occurs when there are high dissolved salts in water and high water evapotranspiration rates. With loss of $H_2O$ from the water through evapotranspiration, salts can concentrate and precipitate in the soil. As soil salinity increases, degradation of soils occurs with detrimental effects on plant growth and crop yield.

Salinization is a worldwide problem that occurs naturally as well as from irrigation practices. Given in **Table 14.8** is the amount of land in the regions listed that are affected by soil salinization. The ions responsible for salinization are primarily $Na^+$ and $Cl^-$ but $K^+$, $Ca^{2+}$, and $Mg^{2+}$ can also be important.

As an example of salinization, consider the cross-section through the subsurface of the San Joaquin Valley, California shown in **Figure 14.20**. The valley has a long growing season with lots of sun and therefore lots of potential for soil water evapotranspiration. The elevated Coast Ranges to the west of the valley and the dominant westerly wind direction causes clouds from the Pacific Ocean to rise and cool over the Coast Ranges. This leads to precipitation in these mountains. This water flows downward toward the San Joaquin Valley underground because of the presence of permeable

*Table 14.8* EXTENT OF SALINIZATION IN THE INDICATED REGION

| REGION | AREA ($10^4$ km$^2$) |
|---|---|
| Africa | 69.5 |
| Asia and Far East | 19.5 |
| Australia | 84.7 |
| Europe | 20.7 |
| Latin America | 59.4 |
| Near and Middle East | 53.1 |
| North America | 16.0 |

*Source:* Data from Brinkman, R., 1980.

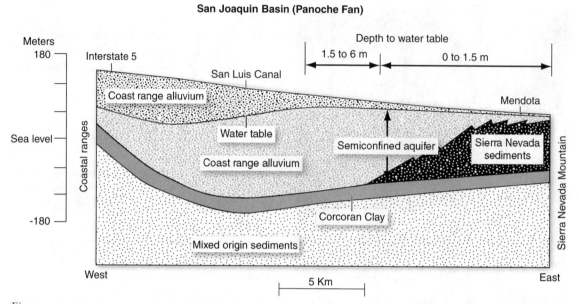

**San Joaquin Basin (Panoche Fan)**

*Figure 14.20* East-west cross-section through sediments in the San Joaquin Basin, CA at the Panoche Fan showing the extent of the semiconfined surface aquifer above the Corcoran clay. (Modified from Belitz, K., 1988, *Character and evolution of the ground-water flow system in the central part of the western San Joaquin Valley, California*, Open file report 87–573, U.S. Geological Survey, Sacramento, Calif., 34p.)

mountain soils due to the presence of Coast Range alluvium. The groundwater picks up salts along the way, particularly from the dissolution of gypsum in the soils. At the base of the Coast Ranges near Interstate 5 it is pumped out of the ground to irrigate crops in the valley to the east where soil is less permeable. The water remains in the surface layer for a long time because the Corcoran clay is a confining layer that does not allow the downward passage of water. Therefore, the water table is at or very near the surface on the eastern side of the valley. This causes extensive evaporation that is added to the substantial transpiration from the growing plants. Salt accumulates in the soil. Salinization is presently reducing the productivity of agricultural lands in the San Joaquin Valley and increasing the cost of purifying urban drinking water and wastewater treatment.

# Wetlands

*Wetlands* are areas of land where the soil is saturated with moisture or ice seasonally or permanently. The wetlands of the world are outlined in **Figure 14.21**. The water can be fresh, saltwater, or brackish in nature. Often, shallow pools of water cover wetlands. Wetlands include bogs, marshes, swamps, fens, and permafrost regions (**Table 14.9**). *Permafrost* soil is soil that has been below the freezing point of water (0°C) for at least two years.

The largest continuous nonpermafrost wetland in the world is the tropical floodplain Pantanal, located mainly in south-central Brazil (**Figure 14.22**). The uplift of the Andean Mountains created a depression to the east that filled with alluvial sediments. These sediments receive runoff from the mountains and the water only drains slowly to the south from the area through the Paraguay River. It sprawls over an area estimated to be between 140 and 200 thousand km$^2$ along the banks of the upper Paraguay River. The Everglades of Florida is the largest wetland in the U.S. and is estimated to have been 10 thousand km$^2$ before its development (**Figure 14.23**).

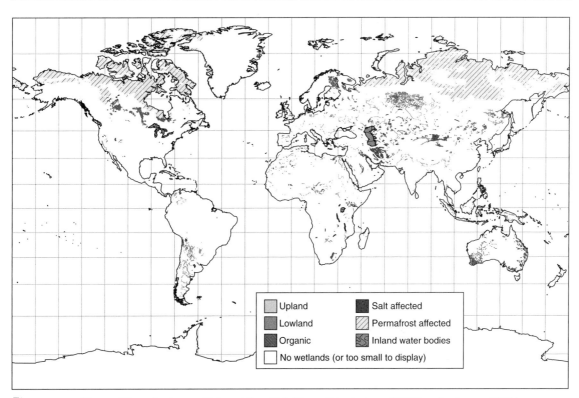

*Figure 14.21* The world's wetland areas. (Adapted from U.S. Dept. of Agriculture, NRCS wetlands map, 1997.)

*Table 14.9* TYPES OF WETLANDS

| TYPE | CHARACTERISTICS |
|------|-----------------|
| Bog | Waters are acidic and contain a thick carpet of spongy sphagnum moss peat. |
| Fen | Peat forming wetlands that receive nutrients from sources other than precipitation, such as upslope mineral soils. Fens are less acidic and have higher nutrient levels than bogs. |
| Marsh | Wetlands dominated by herbaceous or nonwoody plants occurring in a transition zone between land and water. Tidal marshes occur along protected coastlines in middle and high latitudes worldwide. Nontidal marches occur along streams in poorly drained depressions, and in the shallow water along the edges of lakes, ponds, and rivers. |
| Swamp | Any wetland dominated by woody plants, either trees or shrubs. |
| Permafrost | Lands with water that remains permanently frozen year round, typical of Arctic and sub-Arctic areas. |

Wetlands are considered the most biologically diverse of all ecosystems. They serve as natural wastewater purification systems for many urban communities and are a sink for carbon from the atmosphere in their peat deposits. Historically, wetlands have been drained for real estate development or flooded to create lakes. Half the world's wetlands were destroyed by 1993. With recognition of their usefulness and the passing of laws to protect them, the destruction of wetlands has slowed dramatically in recent years.

*Figure* 14.22 South America showing the major rivers and location of the Pantanal wetlands.

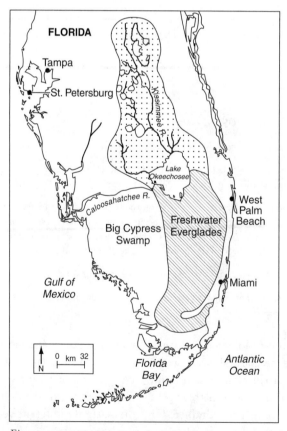

*Figure* 14.23 Florida everglades before development.

## SUMMARY

Soil is a dynamic layer on the earth's land surface. Its texture depends on the amount of sand, silt, and clay it contains. The way soil particles are aggregated together is the soil structure and is described as granular, platy, prismatic, columnar, or blocky. The color of soil can be described with the Munsell soil color index. Terms such as loose, hard, friable, plastic, sticky, and firm are used to describe soils. The soil pH is often a helpful parameter which helps determine the soils ability to hold or release nutrients and other soil water constituents.

The Comprehensive Soil Classification System is used to characterize and classify soils. The most general divisions are the 12 soil orders with the most specific being the >15,000 soil series. The distribution of soil orders on earth is controlled to a large extent by climatic conditions.

Soils form on rocks that are ruptured by stresses that occur on uplift, cyclic freezing and thawing, temperature changes, abrasion, and root wedging. Some primary silicates in rocks become unstable and dissolve in the water in rock cracks and pores with the formation of secondary minerals, commonly clays. Other more resistant minerals such as quartz are disaggregated from the rock. Layers form in a soil due to differential solubility and adsorption of mineral components from the soil water in a process called chromatographic separation.

Water in soils first becomes acidic and then reacts with minerals that neutralize the acidity. Rates of these reactions depend on the presence of water, its pH, temperature, and the strength of bonding between atoms in the soil minerals.

About 16% of the world's cropland is irrigated but it produces 30% of the world's food. Soil loss by both wind and water is an important environmental problem. The soil water can suffer because of excess accumulation of salts leading to salinization. Wetlands occur mainly where there is permafrost, but large tracks of wetlands also occur in tropical river floodplains.

## KEY TERMS

allophanes
bioturbation
cation exchange capacity
chromatographic separation
fulvic acid
humic acid
montmorillonite
Munsell soil color index

osmotic pressure
ped
permafrost
rill
salinization
soil deflation
spalling
wetlands

## PROBLEMS

1. If the average composition of O, C, H, N, and P in terrestrial higher plants is given by:

| ELEMENT | CONCENTRATION g per kg |
|---------|------------------------|
| O | 463 |
| C | 412 |
| H | 54.0 |
| N | 32.5 |
| P | 3.0 |

   Determine the chemical formula of an average higher plant in terms of these elements.

2. Write a balance reaction for the dissolution of the Mg-olivine forsterite ($Mg_2SiO_4$) by hydrolysis.

3. In humid tropical conditions orthoclase ($KAlSi_3O_8$) can weather to kaolinite. Write the balanced chemical reaction that describes this weathering.

4. Humic organic material in aqueous solutions can have acid groups on the framework of its structure. The reaction written to show a carboxyl acidic group that disassociates to give $H^+$ is:

$$\text{R-COOH} = \text{R-COO}^- + H^+$$

   where R is the humic framework. Write the equilibrium constant expression for the reaction. If the equilibrium constant is 0.00012 at what pH will the concentration, that is, activity of R-COOH and R-COO⁻ be equal?

5. Considering Figure 14.1 what is the textural class of soils with 30% sand, 35% clay, and 35% silt?

6. What kind of soil is under your "rents" house? (If you don't know what "rents" are consult the urban dictionary at www.urbandictionary.com/define.php?term=rents.)

   To find the soil open the website:
       http://websoilsurvey.nrcs.usda.gov/app/.
   Start **WSS** and enter the "rents" address.
   On the map use the AOI tool to specify the area of interest.
   Open the **Soil Map** tab on the top of the page.
   Under the **Map unit legend** click on the soil type to get a description of the soil.
   Attach a copy.

# REFERENCES

Bagley, W. C. 1942. *Soil exhaustion and the Civil War.* Washington, DC: American Council on Public Affairs.

Belitz, K. 1988. *Character and evolution of the ground-water flow system in the central part of the western San Joaquin Valley, California.* Open file report 87-573, U.S. Geological Survey, Sacramento, CA.

Brady, P.V. and Walther, J.V. 1989. Controls on silicate dissolution rates in neutral and basic pH solutions at 25°C. *Geochim Cosmochim Acta* 53:2823–2830.

Brinkman, R., 1980. Saline and sodic soils. in: Land reclamation and water management, p. 62–68. International Institute for Land Reclamation and Improvement (ILRI), Wageningen, The Netherlands.

Carroll, D. 1959. Ion exchange in clays and other minerals. *Geological Society of America Bulletin* 70(6): 749–780.

Cunningham, W. P., Cunningham, M. A., and Saigo, B. W. 2003. *Environmental science: A global concern,* 7th ed. New York: McGraw–Hill.

Earth Observing Systems Project, U.S. National Aeronautics and Space Administration (NASA), Photo: Betsiboka Estuary, Madagascar, April 12, 2004. Available online at URL: http://earthobservatory.nasa.gov/IOTD/view.php?id=4388 Accessed 01 Aug 2012.

Munsell, A. H. 1929. *Munsell® Soil Color Book,* 2009 Revised Edition. Grand Rapids, MI: X-Rite Publisher.

Montgomery, D. R. 2007. *Dirt: The erosion of civilizations.* Berkeley: University of California Press.

Pimentel, D., Harvey, C., Resosudarmo, P., Sinclair, K., Kurz, D., McNair, M., Crist, S., Shpritz, L., Fitton, L., Saffouri, R., and Blair, R. 1995. Environmental and economic costs of soil erosion and conservation benefits. *Science* 267(24):1117–1122.

Sposito, G. 1989. *The chemistry of soils.* New York: Oxford University Press.

U.S. Dept. of Agriculture, 1997, *Global Distribution of Wetlands,* Natural Resources Conservation Service map. (Available online at URL: http://soils.usda.gov/use/worldsoils/mapindex/wetlands.html.)

U.S. Department of Agriculture, 1998, *Global Desertification Vulnerability,* Natural Resources Conservation Service map. (Available online at URL: http://soils.usda.gov/use/worldsoils/mapindex/desert.html.)

U.S. Department of Agriculture, 2005, *Global Soil Regions,* Natural Resources Conservation Service map. (Available online at URL: http://soils.usda.gov/use/worldsoils/mapindex/order.html.)

U.S. Department of Agriculture, 2011, *Major Uses of Land in the United States* by Cynthia Nickerson, Robert Ebel, Allison Borchers, and Fernando Carriazo, EIB-89, USDA, Economic Research Service, December 2011. (Available online at URL: www.ers.usda.gov/publications/eib89/.)

Walther, J.V. 2009. *Essentials of geochemistry,* 2nd ed. Sudbury, MA: Jones and Bartlett Learning.

# Prefix Multipliers
# for Metric Quantities

| FACTOR | PREFIX | SYMBOL | FACTOR | PREFIX | SYMBOL |
|--------|--------|--------|--------|--------|--------|
| $10^{-1}$ | deci- | d | $10^{1}$ | deca- | da |
| $10^{-2}$ | centi- | c | $10^{2}$ | hecto- | h |
| $10^{-3}$ | milli- | m | $10^{3}$ | kilo- | k |
| $10^{-6}$ | micro- | $\mu$ | $10^{6}$ | mega- | M |
| $10^{-9}$ | nano- | n | $10^{9}$ | giga- | G |
| $10^{-12}$ | pico- | p | $10^{12}$ | tera- | T |
| $10^{-15}$ | femto- | f | $10^{15}$ | peta- | P |
| $10^{-18}$ | atto- | a | $10^{18}$ | exa- | E |

# Common Ore Minerals

There are over 500 different ore minerals. Given on the next page is a list of some of the more common ore minerals that are helpful to know based on their composition so their distribution in the earth can be put into context. They are arranged by the nature of their negative anion species.

## COMMON ORE MINERAL

### Native metals

Bismuth—Bi

Copper—Cu

Gold—Au

Mercury—Hg

Palladium—Pd

Platinum—Pt

Silver—Ag

### Native nonmetals

Diamond—C

Graphite—C

Sulfur—S

### Halides (Halogens of F$^-$, Cl$^-$, I$^-$, or Br$^-$ as the anion)

Atacamite—$Cu_2Cl(OH)_3$

Chlorargyrite—AgCl

Fluorite—$CaF_2$

Halite—NaCl

Sylvite—KCl

### Sulfides (Reduced sulfur compounds with S$^{2-}$ as the anion)

Argentite—$Ag_2S$

Bornite—$Cu_5FeS_4$

Chalcocite—$Cu_2S$

Chalcopyrite—$CuFeS_2$

Cinnabar—HgS

Colbaltite—CoAsS

Covellite—CuS

Galena—PbS

Millerite—NiS

Pentlandite—$(Fe,Ni)_9S_8$

Pyrrhotite—$Fe_{1-x}S$ (x = 0.1–0.2)

Sphalerite—ZnS

Stibnite—$Sb_2S_3$

### Sulfur pair anion $\left(S_2{}^{2-}\right)$ covalently bonded together

Molybdenite—$MoS_2$

Pyrite—$FeS_2$

### Oxides (O$^{2-}$ as the anion)

Cassiterite—$SnO_2$

Chromite—$FeCr_2O_4$

Cuprite—$Cu_2O$

Hematite—$Fe_2O_3$

Ilmenite—$FeTiO_3$

Magnetite—$Fe_3O_4$

Pyrolusite—$MnO_2$

Rutile—$TiO_2$

Scheelite—$CaWO_4$

Uraninite—$UO_2$

Wolframite—$(Fe,Mn)WO_4$

### Hydroxides (OH$^-$, hydroxal, as the anion)

Gibbsite—$Al(OH)_3$

### Oxide-hydroxides (Both oxygen and hydroxal as the anions)

Boehmite—AlO(OH)

Diaspore—AlO(OH)

Goethite—FeO(OH)

### Carbonates (CO$_3{}^{2-}$ as the anion)

Azurite—$Cu_3(OH)_2(CO_3)_2$

Calcite—$CaCO_3$

Dolomite—$CaMg(CO_3)_2$

Malachite—$Cu_2(OH)_2CO_3$

Siderite —$FeCO_3$

Trona—$Na_3H(CO_3)_2 \cdot 2H_2O$

**Sulfates** ($SO_4^{2-}$ as the anion)

Anhydrite—$CaSO_4$

Barite—$BaSO_4$

Gypsum—$CaSO_4 \cdot 2H_2O$

**Phosphates** ($PO_4^{2-}$ as the anion)

Monazite—$(Ce,La,Th)PO_4$

**Silicates** (Si-bearing mineral with tetrahedral $SiO_4^{4-}$ groups as anion)

Beryl—$Be_3Al_2(SiO_3)_6$

Kaolinite—$Al_4Si_4O_{10}(OH)_8$

Quartz—$SiO_2$

# Energy and Power Conversion Factors

The most widely used systems of measurements are S.I. units (Le Système International d'Unités) of meter-kg-second and CGS units of cm-gram-second. Given on the next page are conversion factors between the two systems and some other common energy and power units.

## FORCE = MASS × ACCELERATION

| SYSTEM | FORCE UNIT |
|---|---|
| CGS | dyne = force to accelerate a mass of 1 g by 1 cm s$^{-2}$ |
| S.I. (MKS) | newton = force to accelerate a mass of 1 kg by 1 m s$^{-2}$ |

## ENERGY = FORCE × DISTANCE OR PRESSURE × VOLUME

| ENERGY UNIT | ERG | JOULE | THERMOCHEMICAL CALORIE | BTU |
|---|---|---|---|---|
| CGS: 1 erg = dyne cm | 1 | 10$^{-7}$ | 2.38901 × 10$^{-8}$ | 9.4782 × 10$^{-11}$ |
| S.I.: 1 joule = newton meter | 10$^7$ | 1 | 0.23901 | 9.4782 × 10$^{-4}$ |
| 1 calorie | 4.194 × 10$^7$ | 4.1840 | 1 | 3.9657 × 10$^{-3}$ |
| 1 Btu = British thermal unit | 1.055056 × 10$^{10}$ | 1.055056 | 252.164 | 1 |
| 1 kilowatt hour | 3.600 × 10$^{13}$ | 3.600 × 10$^7$ | 8.6042 × 10$^5$ | 3,412.1 |

## POWER = ENERGY PER UNIT TIME

| POWER UNIT | WATT | BTU HOUR$^{-1}$ | HORSEPOWER (ELECTRICAL) | TON (REFRIGERATION) |
|---|---|---|---|---|
| S.I.: 1 watt = joule s$^{-1}$ | 1 | 3.41443 | 1.340 × 10$^{-3}$ | 2.8435 × 10$^{-4}$ |
| Btu per hour | 0.292875 | 1 | 3.9259 × 10$^{-4}$ | 8.3278 × 10$^{-5}$ |
| Horsepower (electrical) | 746 | 2,547.2 | 1 | 0.21212 |
| Ton (refrigeration)* | 3,516.9 | 12,000 | 4.7143 | 1 |

* Ton of refrigeration is approximately the energy removal rate that will freeze 2,000 lb of water at 0°C in 24 hours.

# U.S. Primary Drinking Water Standards

The table starting on the following page lists drinking water contaminants and their maximum contaminant level (MCL) for microorganisms, disinfectants, disinfection by-products, inorganic chemicals, organic chemicals, and radionuclides. MCL standards are the highest level of a contaminant that is allowed in drinking water in the U.S. and are enforceable standards. MCLs are set as close to maximum contaminant level goals (MCLG) as feasible using the best available treatment technology and taking cost into consideration. MCLGs are the level of a contaminant in drinking water below which there is no known or expected risk to health. Treatment technique (TT) is a process intended to reduce the level of a contaminant and is required for all drinking water. Units are in milligrams per liter (mg/L) unless otherwise noted. Milligrams per liter are equivalent to parts per million by weight (ppm).

| MICROORGANISM | | | | |
| --- | --- | --- | --- | --- |
| CONTAMINANT | MCLG (mg/L) | MCL (mg/L) OR TT | POTENTIAL EFFECTS FROM WATER INGESTION | SOURCES OF CONTAMINANT IN DRINKING WATER |
| *Cryptosporidium* | 0.0 | TT | Gastrointestinal illness (e.g., diarrhea, vomiting, cramps) | Human and animal fecal waste |
| *Giardia lamblia* | 0.0 | TT | Gastrointestinal illness (e.g., diarrhea, vomiting, cramps) | Human and animal fecal waste |
| Heterotrophic plate count (HPC) | n/a | TT | HPC has no health effects; it is an analytic method used to measure the variety of bacteria that are common in water. The lower the concentration of bacteria in drinking water, the better maintained the water system is. | HPC measures a range of bacteria that are naturally present in the environment |
| *Legionella* | 0.0 | TT | Legionnaire's Disease, a type of pneumonia | Found naturally in water; multiplies in heating systems |
| Total coliforms (including fecal coliform and *E. coli*) | 0.0 | 5.0% | Not a health threat in itself; it is used to indicate whether other potentially harmful bacteria may be present | Coliforms are naturally present in the environment; as well as feces; fecal coliforms and *E. coli* only come from human and animal fecal waste |
| Turbidity | n/a | TT | Turbidity is a measure of the cloudiness of water. It is used to indicate water quality and filtration effectiveness (e.g., whether disease-causing organisms are present). Higher turbidity levels are often associated with higher levels of disease-causing microorganisms such as viruses, parasites, and some bacteria. These organisms can cause symptoms such as nausea, cramps, diarrhea, and associated headaches. | Soil runoff |
| Viruses (enteric) | 0.0 | TT | Gastrointestinal illness (e.g., diarrhea, vomiting, cramps) | Human and animal fecal waste |

## DISINFECTION BY-PRODUCTS

| CONTAMINANT | MCLG (mg/L) | MCL (mg/L) | POTENTIAL HEALTH EFFECTS FROM WATER INGESTION | SOURCES OF CONTAMINANT |
|---|---|---|---|---|
| Bromate | 0.0 | 0.010 | Increased risk of cancer | By-product of drinking water disinfection |
| Chlorite | 0.8 | 1.0 | Anemia; infants and young children: nervous system effects | By-product of drinking water disinfection |
| Haloacetic acids (HAAs) | n/a | 0.060 | Increased risk of cancer | By-product of drinking water disinfection |
| Total trihalomethanes (TTHMs) | n/a | 0.080 | Liver, kidney, or central nervous system problems; increased risk of cancer | By-product of drinking water disinfection |

## DISINFECTANTS*

| CONTAMINANT | MRDLG (mg/L) | MRDL (mg/L) | POTENTIAL HEALTH EFFECTS FROM WATER INGESTION | SOURCES OF CONTAMINANT IN DRINKING WATER |
|---|---|---|---|---|
| Chloramines (as $Cl_2$) | 4 | 4.0 | Eye/nose irritation; stomach discomfort, anemia | Water additive used to control microbes |
| Chlorine (as $Cl_2$) | 4 | 4.0 | Eye/nose irritation; stomach discomfort | Water additive used to control microbes |
| Chlorine dioxide (as $ClO_2$) | 0.8 | 0.8 | Anemia; infants and young children: nervous system effects | Water additive used to control microbes |

* MRDL = Maximum Residual Disinfectant Level; MRDLG = Maximum Residual Disinfectant Level Goal.

## INORGANIC CHEMICALS

| CONTAMINANT | MCLG (mg/L) | MCL (mg/L) OR TT | POTENTIAL HEALTH EFFECTS FROM WATER INGESTION | SOURCES OF CONTAMINANT IN DRINKING WATER |
|---|---|---|---|---|
| Antimony | 0.006 | 0.006 | Increase in blood cholesterol; decrease in blood sugar | Discharge from petroleum refineries; fire retardants; ceramics; electronics; solder |
| Arsenic | 0 | 0.010 | Skin damage or problems with circulatory systems, and may have increased risk of getting cancer | Erosion of natural deposits; runoff from orchards, runoff from glass and electronics production wastes |

## INORGANIC CHEMICALS (CONTINUED)

| CONTAMINANT | MCLG (mg/L) | MCL (mg/L) OR TT | POTENTIAL HEALTH EFFECTS FROM WATER INGESTION | SOURCES OF CONTAMINANT IN DRINKING WATER |
|---|---|---|---|---|
| Asbestos (fiber > 10 µm) | 7 million fibers per liter (MFL) | 7 MFL | Increased risk of developing benign intestinal polyps | Decay of asbestos cement in water mains; erosion of natural deposits |
| Barium | 2 | 2 | Increase in blood pressure | Discharge of drilling wastes; discharge from metal refineries; erosion of natural deposits |
| Beryllium | 0.004 | 0.004 | Intestinal lesions | Discharge from metal refineries and coal-burning factories; discharge from electrical, aerospace, and defense industries |
| Cadmium | 0.005 | 0.005 | Kidney damage | Corrosion of galvanized pipes; erosion of natural deposits; discharge from metal refineries; runoff from waste batteries and paints |
| Chromium (total) | 0.1 | 0.1 | Allergic dermatitis | Discharge from steel and pulp mills; erosion of natural deposits |
| Copper | 1.3 | TT Action Level =1.3 | Short-term exposure: Gastrointestinal distress Long-term exposure: Liver or kidney damage. People with Wilson's Disease should consult their personal doctor if the amount of copper in their water exceeds the action level. | Corrosion of household plumbing systems; erosion of natural deposits |
| Cyanide (as free cyanide) | 0.2 | 0.2 | Nerve damage or thyroid problems | Discharge from steel/metal factories; discharge from plastic and fertilizer factories |
| Fluoride | 4.0 | 4.0 | Bone disease (pain and tenderness of the bones); children may get mottled teeth. | Water additive which promotes strong teeth; erosion of natural deposits; discharge from fertilizer and aluminum factories |

| | | | | |
|---|---|---|---|---|
| Lead | 0.0 | TT Action Level =0.015 | Infants and children: Delays in physical or mental development; children could show slight deficits in attention span and learning abilities. Adults: Kidney problems; high blood pressure | Corrosion of household plumbing systems; erosion of natural deposits |
| Mercury (inorganic) | 0.002 | 0.002 | Kidney damage | Erosion of natural deposits; discharge from refineries and factories; runoff from landfills and croplands |
| Nitrate (measured as nitrogen) | 10 | 10 | Infants below the age of 6 months who drink water containing nitrate in excess of the MCL could become seriously ill and, if untreated, may die. Symptoms include shortness of breath and blue-baby syndrome. | Runoff from fertilizer use; leaching from septic tanks, sewage; erosion of natural deposits |
| Nitrite (measured as nitrogen) | 1 | 1 | Infants below the age of 6 months who drink water containing nitrite in excess of the MCL could become seriously ill and, if untreated, may die. Symptoms include shortness of breath and blue-baby syndrome. | Runoff from fertilizer use; leaching from septic tanks, sewage; erosion of natural deposits |
| Selenium | 0.05 | 0.05 | Hair or fingernail loss; numbness in fingers or toes; circulatory problems | Discharge from petroleum refineries; erosion of natural deposits; discharge from mines |
| Thallium | 0.0005 | 0.002 | Hair loss; changes in blood; kidney, intestine, or liver problems | Leaching from ore-processing sites; discharge from electronics, glass, and drug factories |

## ORGANIC CHEMICALS

| CONTAMINANT | MCLG (mg/L) | MCL (mg/L) OR TT | POTENTIAL HEALTH EFFECTS FROM WATER INGESTION | SOURCES OF CONTAMINANT IN DRINKING WATER |
|---|---|---|---|---|
| Acrylamide | 0.0 | TT | Nervous system or blood problems; increased risk of cancer | Added to water during sewage/wastewater treatment |
| Alachlor | 0.0 | 0.002 | Eye, liver, kidney, or spleen problems; anemia; increased risk of cancer | Runoff from herbicide used on row crops |
| Atrazine | 0.003 | 0.003 | Cardiovascular system or reproductive problems | Runoff from herbicide used on row crops |
| Benzene | 0.0 | 0.005 | Anemia; decrease in blood platelets; increased risk of cancer | Discharge from factories; leaching from gas storage tanks and landfills |
| Benzo(a)pyrene (PAHs) | 0.0 | 0.0002 | Reproductive difficulties; increased risk of cancer | Leaching from linings of water storage tanks and distribution lines |
| Carbofuran | 0.04 | 0.04 | Problems with blood, nervous system, or reproductive system | Leaching of soil fumigant used on rice and alfalfa |
| Carbon tetrachloride | 0.0 | 0.005 | Liver problems; increased risk of cancer | Discharge from chemical plants and other industrial activities |
| Chlordane | 0.0 | 0.002 | Liver or nervous system problems; increased risk of cancer | Residue of banned termiticide |
| Chlorobenzene | 0.1 | 0.1 | Liver or kidney problems | Discharge from chemical and agricultural chemical factories |
| 2,4-D | 0.07 | 0.07 | Kidney, liver, or adrenal gland problems | Runoff from herbicide used on row crops |
| Dalapon | 0.2 | 0.2 | Minor kidney changes | Runoff from herbicide used on rights of way |
| 1,2-Dibromo-3-chloropropane (DBCP) | 0.0 | 0.0002 | Reproductive difficulties; increased risk of cancer | Runoff/leaching from soil fumigant used on soybeans, cotton, pineapples, and orchards |
| o-Dichlorobenzene | 0.6 | 0.6 | Liver, kidney, or circulatory system problems | Discharge from industrial chemical factories |
| p-Dichlorobenzene | 0.075 | 0.075 | Anemia; liver, kidney, or spleen damage; changes in blood | Discharge from industrial chemical factories |

| | | | | |
|---|---|---|---|---|
| 1,2-Dichloroethane | 0.0 | 0.005 | Increased risk of cancer | Discharge from industrial chemical factories |
| 1,1-Dichloroethylene | 0.007 | 0.007 | Liver problems | Discharge from industrial chemical factories |
| cis-1,2-Dichloroethylene | 0.07 | 0.07 | Liver problems | Discharge from industrial chemical factories |
| trans-1,2-Dichloroethylene | 0.1 | 0.1 | Liver problems | Discharge from industrial chemical factories |
| Dichloromethane | 0.0 | 0.005 | Liver problems; increased risk of cancer | Discharge from drug and chemical factories |
| 1,2-Dichloropropane | 0.0 | 0.005 | Increased risk of cancer | Discharge from industrial chemical factories |
| Di(2-ethylhexyl) adipate | 0.4 | 0.4 | Weight loss, liver problems, or possible reproductive difficulties. | Discharge from chemical factories |
| Di(2-ethylhexyl) phthalate | 0.0 | 0.006 | Reproductive difficulties; liver problems; increased risk of cancer | Discharge from rubber and chemical factories |
| Dinoseb | 0.007 | 0.007 | Reproductive difficulties | Runoff from herbicide used on soybeans and vegetables |
| Dioxin (2,3,7,8-TCDD) | 0.0 | 0.00000003 | Reproductive difficulties; increased risk of cancer | Emissions from waste incineration and other combustion; discharge from chemical factories |
| Diquat | 0.02 | 0.02 | Cataracts | Runoff from herbicide use |
| Endothall | 0.1 | 0.1 | Stomach and intestinal problems | Runoff from herbicide use |
| Endrin | 0.002 | 0.002 | Liver problems | Residue of banned insecticide |
| Epichlorohydrin | 0.0 | TT | Increased cancer risk, and over a long period of time, stomach problems | Discharge from industrial chemical factories; an impurity of some water treatment chemicals |
| Ethylbenzene | 0.7 | 0.7 | Liver or kidneys problems | Discharge from petroleum refineries |
| Ethylene dibromide | 0.0 | 0.00005 | Problems with liver, stomach, reproductive system, or kidneys; increased risk of cancer | Discharge from petroleum refineries |
| Glyphosate | 0.7 | 0.7 | Kidney problems; reproductive difficulties | Runoff from herbicide use |

## ORGANIC CHEMICALS (CONTINUED)

| CONTAMINANT | MCLG (mg/L) | MCL (mg/L) OR TT | POTENTIAL HEALTH EFFECTS FROM WATER INGESTION | SOURCES OF CONTAMINANT IN DRINKING WATER |
|---|---|---|---|---|
| Heptachlor | 0.0 | 0.0004 | Liver damage; increased risk of cancer | Residue of banned termiticide |
| Heptachlor epoxide | 0.0 | 0.0002 | Liver damage; increased risk of cancer | Breakdown of heptachlor |
| Hexachlorobenzene | 0.0 | 0.001 | Liver or kidney problems; reproductive difficulties; increased risk of cancer | Discharge from metal refineries and agricultural chemical factories |
| Hexachlorocyclopentadiene | 0.05 | 0.05 | Kidney or stomach problems | Discharge from chemical factories |
| Lindane | 0.0002 | 0.0002 | Liver or kidney problems | Runoff/leaching from insecticide used on cattle, lumber, gardens |
| Methoxychlor | 0.04 | 0.04 | Reproductive difficulties | Runoff/leaching from insecticide used on fruits, vegetables, alfalfa, livestock |
| Oxamyl (Vydate) | 0.2 | 0.2 | Slight nervous system effects | Runoff/leaching from insecticide used on apples, potatoes, and tomatoes |
| Polychlorinated biphenyls (PCBs) | 0.0 | 0.0005 | Skin changes; thymus gland problems; immune deficiencies; reproductive or nervous system difficulties; increased risk of cancer | Runoff from landfills; discharge of waste chemicals |
| Pentachlorophenol | 0.0 | 0.001 | Liver or kidney problems; increased cancer risk | Discharge from wood-preserving factories |
| Picloram | 0.5 | 0.5 | Liver problems | Herbicide runoff |
| Simazine | 0.004 | 0.004 | Problems with blood | Herbicide runoff |
| Styrene | 0.1 | 0.1 | Liver, kidney, or circulatory system problems | Discharge from rubber and plastic factories; leaching from landfills |
| Tetrachloroethylene | 0.0 | 0.005 | Liver problems; increased risk of cancer | Discharge from factories and dry cleaners |
| Toluene | 1 | 1 | Nervous system, kidney, or liver problems | Discharge from petroleum factories |
| Toxaphene | 0.0 | 0.003 | Kidney, liver, or thyroid problems; increased risk of cancer | Runoff/leaching from insecticide used on cotton and cattle |

| | | | | |
|---|---|---|---|---|
| 2,4,5-TP (Silvex) | 0.05 | 0.05 | Liver problems | Residue of banned herbicide |
| 1,2,4-Trichlorobenzene | 0.07 | 0.07 | Changes in adrenal glands | Discharge from textile finishing factories |
| 1,1,1-Trichloroethane | 0.20 | 0.2 | Liver, nervous system, or circulatory problems | Discharge from metal degreasing sites and other factories |
| 1,1,2-Trichloroethane | 0.003 | 0.005 | Liver, kidney, or immune system problems | Discharge from industrial chemical factories |
| Trichloroethylene | 0.0 | 0.005 | Liver problems; increased risk of cancer | Discharge from metal degreasing sites and other factories |
| Vinyl chloride | 0.0 | 0.002 | Increased risk of cancer | Leaching from PVC pipes; discharge from plastic factories |
| Xylenes (total) | 10 | 10 | Nervous system damage | Discharge from petroleum factories; discharge from chemical factories |

## RADIONUCLIDES

| CONTAMINANT | MCLG (mg/L) | MCL (mg/L) OR TT | EFFECTS FROM WATER INGESTION | SOURCES OF CONTAMINANT IN DRINKING WATER |
|---|---|---|---|---|
| Alpha particles | none | 15 picocuries per liter (pCi/L) | Increased risk of cancer | Erosion of natural deposits of certain minerals that are radioactive and may emit a form of radiation known as alpha radiation |
| Beta particles and photon emitters | none | 4 millirems per year | Increased risk of cancer | Decay of natural and man-made deposits of certain minerals that are radioactive and may emit forms of radiation known as photons and beta radiation |
| $^{226}$Radium and $^{228}$Radium (combined) | none | 5 pCi/L | Increased risk of cancer | Erosion of natural deposits |
| Uranium | 0.0 | 30 ug/L as of 12/08/03 | Increased risk of cancer, kidney toxicity | Erosion of natural deposits |

*Source for all tables*: U.S. Environmental Protection Agency, *National Primary Drinking Water Regulations*, 2009. Available to download at: www.epa.gov/ogwdw/consumer/pdf/mcl.pdf.

# Glossary

**Achaean.** A geologic time period from the end of the Hadean eon (3.8 Ga) to the beginning of the Proterozoic eon (2.5 Ga) where Ga = billions of annum.

**acid rain.** Rain or snow that contains acids which cause its pH to be below 5.6 of pure rain.

**actinides.** The 14 elements with atomic numbers from 90 to 103, that is, Th, Pa, U, Np, Pu, Am, Cm, Bk, Cf, Es, Fm, Md, No, and Lr. They are all ratioactive and, except for Th and U, are human-made.

**adit.** A nearly horizontal passageway from the outside into a mine.

**adit portal.** An outside entrance into a nearly horizontal passageway into a mine.

**algae.** A large group of mainly aquatic, nonflowering plants, many microscopic, which includes seaweed.

**Algoma-type BIF.** Rock composed of alternating fine layers of silica and iron minerals, typically hematite or magnetite (banded iron formation). The individual layer lacks the lateral continuity of a Lake Superior-type banded iron formation. These deposits are thought to form as a result of submarine volcanism.

**allophane.** A soft clay mineraloid, a hydrous silicate of aluminum, of varying composition and color, typically occurring as incrustations in chalk and sandstone.

**alluvial fan.** A fan-shaped deposit produced where a stream's velocity decreases abruptly as it flows out of a steep mountain valley onto a plain and deposits its sediments.

**alluvium.** Sediment that has been deposited by river water as in a river bed, on a floodplain, or in a delta.

**alpha particle.** Two protons plus two neutrons bond together into a nuclear particle.

**amalgam.** An alloy of mercury with another metal, including silver used to fill cavities in teeth.

**anaerobic.** Environment without oxygen.

**annulus.** In drilling wells the annulus of the well refers to any void space between the drill pipe and the walls of the well.

**anodize.** To produce a protective oxide layer on the surface of metal parts by having the metal be the anode (positive electrode) of an electrical circuit. Oxygen is released forming an oxide coating on the surface.

**anorthosite.** An uncommon igneous rock made up of 90% plagioclase feldspar crystals with the remainder a mafic component of pyroxene, ilmenite, magnetite, and/or olivine.

**anthracite.** Hard glossy coal with high carbon content of 92% to 98 wt%.

**anthropogenic.** Related to or influenced by human beings.

**apatite.** A phosphate-rich mineral with chemical formula $Ca_{10}(PO_4)_6(OH)_2$. Often $F^-$, $Cl^-$, or $Br^-$ substitute for some $OH^-$ in the structure.

**API (American Petroleum Institute) gravity.** Gives how heavy or light petroleum liquid is given in degrees. API° is determined from $141.5/\rho - 131.5$ where $\rho$ = $g/cm^3$ of the petroleum.

**aquiclude.** A layer above or below an aquifer that is impermeable to the flow of water.

**aquifer.** A sediment or rock that has porosity and allows the transport of water through its porosity.

**asbestos.** Six naturally occurring silicate minerals that occur with a fibrous crystal habit.

**ashlar.** Rectangular building blocks of trimmed stone.

**baby boomers.** The generation born between 1946 and 1964 during the population boom after the end of World War II.

**banded iron formation (BIF).** Rocks composed of fine (0.5 to 2.5 cm) layers of cherty silica alternating with layers of hematite ($Fe_2O_3$), magnetite ($Fe_3O_4$), or siderite ($FeCO_3$).

**barrier island.** A coastal landform composed of long relatively narrow offshore strips of sand that form islands that parallel the coast line.

**baseflow.** The portion of stream flow that comes from subsurface water flow.

**base-load power.** The average demand for power that an electrical utility must produce. This is contrasted to peak power when there is a spike in demand for power.

**base metals.** In mining and industrial usages these are nonferrous, nonprecious metals, which include copper, lead, zinc, tin, mercury, and cadmium.

**batholith.** A greater than 100 km² exposure of intrusive igneous rock that has solidified within the earth, typically with a depth of about 10–30 kilometers.

**bauxite.** A naturally occurring rock with high concentrations of aluminum + oxygen and hydrogen produced from weathering. It is the most important aluminum ore and can contain the minerals gibbsite [$Al(OH)_3$], boehmite [$\gamma$-$AlO(OH)$], and diaspore [$\alpha$-$AlO(OH)$] together with iron oxides and kaolinite.

**basic oxygen steelmaking (BOS) furnace.** A production receptacle used to make steel from carbon-rich molten pig iron. Oxygen is blown through the heated molten pig iron to lower its carbon content producing low-carbon steel. Because the refractories that line the production receptacle are of high pH material they are consider to be basic.

**becquerel (Bq).** The S.I. unit of radioactivity. One Bq equals the quantity of radioactive material for which one nucleus decays per second.

**bentonite.** A water-absorbing aluminum silicate clay formed from volcanic ash.

**benzene-like rings.** Benzene is a ring of six alternating single- and double-bonded carbon atoms with attached hydrogens. Much organic matter includes compounds with the six member carbon rings but with different attached functional groups in place of the hydrogens.

**Besshi-type massive sulfide deposit.** Volcanogenic stratabound deposit containing massive accumulations of sulfides in thin sheets typically associated with terrigenous clastic rocks or mafic volcanic rocks or their equivalents.

**beta particle, $\beta^-$.** High speed electron emitted from a nucleus during radioactive decay when a neutron becomes unstable and is transformed into a proton. There are also less common $\beta^+$ particles, positrons, emitted when a proton is transformed to a neutron.

**binary cycle boiling.** In energy production a type of geothermal power plant where the geothermal water is run through a heat exchanger to boil a secondary fluid to drive a turbine.

**binding energy.** When considering a nucleus, the energy given through Einstein's relationship: nuclear binding energy = $\Delta mc^2$ where $\Delta m$ is the mass difference between unbound protons and neutrons and how they exist in the nucleus and $c$ = speed of light.

**bioaccumulate.** To increase the concentration of a substance in a organism by continual input without significant loss. Usually refers to toxins such as pesticides.

**bioturbation.** The reworking of soils and sediments by animals or plants.

**bitumen.** A mixture of organic compounds that are highly viscous, black, sticky, entirely soluble in carbon disulfide, and composed primarily of cyclic hydrocarbons.

**bituminous.** Soft coal containing bitumen, a highly viscous tar-like substance.

**Black Death plague.** An outbreak of bubonic plague that spread throughout Europe and much of Asia in the fourteenth century.

**blackout.** When referring to electrical power, the complete loss of power.

**blast furnace.** A furnace used to smelt ore. Fuel plus ore are introduced through the top of the furnace, and air or oxygen is blown into the bottom of the furnace.

**boiling water reactor (BWR).** A device that uses nuclear fuel to boil water and sends the steam produced through a turbine to power an electric generator. After use the steam is cooled to liquid water and returned to the reactor core, completing the loop.

**borax.** Short for sodium borate. A boron-containing mineral, a salt of boric acid with the chemical formula $Na_2[B_4O_5(OH)_4] \cdot 8H_2O$.

**brass.** An alloy of copper and zinc.

**breeder reactor.** A type of nuclear reactor that generates new reactor fuel at a greater rate than it consumes it.

**British thermal unit (Btu).** A measure of heat energy necessary to raise the temperature of one pound of air-free water from 60°F to 61°F at 1 atm. 1 Btu = 1,054.68 joules.

**bronze.** A metal alloy made up primarily of copper with added tin. Modern bronze is about 88 wt% copper and 12 wt% tin.

**brown coal.** A term to denote low-grade coal that is brown in color and consists of lignite plus subbituminous coal.

**brownout.** A temporary power reduction in which a utility decreases the voltage on the power lines, so customers receive lower electric current.

**Bushveld Complex.** A very large igneous intrusion in South Africa composed of layered ultramafic and mafic rocks topped by felsic rocks. It covers an area of 66,000 km² and is up to 9 km thick.

**by-product.** A secondary or additional product generated in the process of producing the main product.

**caliche.** A sedimentary rock of gravel, sand, clay, and silt cemented together by calcium carbonate or when ore deposits are considered the cement is nitrate ($NaNO_3$).

**cambium.** Layer of living cells just under the bark from which new plant growth develops.

**capillary fringe.** A layer above the water table where groundwater seeps up by capillary action to fill pores.

**capital costs.** Costs of land, buildings, and equipment and not consumable supplies or salaries.

**caprock.** In the oil industry an impervious layer of rock that traps hydrocarbons in rock below it and keeps the hydrocarbon from rising to the earth's surface.

**carbon steel.** Iron alloyed with carbon.

**carbonate.** Mineral that contains the carbonate anion $CO_3^{2-}$, balancing its cation charge.

**carbonatite.** An unusual intrusive or extrusive igneous rock consisting of greater than 50%

carbonate minerals. It is often associated with alkaline intrusions and is a source of rare minerals.

carcinogen. Substance that is involved in causing cancer.

carrying capacity. The amount that can be supported indefinitely.

cation exchange capacity (CEC). The maximum quantity of cations that a soil is capable of holding, at a given pH, which are available from the soil solution.

celestite. A mineral of strontium sulfate ($SrSO_4$).

cereal crop. A grass plant including wheat, oats, and corn. The starchy grains of the plant are harvested for food.

Chernobyl Nuclear Power Plant. A nuclear reactor situated outside Chernobyl, Ukraine, which suffered a catastrophic meltdown on April 26, 1986. The accident raised concerns about the safety of nuclear power plants.

chromatographic separation. Mixtures of various compounds in a fluid are separated based on their relative attraction to stationary solids as opposed to being retained in the mobile fluid phase passed through the solids.

coal bed methane. Natural gas, primarily made up of methane that occurs in the fractures and matrix of coal beds. It is formed during maturation of the coal.

coke or coking coal. Porous solid carbon material derived from destructive distillation of pure bituminous coal.

comminution. The grinding of a material to form smaller particles, generally a powder in order to liberate the desired mineral from the gangue material.

completed. When referring to an oil or gas well the procedure where small holes called perforations are made in a portion of the casing that passes through the production zone where the hydrocarbons reside or a pack of sand or gravel is installed in the bottom of the drillhole (open-hole completion).

concrete. A strong building material made up of aggregate, sand, and portland cement in water that solidifies and hardens with time.

conduction. For heat, the transfer of energy through a substance by molecule to molecule interaction due to a temperature gradient.

cone of depression. When referring to aquifers, a cone of depressed water levels in a water-table aquifer due to pumping from a well. In a confined aquifer the cone is of reduced water pressure around the extraction point.

connate. Refers to water that is trapped in pores of sediments as a rock is formed. These are usually expelled as the sediments are lithified.

control rods. Rods made of elements capable of absorbing many neutron without themselves fissioning. They control the rate of fission in nuclear reactors.

Copper Age. A period of some human cultures between a Neolithic (New Stone) Age and a Bronze Age, where copper tools and weapons are present in artifacts.

Coriolis effect. An apparent deflection of moving objects when they are viewed from a rotating reference frame such as occurs when viewing the direction of air masses on the earth.

cracking. The process where large molecules are broken down into hydrocarbons of lower molecular weight. Used extensively in oil-refining processes.

craton. The part of continental crust that is stable and has a Precambrian age.

critical mass. The smallest mass of fissionable material need to sustain a nuclear chain reaction.

crude helium. Helium found in natural gas before it is purified.

crude oil. Oil that has been extracted from the ground but has not yet been refined into useful compounds. Also referred to as petroleum.

Cryptosporidium. A parasite in contaminated water, spread through fecal-oral contact, that causes an acute short-term infection generally manifested as diarrhea.

cubic feet (cf). A measure of volume that is also used for natural gas at standard pressure and temperature [corresponding to 60°F (15.56°C) and 14.73 psi] to report a mass of gas. The typical heat energy of natural gas is ~ 1,000 Btu per cubic foot.

**curie (Ci).** A unit of radioactivity equal to $3.7 \times 10^{10}$ decays per second.

**Cyprus-type massive sulfide deposit.** Ophiolite, that is ocean floor, hosted deposits of massive sulfide accumulations as occur in Cyprus. Thought to be produced by mid-ocean ridge black smokers.

**cryolite.** A mineral of composition $Na_3AlF_6$. Because natural cryolite has become so rare, synthetic sodium aluminum fluoride is now produced from fluorite ($CaF_2$) for use in the production of Al metal.

**decommissioning.** A term used to denote the process of removing something from active service.

**deep-sea nodules.** Small hand-sized, irregular concretions found on the deep-sea floor in some locations formed of concentric layers of iron and manganese hydroxides. Also referred to as manganese nodules.

**demographic transition.** The change in population that generally takes place as a country develops. The birth and death rates of its population tend to eventually fall as per capita income rises.

**desertification.** Degradation of land and loss of water in arid climates generally through human activities.

**deuterium.** A stable isotope of hydrogen whose nucleus contains a neutron as well as a proton. Also called heavy hydrogen.

**Dewar tube.** A vacuum tube (thermos) that provides thermal insulation by creating a partial vacuum between the inside and the outside environment.

**diabase.** Fine-grained mafic igneous rock whose composition is the same as basalt. It is typically found as dikes and sills at shallow depths in the earth's crust.

**diatomite.** A lightweight material made up of siliceous microscopic single-celled algae remains, called diatoms. It is generally used as a filtering material and is also called diatomaceous earth.

**dioctahedral.** When referring to clays, the situation where two of the three octahedral sites in the structure are occupied by cations.

**dioxin.** One of a number of toxic chemical compounds composed of two six-member carbon rings fused together with connecting oxygen atoms and with some of the hydrogen atoms replaced by chlorine atoms.

**distillation.** A method of separating compounds in mixtures of liquids that occurs when heated. They become vapors, that is, boil at different temperatures.

**Doctrine of Prior Appropriation.** Law which recognizes water rights unconnected to land ownership that can be sold like other property. The first person to use a quantity of water for a beneficial purpose from a water source has the right to continue to use that quantity of water for that purpose.

**doping.** In the semiconductor industry this refers to the intentional introduction of small amount of material into an extremely pure semiconductor to change its electrical properties.

**drainage basin.** A watershed. The area of land where surface water converges to a single place at the mouth of a river and joins another body of water.

**drilling mud.** A mixture of clay, chemicals, and water pumped down a drill stem to lubricate and cool the drilling bit and to flush out the cuttings and control downhole pressure.

**dry steam.** Steam that does not contain any liquid water.

**Duluth Gabbro.** One of the largest intrusions of gabbro on earth located in Minnesota. Gabbro is a coarse-grained igneous rock primarily consisting of plagioclase feldspar and pyroxene. It is the intrusive (plutonic) equivalent of basalt.

**early diagensis.** The combination of biological, chemical, and physical processes occurring in the upper several hundred meters of sediments after they are deposited and soils after they are buried.

**economic stagnation.** A prolonged period (years) of no or very slow growth of production of goods.

*Eh.* A scale of oxidation-reduction electrical potential (that is voltage, *E*) relative to the hydrogen half reaction (*h*) used to measure the tendency of a chemical species to lose electrons or oxidize in a reaction.

**electrical power grid.** An interconnected network that delivers electricity from suppliers to consumers.

**electric arc furnace (EAF).** A furnace that uses an electric arc produced by an inserted electrode to melt material.

**electricity.** The movement of an electromagnetic field (electric force + magnetic force field) produced by the moving of charged electrons.

**electrolysis.** The use of direct electric current to drive an otherwise nonspontaneous chemical reaction. It is used to separate elements from ores in an electrolytic cell.

**electrolyte.** A substance that contains free ions and therefore readily conducts electricity.

**electromagnetic force field.** An electric and magnetic field of potential energy produced by electrically charged objects.

**electroplating.** A process where electrical current is used to reduce cations of a material in solution so they coat a conductive surface with a thin layer of the material.

**electrowinning.** The deposition, by passing current to electrodes, of metals from ores that have been put in solution.

**electrum.** A naturally occurring alloy of gold with at least 20% silver.

**El Niño event.** An interannual variability in the tropical atmosphere where the normally warm waters and associated low pressure atmosphere over the western Pacific move toward the eastern Pacific.

**endoskarn.** Alteration at country rock–igneous rock contacts that occur within the igneous rock itself, typically a granite.

**energy.** Heat or work, as in a source of usable power.

**energy convection.** The transfer of heat through a fluid by motion of molecules (mass).

**Energy Watch Group.** A group of independent scientists and experts who investigate sustainable concepts for global energy supply.

**entropy.** A measure of the amount of disorder or randomness in an isolated system.

**eolian.** Refers to a process involving wind.

**epicontinental sea.** A shallow ocean that in times past has extended over a significant part of continental crust due to a rise in sea level.

**epigenetic.** When referring to ore deposits those that have been produced or formed at or near the earth's surface after the rock has formed. (The opposite term is syngenetic. Mineralization formed at the same time as the host rocks.)

**epithermal.** Deposits of minerals, typically in veins, precipitated from 50° to 200°C aqueous solutions within about 1 km of the earth's surface.

**eutectic temperature.** The temperature at fixed pressure for a two-component mixture where a liquid solution and both pure component solids exist.

**eutrophication or eutrophic.** Addition of excessive concentrations of nutrients in a body of water that causes a dense growth of plant life and depletion of the oxygen in the water when the plants decay.

**evapotranspiration.** The transfer of water from the earth's surface to the atmosphere by evaporation from wet surfaces and by transpiration from plants.

**excipient.** A chemically inactive substance used as a carrier for the active ingredients in a medication.

**exfoliation.** The process of producing thin sheets of material from a solid.

**exoskarn.** Alterations of country rock at country rock–igneous rock contacts, typically granites.

**fast neutron reactor.** A type of nuclear reactor where fast neutrons are used to sustain the fission chain reaction. It uses no neutron moderator but needs more enriched fuel than used in a thermal reactor.

**feldspathoid.** A group of silicate minerals similar to feldspars but with a lower silica content.

**feller buncher.** A type of logging machine that cuts trees with shears or a saw and places them in a stack.

**felsification.** Assimilation by melting of feldspar components of country rock by a magma.

**Ferrel cell.** The mid-latitude segment of the earth's wind circulation system. It balances the transport in the Hadley and Polar cells. Air motion is opposite to planetary rotation. At the surface this produces the westerlies.

**ferro-alloy metals.** Metals that alloy with iron. These typically include chromium, vanadium, nickel, molybdenum, cobalt, and tungsten.

**ferromagnetic.** Substances such as iron, nickel, and cobalt that can become permanent magnets.

**ferrous.** Bivalent iron, $Fe^{2+}$ (+2 oxidation state).

**fertility rate.** In a particular population the average number of children born to a woman over her lifetime.

**fixed nitrogen.** Atmospheric nitrogen whose oxidation is changed so that it can exist in compounds, such as ammonia ($NH_3$), by either natural or industrial processing.

**flash steam.** Steam produced by lowering the pressure on hot water so it boils.

**flood basalts.** Vast sheets of basalt that spread from a volcanic vent over a large surface of land or ocean floor.

**floodplain.** A plain lower than the banks of an adjacent river that is produced by deposition of river sediment during flooding.

**flux.** The flow of energy or mass across a specified surface. In metallurgy a flux is a substance used to promote fusion.

**fossil fuels.** Organic remains of plants and animals deposited in sediments at an earlier geological time. They develop into the energy sources oil, natural gas, and coal upon burial in the earth.

**foundry.** A factory that produces metal castings.

**fracking.** In the oil industry short for hydrofracturing, a technique where a mixture of water, sand, and proprietary chemicals is pumped down a well at elevated pressure in order to create micro-fractures and release gas and oil from the reservoir rock.

**Frasch process.** A method for extracting sulfur from underground deposits by using superheated steam to melt the sulfur.

**fuel cell.** A cell that produces electricity by a chemical reaction of fuel and oxygen.

**fulvic acid.** One of the constituents of humus, the dark-brown organic layer of soil. It has the ability to transport nutrients and minerals to cells and to help cells digest minerals.

**galvanizing.** An electrochemical process that adds a thin layer of another metal to a steel surface in order to prevent rusting.

**gangue.** Worthless rock that is mixed with wanted ore in a deposit.

**generator.** When considering electricity, a device that converts mechanical energy to electrical energy.

**genetic engineering.** The deliberate alteration of the characteristics of an organism by manipulating its DNA.

**geochemical cycle.** A set of reservoirs or concentrations of some kind of mass (e.g., water) on the earth and the fluxes of this mass between these reservoirs.

**geopressured.** A fluid found at depth in the earth whose pressure is greater than that produced by a standing column of fluid to that depth. It can approach the greater pressure of a column of rocks to that depth.

**geothermal gradient.** The rate at which the earth's temperature increases with depth.

**geothermal heat pump.** A heating and/or cooling system that controls temperature by pumping heat to or from the ground.

*Giardia lamblia.* A parasite with flagella found in contaminated water that colonizes and reproduces in the small intestine. It causes a diarrheal illness known as giardiasis.

**glauconite.** A greenish mica mineral, $(K,Na)(Fe^{3+},Al,Mg)_2(Si,Al)_4O_{10}(OH)_2$, found most often in marine sand deposits.

**gossan.** Highly weathered rust-colored oxidized rock containing altered Fe sulfides that caps an ore deposit.

**grade.** When referring to mineral deposits the average amount of an element of interest (ore) contained in a rock expressed in ounces per metric ton.

**granodioritic.** Pertaining to an intrusive igneous rock that is similar to granite, but contains less quartz and therefore Si and more plagioclase and mafic minerals.

**gray (Gy).** The S.I. unit of absorbed radiation dose equal to 1 joule of radiation energy absorbed by 1 kilogram of matter.

**Great Dyke of Zimbabwe.** A narrow linear north-south trending 550 km long ultramafic lopolith that is exposed through the center of Zimbabwe.

**greenhouse effect.** The process where shorter wavelengths of visible light from the sun pass through the transparent atmosphere. They are absorbed by the earth's surface producing longer wavelength infrared (heat) energy. This longer wavelength energy is unable to pass back through that atmosphere. Instead it heats the atmosphere.

**greenstone.** Low temperature and pressure metamorphosed basalt resulting in a rock with a greenish-black color due to the formation of chlorite and epidote.

**Grenville orogeny.** A long-lived mountain-building and magmatism event from 1250 to 980 million years ago occurring in eastern North America and the southern U.S.

**Haber-Bosch process.** A nitrogen fixation process that takes $N_2$ gas and $H_2$ gas produced from methane gas ($CH_4$) and reacts them over a catalyst like enriched iron or ruthenium to produce ammonia ($NH_3$) gas.

**Hadley cell.** An atmospheric circulation pattern where rising air near the equator moves poleward at 10–15 kilometers above the earth's surface. The air descends at ~30° latitude and moves toward the equator at the surface completing the cell.

**half-life.** The time it takes for something to decay to half its initial value.

**hardwood.** Wood from broad-leaved trees either deciduous or evergreen as opposed to the softwood of conifers.

**harvester.** When referring to trees, a large vehicle that chops, trims, and debarks trees.

**heavy metal.** A metal of density of greater than $5$ g cm$^{-3}$ and of high atomic weight.

**heavy water.** A form of water where the hydrogen in the water molecule has a neutron in its nucleus and is referred to as deuterium. The extra mass of the neutron makes it heavy.

**heliostat.** A device, typically a mirror, that tracks the movement of the sun to concentrate sunlight.

**high-level waste.** A type of nuclear waste created from spent nuclear fuel.

**humic acid.** A principal component of humic substances, which are the major organic constituents of soil (humus), peat, coal, many rivers, still-water lakes, and ocean water. It is produced by biodegradation of dead organic matter.

**hydraulic conductivity.** The ease of flow of water through a sediment or rock measured as a velocity.

**hydrogenation.** A chemical process for adding hydrogen atoms to an oil molecule that is unsaturated with hydrogen.

**hydrograph.** A graph giving changes in the discharge or height of a river over a period of time at a particular location.

**hydrothermal ore deposits.** Ore deposits that are produced by ore minerals being precipitated out of hot water, typically from a crystallizing magma.

**hydroxides.** Compounds of metallic elements containing hydroxyls (OH) such as $Fe(OH)_3$.

**humic coal.** Low hydrogen content coals formed from peat swamps.

**igneous rock.** A rock formed by the crystallization of molten liquid rock (magma).

**index of refraction.** A measure of how much the velocity of a light wave is reduced and therefore bends inside the material.

**infrared radiation.** Electromagnetic radiation also referred to as heat that has wavelengths longer than visible light but shorter than radio waves.

**ironstone.** An ore of Fe made up of iron oxides and/or carbonate. Also contains some clay, and sometimes calcite and quartz.

**isotope.** Atoms of the same element that differ by having different numbers of neutrons in their nucleus.

**itai-itai disease.** Syndrome consisting of severe pain in the joints and spine caused by cadmium poisoning. Also called ouch-ouch sickness.

**karst.** An area of irregular eroded limestone formed by the percolation of acidic groundwater where fissures, sinkholes, and underground caverns are produced.

**karst deposits.** Deposits in karst terrains or fossil karst terrains.

**kerogen.** A mixture of organic compounds that make up a portion of the organic matter in sedimentary rocks. It is insoluble in normal organic solvents because of the large molecular weight of its components.

**kimberlite.** A rare type of peridotite found in pipe-like structures containing a high concentration of $CO_2$ that sometimes contains diamonds.

**Kuroko-type massive sulfide deposit.** An accumulation of sulfide minerals precipitated during or after rhyolitic to rhyodacitic volcanism within a caldera or found as a layer in felsic-rich sediments.

**Le Chatelier's principle.** If a chemical system at equilibrium is disturbed a reaction occurs to counteract the imposed change and a new equilibrium is established.

**laissez-faire policy.** A set of principles letting people act without interference or direction from the government.

**Lake Superior-type BIF.** A type of banded iron formation deposited in shallow waters of continental shelves or in ancient sedimentary basins as opposed to Algoma-type BIF.

**lamproite.** High potassium and magnesium mantle derived volcanic rocks.

**La Niña event.** An interannual variability in the tropical atmosphere that is the counterpart of an El Niño. It is characterized by unusually cold surface ocean temperatures in the equatorial Pacific.

**lanthanides.** The rare earth series of 15 elements with atomic numbers 57 through 71, from lanthanum to lutetium. Also called lanthanoids.

**laterites.** Red iron-rich soils that also contain high concentrations of aluminum formed by extensive weathering of rocks in tropical and subtropical regions.

**Leadership in Energy and Environmental Design (LEED).** Green building certification system developed by the U.S. Green Building Council.

**levee.** A natural or artificial raised embankment along the banks of a river.

**lignite.** A soft brown coal with relatively high moisture ranked in energy content between peat and subbituminous.

**limiting nutrient.** A nutrient that is in low proportion to the others that will limit the growth of organisms. For seawater it is typically nitrogen and for freshwater phosphorous.

**limonite.** A mixture of ferric iron ($Fe^{3+}$), oxides, and hydroxides of varying composition. It generally has a large amount of the mineral goethite [$FeO(OH)$].

**liquid immiscibility.** The property that a liquid can separate into two liquids of different compositions on cooling or that two liquids are incapable of being mixed to produce a homogeneous mixture (e.g., oil and water).

**lithostatic pressure.** A pressure in the earth equal to that imposed by an overlying column of rocks to the earth's surface.

**lode.** A concentration of ore that fills a fissure as in a vein in a rock.

**lopolith.** A large igneous intrusion that is lenticular in shape and formed by injection of magma between beds or layers of existing rock.

**maceral.** Organic component of coal or oil shale. Different macerals are identified by their color, reflectance, and morphology.

**mafic.** A rock, magma, or mineral that is rich in magnesium and iron. This includes the rocks termed basalt and gabbro. They tend to be dark in color.

**magma.** Molten rock formed at depth in the earth that produces lavas when it is erupted on the earth's surface.

**magmatic massive sulfide deposit.** Large accumulations of almost pure sulfide ore minerals found in mafic to ultramafic solidified magmas.

**magnetic hysteresis.** When an external magnetic field is applied to a substance that can be magnetized, like Fe, the atomic magnetic dipoles in the substance align themselves with the external field. When the external field is removed, some of the field is retained and the substance is slightly magnetized.

**manganese nodules.** Nodules found on the seafloor, often partly buried, that are composed of concentric layers of iron and manganese hydroxides. They vary in size from small pellets to greater than 20 cm across. They also contain significant amount of nickel, copper, and cobalt.

**mantle wedge.** The triangular section of the earth's mantle in the asthenosphere above a downgoing lithospheric slab and the lithospheric plate above.

**marsh gas.** Methane gas produced when bacteria decompose vegetation in water.

**maximum sustainable yield.** The largest amount of a renewable resource that can be taken without impairing its population or concentration.

**metal.** An element that can conduct heat and electricity. It forms cations in solution and ionic bonds with nonmetals.

**metallic bonds.** Bonds between atoms formed by interactions of delocalized electrons. This sea of electrons is shared amongst the metal nuclei in the substance.

**metalloid.** An element with properties intermediate between those of metals and nonmetals. They include arsenic, antimony, boron, germanium, silicon, and tellurium.

**metamorphic rock.** Rock formed from preexisting rocks through recrystalization during heating and compression as it is buried in the earth.

**metasomatic.** The chemical alteration of rock by fluids.

**meteoric.** Pertaining to atmospheric phenomena particularly atmospheric water.

**methane clathrate.** Cage-like structures of cubic ice with a methane gas atom within the cage.

**metric ton.** A mass equivalent to 1,000 kilograms.

**Millennials.** An age group of population following Generation X sometimes called Generation Y, born ~1980–1997.

**Mississippi Valley-type (MVT) deposits.** Epigenetic hydrothermal (50°–200°C) ore deposits that are not associated with igneous activity, which form predominantly in dolostone where lead and zinc sulfides (galena and sphalerite) are the major ore minerals.

**moderator.** When considering nuclear reactions a substance that reduces the speed of fast neutrons turning them into thermal neutrons.

**monsoon.** A regional wind pattern that reverses direction seasonally from onshore to offshore and influences climate.

**montmorillonite.** Any of a group of aluminum-rich clay minerals characterized by the ability to expand when they absorb large quantities of water.

**Mohs hardness.** A scale that characterizes the scratch resistance of minerals from talc with a hardness of 1 to diamond with a hardness of 10.

**mud logging.** Creation of a record (well log) of the nature of rocks in a borehole by examining the bits of rock and sediments brought to the surface by the circulating drilling fluid (mud).

**Munsell soil color index.** The official color system for soil research. The system has three components: hue (a specific color), value (lightness and darkness), and chroma (color intensity) that are arranged in books of color chips.

**norite.** A coarsely crystalline mafic plutonic igneous rock composed dominantly of the plagioclase, labradorite, with hypersthene and olivine.

**nuclear fission.** A nuclear reaction where the nucleus of an atom splits into smaller parts.

**nuclear fusion.** A nuclear reaction where multiple atomic nuclei bond together to form a single heavier nucleus.

**octahedral sheet.** Octahedron made up of a center atom of Al, Fe, or Mg surrounded by six oxygen atoms. These are bonded together in a layer by sharing some oxygen atoms between octahedron.

**octane rating.** A measure of the resistance of petroleum fuels to auto-ignition in an internal combustion engine. A high-octane fuel will more likely burn in a controlled manner, rather than exploding uncontrollably.

**oil window.** The depth region over which oil is produced in the subsurface in a particular locality.

**oligotropic.** A body of water that is low in plant nutrients and contains abundant oxygen.

**oolites.** Spherical grains 0.25 to 2 mm in diameter composed of concentric growth layers.

**open pit mine.** A mine where rock or minerals are extracted from the earth from an open cavity or burrow.

**ophiolites.** Sections of oceanic crust and part of the underlying upper mantle exposed on continents.

**optical reflectivity.** The ratio of the energy of a wave reflected from a surface to the energy of the wave striking the surface.

**ore deposit.** A mass of rock in the earth containing metal or other elements of value (e.g., diamonds) of sufficient abundance to be extracted at a profit.

**osmotic pressure.** The pressure exerted by the transport of water through a semipermeable membrane separating two solutions with different concentrations of solute.

**oxidant.** A substance that is an oxidizing agent by donating electrons to another compound.

**oxidation-reduction reaction.** A reaction where electrons have been transferred between species so that their oxidation state has changed (e.g., $Fe^{2+}$ to $Fe^{3+}$). The species that loses electrons is oxidized, while the species that gains electrons is reduced. Redox for short.

**oxidation state.** The state of an atom in a compound or species that refers to the number of electrons it possesses in its outer electron shell relative to the number in the neutral element.

**oxide.** A binary compound of an element together with oxygen such as $Al_2O_3$.

**panning.** To obtain particles of gold or other precious metal by washing away less dense sediments in a pan.

**parabolic trough collector.** When considering sunlight, a parabolic mirrored trough that focuses sunlight on a tube that runs down the length of the trough.

**peak coal.** When plotting time versus production, the time when a peak in the amount of coal use occurs. The peak is typically attributed to the lack of the availability of more coal.

**peak natural gas.** When plotting time versus production, the time when a peak in the amount of natural gas produced occurs. The peak is typically attributed to the lack of the availability of more natural gas.

**peak oil.** When plotting time versus production, the time when a peak in the amount of oil produced occurs. The peak is typically attributed to the lack of the availability of more oil.

**peat.** Compressed partially decomposed plant remains that can be used as a fuel when dried.

**ped.** A naturally formed unit of soil structure.

**pegmatite.** A very coarse-grained, intrusive igneous rock. Generally composed of feldspar, quartz, and mica, and therefore having a granitic composition.

**per capita.** The average per person.

**peridotites.** Intrusive ultramafic igneous rocks that are the main constituent of the earth's mantle.

**perlite.** An amorphous hydrated volcanic glass of obsidian that greatly expands when heated forming coarse lightweight granules.

**permafrost.** Permanently frozen subsoil, occurring throughout the Polar Regions and locally in perennially frigid areas.

**permeability.** A measure of the ability of a rock or sediment to allow a fluid under pressure to flow through it.

**pewter.** A malleable silver-gray metal alloy made of approximately 93% to 98% tin, 1% to 2% copper, and the rest of antimony. Pewter was originally made of tin and lead.

**pH.** A measure of the hydrogen ion, $H^+$, concentration in solution where pH = − (base-10 logarithm of the $H^+$ concentration). At 25°C a solution where pH < 7 is acidic, pH = 7 is neutral, and pH > 7 is alkaline or basic.

**photosynthesis.** The process where plants use light energy to produce carbohydrates and oxygen from $CO_2$ and $H_2O$.

**photovoltaic cell.** A semiconductor device that converts sunlight into electricity.

**phyllosilicates.** Silicate minerals with a crystal structure containing parallel sheets of silica-oxygen tetrahedral.

**pitchblende.** A black amorphous form of crystallized uraninite ($UO_2$).

**pig iron.** Crude (typically 90% pure) Fe metal made by smelting iron ore in a furnace. Originally cast into a row of blocks in earthen holes, having the appearance of a pig feeding her piglets.

**placer deposit.** An accumulation of dense ore minerals deposited when the flow of running water abruptly slows in a river or stream but allows less dense minerals to be carried away.

**plasma.** When referring to atoms, a gas-like state of charged matter where the atoms are ionized by being subjected to such high temperatures that they lose an outer electron.

**plaster.** A mixture of heated gypsum together with other inert fillers, and water that hardens to a smooth solid used for coating walls and ceilings.

**Pleistocene.** A time period between 2.6 million and 12,000 years ago when repeated glaciations occurred.

**podiform.** A descriptive term usually used with chromite ore bodies where the shape of the deposit is a center-expanded tube closed at both ends.

**Polar cell.** An atmospheric circulation pattern where cold dense air sinks as the pole. It is then transported along the surface and warmed. At about 60° latitude it rises 10 to 15 km in the atmosphere and is transported back to the pole completing the cell.

**polyaromatic hydrocarbon (PAH).** Rings of six alternating single- and double-bonded carbon atoms with attached hydrogen that are fused together. They are common atmospheric pollutants produced by burning fossil fuels.

**polychlorinated biphenyl (PCB).** A class of toxic organic compounds with one to 10 chlorine atoms attached to two fused benzene rings given by $C_{12}H_{10-x}Cl_x$.

**population pyramid.** A diagram that separates out the number of people of particular ages in a population with the youngest on the bottom and oldest on top. These diagrams tend to have a triangular type shape indicating a large number of younger people and a small number of older people except in developed countries.

**porphyry copper deposit (PCD).** Copper ore bodies occurring from deposition of copper sulfide minerals from fluids associated with porphyritic intrusive igneous rocks. They may also contain molybdenum, silver, and/or gold.

**porphyritic intrusive igneous rock.** Rock produced from molten magma crystallizing below the earth's surface with minerals with two distinctly different sizes. The larger mineral grains are termed phenocrysts and the finer mineral grains are called matrix or groundmass.

**portland cement.** A construction cement made by heating limestone and clay in a kiln and pulverizing the result. It hardens by chemical reaction with water.

**porosity.** The void spaces in a material given by the volume of voids divided by the total volume of the material.

**potash.** Mined potassium-containing salts. The most common is KCl.

**potentiometric surface.** A hypothetical surface outlining the height to which groundwater would rise if not confined in an aquifer. For a water-table aquifer it is the surface of the water table.

**precious metals.** Rare metals of high economic value, which are not radioactive. Typically they include gold, silver, and the platinum group metals: platinum, palladium, rhodium, iridium, osmium, and ruthenium.

**pressurized water reactor (PWR).** One of two types of light water reactors, the other being a boiling water reactor (BWR). In a PWR a separate loop transfers heat from the reactor core to water under pressure and the other loop, to which the heat is transferred, produces steam to drive a turbine to run a generator to produce electricity. It is the most common type of nuclear power plant.

**Proterozoic.** A geological time period starting at the end of the Archean (2.5 Ga) and lasting until the Phanerozoic starts (0.542 Ga) where Ga is billions of annum.

**pumice.** A light, porous type of solidified frothy lava formed by the expansion of gas in erupting lava that will float on water.

**pumpjack.** An aboveground drive used in a reciprocating piston pump installed in an oil well to bring crude oil to the surface. Also called a rocking horse or nodding donkey.

**pyrobitumen.** A brittle solid hydrocarbon distinguishable from bitumen by being infusible and insoluble. Formed by polymerization of carbon in petroleum over geological time.

**pyroclastic.** Composed of rock fragments ejected from a volcanic vent.

**pyroxenites.** Ultramafic igneous rock consisting primarily of pyroxene.

**quicksilver.** An older name for the element mercury (Hg).

**quartz monzonite.** An intrusive igneous rock that contains 5% to 20% quartz; less than a true granite. Much of the remaining minerals are either alkali or plagioclase feldspar. The ratio of alkali feldspar to total feldspar is between 35% and 65 wt%. Small amounts of biotite and/or hornblende are also present.

**radiation.** Energy transmitted in the form of waves.

**radiation absorbed dose (RAD).** The historical unit used to quantify the energy deposited in a substance by ionizing radiation. 100 RAD = 1 gray, the newer S.I. unit of absorbed radiation.

**rare earth elements (REEs).** Any of the 17 relatively abundant metallic elements of atomic number from 57 through 71. Includes scandium, yttrium, and the 15 lanthanides.

**redox.** Short for oxidation-reduction reaction.

**refactory.** A brick-like substance with a high melting point used to line furnaces.

**reflectance microscope.** A microscope that looks at light reflected from an object's surface.

renewable resource. An asset that will be regenerated or replenished shortly after being used so the amount available does not change significantly with time.

reserve. When referring to resources a concentration of naturally occurring solid, liquid, or gaseous material in or on the earth in such form and amount that its production is currently or potentially feasible under current market conditions.

reserve base. When referring to resources the economically obtainable part of a measured and/or indicated mineral resource under current market conditions.

reservoir. When referring to oil in the subsurface a concentration of hydrocarbons in a particular locality.

residence time. The average amount of time a substance like water spends in a specified reservoir.

resource. An available asset that can be drawn upon when needed.

retrograde solubility. A substance becomes less soluble as temperature increases.

reverse osmosis. A process of purifying a solvent such as water by passing it through a semipermeable membrane where dissolved solute is trapped.

rhyolitic tuff. A rock composed of compacted sand to coarse gravel-sized particles of volcanic ash of rhyolite.

rill. A very small streamlet of running water; a rivulet.

riparian rights. The rights of an owner of a property situated along a river or lake to use the water for irrigation and consumption.

Rock-Eval pyrolysis. A programmed heating and collection of released hydrocarbons used to identify the type and maturity of organic matter and to determine petroleum potential in sedimentary rocks.

roentgen equivalent man (REM). The historical unit of radiation dose equivalent quantifying the biological damage of radiation. The newer S.I. unit is the sievert where 100 REM = 1 sievert.

runoff. The water that runs off a surface as opposed to water that infiltrates or evaporates from the surface.

sabkha. An area of coastal flats typically covered with a salt crust caused by periodic flooding and seawater evaporation. They are common in Arabia and North Africa.

salinization. In a soil of an arid, poorly drained region, the accumulation of soluble salts by the evaporation of the waters brought to the soil by transport of salt-containing water in the subsurface.

saltation. The jumping process of moving particles in fluid over an uneven surface.

salt dome. A diapir of salt formed when a thick bed of salt found at depth intrudes upward into the rock strata above due to its lower density.

saltpeter. Naturally occurring potassium nitrate ($KNO_3$) used in the manufacture of fertilizer and explosives.

sapropelic coal. High hydrogen coals formed in lakes. They are typically thin and form lenticular masses.

scoria. A relatively heavy frothy basaltic rock, full of cavities produced by small gas bubbles trapped in the lava. It is dense enough that it sinks in water.

scram. An automatic emergency shutdown of a nuclear reactor (safety control rod axe man, a term from the first nuclear reactor at the University of Chicago where an axe was used to cut a rope that held a control rode above the reactor core).

sedimentary rock. A rock resulting from the consolidation of loose grains that has accumulated in layers, generally in water.

semiconductor. A substance with electrical properties between a good conductor and a good insulator.

separation work unit (SWU). The energy expended in separating a kilogram $F$ of assay $xf$ into a mass $P$ of product with assay $xp$ and waste of mass $W$ with assay $xw$. SWU = $WV(xw) + PV(xp) - FV(xf)$ where the "value" function, $V(x) = (1 - 2x) \ln ((1 - x)/x)$.

**Siberian Traps.** A very large outpouring of flood basalt covering 2 million km² in Siberia, Russia 251–250 million years ago. The term traps (stairs) stems from the fact that step-like hills were produced by the individual flows of basalt on top of each other.

**sievert (Sv).** The S.I. unit of radiation dose equal to 1 joule per kilogram. It is determined by multiplying the absorbed dose in gray ($J kg^{-1}$) by a dimensionless "quality factor" for each radiation type as well as by a dimensionless N which accounts for other factors.

**silane.** A chemical compound consisting of only Si and H such as monosilane ($SiH_4$).

**sinkhole.** A natural depression in the ground, especially in limestone from dissolution of the rock in water, that provides a route for surface water underground.

**sintering.** A heating process below the substance's melting point that bonds particles in a powder to increase the substance's strength and density.

**skarn.** A metamorphic rock produced at the contact of an igneous intrusive and limestone or dolostone consisting of both metasomatically-altered carbonates and igneous rocks.

**skip.** In mining, a large open-topped container designed for transport of ore to the surface of an underground mine designed for easy loading and unloading.

**slag.** The scum of impurities formed by oxidation at the surface of molten metals.

**smelting.** The process of melting an ore concentrate to extract a metal.

**softwoods.** Wood from conifers. Includes fir, pine, spruce, cedar, larch, hemlock, cypress, redwood, and yew.

**soil deflation.** Erosion of soil as a consequence of sand and dust and loose rocks being removed by the wind.

**solar power.** The generation of electricity from sunlight.

**solid load.** In geology, the solid material that is carried in a river as opposed to the material dissolved in solution.

**solid solution.** A homogeneous mixture of the elements considered in a single mineral.

**slag.** The impurities which result when ore concentrate is smelted.

**spalling.** In geology, the chipping or fracturing with an upward heaving, of rock caused by internal stresses.

**specialty metals.** A set of metals that play an important role in high technology manufacturing. Niobium (Nb), tantalum (Ta), arsenic (As), antimony (Sb), bismuth (Bi), germanium (Ge), gallium (Ga), indium (In), beryllium (Be), and the rare earth elements (REEs) are generally classified as specialty metals.

**standard hydrogen electrode.** A redox electrode where the half reaction $2H^+_{(aq)} + 2e^- \rightarrow H_{2(g)}$ occurs, which forms the basis of the scale of oxidation-reduction potential, *Eh*.

**steady state.** A state where the variables that describe the system are constant so concentrations do not change with time (input = output) but the system is not at equilibrium.

**steel.** A hard and durable yet malleable mixture of iron with small amounts of carbon (between 0.2 wt% and 2.1 wt%) and often with other useful elements depending on the desired properties. It is widely used as a construction material.

**sterling silver.** A silver alloy of 92.5% Ag and 7.5% Cu.

**stock.** An intrusive igneous rock that covers less than 100 km² in area and has steep contacts with the surrounding rocks.

**stopes.** The area in an underground mine from which ore is extracted. They usually occur in a step-like excavation as ore is mined deeper and deeper.

**strategic metal.** A metal that is essential for industrial production and national security, but little or no domestic supply is available. Many countries stockpile these metals.

**stratiform sediment-hosted copper (SSC) deposit.** Sedimentary rock-hosted copper deposits that occur along the bedding of the rock formed by precipitation of copper sulfides out of solution. The copper sulfides occur disseminated to veinlet in the strata.

**stratiform sediment-hosted exhalative (SEDEX) deposit.** Ore deposit that is thought to have formed by release of metal bearing hydrothermal fluid into water, typically the ocean, which results in the precipitation of a layer of ore between sedimentary layers.

**stratiform deposit.** Deposit that is confined to a single strata or bed of rock.

**strip mine.** A mine where the material above the ore is removed (stripped) to expose it for removal.

**subbituminous.** A soft coal with high moisture content and a heat content between bituminous and lignite.

**subsistence living.** Barely consuming enough to keep living.

**sulfide.** A mineral that contains sulfur in its reduced oxidation state like pyrite ($FeS_2$).

**superalloy.** A high-performance metal alloy with superior mechanical strength at high temperatures as well as high corrosion and oxidation resistance.

**superconducting magnet.** An electromagnet constructed with extremely low-resistance superconducting wire. It must be cooled to extremely low temperatures during operation to obtain the low resistance.

**supergene enrichment.** A secondary enrichment of sulfide or metals, at or near the surface, by reaction in acidic groundwater. This water either leaches out non-ore elements or dissolves ore and reprecipitates it at some depth. This upgrades these ores where leaching or reprecipitation has occurred.

**syngenetic.** Mineralization formed at the same time as the host rocks. (The opposite term is epigenetic which implies the mineralization was after the host rocks were formed).

**tailings.** The pile of refuse material resulting from physically processing extracted ore.

**tarmac.** Short for tarmacadam, a paving material used for roadways of tar, bitumen, and crushed stone. Also called blacktop.

**tensile strength.** The resistance of a material to a tearing force.

**terra rosa.** A red clay soil produced by weathering of limestone.

**terminal lake.** A lake where streams and groundwater flow into it but no streams or groundwater flow out of it. Water is lost by evaporation.

**tetrahedral sheet.** Tetrahedron made up of a center atom of Si or Al surrounded by four oxygen atoms. These are bonded together in a layer by sharing some oxygen atoms between tetrahedron.

**thenardite.** Anhydrous sulfate mineral ($Na_2SO_4$), which precipitates from water in arid, evaporite environments.

**thermal reactor.** A nuclear reactor having material to reduce the speed of neutrons released on uranium decay to lower velocity thermal neutrons, which allow greater chance of capture by uranium-235.

**Three Mile Island (TMI) Nuclear Generating Station.** A civilian nuclear power plant located on an island in the Susquehanna River southeast of Harrisburg, Pennsylvania. It is the site of a partial meltdown of a nuclear reactor on March 28, 1979.

**tidal estuary.** A partly enclosed coastal body of water connected to the ocean with a river flowing into it.

**tons of oil equivalent (toe).** Energy released by burning 1 metric ton of crude oil ~42 gigajoules.

**total organic carbon (TOC).** The total amount of organic carbon in a rock given in wt%.

**transuranic waste.** Material contaminated with transuranic elements, These are artificially made, radioactive elements with atomic numbers greater than uranium such as neptunium, plutonium, and americium.

**Transvaal.** The land area north of the Vaal River in South Africa.

**trioctahedral.** When referring to clays having all three octahedral sites in the structure occupied by cations.

**trona.** A carbonate evaporite mineral, $Na_3H(CO_3)_2 \cdot 2H_2O$.

**turbine.** A device where kinetic energy of a moving fluid is converted into mechanical energy of a rotating blade.

**ultramafic.** An igneous rock or magma with very low silica content of < 45 wt% and higher magnesium and iron content than mafic basalt.

**ultraviolet radiation.** Electromagnetic radiation with wave lengths shorter than visible light but longer than X-rays.

**unconventional oil.** Petroleum extracted using techniques other than by a traditional oil well. Unconventional oil production is more expensive and includes extracting hydrocarbons from tar sands and oil shales.

**uraninite.** A mineral with chemical formula $UO_2$ that is the chief ore or both uranium and radium.

**vadose zone.** The region above the water table to the earth's surface where pores contain both air and water. Also called the unsaturated zone.

**valence.** The number of chemical bonds an atom forms with other atoms in a compound by giving up electrons. If the atom attracts electrons its valence is negative.

**Van Krevelen diagram.** A plot of the oxygen:carbon (oxygen index) and hydrogen:carbon (hydrogen index) ratios of crude oils used to understand the oil maturation process.

**vermiculite.** A 2:1 hydrated clay mineral which expands on heating of chemical formula $(Mg, Fe,Al)_3(Al,Si)_4O_{10}((OH)_2 \cdot 4H_2O$.

**vitrinite.** One of the primary components (macerals) of coals and most sedimentary kerogens.

**volatile organic chemical (VOC).** Organic chemical with high enough vapor pressures under normal conditions to form fumes that enter the earth's atmosphere.

**volcanic massive sulfide (VMS) deposit.** An ore deposit of Fe-Cu-Zn-Pb sulfides created by a hydrothermal fluid emanating from a volcanic edifice in a submarine environment.

**Waste Isolation Pilot Plant (WIPP).** A disposal site for U.S. defense-related transuranic radioactive waste located in the Chihuahuan Desert outside Carlsbad, New Mexico.

**water cycle.** Describes how the movement of water on the earth occurs with the total amount of water in the earth system staying constant. Also known as the hydrologic cycle.

**watershed.** An area bounded peripherally by divides where surface water drains into a particular river or lake.

**water table.** The surface in the ground below which all voids are filled completely with water.

**wetlands.** A land area that is saturated with water, either permanently or seasonally, such that it takes on the characteristics of a distinct ecosystem.

**xenolith.** A foreign rock fragment or crystal that is incorporated into an igneous rock.

**yellowcake.** A concentrated uranium powder of uranium oxide ($U_3O_8$) obtained from leaching of milled uranium ore.

**Yucca Mountain, Nevada.** A mountain about 160 km northwest of Las Vegas, Nevada where a radioactive waste repository has been considered for long-term storage of civilian nuclear power plant waste.

**zeolite.** A hydrated aluminosilicate mineral with an open three-dimensional crystal structure in which the water molecules are held.

**zooplankton.** Tiny microscopic animals that drift in seawater.

# Index

*Boxes, figures, and tables are indicated by b, f, and t following page numbers.*

# Photo Credits

## Chapter 6

6.F04 © William D. Bachman/Photo Researchers, Inc.; 6.F12 Courtesy of USGS; 6.F14 © Neil Overy/Getty Images; 6.F15A © Charles D. Winters/Photo Researchers, Inc.; 6.F15B © Institute of Oceanographic Sciences/NERC/Photo Researchers, Inc.

## Chapter 7

7.F03 Courtesy of Kevin Walsh.

## Chapter 8

8.F21 © GC Minerals/Alamy.

## Chapter 10

10.F02 © All Canada Photos/Alamy; 10.F03 © Alexandr Vlassyuk/ShutterStock, Inc.; 10.F06 © Artography/ShutterStock, Inc.; 10.F07 © Katia/ShutterStock, Inc.; 10.F11 © Turnervisual/iStockphoto; 10.F12 © Turnervisual/ShutterStock, Inc.; 10.F13 © Biophoto Associates/Photo Researchers, Inc.; 10.F16 © Philippe Psaila/Science Photo Library; 10.F17 Courtesy of Ed Jackson; 10.F18 © TonyV3112/ShutterStock, Inc.; 10.F19 © farbled/ShutterStock, Inc.; 10.F21A © Graça Victoria/ShutterStock, Inc.; 10.F21B Courtesy of Dan Freeman, Grip-Rite Sports.

## Chapter 11

11.F02 Courtesy of Chris Arp, University of Alaska, Fairbanks; 11.F03 Courtesy of Death Valley National Park/NPS; 11.F04 Courtesy of Claus Ableiter; 11.F06 © Jintana Limpongsa/ShutterStock, Inc.; 11.F07 © Martin Valent/ShutterStock, Inc.; 11.F08 © National Geographic Image Collection/Alamy; 11.F10 © gkuna/ShutterStock, Inc.; 11.F15 Courtesy of Linde Engineering Division.

## Chapter 12

12.F01 © Hemera/Thinkstock; 12.F16 © Reuters/Landov; 12.F27 © Thomas Barrat/ShutterStock, Inc.; 12.F28 © jorisvo/ShutterStock, Inc.; 12.F34 © Tim Roberts Photography/ShutterStock, Inc.

## Chapter 13

13.F02 Courtesy of Wayne Wurtsbaugh, Association for the Sciences of Limnology and Oceanography; 13.F03 Courtesy of SIDEM; 13.F06A Courtesy of Janice Carr/CDC; 13.F06B © Biophoto Associates/Photo Researchers, Inc.; 13.F11 © John Flesher/AP; 13.F15 Courtesy of Jeff Vanuga /USDA NRCS; 13.F16 © MartinMaritz/ShutterStock, Inc; 13.F18 © Jean-Paul Pelissier/Reuters/Landov.

## Chapter 14

14.F04 Courtesy of Bruce Jensen; 14.F05 © Alan Heartfield/ShutterStock, Inc.; 14.F12 © B Brown/ShutterStock Inc.; 14.F15 Courtesy of Jack Dykinga/USDA ARS; 14.F17 Courtesy of Jeff Vanuga /USDA NRCS; 14.F18 Courtesy of Tim McCabe/USDA NRCS; 14.F19 © raphme/ShutterStock, Inc.